Advanced Structured Materials

Volume 113

Series Editors

Common engineering materials reach in many applications their limits and new developments are required to fulfil increasing demands on engineering materials. The performance of materials can be increased by combining different materials to achieve better properties than a single constituent or by shaping the material or constituents in a specific structure. The interaction between material and structure may arise on different length scales, such as micro-, meso- or macroscale, and offers possible applications in quite diverse fields.

This book series addresses the fundamental relationship between materials and their structure on the overall properties (e.g. mechanical, thermal, chemical or magnetic etc.) and applications.

The topics of *Advanced Structured Materials* include but are not limited to

- classical fibre-reinforced composites (e.g. glass, carbon or Aramid reinforced plastics)
- metal matrix composites (MMCs)
- micro porous composites
- micro channel materials
- multilayered materials
- cellular materials (e.g., metallic or polymer foams, sponges, hollow sphere structures)
- porous materials
- truss structures
- nanocomposite materials
- biomaterials
- nanoporous metals
- concrete
- coated materials
- smart materials

Advanced Structured Materials is indexed in Google Scholar and Scopus.

More information about this series at http://www.springer.com/series/8611

Andreas Öchsner · Holm Altenbach
Editors

Engineering Design Applications II

Structures, Materials and Processes

Springer

Editors
Andreas Öchsner
Faculty of Mechanical Engineering
Esslingen University of Applied Sciences
Esslingen, Baden-Württemberg, Germany

Holm Altenbach
Lehrstuhl für Technische Mechanik
Institut für Mechanik
Faculty of Mechanical Engineering
Otto-von-Guericke-Universität Magdeburg
Magdeburg, Sachsen-Anhalt, Germany

ISSN 1869-8433 ISSN 1869-8441 (electronic)
Advanced Structured Materials
ISBN 978-3-030-20800-4 ISBN 978-3-030-20801-1 (eBook)
https://doi.org/10.1007/978-3-030-20801-1

This Springer imprint is published by the registered company Springer Nature Switzerland AG
The registered company address is: Gewerbestrasse 11, 6330 Cham, Switzerland

Preface

Different engineering disciplines such as mechanical, materials, computer and process engineering provide the foundation for the design and development of improved structures, materials and processes. The modern design cycle is characterized by an interaction of different disciplines and a strong shift to computer-based approaches where only a few experiments are performed for verification purposes. A major driver for this development is the increased demand for cost reduction, which is also connected to environmental demands. In the transportation industry (e.g. automotive or aerospace), this is connected with the demand for higher fuel efficiency, which is related to the operational costs and the lower harm for the environment. One way to fulfil such requirements is lighter structures and/or improved processes for energy conversion. Another emerging area is the interaction of classical engineering with the health and medical sector. This volume gives an update on recent developments in the mentioned areas of modern engineering design application.

We would like to express our sincere appreciation to the representatives of Springer, who made this volume possible.

Esslingen, Germany — Prof. Dr.-Ing. Andreas Öchsner, D.Sc.
andreas.oechsner@gmail.com

Magdeburg, Germany — Prof. Dr.-Ing. habil. Dr. h. c. mult. Holm Altenbach
holm.altenbach@ovgu.de

Contents

Investigation of the Damage Behavior of Polyurethane in Stress Relaxation Experiments and Estimation of the Stress-at-Break σ_b with a Failure Envelope 1
Selina Neuhaus, Henning Seibert and Stefan Diebels

Identification of Inelastic Parameters of the AISI 304 Stainless Steel . . . 17
Miguel Vaz Jr., E. R. Hulse and M. Tomiyama

Effect of Fiber Volume Fraction on Mechanical Properties of Type E-Glass in Composite Materials 37
Zamzam A. Elsharif and Bashir M. Gallus

Experimental Study of Temperature Effect on the Mechanical Properties of GFRP and FML Interface 47
Z. P. Chow, Z. Ahmad and K. J. Wong

An Alternative Method for Modelling the Degradation of Hyperelastic Materials Within the Framework of Finite-Strain Elastoplasticity 59
L. Écsi and P. Élesztős

Geometric Nonlinear Numerical Analysis of Shells in the Form of a Hyperboloid of Revolution 83
Ismael Taha Farhan Farhan, Gil-oulbé Mathieu and Timur Soibnazarovich Imomnazarov

Numerical Experiments on Helicoids Stress-Strain State 95
Marina Rynkovskaya, A. S. Markovich, M. Gil-Oulbe and S. V. Strashnov

Transversely Cracked Beams with Quadratic Function's Variation of Height 105
Matjaž Skrinar

Effect of Vapour-Grown Carbon Nanofibres on Thermo-Mechanical Properties of High-Functionality Based Resin Used in CFRP Strengthening System Subjected Severe Service Conditions 117
R. Al-Safy, R. Al-Mahaidi, G. P. Simon and J. Habsuda

Study on Closed-Die Forging in Stainless Steel as Standard ASTM F138 in Grain Size Function 133
Givanildo Alves dos Santos, Wagner Figueiredo, Ylich Peter Schmitt, Maurício Silva Nascimento, Fabio Miranda, Gilmar Ferreira Batalha and Antonio Augusto Couto

Contribution to an Electrical Transport in Montmorillonite/Polyaniline Composite 155
S. Rusnáková, K. Karvanis, P. Koštial, Z. Koštialová-Jančíková and A. Zimula

Chosen Physical Properties of Menzolit BMC 3100 167
S. Rusnáková, K. Karvanis, P. Koštial, Z. Koštialová Jančíková and A. Zimula

Structural Analysis by Finite Element Method in Ball Valves to Improve Their Mechanical Properties 175
M. Egure-Hidalgo, J. M. Aburto-Barrera, C. R. Torres-San Miguel, J. Martinez-Reyes and B. Romero-Ángeles

Modelling of Cavitation Around Hydrofoils with Included Bubble Dynamics and Phase Changes 187
Galina Ilieva and Christo Pirovsky

Implementation of Beamforming Codes in 3D CFD Simulations for the Localization and Visualization of Rotating Sound Sources 201
Christian Maier and Wolfram Pannert

Signal Analysis and Numerical Method Roles in Neural Interfaces Development 215
I. Kuzmanić, I. Vujović, J. Šoda and M. Rogić Vidaković

Optimization of the Head Geometry for a Cable Car Passing over a Support 231
M. Wenin, A. Windisch, S. Ladurner, M. L. Bertotti and G. Modanese

Comparison in Performance of Hybrid and Marvel NoKH Okra/*Abelmoschus esculentus* Fibre Reinforced Polymer Composites Under Tensile Load 243
Nadendla Srinivasababu

Analysis of Stress-Strain State of Multi-wave Shell on Parabolic Trapezoidal Plan 257
V. N. Ivanov, Timur Soibnazarovich Imomnazarov, Ismael Taha Farhan Farhan and Daou Tiekolo

General Boundary Element Method for the Dual-Phase Lag Equations Describing the Heating of Two-Layered Thin Metal Films 263
Ewa Majchrzak

Numerical and Experimental Analysis of a Personalized Prosthesis for a Patient with Unilateral Hip Osteoarthritis 279
Juan Alfonso Beltrán-Fernández, Omar Rolando Ruiz-Muñoz, Luis Héctor Hernández-Gómez, Alejandro González-Rebattú y González, Itzel Bantle-Chávez, Carolina Alvarado-Moreno, Pablo Moreno-Garibaldi and Nefi Pava Chipol

Biomechanical Fundamentals for Designing Machines that Allows Wear Testing 317
Iván González-Uribe, Liliana Gutiérrez-Lonche, Diego Ivan Islas-Jiménez, Guillermo Manuel Urriolagoitia-Calderón and Guillermo Urriolagoitia-Sosa

Mechanical Design and Numerical Analysis of a Femorotibial Implant in Patients with Medial Knee Osteoarthritis 323
José María Aburto-Barrera, Mildred Egure-Hidalgo, Christian Díaz-León, Juan Alejandro Vázquez-Feijoo and Guillermo Urriolagoitia-Sosa

Design and Comparative Numerical Analysis of Designs of Intramedular Telescopic Systems for the Rehabilitation of Patients with Osteogenesis Imperfecta (OI) Type III 333
J. L. Rueda, C. R. Torres, V. Ramírez, L. Martínez and G. Urriolagoitia

Finite Element Analysis of 3D Models of Upper and Lower Limbs of Mexican Patients with Osteogenesis Imperfecta (OI) Type III 343
V. Ramírez, C. R. Torres, J. L. Rueda, L. Martínez, B. Romero and G. M. Urriolagoitia

Manufacturing of a Human's Hand Prosthesis with Electronic Movable Phalanges Based on a CT Image: An Amputation Case 355
Juan Alfonso Beltrán-Fernández, Itzel Alejandrina Aguirre Hernández, Itzel Bantle-Chávez, Carolina Alvarado-Moreno, Luis Héctor Hernández-Gómez, Pablo Moreno-Garibaldi, Nefi Pava Chipol, Juan Carlos Hermida Ochoa, Adolfo Lopez Lievano and Guillermo Manuel Urriolagoitia-Calderón

Reconstruction and Finite Element Evaluation of a Calcaneous Implant by Stereolithographic 3D Printing Technique 397
Juan Alfonso Beltrán-Fernández, Juan Carlos Hermida-Ochoa, Adolfo López-Lievano, Luis Héctor Hernández-Gómez, Berenice Uribe-Cortes, Pablo Moreno-Garibaldi and Nefi Pava-Chipol

Development of an Auxiliary Biomechanical Assistance System for Blindness or Partial Vision 407
L. H. Hernández-Gómez, J. A. Beltrán-Fernández, J. Martínez-Paredes, J. Medel-Ortiz, Itzel Bantle-Chávez, Carolina Alvarado-Moreno, A. Vega-López, D. Villaseñor-Chávez and J. O. Miranda-Peña

Investigation of the Damage Behavior of Polyurethane in Stress Relaxation Experiments and Estimation of the Stress-at-Break σ_b with a Failure Envelope

Selina Neuhaus, Henning Seibert and Stefan Diebels

Abstract In stress relaxation experiments the investigated polyurethane exhibits an unexpected, but repeatable failure during the relaxation period. Images, taken by a camera and a high speed camera, displayed crack initiation several minutes before rupture occurred. The crack growth rate then accelerates and leads to failure rapidly. The present investigation indicates different methods for analyzing the damage evolution with focus on the appearance of damage at low strains, accumulation of damage processes at higher strains, the influence of time on damage evolution and the identification of recovery phenomena during unloading and in the unloaded state. The results show that not only long times at high strains but also the loading history, especially loading and unloading processes, accelerate damage evolution. In a preliminary study the characterization of the ultimate tensile properties by a failure envelope appears to provide an adequate method for estimating the stress-at-break σ_b and the time-to-break t_b in stress relaxation experiments.

Keywords Digital image correlation · Crack initiation · Crack growth · Damage evolution · Predeformation · Failure envelope

1 Introduction

Stress relaxation experiments are usually used to characterize the viscoelastic behavior of polymers [1–4]. During the relaxation period a stress decay due to the viscoelasticity is measurable. In general, damage rises with increasing stress [5]. In spite of the decrease of stress in stress relaxation experiments, the investigated polyurethane ruptures. Friedrich [6] also observed this damage behavior on an ethylene-propylene-diene monomer rubber. Especially in technical use of these materials such an unexpected failure has to be avoided. For this purpose, it is important to characterize the fracture process and to study the damage evolution as well as its dependencies from testing conditions. Friedrich [6] already showed, that the time-to-break during

S. Neuhaus (✉) · H. Seibert · S. Diebels
Applied Mechanics, Saarland University, Campus A4.2, 66123 Saarbrücken, Germany
e-mail: selina.neuhaus@uni-saarland.de

A. Öchsner and H. Altenbach (eds.), *Engineering Design Applications II*,
Advanced Structured Materials 113, https://doi.org/10.1007/978-3-030-20801-1_1

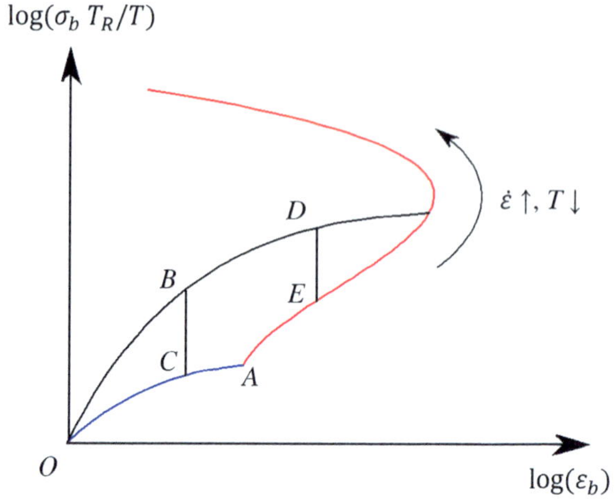

Fig. 1 Schematic representation of the failure envelope which connects rupture points from different tensile tests, used to estimate rupture in stress relaxation experiments

stress relaxation depends on the temperature, the strain rate and the maximal applied strain. In his experiments damage seemed to become minimal at a specific intermediate strain rate. Furthermore, he defined an upper threshold strain value ε_{oG} at which samples already break during loading, and a lower threshold strain value ε_{uG}, at which the specimens do not fail over the entire testing period.

Smith and Stedry [7] also observed rupture during the stress relaxation of a styrene-butadiene rubber. They developed a method to use the stress-strain data from tensile tests to estimate the stress-at-break σ_b, at which rupture occurs in stress relaxation and creep experiments. The ultimate properties are considered in terms of stress-strain curves to rupture measured at various temperatures and strain rates. In a diagram these rupture points $(\varepsilon_b|\sigma_b)$ are connected and build the so-called failure envelope (see Fig. 1). The plot of $\log(\sigma_b\, T_R/T)$ versus the logarithm of the strain-at-break ε_b, where T is the testing temperature and T_R is an arbitrary reference temperature, considers that the elastic retractive force in a specimen at a fixed extension ratio increases in direct proportion to the absolute temperature [8]. Provided time-temperature superposition is applicable, the failure envelope is independent of temperature, strain rate and testing method [7–9]. In Fig. 1 the curve $\overline{OA}$ represents stress-strain values under nearly equilibrium conditions in the absence of viscous effects. If a specimen is stretched along curve $\overline{OB}$ and the strain is held constant, stress will relax until it reaches C on the equilibrium stress-strain curve. But if the sample is elongated along curve $\overline{OD}$ and thereafter the strain is held constant, stress decreases and the specimen breaks by intersecting the failure envelope at E. Indeed, rupture can also occur along the equilibrium curve. Even though there is no viscous relaxation, the ultimate properties of tensile tests with nearly equilibrium conditions vary considerably with temperature and strain rate. Hence, A is not a sharp corner but a fluent transition from the equilibrium curve to the failure envelope [8].

2 Experimental

First the sample preparation, the sample geometry and the testing device are explained. Then the different experiments and the results are presented. In conventional stress relaxation experiments the relaxation behavior and the fracture process are investigated. Afterwards, different modifications of the stress relaxation experiment are outlined to study the time and loading history dependence of the damage evolution. At the end in a preliminary study the method of Smith and Stedry [7] appears to be an adequate possibility to estimate the stress-at-break σ_b in stress relaxation experiments.

2.1 *Material and Apparatus*

The investigated material is a cross-linked polyurethane. The monomer components used to prepare the samples were supplied by the manufacturer Covestro AG. A mixture of polyether-based diol (Desmophen® 3600Z) and polyether-based triol (Baygal® K 55) is used as the resin component with 90% of the hydroxyl groups of the diol and 10% of the triol. Hence, the resulting polyurethane is called PU90/10. The crosslinking agent is a diphenylmethane diisocyanate isomer mixture (Desmodur® VP.PU 1806) that is stoichiometrically mixed with the resin. To avoid undesirable reactions with water from the air, the total sample preparation was performed in a glove box. The polyurethane was then injected into individual grooves milled into a Teflon mold and allowed to cure a week at room temperature in dried air and afterwards a week at 60 °C in dried air. After this curing time the samples were removed from the Teflon mold. The thereby introduced internal stresses were reduced by a heat treatment at 60 °C for 20 min. The samples were then stored in a dry box.

The material was tested displacement-controlled with a uniaxial testing device at 30 °C in a heat chamber. In order to obtain the desired uniform temperature, the samples were preheated in the unstrained state for 15 min before they were elongated. In previous tensile tests the relation between the applied machine displacement and the strain in loading direction ε_{xx} and in cross direction ε_{yy} was determined via digital image correlation DIC. Hence, it was possible to execute the stress relaxation experiments without DIC using the calibration curves. The Cauchy-stress was then calculated by the ratio of the measured load F and the current cross-section ($(1 + \varepsilon_{yy})^2\ bd$) with an initial sample width b and an initial sample thickness d

$$\sigma = \frac{F}{(1+\varepsilon_{yy})^2} bd. \tag{2.1}$$

The results of the DIC also showed the compressible stress-strain behavior of PU90/10. The Poisson's ratio was determined to be approximately 0.39. A special waisted sample geometry (see Fig. 2) realizes a smallest cross-section area in the

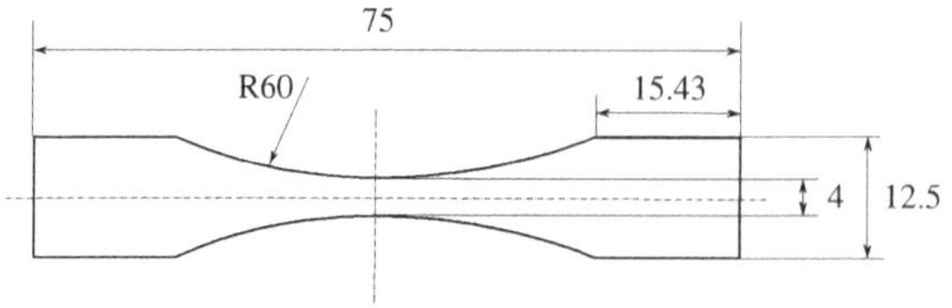

Fig. 2 Sample geometry with a sample thickness of 2 mm

center of the specimen with highest stresses and generates a replicable predetermined breaking point.

2.2 Experiments and Results

2.2.1 Conventional Stress Relaxation Experiments

Conventional stress relaxation experiments to rupture were carried out to get an overview of the relaxation behavior and the fracture process. Ten experiments provided a mean value of the time-to-break t_b of about 1040.5 s with a broad scattering (see Fig. 6) and a standard deviation of 887.1 s. t_b is the elapsed time from the beginning of the stress relaxation period until rupture occurs. All tests were conducted at 30 °C and with a rate of extension of 0.1 mm/s to a machine displacement of 50 mm. The maximal axial strain reaches about 200%. The elongation was held constant until the specimens broke. With these parameters PU90/10 shows a significant viscoelastic behavior in combination with a comparatively short time-to-break.

A DIC-picture (see Fig. 3) shows that the deformation is homogeneous in the middle of the sample. There is no strain concentration at one side, so the testing machine was axial well aligned. In one extra test, images were taken using a camera during the loading and the relaxation period to study the crack initiation and growth. The last seconds before rupture was recorded with a high speed camera to investigate the fracture process. Figure 4 shows the first frame of the high speed camera which displays the cracked sample.

In this investigation the crack appears during loading (see star symbol in Fig. 5) at 43 mm machine displacement. In some subsequent tests cracks just became visible during stress relaxation. In all cases the incipient crack appears long time before failure. In some samples the crack did not appear in the center of the sample which indicates pores inside the specimen. The results of these tests were omitted. The crack length in the pictures of both the camera (see square symbols in Fig. 5) and the high speed camera (see triangle symbols in Fig. 5) was measured with the software National Instruments Vision Assistant 2015. At the beginning of the relaxation period the crack grows barely. Predominantly after a longer time of about 3830 s crack growth increases markedly and the crack tip sharpens while necking in the fracture

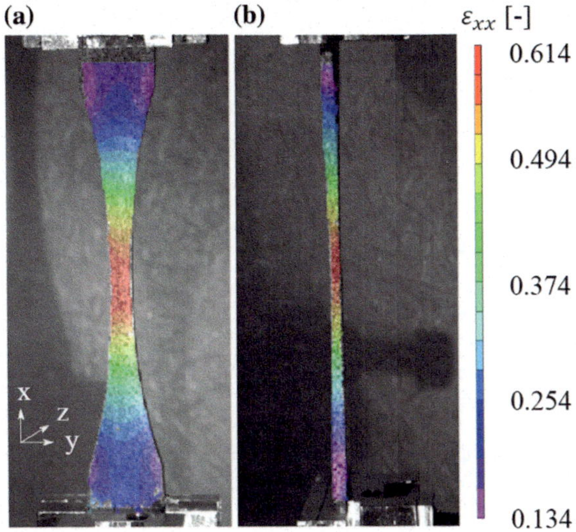

Fig. 3 DIC of a sample in (**a**) front view and (**b**) side view and the determined strain ε_{xx} in tensile direction

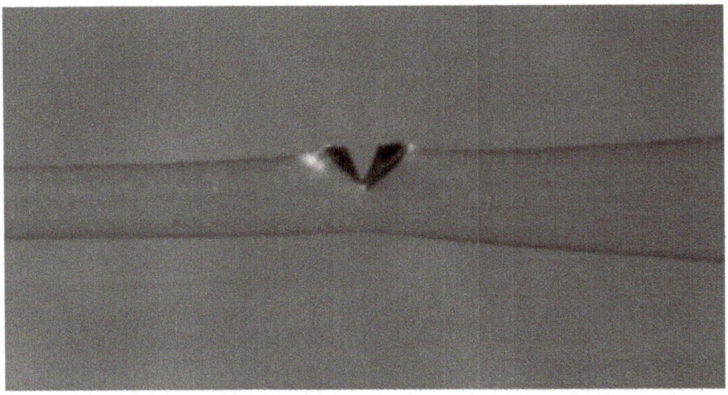

Fig. 4 Cracked sample 2.876 s before rupture occurred, picture taken with a high speed camera with 3000 frames per second

region and bending towards the crack arise. Total rupture occurred after 4995 s testing time.

Sometimes samples slipped slightly out of the clamps during testing. Hence, the effective applied strain and the achieved stress was less than expected. This also affects the strain rate. The local stress-strain behavior appeared to be independent of the slippage, so the reliability and the reproducibility of the results are still guaranteed. After every experiment the maximal stress σ_{max} at the beginning of the relaxation period when the maximum axial displacement is reached and the stress-at-break σ_{min} as the mean value of the last 5 recorded measurements before rupture were

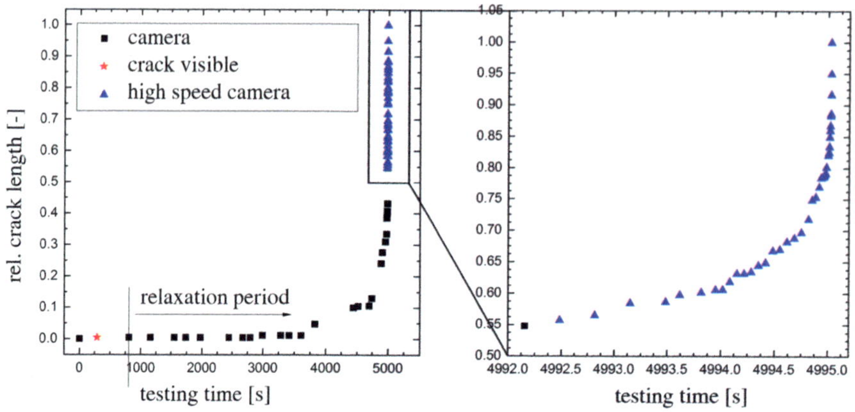

Fig. 5 Increasing crack length related to the initial sample thickness in a stress relaxation experiment

Fig. 6 Normalized stress difference of ten conventional stress relaxation experiments to rupture

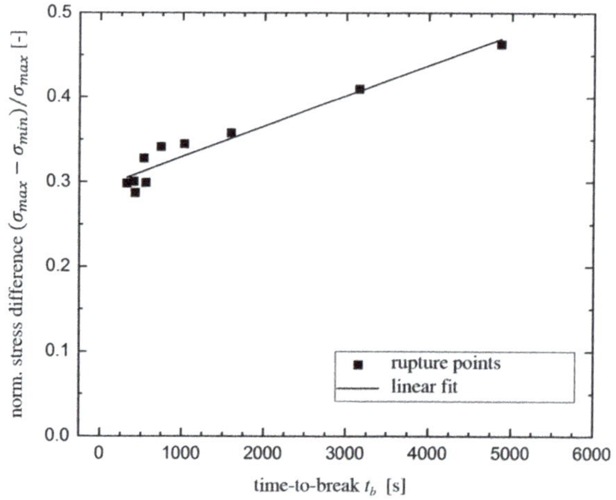

determined. The difference between σ_{max} and σ_{min} was calculated to compare the stress decays of different tests, to characterize the intrinsic material behavior and to quantify the sustained load. However, the less the maximal stress because of slippage of the sample out of the clamps, the slower the initial stress decay in the relaxation period [10, 11]. To reduce this effect, the calculated stress difference was divided by the maximal stress. The dependence of the so obtained normalized stress differences of the ten conventional stress relaxation experiments on the time-to-break t_b can be described by a straight line (see Fig. 6). This relation can be used as a failure criterion in a damage mechanics model.

2.2.2 Discontinuous Relaxation Experiments

To investigate the dependency of the damage evolution on the test duration and the overall time at high strains, in the second test series relaxation experiments were interrupted by several unloading processes and subsequent waiting times, at which the sample remains unstretched. After these waiting times the sample was reloaded up to a machine displacement of 50 mm and again stress relaxation with the same holding time than before was conducted. A waiting time of one day between each cycle of loading, stress relaxation and unloading was chosen for fifteen samples. Fifteen further samples were stored two weeks before the next cycle was conducted. At least, the cycles were performed one after another without waiting time (0 s) for fifteen specimens as well. The maximal waiting time of two weeks after one cycle of loading, stress relaxation and unloading was chosen because the specimen's length regains then its initial value. Hence, time-dependent internal stresses are mostly relieved after two weeks and the PU appears to deform viscoelastically. Therefore, the plastic part of deformation can be neglected. These experiments compared to those without a waiting time between the single cycles also allow to find out if there are recovery processes that reduce or at least retard damage. In the discontinuous relaxation experiments the samples with one day or two weeks of waiting time performed maximal five cycles provided they do not break before. However, those without waiting time were tested until rupture occurred also if more than five cycles were necessary.

The holding time, that defines how long the sample is exposed to the maximal strain, is another parameter, which was varied between 300, 500 and 800 s. Hence, the stress relaxation of five of the fifteen samples with a waiting time of one day (or two weeks or 0 s) was 300 s in every cycle. One half of the remaining ten specimens was exposed to 500 s and the other half to 800 s stress relaxation in each cycle. The mean time-to-break of 1040.5 s of the ten conventional relaxation experiments is the reason for this choice of holding times. Assuming that the mean time-to-break of conventional experiments is comparable to the overall time at maximal strain of all cycles of an interrupted relaxation experiment, rupture is expected to occur in the fourth cycle if the holding time in each cycle is 300 s. Analogously, with a holding time of 500 s or 800 s the sample is anticipated to break in the third or the second cycle. 300 s is the smallest holding time, because in conventional relaxation experiments no specimen failed in this time.

Besides, five other samples were tested in a cyclic test (without waiting time and without holding time) with a mean displacement of 25 mm and a maximal displacement of 50 mm. So, a total number of 50 discontinuous relaxation experiments were performed. Combining the different holding times with the mentioned values for the waiting times, ten interrupted relaxation experiments, including the cyclic tests, can be distinguished, which were each repeated with five samples. Figure 7 illustrates exemplary the five experiments with two weeks waiting time and 800 s holding time. The tests were carried out in a displacement-controlled manner, which cause small compressive stresses after unloading.

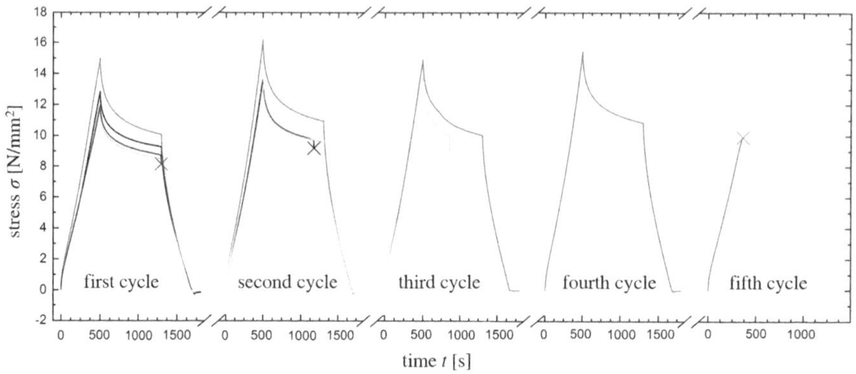

Fig. 7 Stress-time-diagrams of the five discontinuous relaxation experiments with two weeks waiting time (marked with breaks //) and 800 s holding time: crosses mark the rupture points, different gray scales represent distinct samples

Especially in the case of the sample with the highest maximal stress in every cycle an increase of stress in the second cycle compared to the first cycle can be observed. This effect just emerges in the experiments with a waiting time of one day or two weeks because in these tests the specimens were removed from the clamps and again stored in the dry box between the single cycles. When clamping the sample once again to conduct the next cycle, the preload and the clamping is not 100% identical to those in the previous cycle. Therefore, the test starting conditions differ in a series of discontinuous relaxation test. Also the measurement inaccuracy due to slippage is more pronounced, because it varies not only from sample to sample but also from cycle to cycle. The better the clamping, the less the slippage of the sample and the higher the achieved stress. The discontinuous relaxation experiments with waiting times of one day or two weeks provide comparable results. So, the relaxation processes between one day and two weeks appear not to significantly influence the results.

The assumption of damage being reduced or retarded by recovery effects is not confirmed. On the contrary, the samples in tests without waiting time tend to withstand the maximal loading for longer. A conceivable reason may be the fact, that between the cycles in these tests, the samples were not removed from the clamps. Hence, the results of the discontinuous relaxation experiments without waiting time always show a decrease of the maximal stress from cycle to cycle due to the viscoelastic behavior of the polyurethane. In the interrupted relaxation experiments with waiting times viscoelasticity leads to relaxation processes which relieve time-dependent stresses. Thus, in the next cycle the maximal stress is higher. But with increasing maximal stress, the lifetime of the specimen tends to decrease. Besides, the normalized stress difference is higher.

In consequence of the influence of the viscoelasticity, the clamping and the slippage on the results, together with the broad scattering of the time-to-break, comparability between experiments with and without waiting time is reduced. On this basis, it is not possible to get information about the existence of recovery effects during the waiting time and their influence on damage evolution.

To guarantee better comparability of the results of the experiments with and without waiting time, in further investigations samples could also be removed from the clamps in the latter case. On the other hand, the first cycle of these experiments without waiting time can be conducted as shown above and its maximal stress can be preset as the beginning of the stress relaxation in all further cycles. Then DIC will be necessary with the maximal strain differing from cycle to cycle.

In the discontinuous stress relaxation experiments the PU90/10 specimen tended to sustain more cycles before rupture than expected regarding the mean time-to-break in conventional relaxation tests. This is shown by three diagrams in Fig. 8. The ordinate displays the number of cycles within which the begin of the relaxation period was reached before failure occurred. So, the breaking point of a sample which broke during loading in the third cycle is plotted at a number of cycles of 2. The gray scale illustrates the relative number of broken samples at the considered breaking point. This is used as an estimation for the probability of fracture with increasing time. Since each of the ten different discontinuous relaxation experiments was repeated with five samples, a probability of fracture increases from 0.2 to 1 between the first and the fifth breaking point. The abscissa depicts the total testing time including loading, stress relaxation and unloading process of every cycle. The experimental results are divided into three diagrams according to the waiting time.

The fact, that the investigated material sustained more cycles than expected, indicates, that damage depends not only on the endured time at high deformation but also on the loading history, meaning the foregoing loading and unloading paths. Loading and unloading appear to be correlated with damage because samples also break in a cyclic test without waiting time and holding time. Besides, the maximal time-to-break in the cyclic tests is only slightly higher than in discontinuous relaxation experiments with 800 s holding time.

In the tests with waiting times of one day or two weeks, the samples tendentially sustain the least cycles in the experiments with 500 s holding time instead of those with relaxation periods of 800 s, although the percentage of overall time at maximal strain with respect to the total testing time decreases with decreasing holding time. With holding times of 300 s or 500 s the samples also appear to fail earlier than in tests with 800 s. A reduction of the holding time makes the loading and unloading paths all the more significant. Therefore, both the loading history and the overall time at high strains initiate damage processes and lead to failure.

2.2.3 Relaxation Experiments with Preceding Deformation

The light gray curves in Fig. 9 demonstrate the two different kinds of deformation, that were carried out two weeks before a stress relaxation experiment at 50 mm maximal machine displacement to rupture was conducted. On the one hand samples were elongated to 50 mm machine displacement with 0.1 mm/s and immediately unloaded with the same machine rate (see "peak"' in Fig. 9(a)). The previous exper-

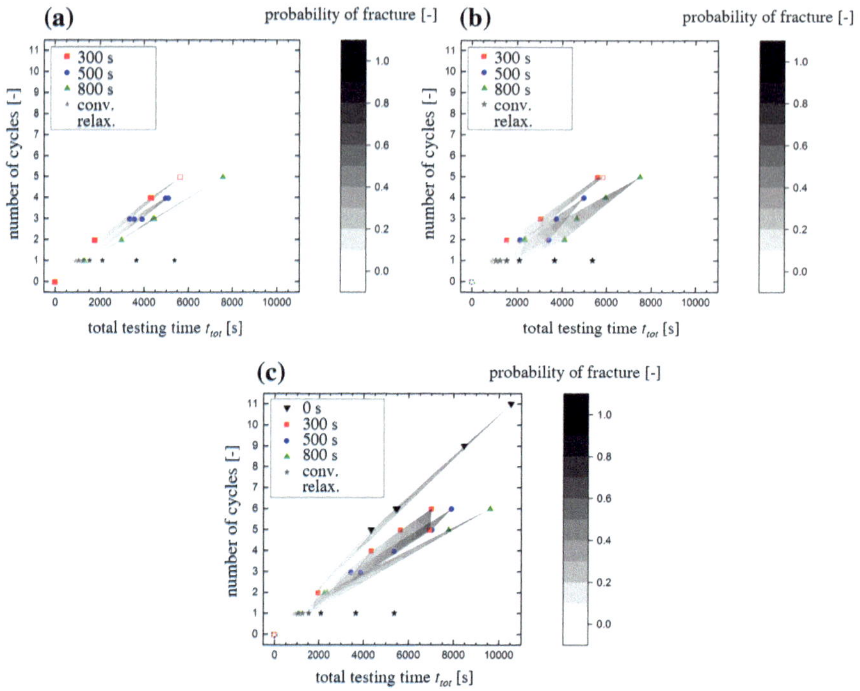

Fig. 8 Probability of fracture depending on the total testing time and the number of sustained cycles split in three diagrams according to the waiting time: (**a**) two weeks, (**b**) one day and (**c**) without waiting time

iments showed that damage depends on time at high strains and loading history. The preceding deformation should clarify, if damage processes already proceed during loading or if they mostly need enough time to form. In the first case damage should be measurable in a following loading process in form of a decrease of the time-to-break of the polyurethane in the relaxation period.

On the other hand, a stress relaxation experiment with a holding time of two hours at 10 mm machine displacement was conducted to see if damage also appears at low strains. 10 mm appeared to be less than the lower threshold strain value ε_{uG} because even after 18 h holding time at 10 mm the sample did not have any visible cracks. To avoid an influence of time-dependent viscoelastic effects on the stress-strain behavior a waiting time of two weeks was chosen between the preceding deformation and the stress relaxation experiment. Every kind of predeformation was applied to five samples. In addition, the "peak" treatment was carried out with two further specimens because of the broad scattering of the time-to-break and the achieved maximal stresses.

The reduction of the normalized stress differences in comparison to the results of stress relaxation experiments without predeformation (see Fig. 10) may suggest, that the preceding deformations had an effect on the stress. The maximal stress at the

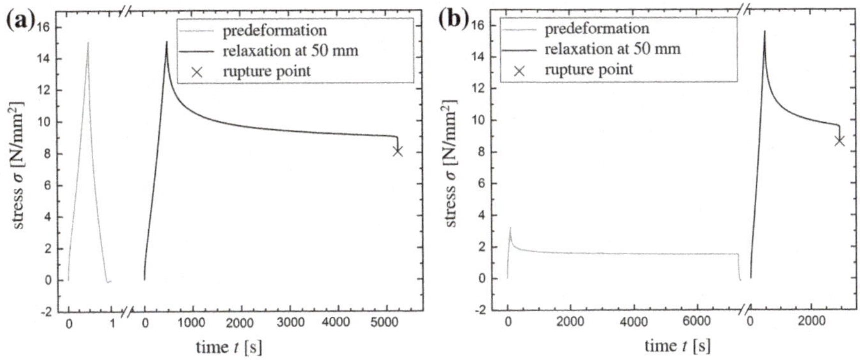

Fig. 9 Stress relaxation experiments with a preceding deformation in terms of (**a**) a 50 mm "peak", (**b**) a stress relaxation at 10 mm machine displacement

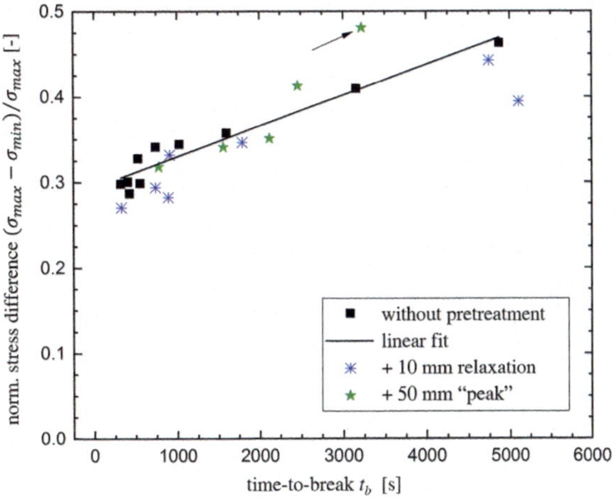

Fig. 10 Normalized stress differences of stress relaxation experiments to rupture with a preceding deformation compared to ten conventional stress relaxation experiments without pretreatment

beginning of the relaxation period was mostly less than that in conventional relaxation experiments without predeformation, whereas the time-to-break remained similar. But the slippage of pretreated samples in stress relaxation experiments was higher than that of specimen without predeformation. Consequently, the difference between the highest and the lowest maximal stress of the stress relaxation experiments with preceding treatment is two to four times higher than in the experiments without predeformation. So, the influence of a decreasing maximal stress on the initial stress decay in the relaxation period was more pronounced and caused a decrease in the normalized stress difference. Thus, this effect superposes the influence of the predeformation on stress. An arrow in Fig. 10 marks the breaking point of a specimen with a maximal stress far less compared to the other samples.

In conclusion, it is not clear, how the different kinds of preceding deformation affect the relaxation behavior and the fracture process. However, these are adequate methods to investigate the time- and strain-dependency of the damage evolution

Table 1 Testing conditions of tensile experiments to rupture (rupture points see black crosses in Fig. 11) to create a failure envelope

T [°C]	25	30	30	30	30	40	40	40	40	40
$\dot{u}$[mm/s]	0.05	0.05	0.1	0.5	10	0.02	0.05	0.1	0.5	1

provided the disturbing effect of slippage is eliminated. This can be realized by choosing a fix stress value for the beginning of the relaxation period, instead of a fix machine displacement value, in combination with DIC.

2.2.4 Failure Envelope

To test the applicability of the previously explained method of Smith and Stedry [7] to the investigated PU90/10, ten tensile tests under different testing conditions (different temperatures and constant machine rates, see Table 1) to rupture were conducted. For the sake of simplicity, tests were carried out without DIC in this preliminary study. Therefore, the measured load-at-break F_b and the achieved displacement were recorded and plotted as crosses on a load-displacement-diagram (see Fig. 11). In contrast to the method of Smith and Stedry [7], in this investigation the influence of temperature on the elastic retractive force was neglected, because the chosen temperature range was small. Hence, the load-at-break F_b is not multiplied by the ratio of a reference temperature to the testing temperature. The resulting failure envelope is estimated and sketched in Fig. 11 as a dashed line. The data of the tensile test at 40 °C and with a machine rate of 0.02 mm/s represents approximately the equilibrium curve. Three of the ten conventional stress relaxation experiments at a maximal machine displacement of 50 mm to rupture and the previously mentioned stress relaxation experiment at 10 mm over 18 h without failure are also displayed in the diagram.

Remarkably, the three samples, that relaxed at 50 mm, did not break until the decreasing load intersected the failure envelope during the relaxation period. However, in the relaxation test at 10 mm the load reached the equilibrium curve and the sample did not break even after 18 h. Therefore, the experiments show the material behavior, that was expected by the theory of Smith and Stedry [7]. The failure envelope appears to be an adequate possibility of estimating the stress-at-break σ_b (here: the load-at-break F_b) in a stress relaxation experiment and a prediction of the time-to-break can be possible, provided that the relaxation rate is known.

3 Discussion

Of course, the interpretation of the shown results should be handled with care. The fracture process and the time-to-break is subjected to significant statistical fluctua-

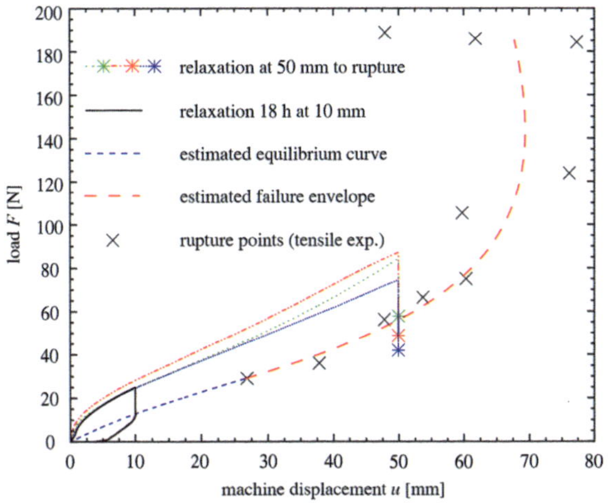

Fig. 11 Rupture points measured in tensile experiments build a failure envelope, which is compared to three conventional stress relaxation experiments to rupture (maximal machine displacement of 50 mm) and one conventional stress relaxation experiment without rupture (18 h at a maximal machine displacement of 10 mm)

tions. Hence, the small number of experiments just allows to observe tendencies. In further investigations the conclusions have to be verified by a larger number of samples per experiment. Besides, the storage of the specimen was humidity controlled but during testing it was not possible to adjust the humidity. So, it is not clear how moisture absorption influences the results.

Further tests should be carried out with optimized clamps, combined with DIC and a modification of the discontinuous relaxation experiments without waiting time. To modify the interrupted experiments, the first cycle can be conducted as shown in the present investigation. But afterwards, the maximal stress achieved in the first cycle determines the beginning of the stress relaxation period in all further cycles in order to reduce the influence of slippage of the samples out of the clamps on the maximal stress. Therefore, the initial stress decay in the relaxation period will be less dependent on slippage. In general, this modification is applicable, because it turned out that the stress-strain behavior is not considerably influenced by slippage. The suggested modification reduces the effect of the viscoelastic properties on the results of experiments without waiting time. These results will be more comparable to those of experiments with waiting time. Consequently, these tests will show, if there are recovery phenomena during unloading and in the unloaded state, that reduce or at least retard damage. Furthermore, it will be possible to study the effect of the predeformation on stress and therefore on damage evolution and on the time-to-break. Relaxation experiments with a foregoing relaxation at 10 mm machine displacement over two hours then will provide information about the appearance of damage at low strains.

In cyclic experiments, samples broke after a certain number of cycles. Combined with the conclusion that damage is not only influenced by the overall time at high strains, but also by the loading history, the question arises, whether damage processes already proceed during loading and unloading, but need some time to fully develop. A preceding deformation in form of one single cycle together with the above explained modification of the experiments will clarify, if damage processes need enough time to form by causing the samples to fail earlier.

In Sect. 2.2.4 the applicability of the method of Smith and Stedry [7] to the investigated PU90/10 is analyzed. Similar to the results of Friedrich [6] there is a lower threshold strain value, which is dependent on temperature and strain rate. This threshold defines a low strain range within which samples do not break in a relaxation experiment. In addition, in Friedrich [6] damage became minimal at an intermediate machine rate. Equally, in the investigations of Smith [12] the strain-at-break reached a maximum at a certain strain rate, so the damage became less. In this preliminary study, it appears to be possible to estimate the stress-at-break σ_b in stress relaxation with the failure envelope. This has to be verified in further investigations combined with DIC to allow the calculation of $\log(\sigma_b\ T_R/T)$ and $\log(\varepsilon_b)$ and to create a diagram similar to that of Smith and Stedry [7]. Besides, more tensile experiments to rupture have to be conducted to improve the approximation of the failure envelope. Also the capability of the method should be proved with a number of different relaxation experiments at different machine rates, temperatures and maximal machine displacements.

4 Conclusion

In spite of the small number of repetition per experiment, the results show the potential of the presented methods for analyzing the damage behavior of the investigated polyurethane. Discontinuous stress relaxation experiments led to the conclusion, that damage not only depends on the overall time at high strains, but is also influenced by the loading history. Cyclic experiments (discontinuous stress relaxation experiments without waiting time and holding time) proved, that damage already proceeds during loading and unloading. However, it should be clear, that the statistical uncertainty in those findings is very large, but the results display qualitative trends.

Moreover, in a preliminary study the method of Smith and Stedry [7] appeared to be adequate to estimate the stress-at-break in a stress relaxation experiment with a given maximal stress. Provided the relaxation behavior of PU90/10 in the relaxation phase is known, the time-to-break can also be predicted. The method holds potential for the prediction of the probability of fracture during the relaxation period and the time-to-break.

Acknowledgements We gratefully acknowledge the support of Prof. Dr. rer. nat. habil. Wulff Possart, Chair for Adhesion and Interphases in Polymers, Saarland University, for providing access to materials and equipment for sample preparation and to the dry box.

References

1. Tobolsky AV (1956) Stress relaxation studies of the viscoelastic properties of polymers. J App Phys 27:673–685
2. Bergström JS, Boyce MC (1998) Constitutive modeling of the large strain time-dependent behavior of elastomers. J Mech Phys Solids 46:931–954
3. Kahn AS, Lopez-Pamies O (2002) Time and temperature dependent response and relaxation of a soft polymer. Int J Plasticity 18:1359–1372
4. Tobolsky AV, Prettyman IB, Dillon JH (1944) Stress relaxation of natural and synthetic rubber stocks. Rubber Chem Technol 17:551–575
5. Kausch HH (2012) Polymer fracture. Springer Science & Business Media, Berlin
6. Friedrich L (2017) Untersuchungen zum Materialverhalten poröser Elastomere während der Relaxation. Bachelor thesis, Chair of Applied Mechanics, Saarland University
7. Smith TL, Stedry PJ (1960) Time and temperature dependence of the ultimate properties of an SBR rubber at constant elongations. J Appl Phys 31:1892–1898
8. Smith TL (1963) Ultimate tensile properties of elastomers. I. Characterization by a time and temperature independent failure envelope. J Polym Sci Part A General Papers 1:3597–3615
9. Smith TL (1964) Ultimate tensile properties of elastomers. II. Comparison of failure envelopes for unfilled vulcanizates. Rubber Chem Technol 4:792–807
10. Lai JS, Findley WN (1968) Stress relaxation of nonlinear viscoelastic material under uniaxial strain. Trans Soc Rheol 12:259–280
11. Wang TT, Klosner JM (1969) Relaxation properties of polyester-based polyurethane under small deformations superposed on large deformations. Trans Soc Rheol 13:193–208
12. Smith TL (1958) Dependence of the ultimate properties of a GR-S rubber on strain rate and temperature. J Polym Sci 32:99–113

Identification of Inelastic Parameters of the AISI 304 Stainless Steel

Miguel Vaz Jr., E. R. Hulse and M. Tomiyama

Abstract A proper choice of inelastic parameters is one of the most important aspects for a successful simulation of metal forming processes. Several issues must be observed when choosing such parameters, amongst which the compatibility between the magnitude of the plastic deformation of the target forming operation and the mechanical test employed to obtain those parameters. Within this context, the present work addresses the suitability of selected phenomenological hardening models and identification of the corresponding inelastic parameters based on curve-fitting strategies (logarithmic-based equations) and optimization methods (non-logarithmic models) for the AISI 304 austenitic stainless steel. Tensile tests were performed using specimens of different sizes. Based on a combined assessment of all types of specimens, it was observed that the curve-fitting technique was able to describe with excellent accuracy deformations up to maximum load. In order to contemplate larger plastic deformations, an inverse problem strategy based on optimization methods was used to account for material response up to macroscopic failure of the specimens. Numerical simulation of the tensile tests shows that the latter technique associated with non-logarithmic hardening equations provided the best approximation to the experimental data.

1 Introduction

Austenitic stainless steels are the most widely used type of corrosion-resistant steels. This class of steels are generally characterized by containing low levels of Carbon and high levels of Chromium and Nickel. In addition to Cr and Ni, other elements

M. Vaz Jr. (✉) · E. R. Hulse · M. Tomiyama
State University of Santa Catarina, Campus Universitário Prof. Avelino Marcante, Joinville 89219-710, Brazil
e-mail: Miguel.Vaz@udesc.br

E. R. Hulse
e-mail: emilio.r.hulse@embraco.com

M. Tomiyama
e-mail: masahiro.tomiyama@udesc.br

A. Öchsner and H. Altenbach (eds.), *Engineering Design Applications II*,
Advanced Structured Materials 113, https://doi.org/10.1007/978-3-030-20801-1_2

may be added in order to confer specific properties and characteristics to the material, e.g. Molybdenum to increase pitting resistance, Phosphorus and Sulphur to improve machinability, Selenium for better machined surfaces, Boron to enhance the steel capacity to absorb neutrons in nuclear thermal reactors, Silicon for higher heat resistance and Copper to improve cold working, amongst other elements [1]. The stainless steel AISI 304 is the most used grade due to its combination of mechanical (formability and weldability) and chemical (corrosion and oxidation resistance) characteristics.

The good formability of the type 304 stainless steel is associated with its relatively low yield stress and high ductility, thereby allowing large plastic deformation before onset of mechanical failure. Such features favour cold working processing [2] (roll forming, deep drawing, etc.), making possible to manufacture with greater efficiency a wide range of products and components, including consumer items, architectural elements and industrial equipment. Therefore, a proper description of mechanical properties is highly significant to industries, especially when using computational packages to simulate metal forming operations (the user must provide actual inelastic parameters). This work is inserted in this context, aiming to contribute to the discussion by assessing the suitability of some selected phenomenological equations to model hardening evolution and alternative strategies to obtain the corresponding parameters.

Form the viewpoint of tool design and prediction of manufacturing costs, including the modern concept of energy accountability, the issues briefly discussed in the previous paragraph highlights the necessity to develop a methodology to determine material parameters. Material parameters describe different aspects of the nature of the material, ranging from mechanical and thermal behaviour to chemical kinetics. The strategy addressed in this work is primarily concerned with mechanical parameters in association with inelastic deformation.

The literature shows an increasing number of works discussing strategies to determine different types of inelastic parameters, which in turn are associated with the material constitutive relation (plastic, viscoplastic, etc.) and purpose of the analysis itself (hardening, damage, etc.). Techniques based on *curve fitting* and *optimization methods* have been largely adopted in this class of problems. For instance, the suitability of selected hardening equations to describe plastic deformation of ductile materials was addressed by Samuel and co-workers [3, 4]. The authors studied several ductile materials, amongst which the AISI 316L stainless steel, and used a *curve-fitting technique* up to instability onset to obtain the hardening parameters. One of the relevant works on the application of *optimization methods* to obtain hardening parameters was presented by Ponthot and Kleinermann [5]. The study is mostly focused on mathematical aspects of several gradient-based optimization schemes. The authors proposed application in cascade of optimization methods to identification of hardening parameters of the ASTM A533, Grade B, Class 1 low alloy steel based on tensile tests.

A discussion on the hardening behaviour and a proposal of a new phenomenological hardening equation were presented by Dimatteo et al. [6] for Dual Phase (DP) 450/600/800/1000 and Transformation Induced Plasticity (TRIP) 800 steels.

A *curve-fitting strategy* associated with tensile tests was adopted to determine the hardening parameters of a new 3rd order logarithmic yield stress curve. The validation procedure indicated excellent correlation of the engineering stress and strain curves up to maximum load. Noticeably, the present work also investigates the suitability of the hardening equation proposed by Dimatteo et al. [6] to describe isothermal hardening evolution for the AISI 304 stainless steel. A parameter identification technique using *optimization methods* was used by Vaz Jr. et al. [7] to obtain hardening and fracture parameters of low Carbon AISI 1020 steel. The authors discussed application of hybrid optimization strategies and described an experimental-numerical scheme to determine inelastic parameters for two damage constitutive relations.

The brief review presented in the previous paragraphs is not exhaustive and intends to illustrate some of the existing approaches regarding the nature of the inelastic parameters and methodology of parameter identification. It is important to mention that other relevant investigations on the subject are discussed later in the appropriate sections. The present work addresses the suitability of selected phenomenological hardening models and identification of the corresponding inelastic parameters based on *curve-fitting strategies* (logarithmic-based equations) and *optimization methods* (non-logarithmic models) for the AISI 304 stainless steel. The chapter is organized as follows: Section 2 introduces a discussion on logarithmic-based phenomenological hardening equations and its intrinsic association with curve-fitting procedures. Section 3 presents alternative, non-logarithmic hardening equations and an optimization-based strategy to obtain the corresponding hardening parameters. Application of both parameter identification techniques is discussed in Sect. 4 based on tensile tests using specimens of different sizes prepared according to two different technical standards. The main conclusions and remarks are summarised in Sect. 5.

2 Phenomenological Hardening Equations and Curve Fitting Strategies

Tensile tests associated with curve fitting strategies have largely been adopted to determine yield stress parameters for metal materials. In addition to its simplicity, this technique makes possible to directly correlate *true stresses* and *true strains*, thereby conferring a straightforward physical significance. For the sake of objectivity, the reader is referred to Davis [8] for a detailed description of the method, including experimental considerations and requirements. Notwithstanding, it is relevant to emphasise that logarithmic true strains and true stresses represent the actual state of the material and are derived by considering the actual area and length of the specimen as well as constant volume during plastic deformation. The assumption of uniform deformation required by this method restricts evaluation of strains and stresses up to necking (instability) onset. The curve fitting technique is particularly significant when determining the material parameters based on specimens of different sizes. This strategy evinces eventual measurement discrepancies: in spite of different levels of tensile loads (owing to different cross-sectional areas), the true stress strain curves for each test must show very close approximation up to maximum load.

The literature shows a wide range of empirical equations to describe the yield stress based on logarithmic, $(\ln\sigma) \times (\ln\varepsilon)$ curves. This work addresses yield stress described by the classical Swift's equation [9] and alternative approaches, such as the double-n Swift [10] and a third-order logarithmic polynomial curve [6] (referred here also as cubic logarithmic equation).

One of the most well-known yield stress equation is due to Swift [9]. The author addressed mathematically the conditions for instability of plastic strains based on the Mises-Henky constitutive model. The instability conditions were demonstrated for a strain-hardening material described by the empirical relation

$$\sigma_Y = k\left(\bar{\varepsilon}_p + \varepsilon_0\right)^n, \tag{1}$$

where σ_Y is the yield stress, $\bar{\varepsilon}_p$ is the equivalent plastic strain, k is known as strength coefficient, n is the strain-hardening exponent, and ε_0 is an initial measure of the plastic state of the material (referred here as initial strain). The fitting constants k and n can be easily determined from double logarithm plot of the experimental true stress-strain data within the envelope comprised by the elastic deformation and onset of plastic instability (maximum load). The initial strain, ε_0, is obtained by the intersection of the uniaxial elastic and plastic curves, $\sigma_0 = E\varepsilon_0 = k\varepsilon_0^n$, so that $\varepsilon_0 = (E/k)^{1/(n-1)}$, where σ_0 is the initial yield stress and E is the Young's modulus.

Further modifications of Eq. (1) have been proposed to accommodate non-linear variations of the hardening evolution. Hertelé et al. [10] highlight the fact that two-stage hardening has been observed in various types of metals, including stainless steels. The authors proposed a strategy based on the Ramberg-Osgood equation and presented results for the DIN 1.4462 duplex stainless steel. Kashyap et al. [11, 12] adopted two and three stages of the Holloman relation, $\sigma_Y = k\varepsilon^n$, to describe the behaviour of the AISI 316L stainless steel under high temperature.

In the present work, a double-n Swift curve is investigated, so that

$$\sigma_Y = \begin{cases} k_1(\bar{\varepsilon}_p + \varepsilon_0)^{n_1} & \text{if} \quad \bar{\varepsilon}_p \leqslant \bar{\varepsilon}_p^T \\ k_2(\bar{\varepsilon}_p + \varepsilon_0)^{n_2} & \text{if} \quad \bar{\varepsilon}_p > \bar{\varepsilon}_p^T \end{cases}, \tag{2}$$

in which k_1 and k_2 are the strength coefficients of each stage, n_1 and n_2 are the corresponding hardening indices, $\varepsilon_0 = (E/k_1)^{1/(n_1-1)}$ is the initial strain, and $\varepsilon_p^T = \varepsilon^T - \varepsilon_0 = \exp[\ln(k_1/k_2)/(n_2 - n_1)] - \varepsilon_0$ is the transitional equivalent plastic strain.

The material parameters (k_1, n_1) and (k_2, n_2) of Eq. (2) can be obtained from a double logarithmic plot of the experimental true stress-true strain data based on the assumption of uniform plastic deformation. The key aspect of the double-n Swift curve is the proper computation of the transition point. In this work, the transition between curves "1" and "2" is determined by minimizing a combination of the fitting errors of both hardening stages when determining parameters (k_1, n_1) and (k_2, n_2) (the technique makes use of a simple line search method which maximizes the average coefficient of determination, $\overline{R}^2(\varepsilon^T) = (R_1^2 + R_2^2)/2$, of curves "1" and "2"). It is also important to note that, contrasting to other empirical equations, the material parameters k and n of the Swift and double-n Swift curves provide a strong physical and technological significance. The strength coefficient, k, directly indicates the magnitude of the forces involved in forming operations, whereas the value of the hardening exponent, n, points out preferred forming processes according to its level.

Within the framework of empirical descriptions of plastic deformation, Dimatteo et al. [6] proposed use of a 3rd order $(\ln \sigma) \times (\ln \varepsilon)$ polynomial fit to DP and TRIP steels in an attempt to predict the influence of alloying elements and heat treatment conditions in a simple manner. This work investigates a possible extension of the applicability range to the AISI 304 stainless steel. In the present study, provision for the elastic envelope is also included, so that

$$\ln(\sigma_Y) = A[\ln(\bar{\varepsilon}_p + \varepsilon_0)]^3 + B[\ln(\bar{\varepsilon}_p + \varepsilon_0)]^2 + C[\ln(\bar{\varepsilon}_p + \varepsilon_0)] + D, \tag{3}$$

where A, B, C, and D are the model parameters to be determined by the curve fitting procedure. Similarly to Swift's equation, the initial strain, ε_0, is obtained by the intersection between the uniaxial elastic and plastic curves and solving the cubic equation, $A[\ln(\varepsilon_0)]^3 + B[\ln(\varepsilon_0)]^2 + (C - 1)[\ln(\varepsilon_0)] + [D - \ln(E)] = 0$, for the initial strain.
Remark (i): The curve fitting procedures seek the best approximation under uniaxial stress conditions. Therefore, a combination of such strategy and the aforementioned hardening equations does not guarantee a priori accurate yield stress predictions beyond the instability point. This fact has important implications when computing the actual forming load for large plastic deformations, as discussed in the following sections.
Remark (ii): It is relevant to mention that the AISI 304 stainless steel is subject to deformation-induced phase transformation of austenite to martensite. The phase transformation mechanisms give rise to complex hardening behaviour according temperature and strain rates. De et al. [13] indicate that the stress-strain curves for low strain rates and temperatures higher than 298 K follow a typical "parabolic" evolution, which, therefore, makes possible to model inelastic deformation using straightforward phenomenological approaches.

3 Alternative Yield Stress Equations and Identification of Material Parameters

It is particularly relevant to acknowledge that phenomenological constitutive models have been largely used (and preferred) in industry. Moreover, many non-logarithmic hardening equations are empirical in nature and have been recommended for individual materials or else portraying specific internal structures (see, for instance, Larour [14] and references therein). However, there have been proposed some yield stress equations derived by using micromechanical concepts that can also be applied within the realm of phenomenological models. This work considers two non-logarithmic hardening equations which can be obtained from physical mechanisms of dislocation theory and yet recover well-known empirical models (without, however, addressing phase transformation).

El-Magd [15] accounts for physical mechanisms of dislocation theory assuming a Taylor-like correlation between the yield stress and total dislocation density associated with Mecking and Kocks' [16] evolution of the dislocation density. A constant assumption of the micromechanical parameters gives rise to a yield stress equation identical with Voce's [17] empirical relation,

$$\begin{aligned}\sigma_Y &= C_1^v + \left(C_2^v - C_1^v\right)\exp\left(-C_3^v\bar{\varepsilon}_p\right)\\ &= \sigma_0^v + \left(\sigma_\infty^v - \sigma_0^v\right)\left[1 - \exp\left(-\delta^v\bar{\varepsilon}_p\right)\right]\end{aligned}, \tag{4}$$

where $C_1^v = \sigma_\infty^v$ and $C_2^v = \sigma_0^v$ are associated with the saturation and initial yield stress, respectively, and $C_3^v = \delta^v$ represents the deformation rate to approach the saturation stress. Depending upon the material, Voce's equation provides good fit only for small plastic strains owing to the fact that the yield stress asymptotically approaches a constant value for large plastic deformations, i.e. $\sigma_Y \to \sigma_\infty^v$. For instance, El-Magd [15] indicates that Eq. (4) presents acceptable approximations for the austenitic steel X6CrNi18-11 at different temperatures in the range of relatively small strain up to 0.2.

A better representation for larger plastic strains can be derived by approximating the effects of the dislocation storage rate as a linear function of plastic strains [15], so that

$$\begin{aligned}\sigma_Y &= C_1^t + C_4^t\bar{\varepsilon}_p + \left(C_2^t - C_1^t\right)\exp\left(-C_3^t\bar{\varepsilon}_p\right)\\ &= \sigma_0^t + \zeta\bar{\varepsilon}_p + \left(\sigma_\infty^t - \sigma_0^t\right)\left[1 - \exp\left(-\delta^t\bar{\varepsilon}_p\right)\right]\end{aligned}, \tag{5}$$

in which $C_1^t = \sigma_\infty^t$ and $C_2^t = \sigma_0^t$ represents the saturation and initial yield stress, respectively, $C_3^v = \delta^t$ is the deformation rate to approach the saturation stress, and $C_4^t = \zeta^t$ portraits the linear hardening coefficient. This equation closely approximates an empirical model proposed by Tome et al. [18]. In addition, hardening laws similar to Eq. (5) have oftentimes been referred as *modified Voce* model. The literature shows increasing use of this hardening equation to describe the yield stress evolution of different materials, such as the 100Cr6 steel [19], X8CrNiMoNb16-16 austenitic steel [15], DP780 and TRIP780 high strength steels [20], and Al-2/4/8/10Si casting alloys [21].

Remark (iii): Voce's and modified Voce hardening equations were not conceived to follow the plastic deformation behaviour naturally represented in the classical logarithmic form. Therefore, this work adopts an optimization strategy to determine parameters of both Eqs. (4) and (5), thereby making possible to evaluate plastic deformation up to final deformation stages (catastrophic failure).

3.1 Identification of Material Parameters—The Optimization Algorithm

Parameter identification can also be regarded as an inverse problem, which consists of finding a set of material parameters which minimizes the differences between *experimental measures* and corresponding *computed responses* with respect to a given norm [22]. In the present study, (i) the *experimental measures* comprise loads and displacements evaluated in tensile tests using specimens of different geometry up to macroscopic failure. (ii) The *computed response* is determined by solving an elastic-plastic boundary value problem, which, in this work, is solved by an in-house finite element code based on a hyperelastic, large deformation formulation [23, 24]. (iii) The norm defines how differences are computed and this study adopts a linear combination of the L_2 error norm for each tensile test.

Unconstrained nonlinear optimization is utilised in the present parameter identification problem, which can be generally defined as

$$\begin{array}{lll} \text{Minimise} & g(\boldsymbol{p}) = \sum_{s=1}^{n_s} \lambda_s g_s(\boldsymbol{p}) & \boldsymbol{p} \in R^{n_p} \\ \text{Such that} & p_i^{\inf} \leq p_i \leq p_i^{\sup} & i = 1, \ldots, n_p \end{array}, \quad (6)$$

in which $g(\boldsymbol{p})$ is the global objective function (global fitness), computed from individual fitness measures, $g_s(\boldsymbol{p})$, where the subscript "s" indicates a tensile test of an individual type of specimen, and n_s is the total number of specimen types. In Eq. (6), $\boldsymbol{p} = \left[p_1\, p_2 \cdots p_i \cdots p_{n_p}\right]^T$ is the design vector containing n_p design variables p_i (material parameters), and $p_i^{\inf}$ and $p_i^{\sup}$ are lateral limits (minimum and maximum values of each material parameter).

The contribution of the tensile tests of each type of specimen to the global objective function is defined by weights, λ_s, so that $\sum_{s=1}^{n_s} \lambda_s = 1$. The individual fitness, $g_s(\boldsymbol{p})$, is evaluated by the L_2 norm of the differences between the experimental, $\boldsymbol{R}_s^{Exp}$, and corresponding computed loads, $\boldsymbol{R}(\boldsymbol{p})_s^{Num}$, of a tensile test "$s$", as

$$\begin{aligned} g_s(\boldsymbol{p}) &= \left\| \boldsymbol{R}_s^{Exp} - \boldsymbol{R}(\boldsymbol{p})_s^{Num} \right\|_{L_2} \\ &= \sqrt{\sum_{j=1}^{N_s} \left[R_{s,j}^{Exp} - R(\boldsymbol{p})_{s,j}^{Num} \right]^2} \end{aligned}, \quad (7)$$

where N_s is the number of experimental points for tensile test "s". The optimization technique adopted in this work uses the gradient-free downhill simplex method, also known as Nelder-Mead (NM) algorithm [25]. *It is important to mention that other optimization methods could have been used in conjunction with the proposed strategy*. Notwithstanding, the NM algorithm was adopted based on the following reasons:

(i) The NM method does not require a continuously differentiable objective function. The combination of individual objective functions and the high nonlinearity of the plasticity problem can cause the sensitivity analysis (derivative of the global objective function with respect to each design variable) of gradient-based schemes to fail, especially after the maximum load, close to the macroscopic failure (see, for instance, the issues associated with the gradient-based Sequential Quadratic Programming (SQP) method discussed in Vaz Jr. et al. [26]). In the present NM implementation, even if computation of one or more individual fitness, $g_s(\boldsymbol{p})$, fail, the scheme naturally purges the corresponding set of material parameters $\boldsymbol{p}$ and the optimization process continues.

(ii) The NM method was shown competitive in this class of problems when compared to some gradient-based techniques and soft-computing algorithms [27]. The authors discussed application of the Broyden-Fletcher-Goldfarb-Shanno (BFGS), Sequential Quadratic Programming (SQP), Globally Convergent Method of Moving Asymptotes (GCMMA), Particle Swarm Optimization (PSO), Genetic Algorithm (GA), Nelder-Mead and Univariate methods to identification of inelastic parameters using tensile tests and deep drawing operations.

(iii) The Nelder-Mead method has been used successfully in parameter identification associated with more elaborate mechanical tests and complex constitutive relations (which indicates its robustness). For instance, Banabic et al. [28] applied the NM scheme to identification of inelastic parameters for a biaxial tensile test, and Pannier et al. [29] used this optimization method to find elastic-plastic constitutive parameters based on the Virtual Fields Method associated to tensile tests. The NM technique was also successfully used in conjunction with different material constitutive relations, such as composite [30] and damaged [7] materials.

Highlights of the Nelder-Mead algorithm: The NM optimization method defines a regular polytope of $n_p + 1$ vertices (for a n_p-dimensional design space corresponding to a n_p number of material parameters), which moves towards the optimum by replacing the worst vertex by a new one selected from pre-defined positions along a given search direction. The reader is referred to Nelder and Mead [25], Luersen and Le Riche [30], and Lagarias et al. [31] for additional details on the NM optimization technique, and Vaz Jr. et al [26] for the NM algorithm used in the present investigation.

4 Numerical Results and Discussions

The previous sections show that inelastic parameters can be determined using tensile tests based on calibration or inverse problem techniques. The former adopts curve fitting strategies of the true stresses and strains in a $(\ln \sigma) \times (\ln \varepsilon)$ plot, whereas the latter uses optimization methods.

Table 1 Chemical composition of the AISI 304 steel (in wt%)

	Cr	Ni	Mn	Si	C	P	S
Nominal	18–20	8–10.5	max 2.0	max 1.0	max 0.08	max 0.045	max 0.03
This work	18.677	8.780	1.9856	0.21897	0.07612	0.01984	0.02767

Experimental investigations shows that tensile tests of ductile materials involve four general stages: (i) uniform elastic deformation; (ii) plastic deformation under uniform stress states; (iii) instability onset and growth of triaxial stress states; and (iv) catastrophic/macroscopic failure. Calibration procedures are performed under uniaxial stress/strain conditions comprising stage (ii) up to instability onset (or necking) to avoid triaxial stress states. Contrastingly, inverse problem techniques can account for elastic-plastic deformations up to macroscopic failure. This study is performed for the AISI 304 stainless steel with chemical composition presented in Table 1.

In the present work, tensile tests were executed using cylindrical specimens prepared according to the American ASTM E8M-09 [32] and Brazilian ABNT NBR ISO 6892 [33] technical standards under displacement control at room temperature (298 K). Extensometers with initial gauge lengths $l_0 = 25$ mm and $l_0 = 50$ mm were used according to the specimen size with maximum crosshead speed $v_c = 3$ mm/min. The specimens used in this work are illustrated in Fig. 1 and referred as follows:

- ASTM #1: initial gauge length $l_0 = 25$ mm and diameter $d_0 = 6.0$ mm,
- ASTM #2: initial gauge length $l_0 = 50$ mm and diameter $d_0 = 12.54$ mm,
- NBR #3: initial gauge length $l_0 = 50$ mm and diameter $d_0 = 10.0$ mm.

The geometrical models (used in the simulations) for all types of specimens consider axisymmetry around the rotation axis $Z - Z'$ and symmetry about the $R - R'$ axis, making possible to model only 1/4 of the specimen. All cases adopt a structured, eight-noded quadrilateral finite element mesh with 200 elements and 661 nodes with progressive refinement towards the specimen $R - R'$ axis. The finite element mesh used for specimen ASTM #2 is presented in Fig. 2. Meshes for ASTM #1 e NBR #3 specimens are geometrically proportional to ASTM #2 with identical element topology. The Young modulus and Poisson's ratio used in the simulations are $E = 205$ GPa and $\nu = 0.29$, respectively.

4.1 *Curve Fitting of Logarithmic-Type Yield Stress Curves*

The curve fitting procedure for the Swift (Eq. 1), double-*n* Swift (Eq. 2) and third-order logarithmic curves (Eq. 3) requires definition of lower and upper limits, i.e. the *elastic limit* and *instability onset*, respectively. The elastic limit corresponds to the lower bound imposed by the elastic curve, $\sigma = E\varepsilon$, whereas the maximum load was assumed as the instability limit. The true strains and true stresses are determined as $\varepsilon = \ln(1 + e)$ and $\sigma = S(1 + e)$, respectively, where $e = \Delta l / l_0$ is the elongation

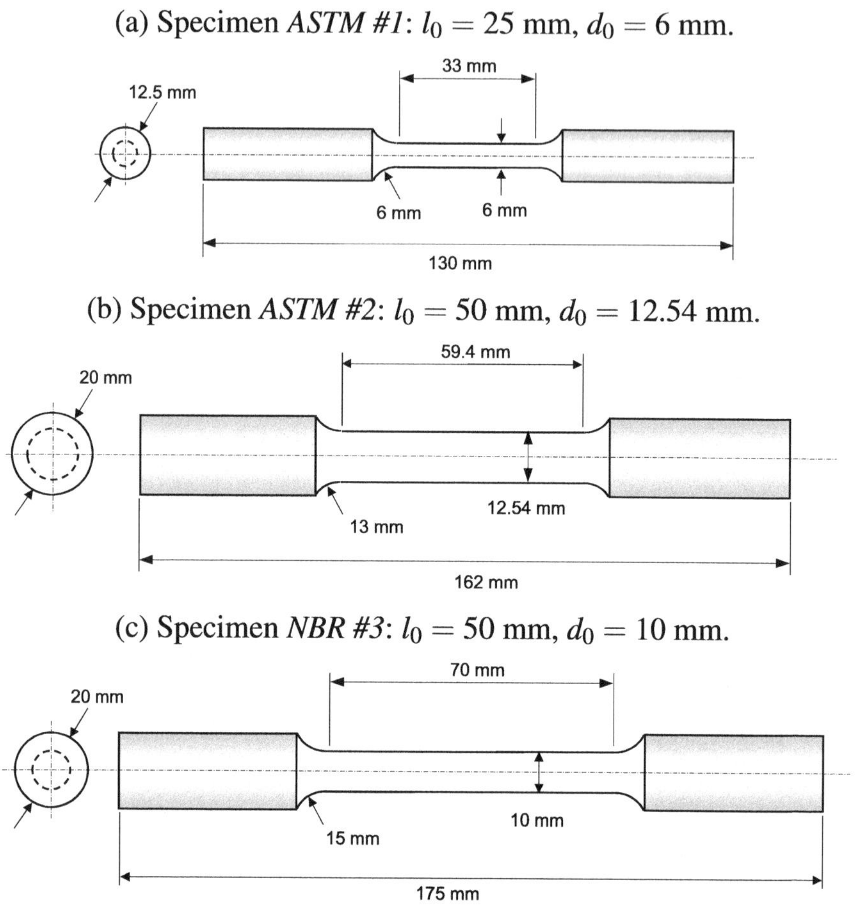

Fig. 1 Geometry of the specimens for the ASTM and ABNT-NBR standards

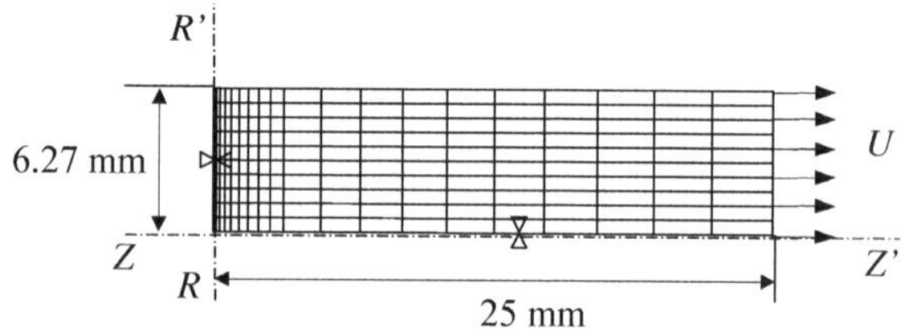

Fig. 2 Finite element mesh for the *ASTM #2* specimen

(engineering strain), $S = R/A_0$ is the engineering stress, R is the axial load, and A_0 is the initial cross-sectional area [8]. It was observed that evolution of the true stress—true strain curve for the ASTM #1, ASTM #2 and NBR #3 specimens presents small variations up to instability onset. Such results indicate a good consistency of the measured data. *Therefore, curve fitting was performed taking into account the true stress–true strain data of all types of specimens, making possible to determine a single set of parameters for each hardening curve, i.e. Swift, double-n Swift and cubic logarithmic curves.*

Figure 3 shows the experimental data and fitted equations in a $(\ln \sigma) \times (\ln \varepsilon)$ plot, and Table 2 presents the corresponding material parameters. For the sake of clarity, not all experimental points are represented in Fig. 3 and forthcoming figures. The best fit was achieved by the cubic logarithmic curve with a coefficient of determination $R^2 = 0.998627$, as represented in Fig. 3 by the thick solid line. The double-n Swift yield stress curve was also able to provide a good approximation to the experimental values with a mean coefficient of determination $\overline{R}^2 = 0.979821$ (solid line in Fig. 3). On the other hand, the Swift curve provided a very poor fitting, clearly indicated by the dashed line in Fig. 3, with a coefficient of determination $R^2 = 0.948860$.

Numerical simulations of the tensile test were conducted for the ASTM #1, ASTM #2 and NBR #3 specimens using the geometrical models previously described. The maximum experimental displacements measured for each specimen are applied to the corresponding geometrical model as: [ASTM #1 ($l_0/2 = 12.5$ mm): $U_{\#1} = 7.30$ mm], [ASTM #2 ($l_0/2 = 25$ mm): $U_{\#2} = 12.88$ mm], and [NBR #3 ($l_0/2 = 25$

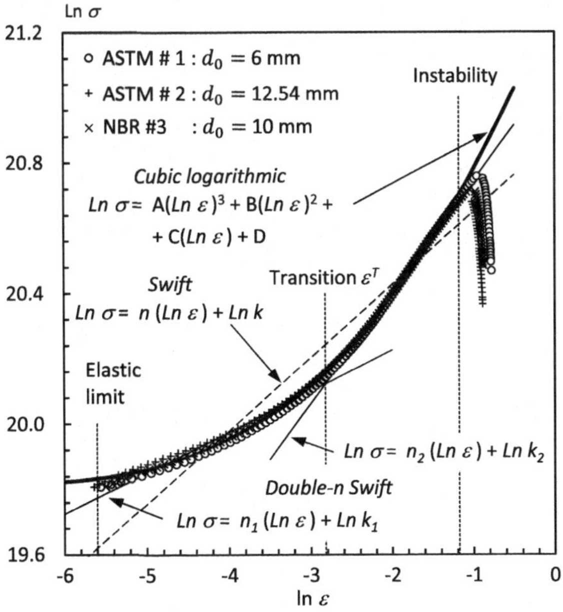

Fig. 3 Curve fitting of the true stresses and strains for the Swift, double-n Swift and cubic logarithmic hardening equations

Table 2 Material parameters for the Swift, double-*n* Swift and cubic logarithmic hardening equations

Model	Parameter	Value
Swift	k	1162.60 kPa
	n	0.223907
	ε_0	1.27526×10^{-3}
Double-*n Swift*	k_1	781.201 kPa
	n_1	0.125434
	k_2	1436.36 kPa
	n_2	0.341271
	ε_0	1.71422×10^{-3}
	ε_p^T	5.95026×10^{-3}
Cubic logarithm	A	3.59939×10^{-3}
	B	8.17363×10^{-2}
	C	0.608688
	D	21.3109
	ε_0	1.97856×10^{-3}

mm): $U_{\#3} = 12.80$ mm]. The material parameters presented in Table 2 for the Swift, double-*n* Swift and cubic logarithmic equations were used in the simulations.

Figure 4a–c present the loading curves for specimens ASTM #1, ASTM #2 and NBR #3, respectively, computed using Swift, double-*n* Swift and cubic logarithmic models. The poor fitting exhibited by the Swift curve (see the dashed line in Fig. 3) is translated into also poor numerical loading response for all specimens: the tensile load is overestimated for small strains and grossly underestimated for large deformations, i.e. the excessive necking towards the end of the tensile test caused an unrealistic decrease of the tensile load. The best data fitting demonstrated by the cubic logarithmic curve is strictly observed up to the maximum load for all specimens (thick solid lines in Fig. 4). However, the hardening parameters of the cubic logarithmic curve obtained for the present tensile tests caused the computed tensile load to steadily increase after the maximum experimental load, i.e. no necking was observed in the simulations up to the maximum applied displacements ($U_{\#1}$, $U_{\#2}$ and $U_{\#3}$) for the respective specimens.

The fitted double-*n* Swift equation provides the best results for the tensile load. Such good approximation (up to the maximum load) for stainless steels has already been established in the literature [10–12]. Figure 4a and c show that this hardening equation was also able to capture the general tendency of the load behaviour after the maximum load. Notwithstanding, the transition between the different slopes of the double-*n* Swift curve (represented by the hardening coefficients n_1 and n_2, as shown in Fig. 3) caused an unrealistic bulge in the loading curve when the elongation approaches $\varepsilon^T \approx 0.06314$ (i.e. when plastic strains approximate ε_p^T), as highlighted in Fig. 5. The maximum differences between experimental and numerical loads in the transition region of the double-*n* Swift curve increase with the cross-sectional area, so that $\Delta R_{\#1} \approx 310$ N ($d_0 = 6$ mm), $\Delta R_{\#3} \approx 1200$ N ($d_0 = 10$ mm), and $\Delta R_{\#2} \approx 2700$ N ($d_0 = 12.54$ mm) for specimens ASTM #1, NBR #3, and ASTM #2, respectively.

(a) ASTM #1 : $d_0 = 6$ mm. (b) ASTM #2 : $d_0 = 12.54$ mm. (c) NBR #3 : $d_0 = 10.00$ mm

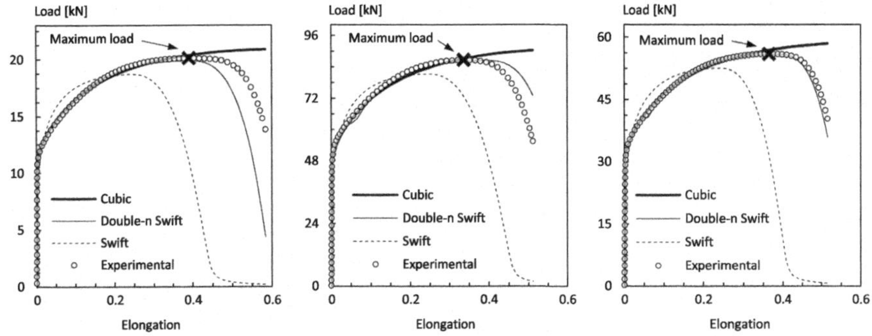

Fig. 4 Loading curve for specimens ASTM #1, ASTM #2 and NBR #3: logarithmic-based yield equations

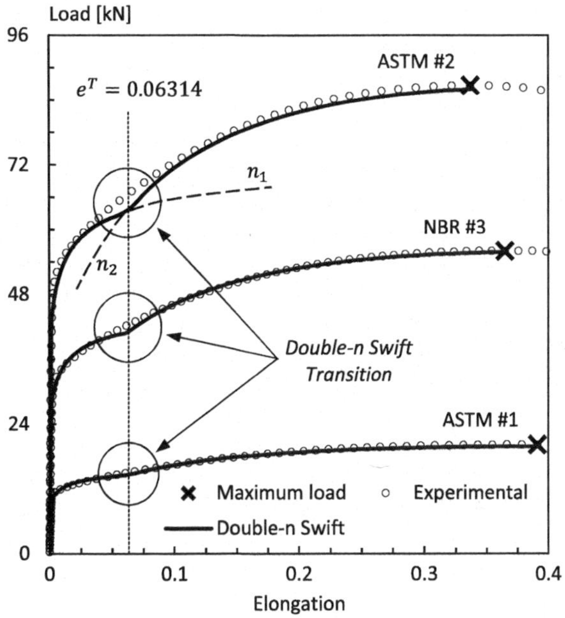

Fig. 5 Loading curves for the *double-n Swift* curve

Remark (iv): the curve-fitting strategy to determine the hardening parameters for the Swift, double-n Swift and cubic logarithmic equations minimizes the differences between experimental and numerical true strains and true stresses based on polynomial least square regression for a $(\ln \sigma) \times (\ln \varepsilon)$ plot. This strategy does not require information *after* the upper limit given by the maximum load. Therefore, it is not expected, a priori, a good approximation of the loading curve for the logarithmic-based hardening equations after necking onset under triaxial stress states.

4.2 Parameter Identification of Alternative Yield Stress Equations

The strategy used to determine the hardening parameters of Voces (Eq. 4) and modified Voce (Eq. 5) yield equations is markedly different from calibration procedures. Instead of using curve fitting techniques, material parameters of both equations are determined by numerically assessing the load response of the specimens submitted to controlled displacements. The final hardening parameters are those which minimize compound differences (evaluated by the L_2 norm) between experimental and numerical loads for specimens ASTM #1, ASTM #2 and NBR #3. The Nelder-Mead optimization algorithm was used in the simulations, which requires an initial estimate of the material parameters, $\boldsymbol{p}_i^{(0)}$, and corresponding minimum and maximum values (lateral constraints, p_i^{inf} and p_i^{sup}), as indicated in Table 3. For the sake of objectivity, assessment of the capacity of the algorithm to solve the problem and other studies of the optimization scheme are omitted here (e.g. convexity, sensitivity to mesh size and to changes of the initial estimates and lateral constraints, influence of the convergence limit, etc.). It is also relevant to emphasise that equal contribution for tensile tests of each specimen ASTM #1, ASTM #2 and NBR #3 was assumed in the identification process, so that $\lambda_1 = \lambda_2 = \lambda_3 = 1/3$ in Eq. (6).

The material parameters determined for the original Voce and modified Voce hardening equations are presented in Table 3, whereas the loading curves for the tensile test of each specimen are shown in Fig. 6. The simulations indicate that both hardening models present good approximation to the experimental data up to the maximum load. Nevertheless, evaluation of the loading curves of all specimens indicates that the modified Voce equation was able to provide a better prediction than the Voces model. It is possible to visually identify in Fig. 6a that, for the smallest specimen ASTM #1 ($d_0 = 6$ mm), the original Voce equation led to large differences towards the end of the loading process. Quantification of such differences is discussed in the following paragraph.

A comparative (quantitative) assessment with respect to load evolution for hardening equations associated with both curve-fitting (Swift, double-*n* Swift and cubic

Table 3 Lateral constraints, initial estimates and final hardening parameters for the *Voce* and *modified Voce* hardening equations

Model	Parameter	p_i^{inf}	p_i^{sup}	$\boldsymbol{p}_i$	Value
Voce Eq. (4)	σ_0^v (MPa)	200	800	500	432.137
	σ_∞^v (MPa)	400	1500	800	1424.73
	δ^v	0	50	25	2.46965
Modified Voce Eq. (5)	σ_0^t (MPa)	200	800	500	406.749
	σ_∞^t (MPa)	400	1200	800	768.524
	ζ^t (MPa)	300	1200	750	746.444
	δ^t	0	50	25	7.04351

(a) ASTM #1 : $d_0 = 6$ mm. (b) ASTM #2 : $d_0 = 12.54$ mm. (c) NBR #3 : $d_0 = 10.00$ mm

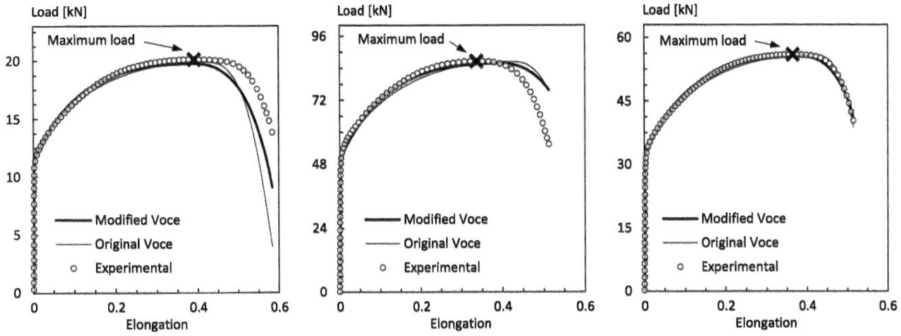

Fig. 6 Loading curve for specimens ASTM #1, ASTM #2 and NBR #3: original Voce and modified Voce yield equations

polynomial) and inverse problem (Voce and modified Voce) techniques was also performed. In order to better perceive eventual discrepancies, specimen-based, $\varphi_s(\boldsymbol{p})$, and global, $\Phi(\boldsymbol{p})$, indices were defined based on the relative L_2 error norm between experimental and numerical tensile loads as

$$\varphi_s(\boldsymbol{p}) = \frac{\left\| \boldsymbol{R}_s^{Exp} - \boldsymbol{R}_s^{Num} \right\|_{L_2}}{\left\| \boldsymbol{R}_s^{Exp} \right\|_{L_2}} \quad \text{and} \quad \Phi(\boldsymbol{p}) = \sum_{s=1}^{3} \varphi_s(\boldsymbol{p}). \tag{8}$$

Table 4 presents both $\varphi_s(\boldsymbol{p})$ and $\Phi(\boldsymbol{p})$, from which one can conclude that the modified Voce equation presents the best global performance, followed by the double-n Swift, original Voce, cubic logarithmic and Swift equations. Noticeably, the global error measure $\Phi(\boldsymbol{p})$ for the modified Voce equation is 30% smaller than the double-n Swift. Table 4 also indicates that differences change for individual specimens, with the modified Voce equation attaining best approximations for ASTM #1 and NBR #3 specimens and the double-n Swift for the ASTM #2 specimen. Figure 7 enables a visual appraisal of the differences between the loading curves for the modified Voce and double-n Swift hardening equations. The relative large differences exhibited by the double-n Swift loading curve after the maximum load for specimens ASTM #1 and NBR #3 (clearly visible in Fig. 7) led to the larger values of the global index, $\Phi(\boldsymbol{p})$. It can also be observed in Fig. 7 and Table 4 that differences for the ASTM #2 specimen are marginal for both hardening equations.

Remark (v): Some reasons might explain the better performance of the modified Voce hardening equation with respect to the double-n Swift model: (i) the optimization technique (minimization of the experimental-numerical differences) utilised for the modified Voce equation accounts for the experimental loading curve up to macroscopic failure, which contrasts to the double-n Swift model that performs curve fitting up to the maximum load. (ii) The unrealistic bulge located in the transition between hardening indices n_1 and n_2 imposes further differences for the double-n Swift equa-

Table 4 Individual, $\varphi_s(\boldsymbol{p})$, and global, $\Phi(\boldsymbol{p})$, error indices for all hardening equations

Technique	Hardening model	$\varphi_s(\boldsymbol{p})$			$\Phi(\boldsymbol{p})$
		ASTM #1	ASTM #2	NBR #1	Global index
Curve fitting	*Swift*	0.5184282	0.4239484	0.4881430	1.430520
	Double-n Swift	0.1481382	**0.0735302**	0.0249953	0.246664
	Cubic logarithmic	0.0981038	0.1419500	0.0900999	0.330154
Optimization	*Original Voce*	0.1498700	0.0871119	0.0161115	0.253093
	Modified Voce	**0.0825237**	0.0805894	**0.0108542**	**0.173967**

The bold value indicates the minimum values of the error indices

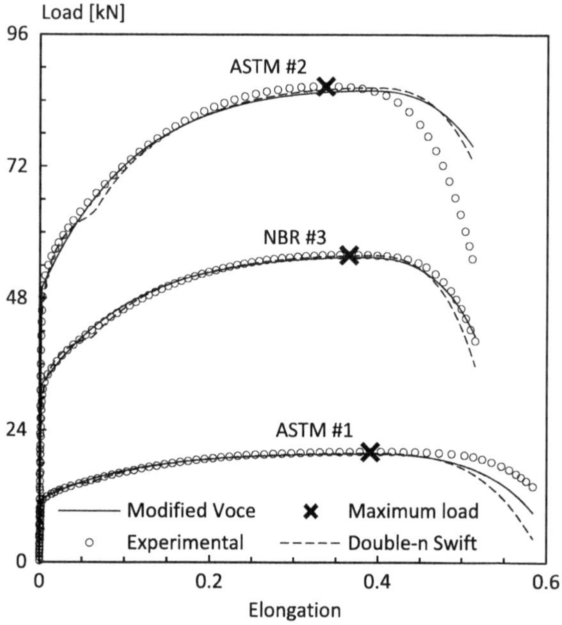

Fig. 7 Loading curve for *double-n Swift* and *modified Voce* hardening equations

tion. (iii) No physical justification holds for double-*n* Swift model since this equation if essentially of empirical character.

5 Concluding Remarks

Austenitic stainless steels have been used in many applications, ranging from consumer items (cutlery, pans, sinks, etc.) and architectural components (ornamental work, decorative panels, poolside fittings and fixings, etc.) to industrial equipment (e.g. pharmaceutical, food and beverage, chemical and mining equipment). The stainless steel AISI 304 constitutes the most used grade owing to its combination of good corrosion/oxidation resistance, weldability, formability, toughness characteristics and ductility (large deformation before failure onset). Due to such high industrial interest, this work discusses suitability of some selected equations to model hardening evolution and alternative strategies to obtain the corresponding parameters within the phenomenological framework.

Tensile tests have long been used to determine hardening parameters of ductile materials. Aiming to confer further credibility to the present analysis, specimens prepared according to the American ASTM E8M-09 [32] (ASTM #1 and ASTM #2) and Brazilian ABNT NBR ISO 6892 [33] (NBR #3) standards were used. Due to simplicity, calibration procedures based on curve-fitting strategies are by far the most common approach adopted to determine hardening parameters. The main drawback of this technique is the requirement of uniform deformation, which restricts its application to relatively small plastic deformations (up to instability onset). This work addresses this method by comparing the following logarithmic-type hardening equations: *Swift* [9] (largely used in industry), *double-n Swift* (previously recommended for austenitic steels [10–12]), and a *3rd order* $(\ln\sigma) \times (\ln\varepsilon)$ *polynomial equation* (recommended for some DP and TRIP steels [6]). The latter provided the best fit of the experimental data up to maximum load, followed by the *double-n Swift* and *Swift* models.

Simulation of metal forming processes, such as extrusion, wire drawing and deep drawing, involves larger plastic deformations than those permitted by curve-fitting strategies. Therefore, this study also addresses identification of material parameters based on inverse problem techniques (by using optimization methods). In such case, *Voce*'s and *modified Voce* equations were adopted to describe material hardening. Despite of phenomenological character, these equations were selected due to their association with physical mechanisms of the dislocation theory.

A comparative assessment of all hardening equations (*Swift, double-n Swift, 3rd order logarithmic, Voce* and *modified Voce*) were performed by numerical simulation of tensile tests for the ASTM #1, ASTM #2 and NBR #3 specimens up to macroscopic failure. The computational model uses an in-house fine element code based on an elastic-plastic, finite strain formulation [23, 24]. The experimental data shows relevant plastic deformation after the maximum load, reaching elongations (engineering strains) up to 50% higher than those observed for uniform deformation. In order to quantify the approximation provided by the aforementioned hardening equations and corresponding parameters, an index was defined based on the relative L_2 error norm between experimental and numerical loads up to macroscopic failure. The simulations indicate that the *modified Voce* equation produced an error index 30% smaller than the *double-n Swift*. Moreover, both equations were able to predict the load behaviour after the instability onset with acceptable accuracy. Contrastingly, *Swift*'s hardening model grossly underestimates tensile loads after maximum load, whereas no instability onset was predicted by the *3rd order polynomial* equation.

Therefore, owing to the discussion summarised in the previous paragraphs, the *modified Voce* equation associated with inverse problem techniques based on optimization methods are recommended to describe isothermal hardening for the austenitic stainless steel AISI 304. The *double-n Swift* equation can also be used, but some care should be exercised.

Acknowledgements The authors acknowledge the financial support provided by the Brazilian funding agency CNPq (National Council for Scientific and Technological Development), Grant no. 303412/2016-0.

References

1. McGuire MF (2008) Stainless steels for design engineers. ASM, Materials Park
2. Subramonian S, Kardes N (2012) Materials for Sheet Forming. In: Altan T, Tekkaya E (eds) Sheet Metal Forming: Fundamentals. ASM, Materials Park
3. Samuel KG, Rodriguez P (2005) On power-law type relationships and the Ludwigson explanation for the stress-strain behaviour of AISI 316 stainless steel. J Mater Sci 40:5727–5731
4. Samuel KG (2006) Limitations of Hollomon and Ludwigson stress strain relations in assessing the strain hardening parameters. J Phys D: Appl Phys 39:203–212
5. Ponthot JP, Kleinermann JP (2006) A cascade optimization methodology for automatic parameter identification and shape/process optimization in metal forming simulation. Comput Meth Appl Mech Eng 195:5472–5508
6. Dimatteo A, Colla V, Lovicu G, Valentini R (2015) Strain hardening behavior prediction model for automotive high strength multiphase steels. Steel Res Int 86:1574–1582
7. Vaz M Jr, Muoz-Rojas PA, Cardoso EL, Tomiyama M (2016) Considerations on parameter identification and material response for Gurson-type and Lemaitre-type constitutive models. Int J Mech Sci 106:254–265
8. Davis JR (2004) Tensile testing, 2nd edn. ASM, Materials Park
9. Swift HW (1952) Plastic instability under plane stress. J Mech Phys Solids 1:1–18
10. Hertelé S, De Waele W, Denys R (2011) A generic stress strain model for metallic materials with two-stage strain hardening behaviour. Int J Non-linear Mech 46:519–531
11. Kashyap BP, McTaggart K, Tangri K (1988) Study on the substructure evolution and flow behaviour in type 316L stainless steel over the temperature range 21–900 °C. Phil Mag A 57:97–114
12. Kashyap BP, Tangri K (1995) On the Hall-Petch relationship and substructural evolution in type 316L stainless steel. Acta Metall Mater 43:3971–3981
13. De AK, Speer JG, Matlock DK et al (2006) Deformation-induced phase transformation and strain Hardening in type 304 austenitic stainless steel. Metall Mater Trans A 37A:1875–1886
14. Larour P (2010) Strain rate sensitivity of automotive sheet steels: influence of plastic strain, strain rate, temperature, microstructure, bake hardening and pre-strain. Dr.-Ing. Dissertation, RWTH Aachen
15. El-Magd E (2004) Modeling and Simulation of Mechanical Behavior. In: Totten GE, Xie L, Funatani K (eds) Modeling and Simulation of Material Selection and Mechanical Design. Dekker, New York
16. Mecking H, Kocks UF (1981) Kinetics of flow and strain hardening. Acta Metall 29:1865–1875
17. Voce E (1948) The relationship between stress and strain for homogeneous deformation. J Inst Metals 74:537–562
18. Tome C, Canova GR, Kocks UF et al (1984) The relation between macroscopic and microscopic strain hardening in f.c.c. polycrystals. Acta Metall 32:1637–1653
19. Gruber M, Lebaal N, Roth S et al (2016) Parameter identification of hardening laws for bulk metal forming using experimental and numerical approach. Int J Mater Form 9:21–33
20. Panich S, Barlat F, Uthaisangsuk V, Suranuntchai S, Jirathearanat S (2013) Experimental and theoretical formability analysis using strain and stress based forming limit diagram for advanced high strength steels. Mater & Design 51:756–766
21. Sugio K, Sasaki G, Tabata J, Fuyama N (2017) In: Proceedings of the International Symposium on Green Manufacturing and Application. The Korean Society for Precision Engineering, Seoul, pp 51–53

22. Vaz M Jr, Cardoso EL, Stahlschmidt J (2013) Particle swarm optimization and identification of inelastic material parameters. Eng Comput 30:936–960
23. de Souza Neto EA, Peric D, Owen DRJ (2008) Computational Methods for Plasticity: Theory and Applications. Wiley, Chichester
24. Vaz M Jr, Lange MR (2017) Thermo-mechanical coupling strategies in elasticplastic problems Continuum Mech. Thermodyn. 29:373–383
25. Nelder JA, Mead R (1965) A simplex method for function minimization. Comput J 7:308–313
26. Vaz M Jr, Luersen MA, Muoz-Rojas PA, Trentin RG (2016) Identification of inelastic parameters based on deep drawing forming operations using a globallocal hybrid Particle Swarm approach. C R Mecanique 344:319–334
27. Vaz M Jr, Cardoso EL, Muñoz-Rojas PA et al (2015) Identification of constitutive parameters – optimization strategies and applications. Mat-wiss Werkstofftech 46:477–491
28. Banabic D, Kuwabara T, Balan T et al (2003) Non-quadratic yield criterion for orthotropic sheet metals under plane-stress conditions. Int J Mech Sci 45:797–811
29. Pannier Y, Avril S, Rotinat R, Pierron F (2006) Identification of elasto-plastic constitutive parameters from statically undetermined tests using the virtual fields method. Exp Mech 46:735–755
30. Luersen MA, Le Riche R (2004) Globalized NelderMead method for engineering optimization Comput & Struct 82:2251–2260
31. Lagarias JC, Reeds JA, Wright MH, Wright PE (1998) Convergence properties of the Nelder-Mead simplex method in low dimensions. SIAM J Optim 9:112–147
32. ASTM E8, E8M–09, (2009) Standard Test Methods for Tension Testing of Metallic Materials. ASTM International, West Conshohocken
33. ABNT NBR-ISO 6892, (2002) Metallic materials – Tensile testing at ambient temperature. ABNT, Rio de Janeiro

Effect of Fiber Volume Fraction on Mechanical Properties of Type E-Glass in Composite Materials

Zamzam A. Elsharif and Bashir M. Gallus

Abstract This chapter presents an experimental study about the effect of the fiber volume fraction on mechanical properties of E-glass types with polyester in composite laminate. This requires to carry out experimental studies to exanimate the fiber volume fraction, fiber weight, tensile test, hardness test, impact test. The prepared samples of woven roving at $[0^\circ/90^\circ/\pm 45^\circ]_s$,$[\pm 30^\circ/\pm 60^\circ]_s$ and chopped mat. Experimental results have shown that hardness values increase with the volume fraction of the laminate. $[0^\circ/90^\circ/\pm 45^\circ]_s$ revealed ideal behavior in contrast to $[\pm 30^\circ/\pm 60^\circ]_s$ and chopped mat materials. Fiber volume fraction results of $[0^\circ/90^\circ/\pm 45^\circ]_s$, $[\pm 30^\circ/\pm 60^\circ]_s$ and chopped mat were found to be 44, 41 and 29% at hardness values of 93, 78, and 75, respectively. Toughness values of $[0^\circ/90^\circ/\pm 45^\circ]_s$, $[\pm 30^\circ/\pm 60^\circ]_s$, and chopped mat were obtained to be 25.7 J, 26.14 J, and 25.5 J, respectively. In addition, yield stresses of $[0^\circ/90^\circ/\pm 45^\circ]_s$, $[\pm 30^\circ/\pm 60^\circ]_s$, and chopped mat were obtained to be ideal plastic behavior, 52.5325, 17.3578 MPa, respectively.

Keywords E-glass fiber · Lay-up · Quasi-isotropic laminate · Composite material · Delamination · Fiber breakage · Fiber volume fraction

1 Introduction

A combination of two or more materials (reinforcing elements, fillers, and composite matrix binder), differing in form or composition on a macroscale is called a composite. The constituents retain their identities, that is, they do not dissolve or merge completely into one another although they act together. Normally, the components can be physically identified and exhibit an interface between each other. Examples are cermets and metal-matrix composites [1].

Z. A. Elsharif (✉) · B. M. Gallus
Mechanical Engineering, College of Technical Sciences Misrata, Misrata, Libya
e-mail: zamzamalsharif@hotmail.com

B. M. Gallus
e-mail: bashir.bashir11311@gmail.com

A. Öchsner and H. Altenbach (eds.), *Engineering Design Applications II*,
Advanced Structured Materials 113, https://doi.org/10.1007/978-3-030-20801-1_3

The physical properties of composite materials are generally not isotropic in nature, but rather are typically orthotropic. For instance, the stiffness of a composite panel will often depend upon the directional orientation of the applied forces and/or moments. Panel stiffness is also dependent on the design of the panel. For instance, the fiber reinforcement and matrix used, the method of panel build, thermosetting versus thermoplastic, type of weave, and orientation of fiber axis to the primary force are major factors.

Isotropic materials (for example, aluminium or steel), in standard wrought forms, typically have the same stiffness regardless of the directional orientation of forces and/or moments.

The relationship between forces/moments and strains/curvatures for an isotropic material can be described with the following material properties: Young's modulus, the shear modulus and the Poisson's ratio, in relatively simple mathematical relationships. For an elastic material, it requires the mathematics of a second order tensor and can require up to 21 material property constants. For the special case of orthogonal isotropy, there are three different material property constants for each of Young's modulus, shear modulus and Poisson's ratio for a total of 9 material property constants to describe the relationship between forces/moments and strains/curvatures [1].

Fiber reinforced composite materials can be divided into two main categories normally referred to as short fiber reinforced materials and continuous fiber reinforced materials. Continuous reinforced materials will often constitute a layered or laminated structure. The woven and continuous fiber styles are typically available in a variety of forms, being pre-impregnated with the given matrix (resin), dry, undirectional tapes of various widths, plain weave, harness satins, braided, and stitched.

The short and long fibers are typically employed in compression molding and sheet molding operations. These come in the form of flakes, chips, and random mate (which can also be made from a continuous fiber laid in random fashion until the desired thickness of the ply/laminate is achieved) [1].

Shock, impact, or repeated cyclic stresses can cause the laminate to separate at the interface between two layers, a condition known as delamination. Individual fibers can separate from the matrix e.g. fiber pull-out. Composites can fail on the microscopic or macroscopic scale. Compression failures can occur at both the macro scale or at each individual reinforcing fiber in compression buckling. Tension failures can be net section failures of the part or degradation of the composite at a microscopic scale where one or more of the layers in the composite fail in tension of the matrix or failure the bond between the matrix and fibers. Some composites are brittle and have little reserve strength beyond the initial onset of failure while others may have large deformations and have reserve energy absorbing capacity past the onset of damage. The variations in fibers and matrices that are available and the mixtures that can be made with blends leave a very broad range of properties that can be designed into a composite structure [1].

The use of glass fiber reinforced plastic (GRP) tubes has spread in the chemical, oil, gas and other energy-related industries because of their lightness, corrosion resistance, durability, ease of installation and low through life cost [2, 3].

There are many literature reviews on this point such as, [4] has studied the effect of changing the reinforcement percentage of fibers on the mechanical properties, for composite material consists of conbextra epoxy (EP-10) resin reinforced by biaxial woven roving kevlar fibers) $\left[0^{\circ} - 45^{\circ}\right]$ with density (340 g/cm^3) which included impact strength, tensile strength, flexural strength and hardness. Authors of [5] applied chip and powder copper which are used as the reinforcing phase in polyester matrix to form composites. Mechanical properties such as flexural strength and impact test of polymer reinforcement copper (powder and chip) were done. In addition, the author of [6] has studied the mechanical properties such as, Young's modulus (E), impact strength (I.S), the Brinell hardness (B.H) and compression strength (C.S) of polyester reinforced with 20% (v/v) glass fiber woven randomly E-glass. Also, the authors of [7] studied about composite materials which consisted of polyester as the matrix and BR4RC as the additive material with grain size equal to 25 μm with different weight fractions (10, 20, 30, 40, 50%). This investigation was done into two stages: the first stage was to produce the composite material, while the second stage was to test the new material which included tensile tests, hardness and microstructure evaluations. Also photomicrographs were taken by ordinary microscopes.

In this chapter, studying of effect the fiber volume fraction of composite material will be presented including experimental investigation to evaluate the fiber volume fraction, fiber weight and the mechanical properties, such as tensile strength, toughness and hardness.

2 Experimental Set-up

2.1 Determination of Fiber Volume Fraction and Fiber Weight for E-Glass/Polyester Laminate

The method used for determining the percentage of fiber content in a GRP laminate is the burn-off test [8]. This test is based on finding the mass of fiber by burning a known mass of composite laminate sample in a furnace. The matrix resin will melt and evaporate while the fiber will remain. By weighing the remaining fiber, and dividing by the initial mass of the laminate one obtains the fiber mass fraction. Square test samples of approximately (20×20 mm^2), were cut from the manufactured laminate.

In order to determine the volume fraction and fiber weight, a number of small crucibles were weighed, both empty and with samples. The crucibles with samples were then placed inside an electric furnace for one hour at temperature of 580–600 °C [8] as shown in Fig. 1. Before placing them in the furnace, each sample and crucible was labeled for identification after burning. After the burning was completed, the mass of the crucible and the remaining glass fibers were weighed. The mass of fiber was determined by subtracting the mass of the empty crucible from the total mass of

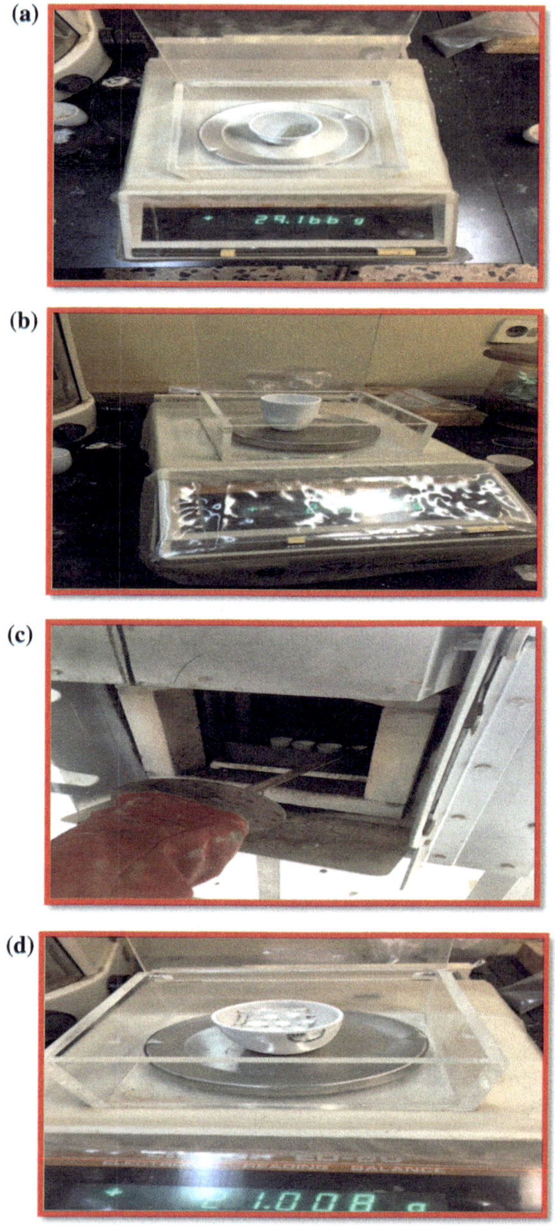

Fig. 1 **a** Weight sample with crucible, **b** weight crucible before burn without sample, **c** samples in a furnace, **d** weight sample with crucible after burn

the crucible and remaining glass fiber. Tables 1, 2 and 3 show the average values of the fiber weight for four samples.

Table 1 Average values of fiber weight of woven roving at $[0^\circ/90^\circ/\pm 45^\circ]_s$

Weight samples (gram)	Weight crucible (gram)	Weight sample with crucible before burn (gram)	Weight sample with crucible after burn (gram)	Weight fiber (gram)	Weight resin (gram)
1.843	20.455	22.304	21.541	1.086	0.757
2.107	20.870	22.976	22.121	1.251	0.856
1.955	20.593	22.548	21.745	1.152	0.803
2.024	19.826	21.850	21.002	1.176	0.846
Ave.1.98225	Ave.20.436	Ave.22.4195	Ave.21.60225	Ave.1.16625	Ave.0.8155

Table 2 Average values of fiber weight of woven roving at $[\pm 30^\circ/\pm 60^\circ]_s$

Weight samples (gram)	Weight crucible (gram)	Weight sample with crucible before burn (gram)	Weight sample with crucible after burn (gram)	Weight fiber (gram)	Weight resin (gram)
1.996	25.960	27.960	27.010	1.05	0.946
2.047	25.647	27.693	26.749	1.102	0.945
1.956	28.088	30.044	29.250	1.162	0.794
2.308	27.149	29.466	28.356	1.207	1.101
Ave.2.0825	Ave.26.711	Ave.28.79075	Ave.27.84125	Ave.1.13025	Ave.0.9465

Table 3 Average values of fiber weight of chopped mat

Weight samples (gram)	Weight crucible (gram)	Weight sample with crucible before burn (gram)	Weight sample with crucible after burn (gram)	Weight fiber (gram)	Weight resin (gram)
3.522	26.871	30.390	28.171	1.30	2.222
3.250	27.846	31.459	29.419	1.573	1.677
3.639	25.499	29.141	26.861	1.362	2.277
4.235	27.042	31.287	28.526	1.484	2.751
Ave.3.6615	Ave.26.8145	Ave.30.56925	Ave.28.24425	Ave.1.42975	Ave.2.2318

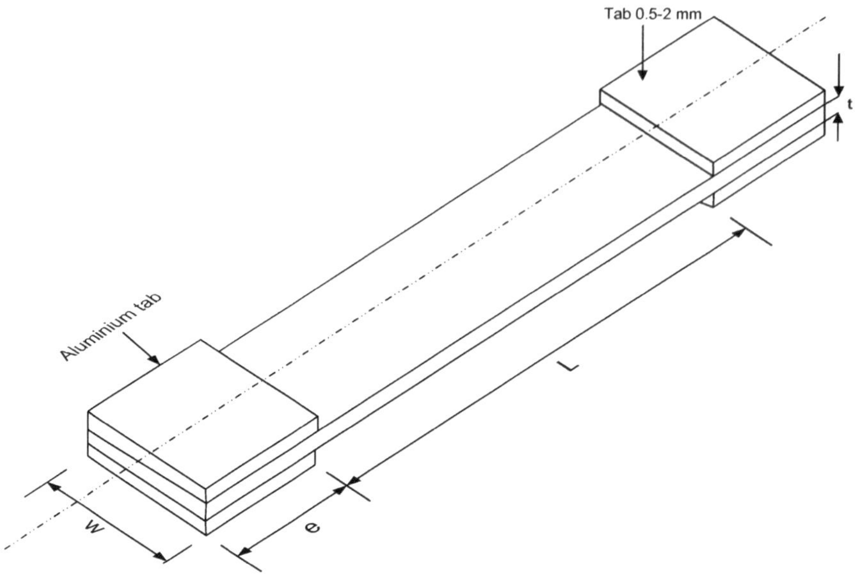

Fig. 2 Schematic diagram of tensile test specimen

2.2 Determination of the Tensile Strength for the Composite Laminate

In order to determine the tensile strength of the composite laminate, three samples were cut from the previous laminate according to the test method of composite laminate given in [8]. The dimensions of the specimen are shown in Fig. 2.

Each specimen was equipped with small rectangular aluminium sheets at its ends for machine grips. The samples must be carefully aligned in the test machine jaws to avoid induced sample bending. All samples were geometrically similar (L = 150 mm, e = 50 mm, w = 26 mm, t = depend on type E-glass fiber).

2.3 Determination of the Hardness of the Composite Laminate

In order to determine the hardness of the composite laminate, two samples were cut from the previous laminate according to the test method of composite laminate by [8]. The dimensions of the chopped mat specimen were 20 mm $\times$20 mm with a thickness of 7 mm, while the dimensions of the woven roving specimen was 20 mm $\times$ 20 mm with a thickness of 5 mm and orientation $\left[0^{\circ}/90^{\circ}/\pm 45^{\circ}\right]_s$, $[\pm 30^{\circ}/\pm 60^{\circ}]_s$ as shown in Fig. 3.

(a)
(b)
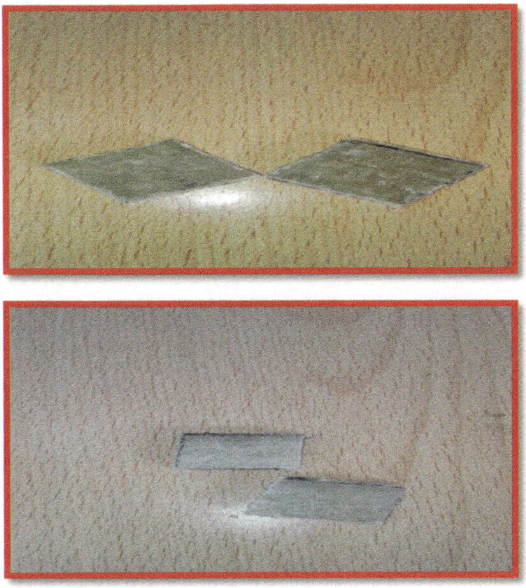

Fig. 3 **a** Photo of the woven roving specimen, **b** photo of the chopped mat specimen

2.4 Determination of Toughness of the Composite Laminate

In order to determine the toughness of the composite laminate, three samples were cut from the previous laminate according to the test method of composite laminate by [8]. The dimensions of the chopped mat specimen were 55 mm × 10 mm with thickness 7.6 mm while, the dimensions of the woven roving specimen were 55 mm × 10 mm with thickness 5.8 mm and orientation $\left[0^\circ/90^\circ/\pm 45^\circ\right]_s$, $[\pm 30^\circ/\pm 60^\circ]_s$ as shown in Fig. 4.

3 Results and Discussion

3.1 Volume Fraction

Tables 4, 5 and 6 show the results of burn-off tests which were carried out on the specimens of composite laminate as described in Sect. 2.1. The average volume fraction of the fiber is shown in the table.

(a)
(b)

Fig. 4 **a** Photo of the chopped mat specimen, **b** photo of the woven roving specimen

Table 4 Results of burn-off tests of woven roving at $[0^{\circ}/90^{\circ}/\pm 45^{\circ}]_s$

Sample No (gram)	Weight samples (gram)	Weight fiber (gram)	Weight resin (gram)	Density sample g/cc	Fiber volume fraction	Resin volume fraction
1	1.843	1.086	0.757	1.9	0.44	0.56
2	2.107	1.251	0.856	1.9	0.44	0.56
3	1.955	1.152	0.803	1.9	0.44	0.56
4	2.024	1.176	0.846	1.9	0.44	0.56
Average	1.98225	1.16625	0.8155	1.9	0.44	0.56

3.2 *Mechanical Properties of Composite Material*

Table 7 shows the experimental results of all tests as described in Sects. 2.1, 2.2, 2.3, 2.4. It can be observed that the increase of hardness value is 93 with an increased volume of fiber fraction of 44% for woven rovings $[0^{\circ}/90^{\circ}/\pm 45^{\circ}]_s$ that it compared with chopped mat which is 17% of volume fraction with a hardness value of 73. In addition, it can be observed at $[0^{\circ}/90^{\circ}/\pm 45^{\circ}]_s$ that after a linear behavior to followed by a steady continue without failure. This is called "ideal plastic" with a

Table 5 Results of burn-off tests of woven roving at $[\pm 30^{\circ}/\pm 60^{\circ}]_s$

Sample No (gram)	Weight samples (gram)	Weight fiber (gram)	Weight resin (gram)	Density sample (g/cc)	Fiber volume fraction	Resin volume fraction
1	1.996	1.05	0.946	1.9	0.41	0.59
2	2.047	1.102	0.945	1.9	0.41	0.59
3	1.956	1.162	0.794	1.9	0.41	0.59
4	2.308	1.207	1.101	1.9	0.41	0.59
Average	2.0825	1.13025	0.9465	1.9	0.41	0.59

Table 6 Results of burn-off tests of chopped mat

Sample No (gram)	Weight samples (gram)	Weight fiber (gram)	Weight resin (gram)	Density sample (g/cc)	Fiber volume fraction	Resin volume fraction
1	3.522	1.30	2.222	1.9	0.29	0.71
2	3.250	1.573	1.677	1.9	0.29	0.71
3	3.639	1.362	2.277	1.9	0.29	0.71
4	4.235	1.484	2.751	1.9	0.29	0.71
Average	3.6615	1.42975	2.2318	1.9	0.29	0.71

yield stress of about 28.856 MPa and the failure stroke was infinite, also woven roving of $[\pm 30^{\circ}/\pm 60^{\circ}]_s$ was 118.586 MPa with fiber volume fraction 41% and hardness value is 78. Nevertheless, the maximum failure stress was 124.363 MPa of chopped mat with fiber volume fraction 29% and hardness value is 75, while chopped mat with largest fiber volume fraction is 50% and hardness value is 84.33.

4 Conclusion

- The study showed that composite materials have good mechanical properties usefully to repair defected metal pipes which are used in oil and gas industry with minimum effective cost.
- Manufacturing of composite material by lay-up can be known by the properties of types of fibre and resin.
- The non-linearity behaviour of woven roving at $[0^{\circ}/90^{\circ}/\pm 45^{\circ}]_s$ continuous without failure at V_f44%.
- The yield stress of woven roving at $[\pm 30^{\circ}/\pm 60^{\circ}]_s$ was the highest, i.e. 52.5323 MPa.
- The maximum ultimate stress was observed at chopped mat 124.363 MPa at V_f 29%.
- The maximum value of hardness was observed at $[0^{\circ}/90^{\circ}/\pm 45^{\circ}]_s$ H.R.C 93 and V_f44%.

Table 7 Mechanical properties of composite materials (E-glass)

Composite material E-glass	Volume fraction $\left(V_f\%\right)$	Energy of Fracture (J)	Hardness (H.R.C)	Yield stress σ_y(MPa)	Ultimate stress σ_{ult}(MPa)
Woven roving $[0^\circ/90^\circ/\pm 45^\circ]_S$	44	274.24	93	–	28.856
Woven roving $[\pm 30^\circ/\pm 60^\circ]_s$	41	273.86	78	52.5325	118.586
Chopped mat	29	274.43	75	17.3578	124.363
Chopped mat	22	274.43	75	N.A	46.33
Chopped mat	17	273	73	N.A	36.23
Chopped mat	50	297.28	84.33	N.A	97.9

- The highest value of energy of fracture was for the chopped mat, i.e. 297.28 J at V_f 50%.
- The highest of toughness was observed the woven roving at $[0^\circ/90^\circ/\pm 45^\circ]_s$ and V_f44%.

References

1. Reference Hand Book.: Introduction to Composite. Fourth Edition. Composites Institute. New York, NY10017. Issue:1-56676-659-1
2. Forst SR (1998) Applications of polymer composites within the oil industry. In: Gibson AG (ed) Seventh international conference on fiber reinforced composites, pp. 84–91. University of Newcastle upon Tyne
3. Gibson AG (2000) Composite Material in Offshore Industry. Chapter 6.23 of Comprehensive Composite Materials. In: Kelly A, Zweben C (eds.) Elsevier
4. AL-Mosawi AI (2009) Study of some mechanical properties for polymeric composite material reinforced by fibers. ALQADISIA J Eng Sci 2(1)
5. Ali HA, Layth WE, Aseel MA (2012) Study of some mechanical properties for a polymer material reinforcement with chip or powder copper. Eng J 18(5)
6. Huda AA (2012) Study of the mechanical properties of unsaturated polyester composite reinforced by randomly woven fiber glass and effect of acidic solutions on some of its physical properties. RAFEDIN Sci J 23(1):114–129
7. Khansaa DS, Sbah N, Ahlam AA, Layth HM (2014) Studying the mechanical properties of unsaturated polyesters-B_4C system. Iraqi J Mech Mater Eng 14(1)
8. Curtis PT (1988) Crag test methods for measurement of the engineering properties of fiber reinforced plastics. Controller HMS **6** London

Experimental Study of Temperature Effect on the Mechanical Properties of GFRP and FML Interface

Z. P. Chow, Z. Ahmad and K. J. Wong

Abstract Interface between laminates has always been the weakest part of bonded materials which is prone to delamination. This is even more prevalent in bonding of two different materials. The research aims to evaluate delamination of dissimilar materials under a range of temperature. This is a part of the experimental study to investigate the potential of fiber metal laminates (FML) to be used in high temperature environment. The mechanical response of interface of hybrid laminate was characterized at temperatures ranging from 30 to 110 °C. Double cantilevered beam (DCB) and end notched flexure (ENF) tests were conducted on glass fiber laminated aluminum specimens to obtain Mode-I and Mode-II delamination properties with use of data reduction. Mode-I fracture toughness (G_{IC}) is significantly degraded by 59.45% at 70 °C and up to 83.65% at 110 °C. Mode-II fracture toughness (G_{IIC}) only slightly degrades by 10.91% at 70 °C but drops rapidly by 82.84% at 110 °C.

Keywords Temperature · Delamination · Fiber metal laminate · Glass fiber · Aluminum

1 Introduction

Abstract This chapter covers the research background on the development of metallic and composite materials. Both materials exhibit their respective advantages and disadvantages. As such, this leads to the development of fiber metal laminates (FML) with superior characteristics. However, FML is not without its own drawbacks primarily with delamination. The focus is drawn on the suitability of using FML at high thermal environments. Brief coverage on literature about previous studies of temperature effects on composite delamination are discussed, which helped to choose type of material and analysis methods of Mode-I and Mode-II delamination. The literature

Z. P. Chow · Z. Ahmad (✉) · K. J. Wong
Computational Solid Mechanics (CSM) Laboratory, School of Mechanical Engineering, Faculty of Engineering, Universiti Teknologi Malaysia, 81310 Johor Bahru, Malaysia
e-mail: azaini@utm.my

A. Öchsner and H. Altenbach (eds.), *Engineering Design Applications II*,
Advanced Structured Materials 113, https://doi.org/10.1007/978-3-030-20801-1_4

helped to elaborate on the research gap of temperature effect on FML delamination. Finally, an objective for this paper is drawn, followed by the rough methodology used to evaluate delamination in both Mode-I and Mode-II.

Following the aviation and aerospace demand for lightweight and high strength materials, advanced composite was developed for its good strength-to-weight ratio, corrosion resistance and long lifespan. Laminated fiber reinforced polymers applications are found in aircraft components due to improved structural and fatigue strength compared to its monolithic counterparts. However, the rigid and high stiffness of composites makes it prone to internal damage such as matrix cracking, fiber fracture and delamination. These hidden internal damages can lead to severely weakened structure and eventually catastrophic failure [1, 2]. Therefore, a combination of laminated metal and composites plies, known as fiber metal laminates were developed. The resulting combination of both dissimilar materials has improved fatigue, impact and damage tolerance that extends lifespan and reduce cost [3–5]. However, the dissimilar material bonding causes even more risk of failure in terms of delamination as FML that undergoes delamination has much lower damage tolerance [6]. Hence, it is important to evaluate the interface strength under different environmental conditions and circumstances.

Furthermore, fiber metal laminates constituting of glass fiber reinforced polymers (GFRP) offer great heat resistance due to multiple layers and high melting point of GFRP layers, hence have potential to be used at high temperatures [3–5]. The temperature resistance and structural integrity of composites at elevated temperature depends on the resin that forms the matrix. Polyether ether ketone (PEEK) type composite possesses very high melting point and its interlaminar strength was studied [7]. Temperature effects on delamination were foremost explored on pure composites. A study by Czabaj and Davidson [8] on woven graphite/polyimide composite at room temperature and 300 °C found Mode-I and mixed-mode toughness to have improved at higher temperature. On the other hand, Mode-II properties decreased. Such findings proposed some modifications on DCB, ENF and MMF tests that suit high temperature applications. Another research was performed by Charalambous et al. [9] on mixed-mode delamination of carbon-fibre/epoxy at temperatures of −50, 20, 50 and 80 °C under quasi-static and fatigue loading. Depending on the rate of loading, the fracture toughness of carbon epoxy shows different responses at different temperatures. At 80 °C, the interlaminar fracture toughness is much higher due to softening of the composite matrix. From open literature, it is found that there has been very limited research on the temperature effects of FML delamination.

In this paper, the temperature effects on Mode-I and Mode-II delamination of glass fiber reinforced polymer laminated with aluminum alloy is investigated experimentally. Firstly, experimental tests are carried out in accordance with ASTM standards to achieve delamination of respective modes. Double cantilever beam and 3-point end notched flexure tests are conducted at steady state temperature of 30, 70 and 110 °C to obtain the load-displacement curves. From the load-displacement, the stiffness and peak loads are acquired for data reduction to obtain Mode-I fracture toughness and Mode-II fracture toughness. Relationship between modulus and fracture toughness with temperature of both modes are compared.

2 Experimental Program

Abstract The main methodology to evaluate Mode-I and Mode-II delamination using experimental method is presented. The source of material acquisition of both aluminum 2024-T3 and GFRP and their respective specification are shown. Next, the fabrication processes and bonding of aluminum to GFRP layers to achieve FML specimens are elaborated. The bonding between aluminum and GFRP used a different epoxy compared to GFRP. FML tests are prepared in reference to ASTM, hence two different dimensions of specimens for both double cantilever beam and end notched flexural three point bending tests. The setups of both tests together with their respective relative dimensions are illustrated, including the pre-crack length and span length. Shimadzu universal testing machine with a thermostatic chamber is used to achieve uniform temperature that is below the glass transition temperature of the epoxy.

This section describes the acquirement and preparation of specimens, followed by the experimental setup and testing of Mode-I delamination under double cantilever beam (DCB) test and Mode-II delamination under 3-point end notched flexure (ENF) test of glass fiber laminated aluminum.

Raw material in bulk form aluminum 2024-T3 alloy and S-2 glass fiber are acquired from X Plas Singapore. Two millimeter thick Aluminum 2024-T3 alloy is sand blasted with 150 grids, and then gloss anodized with industrial grade sulfuric acid. GFRP is fabricated from glass/epoxy prepregs of unidirectional S2-type glass fibers with glass transition temperature (T_g) of the epoxy of 125 °C, fiber volume fraction of 0.5 and individual ply thickness of 0.15 mm. 14 plies are cured to form a total thickness of 2.2 mm. The GFRP is fabricated using hot press to form each part, then allowing the prepregs to bind using hot oven at temperature of around 120 °C.

DCB test is prepared based on ASTM D5528-13 for Mode-I while ENF test is prepared based on ASTM D7905-14 for Mode-II interlaminar fracture toughness of FRP composites under 3-point bending [10, 11]. The fiber directions are aligned to be 0° along the specimen length. To prepare the DCB and ENF specimens, the supplied bulk plate of aluminum with 2 mm thickness and GFRP with 2.2 mm thickness are first sanded using 1200 grit sandpaper, then applied with acetone to ensure the bonding surfaces are free of oil and dirt. Non-adhesive Teflon film of thickness 13 μm are then inserted between the plates to act as the pre-crack. ProAssure Wrap epoxy resin supplied from IO Setia Ventures with a T_g of 130 °C is used for the bonding of aluminum and GFRP. The resin is bonded at room temperature for at least 4 days. DCB specimens are then cut into dimensions of length 125 mm and width $B = 20$ mm while ENF specimens are cut into dimensions of 160 mm and $B = 20$ mm using band saw machine. DCB and ENF specimens have fixed crack length, a_o of 40 and 25 mm respectively. According to ASTM, the pre-crack has to be at the middle point, however in this case, GFRP is slightly thicker with 2.2 mm compared to aluminum with 2 mm. Hence, each specimen has total thickness, 2 h of around 4.2 mm.

Commercial hasp hinges are bonded onto both top and bottom parts of DCB to allow Mode-I peeling by control of crosshead movement, and to ensure strong bonding between hinges and specimen surface. Araldite AW 4859 with high strength, high toughness and T_g of 140 °C is used as adhesive. The DCB specimens are cured at room temperature for 24 h and then post-cured at 150 °C for 2 h to achieve optimum strength.

DCB and ENF tests are subjected under quasi-static loading rates of 1.0 and 0.5 mm/min respectively using Shimadzu AG-X plus 100kN machine. TCE-N300 thermostatic chamber is used for achieving and maintaining control of steady state temperatures throughout the duration of the tests. Load and displacement data is processed and analyzed using Trapezium X. For DCB test, the top hinge is clamped by the crosshead and pulled upwards while the bottom hinge is fixed, as shown in Fig. 1. Meanwhile, ENF test setup is shown in Fig. 2, where two bottom rollers support the specimen in place while the top roller moves downwards. At least five samples are used for each temperature of 30, 70 and 110 °C for each test. Three samples with the best results are chosen to be reported here. All ENF specimens have same span length, L of 50 mm.

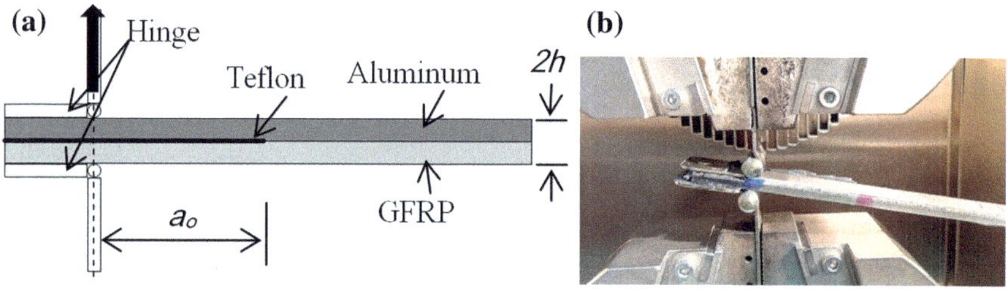

Fig. 1 DCB test setup in **a** sketched configurations and **b** specimen test

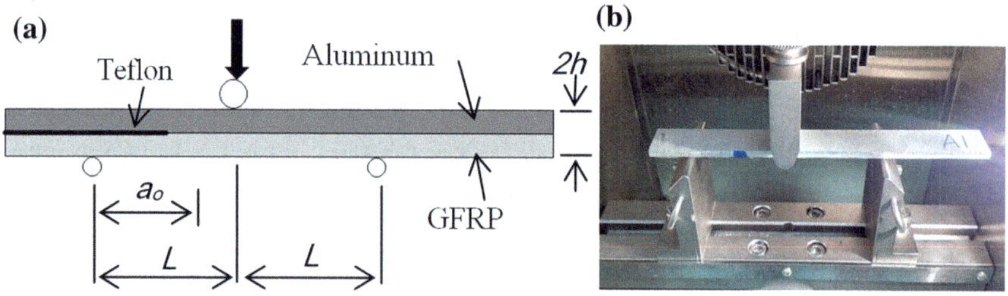

Fig. 2 ENF test setup in **a** sketched configuration and **b** specimen test

3 Experimental Results

Abstract Response of both DCB and ENF tests are displayed in load-displacement curves. Both Mode-I and Mode-II delamination are described separately in subsections and the overall trend are explained with respect to the delamination behavior and response. Differences between DCB and ENF profile are foremost elaborated, where the characteristics of loading and unloading of DCB are described based on literature. The dissimilarity of curve profile and magnitude at each temperature of 30, 70 and 110 °C are addressed and discussed. Values of stiffness, k and peak load, F_P are extracted from the load-displacement curves and plotted against temperature to acquire the respective trends for the subsequent chapters.

3.1 DCB

For both DCB and ENF tests, the specimens of same dimensions, crack length and setup were tested to analyze their response at different temperatures. Representative load-displacement curves of DCB and ENF tests are shown in Fig. 3. Figure 3a shows the load curves of DCB tests for Mode-I delamination at three temperatures. Generally, all three load curves firstly follow a pattern of linear elastic region until reaching peak load, where delamination initiates. Afterwards, the load begins to decrease, indicated by the drop of load. By comparing between the load curves, it is very clear that temperature decreases the peak load, especially when reaching higher temperature range (110 °C). Referring to Table 1, the loading stiffness, k has little change when comparing 70 to 30 °C, yet drops to 57.7% at 110 °C. This can be attributed to the softening of the cohesive region when exceeding 70 °C. Peak load, F_P on the other hand decreases from 34.51 N to 22.04 N even at 70 °C and further to 10.55 N at 110 °C.

It is worth noting that the delamination onset at 30 and 70 °C after the peak load exhibits a rapid load decline, followed by a repeating decreasing zigzag pattern of overall drop of load and is in tandem as observed by other previous researchers [12–14]. It can be explained that the pattern is due to regions across the crack with inconsistent resin toughness. Stress concentration builds up at the region with higher strength, and when the stress exceeds the local interlaminar strength, the energy released is then much higher than region that follows with weaker strength [14]. Hence, the load drops rapidly, and the cycle continues, which explains the zig-zag profile. However, this pattern does not occur at 110 °C where the profile is much smoother. A much higher temperature caused the crack to propagate at a steady pace as the interface is softened by the heat and much more ductile.

3.2 ENF

For Mode-II delamination, the load-displacement curves at 30, 70 and 110 °C are shown in Fig. 3b. Similar to Mode-I, ENF tests display linear elastic region until reaching peak load. The peak load of Mode-II is obviously much higher compared to Mode-I. However, unlike Mode-I, the stiffness k decreases to 56.0% at 70 °C and further decreases to 38.3% of the original value at 110 °C. F_P which is 297.3 N at 30 °C drops significantly to 208.32 N and 74.97 N at 70 and 110 °C respectively. It should be noted that the displacement until initial peak load was actually larger when comparing 70 to 30 °C, as the increase in temperature prolonged the pre-crack initiation. Nevertheless, the initial peak load for 110 °C was at much smaller displacement as the interface is considerably weaker.

In all three temperature cases, Mode-II load curve demonstrates a load drop, followed by a gradual increment of load that persists, which is different from Mode-I. However, higher temperature changes the profile from sudden drop to a gradual one, and at the same time the amplitude of decrease is also significantly less. This once again shows softening of interlaminar surface due to higher temperature, which makes the delamination occur much steadily. For 70 °C, the extension point for initial peak load is actually extended when compared to 30 °C. This implies that at medium temperature range, higher shear displacement between the aluminum and GFRP is possible before crack initiation occurs.

4 Data Reduction

Abstract This chapter focuses on the method of converting and calculating existing stiffness and peak load into fracture toughness. Both Mode-I and Mode-II delami-

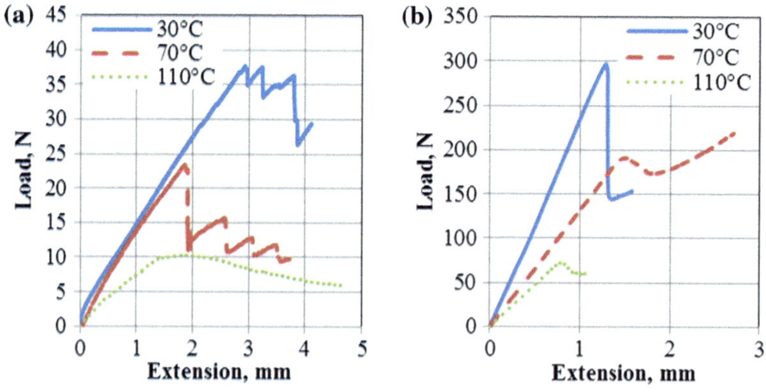

Fig. 3 Representative load versus displacement curve at each temperature from **a** DCB test, **b** ENF test

Table 1 Effect of temperature on Mode-I and Mode-II delamination properties

–	Mode-I	–	–	Mode-II	–	–
–	30 °C	70 °C	110 °C	30 °C	70 °C	110 °C
k (N/mm)	12.57 ± 0.15	12.47 ± 0.46	7.25 ± 0.08	239.97 ± 7.10	134.50 ± 1.91	91.97 ± 2.59
F_P (N)	34.51 ± 2.42	22.04 ± 1.23	10.55 ± 0.16	297.3 ± 4.63	208.32 ± 10.48	74.97 ± 2.34

nation results are analyzed using back-calculated modulus and Irwin-Kies equation. Back-calculated modulus is used together with compliance C, which is the inverse of stiffness k to get the flexural modulus E. The value of E is then subsequently used with peak load F_p, initial crack length a_o, and dimensions width b and single arm thickness h within the Irwin-Kies equations to get Mode-I fracture toughness G_{IC} and Mode-II fracture toughness G_{IIC}. Each values of G_{IC} and G_{IIC} at each temperature are calculated separately where they are plotted in the next chapter against temperature to acquire the general trend.

To obtain the fracture energy of the Mode-I and Mode-II results, analytical model based on Irwin-Kies equation are implemented. The Mode-I fracture energy is expressed as:

$$G_{IC} = \frac{12F_p^2 \cdot a_0^2}{E \cdot b^2 \cdot h^3} \tag{1}$$

where G_{IC} is the Mode-I fracture toughness, F_p is the peak load, a_o is the initial crack length, E is the flexural modulus, b is the width and h is the single arm thickness. The modulus E is calculated from back-calculated modulus method, as shown in Eq. (2):

$$E = \frac{8a_0^3}{C \cdot b \cdot h^3} \tag{2}$$

where C is the compliance, which is the inverse of stiffness, k. The Mode-II fracture energy is defined as:

$$G_{IIC} = \frac{9F_p^2 \cdot a_0^2}{16E \cdot b^2 \cdot h^3} \tag{3}$$

where G_{IIC} is the Mode-II fracture toughness. Similar to Mode-I, the modulus, E is also calculated using back-calculated modulus as shown in Eq. (4):

$$E = \frac{3a_0^3 + 2L^3}{8C \cdot b \cdot h^3} \tag{4}$$

5 Discussion

Abstract The values of back calculated modulus and fracture toughness calculated from previous data reduction are plotted against temperature. Two separate plots of DCB and ENF are compared and discussed based on the specimen behavior. The standard deviations of each result are shown in figure and the points of significant scatter are described. Based on the percentage of degradation, the delamination properties of both Mode-I and Mode-II on the FML application is demonstrated. It is noted that Mode-I fracture toughness is temperature dependent even at medium temperature. On the other hand, Mode-II fracture toughness is only significantly affected

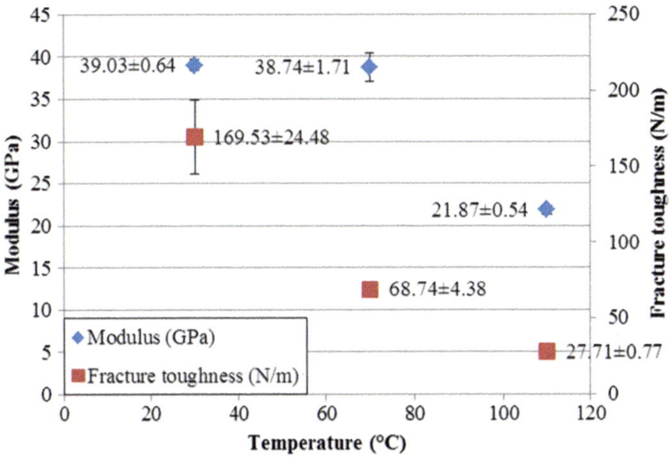

Fig. 4 Mode-I modulus and fracture toughness versus temperature (with standard deviation)

at high temperature range. However, at high temperature, the degradation of both delamination modes is very drastic.

Figure 4 shows the average value of Mode-I modulus and G_{IC} plotted at each temperature with standard deviation. With similarity to the stiffness, Mode-I modulus dropped slightly from 39.03GPa at 30 °C to 38.74GPa at 70 °C while the value is almost halved at 110 °C with 21.87GPa. However, the G_{IC} is greatly reduced from 169.53 N/m to 68.74 N/m only at medium temperature and drops further to 27.71 N/m at 110 °C. The data reduction value demonstrates the major influence of temperature on the Mode-I delamination resistance of glass fiber laminated aluminum, where the G_{IC} is significantly weaker with the increase in temperature. It is worth indicating that the G_{IC} at 30 °C shows relatively large deviation, due to scatter in the peak load values from the tests where its value squared in data reduction. It is suspected that Mode-I crack initiation is sensitive to any minor adhesive flaws, variable thickness across the crack tip and/or experimental factors when taking into account the misalignment or friction of the hasp hinges which impair the test repeatability.

Data reduced Mode-II properties are plotted against temperature in Fig. 5. Interestingly, contradicting Mode-I, the modulus of Mode-II drops significantly by 41.7% at medium temperature range while the G_{IIC} is only marginally reduced by 10.9% from 165.04 to 147.03 N/m. Furthermore, from 70 to 110 °C, modulus property degraded more steadily while G_{IIC} plummeted by 80.7%. The reasoning behind retaining of G_{IIC} at 70 °C can be attributed to Fig. 3b where although the peak load is lower than 30 °C, the extension is longer. Similar to Mode-I, the G_{IIC} at 30 and 70 °C exhibits a certain amount of scatter, though less than G_{IC}. The peak load governed by crack initiation that plays a role in controlling the deviation of G_{IIC} is more controlled in shear delamination.

In summary, G_{IC} is similar to G_{IIC} at 30 °C, which opposes findings by Pan et al. [15]. However, at 70 °C, G_{IIC} becomes extensively higher in comparison such that

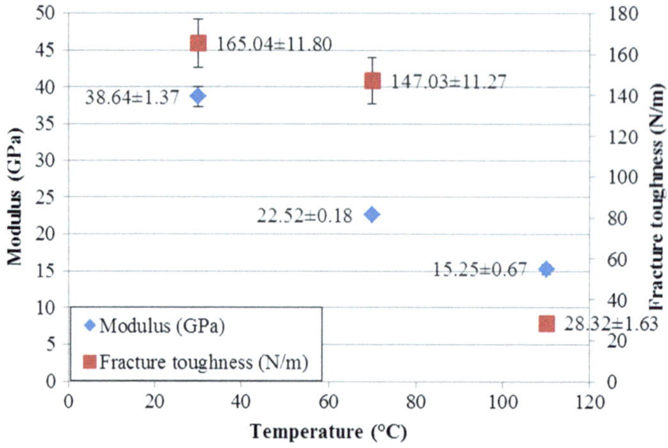

Fig. 5 Mode-II modulus and fracture toughness versus temperature (with standard deviation)

G_{IC} is substantially reduced at 70 °C while G_{IIC} remains mainly unaffected. Finally at 110 °C, both G_{IC} and G_{IIC} are only around 16-17% of original fracture toughness with G_{IIC} slightly higher.

6 Conclusion

Abstract In the previous chapters, the experiments were planned and carried out, then followed by results and calculations, and the main discussion was displayed. As summary, this chapter shows the important parts of the findings concerning the main objective and filling of the research gap. Generally, the modulus and fracture toughness of Mode-I and Mode-II delamination of FML at 3 different temperatures of 30, 70 and 110 °C were shown. The degree of degradation for Mode-I delamination was consistently acute even from medium temperature of 70 °C. On the other hand, degradation of Mode-II is minimal at 70 °C but drastically falls when reaching 110 °C. The applications of the FML at elevated temperature based on delamination modes are elaborated.

The present study scrutinized effect of temperature of 30, 70 and 110 °C on the Mode-I and Mode-II delamination of fiberglass laminated aluminum via DCB and ENF experimental tests. The experimental findings can be summarized as follows.

- Elevated temperature imposed a significant degradation on the G_{IC} both on medium (59.5%) and high temperature range (83.7%).
- At 70 °C, although the Mode-II modulus is much lower compared to 30 °C, the G_{IIC} is only slightly degraded (10.9%). At 110 °C, G_{IIC} sharply dropped by a great margin with relation to 30 °C (82.8%).

- Therefore, fiberglass laminated aluminum can be appropriately used under Mode-II shear loading at 70 °C but not fitting for Mode-I peel type loading. But at 110 °C, exceeding degradation of both Mode-I and Mode-II loading means unacceptable use at temperature close to T_g.

Acknowledgements This project is supported by the Ministry of Higher Education (MOHE) Malaysia under Grant Vote No. R.J130000.7824.4F248, Grant Vote No. Q.J130000.2424.03G71 and Contract Research Grant R.J130000.7624.4C089. Sincere appreciation and acknowledgement also goes to Universiti Teknologi Malaysia (UTM) for the continuous support in completing this project.

References

1. Karakuzu R, Erbil E, Aktas M (2010) Impact characterization of glass/epoxy composite plates: an experimental and numerical study. Compos Part B-Eng 41(5):388–395. https://doi.org/10.1016/j.compositesb.2010.02.003
2. Sayer M, Bektaş NB, Sayman O (2010) An experimental investigation on the impact behavior of hybrid composite plates. Compos Struct 92(5):1256–1262. https://doi.org/10.1016/j.compstruct.2009.10.036
3. Asundi A, Choi AYN (1997) Fiber metal laminates: an advanced material for future aircraft. J Mater Process Tech 63(1–3):384–394. https://doi.org/10.1016/S0924-0136(96)02652-0
4. Sinmazçelik T, Avcu E, Bora MÖ, Çoban O (2011) A review: fibre metal laminates, background, bonding types and applied test methods. Mater Des 32(7):3671–3685. https://doi.org/10.1016/j.matdes.2011.03.011
5. Vogelesang LB, Vlot A (2000) Development of fibre metal laminates for advanced aerospace structures. J Mater Process Tech 103(1):1–5. https://doi.org/10.1016/S0924-0136(00)00411-8
6. Pärnänen T, Vänttinen A, Kanerva M, Jokinen J, Saarela O (2016) The effects of debonding on the low-velocity impact response of steel-CFRP fibre metal laminates. Appl Compos Mater 23(6):1151–1166. https://doi.org/10.1007/s10443-016-9505-4
7. Fracasso R, Rink M, Pavan A, Frassine R (2001) The effects of strain-rate and temperature on the interlaminar fracture toughness of interleaved PEEK/CF composites. Compos Sci Technol 61(1):57–63. https://doi.org/10.1016/S0266-3538(00)00153-6
8. Czabaj MW, Davidson BD (2015) Determination of the mode I, mode II, and mixed-mode I-II delamination toughness of a graphite/polyimide composite at room and elevated temperatures. J Compos Mater 50(16):2235–2253. https://doi.org/10.1177/0021998315602945
9. Charalambous G, Allegri G, Hallett SR (2015) Temperature effects on mixed mode I/II delamination under quasi-static and fatigue loading of a carbon/epoxy composite. Compos Part A-Appl S 77:75–86. https://doi.org/10.1016/j.compositesa.2015.05.016
10. ASTM D5528-13 (2013) Standard test method for Mode I interlaminar fracture toughness of unidirectional fiber-reinforced polymer matrix composites. ASTM International, West Conshohocken, PA
11. ASTM D7905 / D7905 M–14 (2014) Standard test method for determination of the Mode II interlaminar fracture toughness of unidirectional fiber-reinforced polymer matrix composites. ASTM International, West Conshohocken
12. De Baere I, Jacques S, Van Paepegem W, Degrieck J (2012) Study of the Mode I and Mode II interlaminar behaviour of a carbon fabric reinforced thermoplastic. Polym Test 31(2):322–332. https://doi.org/10.1016/j.polymertesting.2011.12.009
13. Johar M, Wong KJ, Tamin N (2017) Mixed-mode delamination failures of quasi-isotropic quasi-homogeneous carbon/epoxy laminated composite. https://doi.org/10.5772/intechopen.69440

14. Reis PNB, Ferreira JAM, Antunes FV, Costa JDM (2015) Initial crack length on the interlaminar fracture of woven carbon/epoxy laminates. Fiber Polym 16(4):894–901. https://doi.org/10.1007/s12221-015-0894-z
15. Pan Y, Wu G, Cheng X, Zhang Z, Li M, Ji S, Huang Z (2016) Mode I and Mode II interlaminar fracture toughness of CFRP/magnesium alloys hybrid laminates. Compos Interface 23(5):453–465. https://doi.org/10.1080/09276440.2016.1144911

An Alternative Method for Modelling the Degradation of Hyperelastic Materials Within the Framework of Finite-Strain Elastoplasticity

L. Écsi and P. Élesztős

Abstract In this paper an alternative method for modeling the degradation of hyperelastic materials within the framework of finite-strain elastoplasticity is presented. The material model is based on the first nonlinear continuum theory for finite deformations for elastoplastic media which allows for the development of objective and thermodynamically consistent material models. Therefore the model and its results, when used in numerical analyses, are independent of the model description and the particularities of the mathematical formulation. Moreover, the model allows for the study of the body's behavior from a thermodynamic aspect as well, using clear physical interpretations of the plastic flow and "normality rules" in all configurations of the body and all stress spaces, as it enables the expression of the rate of change of internal mechanical energy accumulated in the body in terms of internal mechanical power conjugate stress measures and strain rates. As a result, the internal power density of the model can be related directly to the internal power density of the specimen used in the tensile test of the modelled material. In this paper natural rubber behavior is studied using modified Mooney-Rivlin and neo-Hookean material models.

Keywords Nonlinear continuum theory for finite deformations of elastoplastic media · Thermodynamically consistent formulation · Hyperelastic material · Mooney-Rivlin and neo-Hookean material models

1 Introduction

Hyperelastic materials, such as natural rubber, rubber-like solids, various polymers, vulcanized elastomers, metallic foams, biomaterials, etc. have a wide range of applications in industry, medicine, chemistry and they play essential and ubiquitous roles in our everyday lives as well. Mechanically damaged parts of constructions, failure of safety-critical components manufactured from these materials can cause serious

L. Écsi (✉) · P. Élesztős
Faculty of Mechanical Engineering, Slovak University of Technology in Bratislava, Námestie slobody 17, 812 31 Bratislava 1, Slovakia
e-mail: ladislav.ecsi@stuba.sk; ecsi_ladislav@mail.t-com.sk

A. Öchsner and H. Altenbach (eds.), *Engineering Design Applications II*, Advanced Structured Materials 113, https://doi.org/10.1007/978-3-030-20801-1_5

accidents. It is therefore crucial that their behaviour can be predicted accurately under various forms of mechanical loading.

Currently multiplicative plasticity models are used to model material degradation of hyperelastic materials within the framework of finite-strain elastoplasicity. The flow plasticity theories employed in the models date back to the 1960s and since then have been considered by Lee and Liu [1], Lee [2], Kroner and Teodosiu [3], Mandel [4–6], Kratochvil [7], Sidoroff [8], Nemat-Nasser [9, 10], Dafalias [11, 12], Lubarda and Lee [13], Agah-Tehrani et al. [14], Lubliner [15, 16], Simo and Ortiz [17], Simo [18, 19] and others. The basic ideas, implemented in the models, however originate from the fundamental work of Taylor and his associates, e.g. Taylor and Elam [20, 21], Taylor [22], which have been expanded upon in Hill [23], Rice [24], Hill and Rice [25], Asaro and Rice [26], Asaro [27] which would later on be developed into a coherent theory of single-crystal plasticity by Asaro [28], Peirce et al. [29], Peirce [30], allowing for the micromechanical description of the plastic flow in its current form.

Although multiplicative plasticity models are now considered to be continuum based, in reality they are not, if one studies them strictly from the standpoint of nonlinear continuum theory. They may well be disregarding the basic principles of nonlinear continuum mechanics when describing the kinematics of motion during elastoplastic deformations of the body. Despite appearances, it is not our intention to criticise multiplicative plasticity theories and their related material models, though we cannot avoid pointing out a few of their disadvantages, which are not quite consistent with the theory of nonlinear continuum mechanics.

In continuum mechanics the motion, the displacement field and the deformation gradient have exact mathematical formulations, where each of them has an undeniable physical meaning [31–33]. The motion, for instance, from the mathematical point of view is a vector field or a vector function, which maps each material point of the body from one of its configuration to the other. The function must exist whenever the body moves and it determines the position vector of the material point at each time instant. The vector field that connects any pairs of the position vectors of the material point is a displacement field. It must have both Lagrangian and Eulerian forms, otherwise a physical phenomenon expressed in Lagrangian form cannot be re-expressed in Eulerian form and vice versa. The deformation gradient then becomes the derivative of the position vector of the material point after motion with respect to the position vector of the material point before motion, or simply the gradient of the vector function describing the motion. If the motion takes place in several stages, so that it can be decomposed into several parts, while the body traverses from its initial configuration to its current configuration through several intermediate configurations, the above definitions apply between any pairs of configurations of the body.

It is quite clear from the above that contemporary multiplicative plasticity theories treat the motion consistently with the theory of nonlinear continuum mechanics between the initial and current configurations of the body only, where the motion, the displacement field and the deformation gradient are considered properly, but not between configurations where one of the configurations is an intermediate configuration. Here the motion and the displacement field are disregarded and as a result, the deformation gradient loses its physical meaning. Moreover, multiplicative plasticity theories assume that the intermediate configuration of the body is 'stress-free' [34], or at least locally unstressed [35]. The assumption is however not compatible with the theory of nonlinear continuum mechanics, because according to the theory proper stress transformations must hold true between all pairs of stress measures in all configurations of the body. As a result, whenever the stress tensor is determined at a material point in one stress space, in any configuration of the body, it is then determined for all stress spaces and all configurations of the body. This means, at a material point, the Lagrangian forms of the multiplicative split for the deformation gradient $\mathbf{F} = \mathbf{F}^{el} \cdot \mathbf{F}^{pl}$ and the 2nd Piola-Kirchhoff stress tensor $\mathbf{S}$ respectively are given, the corresponding Cauchy's stress measure at the spatial point in the intermediate configuration is

$$\sigma^{\text{int}} = \frac{1}{\det\left(\mathbf{F}^{pl}\right)} \cdot \mathbf{F}^{pl} \cdot \mathbf{S} \cdot \left(\mathbf{F}^{pl}\right)^{T}. \tag{1}$$

Therefore a global/local stress-free state in the intermediate configuration should not exist, unless the material point is unstressed in all configurations and all stress spaces of the body. The assumption of an unstressed intermediate configuration is used as an excuse for disregarding the proper kinematics of motion between any two configurations of the body involving an intermediate configuration by declaring that "compatible unstressed configurations of multiaxially stretched elastoplastic bodies do not exist in general; that is, it is generally not possible to find a deformation $\mathbf{\Phi}^{pl} = \mathbf{\Phi}^{pl}(\mathbf{X}, t)$ such that" [35].

$$\mathbf{F}^{pl} = \frac{\partial \mathbf{\Phi}^{pl}}{\partial \mathbf{X}} \tag{2}$$

In addition to these, multiplicative plasticity theories only offer a solution to problems with a particular order of deformations, namely when the body undergoes plastic deformations first and elastic deformations afterwards at each of its constituent. Virgin materials just do not deform in this way, since they cannot undergo plastic deformations until they have undergone elastic deformation. As a result, the theories fail to even imitate the true physics of the deformation process. Proving that the theories are thermodynamically consistent may also pose a problem and according to the authors this may not be possible until the deformation gradient and its parts are somehow related to the 'missing' displacement fields in the theories.

Proper mathematical formulation of a physical phenomenon has to be independent of the description and the particularities of the mathematical model formulation.

The aim of this paper is to present an alternate way of modelling degradation of hyperelastic materials within the framework of finite-strain elastoplasticity. The model is based on the first nonlinear continuum theory for finite deformations of elastoplastic media which allows for the development of objective and thermodynamically consistent material models. Therefore the model and its results, when used in numerical analyses, are independent of the model description and the particularities of the mathematical formulation. By the description we mean total or updated Lagrangian description and by the particularities of the model's mathematical formulation, the ability to work in various stress spaces in whatever configuration of the body employing appropriate pairs of stress measures and strain rates in the formulation, which are conjugate with respect to the internal mechanical power. The later also means that the description of the plastic flow for the first time becomes thermodynamically consistent, as it can be expressed in terms of various definitions of the yield surface and corresponding 'normality rules' in all stress spaces and all configurations of the body, while the plastic power density of the model can be related directly to the plastic power density of the specimen coming from the uniaxial tensile test of the modelled material. In this paper natural rubber material degradation is studied using the Mooney-Rivlin and the neo-Hookean material models, which have been modified in accordance with the nonlinear continuum theory for finite deformations of elastoplastic media.

2 Theory

2.1 A Short Overview of the Nonlinear Continuum Theory for Finite Deformations of Elasoplastic Media

Écsi and Élesztős presented the foundation of the nonlinear continuum theory for finite deformations of elastoplastic media which allows for the development of objective and thermodynamically consistent material models within the framework of finite-strain elastoplasticity [36]. The theory was derived on the assumption that the body first undergoes plastic deformations and then elastic deformations at each of its constituents using a stressed intermediate configuration at current time. The most important result of the theory is that the multiplicative split of the deformation gradient is not the final form of the deformation gradient, but the product can still be simplified, so the theory is actually based on the additive decomposition of the Lagrangian displacement field into an elastic part and a plastic part. The deformation gradient then takes its final form

$$\mathbf{F}(\mathbf{X},t)=\mathbf{I}+\frac{\partial\,{}^{0}\mathbf{u}}{\partial\mathbf{X}}={}^{pl-el}\mathbf{F}^{el}(\mathbf{X},t)\cdot{}^{pl-el}\mathbf{F}^{pl}(\mathbf{X},t)=\mathbf{I}+\frac{\partial\,{}^{0}\mathbf{u}^{pl}}{\partial\mathbf{X}}+\frac{\partial\,{}^{0}\mathbf{u}^{el}}{\partial\mathbf{X}},\qquad(3)$$

where ${}^{0}\mathbf{u}={}^{0}\mathbf{u}(\mathbf{X},t)$, ${}^{0}\mathbf{u}^{el}={}^{0}\mathbf{u}^{el}(\mathbf{X},t)$, ${}^{0}\mathbf{u}^{pl}={}^{0}\mathbf{u}^{pl}(\mathbf{X},t)$ denotes the Lagrangian total and elastic and plastic displacement fields respectively. In Eq. (3) the Lagrangian forms of the plastic and the elastic parts of the deformation gradient are defined as ${}^{pl-el}\mathbf{F}^{pl}(\mathbf{X},t)=\partial\,{}^{pl-el}\boldsymbol{\Phi}^{pl}(\mathbf{X},t)/\partial\mathbf{X}$, ${}^{pl-el}\mathbf{F}^{el}(\mathbf{X},t)=\partial\,{}^{pl-el}\boldsymbol{\Phi}^{el}\left({}^{pl-el}\boldsymbol{\Phi}^{pl}(\mathbf{X},t),t\right)/\partial\,{}^{pl-el}\boldsymbol{\Phi}^{pl}(\mathbf{X},t)$. The motion can then be expressed by a composite function $\boldsymbol{\Phi}(\mathbf{X},t)=\partial\,{}^{pl-el}\boldsymbol{\Phi}^{el}\left({}^{pl-el}\boldsymbol{\Phi}^{pl}(\mathbf{X},t),t\right)$, where the left superscripts denote the order of deformations assumed in the theory.

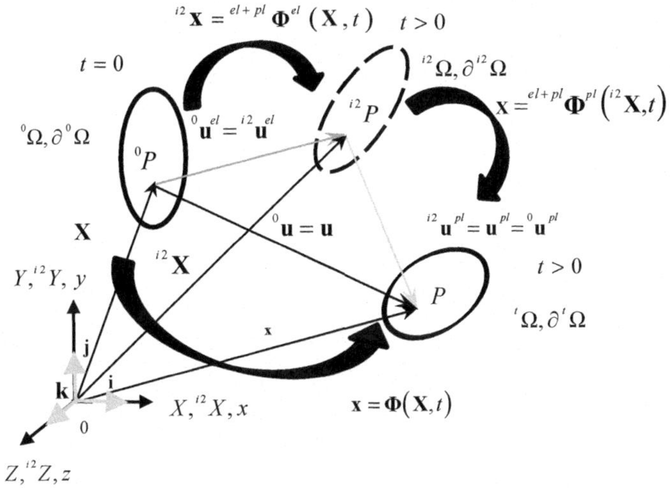

Fig. 1 The kinematics of motion of a deformable body when plastic deformations take place after elastic deformations

It should be noted that the physics of the deformation process is immediately put in order when the kinematics of motion between all configurations of the body is considered with respect to the theory of nonlinear continuum mechanics, because vector addition is commutative and as a result, the theory no longer depends on the order of elastic and plastic deformations. The resulting material models are then capable of imitating the behavior of both, virgin materials and materials with initial plastic deformations. Now let us consider a deformation process during which the order of elastic and plastic deformations is reversed at all material points of the body (Fig. 1).

The motion $\mathbf{\Phi}(\mathbf{X}, t)$ and the Lagrangian displacement field $^{0}\mathbf{u} = {}^{0}\mathbf{u}(\mathbf{X}, t)$ at a material point ^{0}P are then defined as

$$\left\{\mathbf{x}|\mathbf{x} = \mathbf{\Phi}(\mathbf{X}, t), \quad \text{for} \quad \mathbf{X} \in {}^{0}\Omega, \quad \mathbf{x} \in {}^{t}\Omega \, \text{and} \, t \geq 0\right\}, \tag{4}$$

$$^{0}\mathbf{u} = {}^{0}\mathbf{u}(\mathbf{X}, t) = \mathbf{x} - \mathbf{X} = \mathbf{\Phi}(\mathbf{X}, t) - \mathbf{X}, \text{ for } \mathbf{X} \in {}^{0}\Omega, \ \mathbf{x} \in {}^{t}\Omega \, \text{and} \, t \geq 0. \tag{5}$$

In Eqs. (4) and (5) the symbols $^{0}\Omega$ and $^{t}\Omega$ denote the domain and the range of the vector function $\mathbf{\Phi}(\mathbf{X}, t)$ defining the motion, where $^{0}\Omega$ and $^{t}\Omega$ represent the initial and the current volumes of the body. When the motion is decomposed into an elastic part and a plastic part, the material point $^{0}P \in {}^{0}\Omega$ is mapped in succession into the spatial points $^{i2}P \in {}^{i2}\Omega$ and $P \in {}^{t}\Omega$ respectively. Then the elastic motion $^{el-pl}\mathbf{\Phi}^{el}(\mathbf{X}, t)$, the Lagrangian elastic displacement field $^{0}\mathbf{u}^{el}$, the plastic motion $^{el-pl}\mathbf{\Phi}^{pl}(^{i2}\mathbf{X}, t)$, the Eulerian plastic displacement field $^{i2}\mathbf{u}^{pl}$ and the Lagrangian plastic displacement field $^{0}\mathbf{u}^{pl}$ are defined as

$$\left\{ {}^{i2}\mathbf{X} | {}^{i2}\mathbf{X} = {}^{el-pl}\boldsymbol{\Phi}^{el}(\mathbf{X}, t), \text{ for } \mathbf{X} \in {}^{0}\Omega, {}^{i2}\mathbf{X} \in {}^{i2}\Omega \text{ and t} \geq 0 \right\}, \tag{6}$$

$$\begin{gathered} {}^{0}\mathbf{u}^{el} = {}^{0}\mathbf{u}^{el}(\mathbf{X}, t) = {}^{i2}\mathbf{X} - \mathbf{X} = {}^{el-pl}\boldsymbol{\Phi}^{el}(\mathbf{X}, t) - \mathbf{X}, \\ \text{for } \mathbf{X} \in {}^{0}\Omega, {}^{i2}\mathbf{X} \in {}^{i2}\Omega \text{ and } t \geq 0, \end{gathered} \tag{7}$$

$$\left\{ \mathbf{x} | \mathbf{x} = {}^{el-pl}\boldsymbol{\Phi}^{pl}\left({}^{i2}\mathbf{X}, t\right), \text{ for } {}^{i2}\mathbf{X} \in {}^{i2}\Omega, \mathbf{x} \in {}^{t}\Omega \text{ and } t \geq 0 \right\}, \tag{8}$$

$$\begin{gathered} {}^{i2}\mathbf{u}^{pl} = {}^{i2}\mathbf{u}^{pl}\left({}^{i2}\mathbf{X}, t\right) = \mathbf{x} - {}^{i2}\mathbf{X} = {}^{el-pl}\boldsymbol{\Phi}^{pl}\left({}^{i2}\mathbf{X}, t\right) - {}^{i2}\mathbf{X}, \\ \text{for } {}^{i2}\mathbf{X} \in {}^{i2}\Omega, \mathbf{x} \in {}^{t}\Omega \text{ and } t \geq 0, \end{gathered} \tag{9}$$

$$\begin{gathered} {}^{0}\mathbf{u}^{pl} = {}^{0}\mathbf{u}^{pl}(\mathbf{X}, t) = \mathbf{x} - {}^{i2}\mathbf{X} = {}^{el-pl}\boldsymbol{\Phi}^{pl}\left[{}^{el-pl}\boldsymbol{\Phi}^{el}(\mathbf{X}, t), t\right] - {}^{el-pl}\boldsymbol{\Phi}^{el}(\mathbf{X}, t), \\ \text{for } \mathbf{X} \in {}^{0}\Omega, \mathbf{x} \in {}^{t}\Omega \text{ and } t \geq 0. \end{gathered} \tag{10}$$

In Eqs. (6)–(10) the symbol ${}^{i2}\Omega$ stands for the volume of the body in its intermediate configuration when the body undergoes elastic deformations only and the left superscript ${}^{el-pl}(\cdot)$ denotes the order of deformations. Adding up Eqs. (7) and (10) results in the following formula for the Lagrangian displacement field

$$\begin{gathered} {}^{0}\mathbf{u} = {}^{0}\mathbf{u}^{el} + {}^{0}\mathbf{u}^{pl} = \mathbf{x} - \mathbf{X} = {}^{el-pl}\boldsymbol{\Phi}^{pl}\left[{}^{el-pl}\boldsymbol{\Phi}^{el}(\mathbf{X}, t), t\right] - \mathbf{X}, \\ \text{for } \mathbf{X} \in {}^{0}\Omega, \mathbf{x} \in {}^{t}\Omega \text{ and } t \geq 0. \end{gathered} \tag{11}$$

Eqn (11) then implies a composite function for the overall motion $\boldsymbol{\Phi}(\mathbf{X}, t) = {}^{el-pl}\boldsymbol{\Phi}^{pl}\left[{}^{el-pl}\boldsymbol{\Phi}^{el}(\mathbf{X}, t), t\right]$. Using the chain rule of partial differentiation, the Lagrangian multiplicative split of the deformation gradient can be expressed as

$$\begin{aligned} \mathbf{F}(\mathbf{X}, t) &= \frac{\partial \mathbf{x}}{\partial \mathbf{X}} = \frac{\partial \mathbf{x}}{\partial {}^{i2}\mathbf{X}} \cdot \frac{\partial {}^{i2}\mathbf{X}}{\partial \mathbf{X}} = {}^{el-pl}\mathbf{F}^{pl}(\mathbf{X}, t) \cdot {}^{el-pl}\mathbf{F}^{el}(\mathbf{X}, t) \\ &= \mathbf{I} + \frac{\partial {}^{0}\mathbf{u}^{el}}{\partial \mathbf{X}} + \frac{\partial {}^{0}\mathbf{u}^{pl}}{\partial \mathbf{X}}, \end{aligned} \tag{12}$$

where

$${}^{el-pl}\mathbf{F}^{pl}(\mathbf{X}, t) = \frac{\partial \mathbf{x}}{\partial {}^{i2}\mathbf{X}} = \mathbf{I} + \frac{\partial {}^{i2}\mathbf{u}^{pl}}{\partial {}^{i2}\mathbf{X}} = \mathbf{I} + \frac{\partial {}^{0}\mathbf{u}^{pl}}{\partial \mathbf{X}} \cdot \left[{}^{el-pl}\mathbf{F}^{el}(\mathbf{X}, t)\right]^{-1}, \tag{13}$$

$${}^{el-pl}\mathbf{F}^{el}(\mathbf{X}, t) = \frac{\partial {}^{i2}\mathbf{X}}{\partial \mathbf{X}} = \mathbf{I} + \frac{\partial {}^{0}\mathbf{u}^{el}}{\partial \mathbf{X}}. \tag{14}$$

Considering that ${}^{0}\mathbf{u}^{pl} = {}^{i2}\mathbf{u}^{pl}$ [see also Eqs. (9) and (10)] and taking into account the mathematical identity

$$\begin{aligned} \mathbf{I} &= \frac{\partial {}^{i2}\mathbf{X}}{\partial {}^{i2}\mathbf{X}} = \frac{\partial {}^{i2}\mathbf{X}}{\partial \mathbf{X}} \cdot \frac{\partial \mathbf{X}}{\partial {}^{i2}\mathbf{X}} = {}^{el-pl}\mathbf{F}^{el}(\mathbf{X}, t) \cdot \frac{\partial \mathbf{X}}{\partial {}^{i2}\mathbf{X}} \\ &\Leftrightarrow \frac{\partial \mathbf{X}}{\partial {}^{i2}\mathbf{X}} = \left[{}^{el-pl}\mathbf{F}^{el}(\mathbf{X}, t)\right]^{-1}, \end{aligned} \tag{15}$$

one can rewrite $\partial^{i2}\mathbf{u}^{pl}/\partial^{i2}\mathbf{X}$ into a Lagrangian form as follows

$$\frac{\partial^{i2}\mathbf{u}^{pl}}{\partial^{i2}\mathbf{X}} = \frac{\partial^{0}\mathbf{u}^{pl}}{\partial\mathbf{X}} \cdot \frac{\partial\mathbf{X}}{\partial^{i2}\mathbf{X}} = \frac{\partial^{0}\mathbf{u}^{pl}}{\partial\mathbf{X}} \cdot \left[{}^{el-pl}\mathbf{F}^{el}(\mathbf{X}, t)\right]^{-1}. \tag{16}$$

It should be noted in the above that the multiplicative split of the deformation gradient is not the final form of the gradient, because the product can still be simplified to obtain the gradient in its final and simplest form defined by Eq. (3) or Eq. (12) respectively. Alternatively, the deformation gradient can be rewritten as

$$\begin{aligned}
\mathbf{F}(\mathbf{X}, t) &= \frac{\partial\mathbf{x}}{\partial\mathbf{X}} = \mathbf{I} + \frac{\partial^{0}\mathbf{u}}{\partial\mathbf{X}} = \frac{\partial\mathbf{x}}{\partial^{i2}\mathbf{X}} \cdot \frac{\partial^{i2}\mathbf{X}}{\partial\mathbf{X}} \\
&= \frac{\partial^{el-pl}\boldsymbol{\Phi}^{pl}\left[{}^{el-pl}\boldsymbol{\Phi}^{el}(\mathbf{X}, t), t\right]}{\partial^{el-pl}\boldsymbol{\Phi}^{el}(\mathbf{X}, t)} \cdot \frac{\partial^{el-pl}\boldsymbol{\Phi}^{el}(\mathbf{X}, t)}{\partial\mathbf{X}} \\
&= {}^{el-pl}\mathbf{F}^{pl}(\mathbf{X}, t) \cdot {}^{el-pl}\mathbf{F}^{el}(\mathbf{X}, t) = {}^{el-pl}\mathbf{F}^{el}(\mathbf{X}, t) + \frac{\partial^{0}\mathbf{u}^{pl}}{\partial\mathbf{X}} \\
&= \mathbf{I} + \frac{\partial^{0}\mathbf{u}^{el}}{\partial\mathbf{X}} + \frac{\partial^{0}\mathbf{u}^{pl}}{\partial\mathbf{X}},
\end{aligned} \tag{17}$$

or equivalently expressed as

$$\mathbf{F}(\mathbf{X}, t) = \frac{\partial\mathbf{x}}{\partial\mathbf{X}} = \mathbf{I} + \frac{\partial^{0}\mathbf{u}}{\partial\mathbf{X}} = \mathbf{I} + \frac{\partial^{0}\mathbf{u}^{el}}{\partial\mathbf{X}} + \frac{\partial^{0}\mathbf{u}^{pl}}{\partial\mathbf{X}} = \mathbf{F}^{el}(\mathbf{X}, t) + \frac{\partial^{0}\mathbf{u}^{pl}}{\partial\mathbf{X}}, \tag{18}$$

after having dropped the left superscript ${}^{el-pl}(\cdot)$ in the elastic part of the deformation gradient ${}^{el-pl}\mathbf{F}(\mathbf{X}, t)$ in Eq. (17). Using Eq. (3), the deformation gradient then implies the following formulas for the additive decomposition of the material $\dot{\mathbf{E}}$ and spatial $\mathbf{d}$ strain rate tensors [36]:

$$\dot{\mathbf{E}} = \frac{1}{2} \cdot (\dot{\mathbf{F}}^{T} \cdot \mathbf{F} + \mathbf{F}^{T} \cdot \dot{\mathbf{F}}) = \dot{\mathbf{E}}^{el} + \dot{\mathbf{E}}^{pl}, \tag{19}$$

$$\mathbf{d} = \mathbf{d}^{el} + \mathbf{d}^{pl},\ \mathbf{d} = \mathbf{F}^{-T} \cdot \dot{\mathbf{E}} \cdot \mathbf{F}^{-1},\ \mathbf{d}^{el} = \mathbf{F}^{-T} \cdot \dot{\mathbf{E}}^{el} \cdot \mathbf{F}^{-1},\ \mathbf{d}^{pl} = \mathbf{F}^{-T} \cdot \dot{\mathbf{E}}^{pl} \cdot \mathbf{F}^{-1}, \tag{20}$$

where

$$\begin{aligned}
\dot{\mathbf{E}}^{el} &= \frac{1}{2} \cdot \left[\left(\frac{\partial^{0}\dot{\mathbf{u}}^{el}}{\partial\mathbf{X}}\right)^{T} \cdot \mathbf{F} + \mathbf{F}^{T} \cdot \frac{\partial^{0}\dot{\mathbf{u}}^{el}}{\partial\mathbf{X}}\right], \\
\dot{\mathbf{E}}^{pl} &= \frac{\dot{\lambda}}{2} \cdot \left[\left(\frac{\partial^{P}\Psi}{\partial\mathbf{P}}\right)^{T} \cdot \mathbf{F} + \mathbf{F}^{T} \cdot \frac{\partial^{P}\Psi}{\partial\mathbf{P}}\right],
\end{aligned} \tag{21}$$

$$\frac{\partial^{0}\dot{\mathbf{u}}^{pl}}{\partial\mathbf{X}} = \dot{\lambda} \cdot \frac{\partial^{P}\Psi}{\partial\mathbf{P}}, \quad \text{and} \quad \frac{\partial^{P}\Psi}{\partial\mathbf{P}} \neq \left(\frac{\partial^{P}\Psi}{\partial\mathbf{P}}\right)^{T}. \tag{22}$$

Here $\mathbf{X}$ denotes the position vector of a material point and $\mathbf{x} = \mathbf{X} + {}^0\mathbf{u}$ is the position vector of the corresponding spatial point after deformation. The deformation gradient $\mathbf{F} = \mathbf{I} + \partial^0\mathbf{u}/\partial\mathbf{X} = \mathbf{I} + \partial^0\mathbf{u}^{el}/\partial\mathbf{X} + \partial^0\mathbf{u}^{pl}/\partial\mathbf{X}$ in this case can either be expressed as a function of the material displacement field ${}^0\mathbf{u}$ alone or as a function of its elastic ${}^0\mathbf{u}^{el}$ and plastic ${}^0\mathbf{u}^{pl}$ parts. The symbols $\dot{\mathbf{E}}^{el}, \dot{\mathbf{E}}^{pl}/\mathbf{d}^{el}, \mathbf{d}^{pl}$ in the above denote the elastic and the plastic material/spatial strain rate tensors, where in the latter, the plastic flow is defined by Eq. $(22)_1$ as a product of a plastic multiplier $\dot{\lambda}$ and an appropriate yield surface normal $\partial^P\Psi/\partial\mathbf{P}$ defined in terms of the 1st Piola-Kirchhoff stress tensor $\mathbf{P}$. It should be noted that both the elastic and the plastic material/spatial strain-rate tensors have forms similar to the material/spatial strain-rate tensor and that the plastic flow defined by Eq. $(22)_1$ is not constrained, resulting in Eqs. $(21)_2$ and $(20)_4$ respectively, being the only non-degenerated forms of the material and spatial plastic strain rate tensors.

In order to modify the nonlinear continuum theory for finite deformations of elastoplastic media, it is assumed that the yield surface has various definitions in all stress spaces ${}^S\Psi = {}^S\Psi\left[{}^S\sigma_{eq}(\mathbf{S}), \mathbf{q}\right]$, ${}^P\Psi = {}^P\Psi\left[{}^P\sigma_{eq}(\mathbf{P}), \mathbf{q}\right]$, ${}^\tau\Psi = {}^\tau\Psi\left[{}^\tau\sigma_{eq}(\boldsymbol{\tau}), \mathbf{q}\right]$, ${}^\sigma\Psi = {}^\sigma\Psi\left[{}^\sigma\sigma_{eq}(\boldsymbol{\sigma}), \mathbf{q}\right]$ in terms of the 2nd Piola-Kirchhoff stress tensor $\mathbf{S}$, the 1st Piola-Kirchhoff stress tensor $\mathbf{P}$, the Kirchhoff stress tensor $\boldsymbol{\tau}$, the Cauchy stress tensor $\boldsymbol{\sigma}$ and a vector of internal variables $\mathbf{q}$, where ${}^S\sigma_{eq}(\mathbf{S}), {}^P\sigma_{eq}(\mathbf{P}), {}^\tau\sigma_{eq}(\boldsymbol{\tau}), {}^\sigma\sigma_{eq}(\boldsymbol{\sigma})$ are the corresponding equivalent stresses. After changing the physical interpretation of the plastic flow and applying push-forward and pull-back operations to the material gradient of the plastic velocity field Eq. $(22)_1$ as follows

$$\frac{\partial\dot{\mathbf{u}}^{pl}}{\partial\mathbf{x}} = \frac{\partial^0\dot{\mathbf{u}}^{pl}}{\partial\mathbf{X}}\cdot\mathbf{F}^{-1} = \dot{\lambda}\cdot\frac{\partial^\sigma\Psi}{\partial\boldsymbol{\sigma}},\quad \frac{\partial\dot{\mathbf{u}}^{pl}}{\partial\mathbf{x}} = \frac{\partial^0\dot{\mathbf{u}}^{pl}}{\partial\mathbf{X}}\cdot\mathbf{F}^{-1} = \dot{\lambda}\cdot\frac{\partial^\tau\Psi}{\partial\boldsymbol{\tau}},$$
$$\mathbf{F}^T\cdot\frac{\partial^0\dot{\mathbf{u}}^{pl}}{\partial\mathbf{X}} = \dot{\lambda}\cdot\frac{\partial^S\Psi}{\partial\mathbf{S}}, \tag{23}$$

it can be shown that the definitions of the yield surface are not independent of each other, but are related by the following formulas

$$\frac{\partial^P\Psi}{\partial\mathbf{P}}\cdot\mathbf{F}^{-1} = \frac{\partial^\sigma\Psi}{\partial\boldsymbol{\sigma}},\quad \frac{\partial^P\Psi}{\partial\mathbf{P}}\cdot\mathbf{F}^{-1} = \frac{\partial^\tau\Psi}{\partial\boldsymbol{\tau}},\quad \mathbf{F}^T\cdot\frac{\partial^P\Psi}{\partial\mathbf{P}} = \frac{\partial^S\Psi}{\partial\mathbf{S}}. \tag{24}$$

As a result, one of the definitions of the yield surface has to be chosen as a reference in order to define the material model while the rest of them can be calculated by solving the partial differential equations Eq. (24). Moreover, when ${}^\sigma\Psi$ or ${}^\tau\Psi$ is used as the reference definition of the yield surface in the current configuration of the body, the contemporary plasticity models will be recovered. The various definitions of the yield surface ${}^S\Psi, {}^P\Psi, {}^\tau\Psi, {}^\sigma\Psi$ and the corresponding equivalent stresses ${}^S\sigma_{eq}(\mathbf{S}), {}^P\sigma_{eq}(\mathbf{P}), {}^\tau\sigma_{eq}(\boldsymbol{\tau}), {}^\sigma\sigma_{eq}(\boldsymbol{\sigma})$ expressed in the 2nd Piola-Kirchhoff, 1st Piola-Kirchhoff, Kirchhoff and Cauchy's stress spaces moreover have the following properties

$$^{S}\sigma_{eq} = {}^{P}\sigma_{eq} = {}^{\tau}\sigma_{eq} = J \cdot {}^{\sigma}\sigma_{eq}, \tag{25}$$

$$\frac{\partial^{S}\Psi}{\partial \mathbf{S}} : \mathbf{S} = \frac{\partial^{P}\Psi}{\partial \mathbf{P}} : \mathbf{P} = \frac{\partial^{\tau}\Psi}{\partial \boldsymbol{\tau}} : \boldsymbol{\tau} = J \cdot \frac{\partial^{\sigma}\Psi}{\partial \boldsymbol{\sigma}} : \boldsymbol{\sigma}, \tag{26}$$

where Eq. (26) represents a 'normality rule', which from a physical point of view, represents the rate of change of internal plastic work per unit volume, or the internal plastic power density which can also be expressed in the following form

$$dW^{pl} = \dot{\mathbf{E}}^{pl} : \mathbf{S} \cdot dV_0 = \frac{\partial^{0}\dot{\mathbf{u}}^{pl}}{\partial \mathbf{X}} : \mathbf{P} \cdot dV_0 = \mathbf{d}^{pl} : \boldsymbol{\tau} \cdot dV_0 = \mathbf{d}^{pl} : \boldsymbol{\sigma} \cdot dv. \tag{27}$$

Here $J = \det(\mathbf{F})$ is the Jacobian of deformation, dV_0 is an infinitesimal volume element in the initial configuration of the body and $dv = J \cdot dV_0$ is its spatial counterpart.

2.2 *The Kinematics of Motion and the Stress Constitutive Functions of Mooney-Rivlin and Neo-Hookean Hyperelastic Materials Within the Framework of Finite Strain Elastoplasticity*

Hyperelastic materials are nonlinear elastic materials, whose constitutive functions are derived from strain energy density functions. They are commonly used for modelling material behavior of rubber-like solids, polymers, vulcanized elastomers, foams and biomaterials. The stress-strain relationship of such materials is often considered to be non-linearly elastic, isotropic, incompressible and generally independent of strain rate [37]. Without the need for completeness let us just mention a few concrete examples of hyperelastic materials with a widespread use in computational mechanics, such as Ogden material [38, 39], Hencky material [40], Mooney–Rivlin material, neo-Hookean material and their regularized or compressible versions for modelling rubber-like solids and polymers [41]. Later material models for rubber-like solids and polymers have been extended with a few extra terms in their strain energy density functions and were proposed by Yeoh [42], Gent [43], Dill [44] and Bower [45]. The Holzapfel material [32] has specially been designed for modelling carbon-black-filled rubber vulcanizates, while a few earlier models in the class of hyperelastic materials, such as the Blatz-Ko material [46] and the Arruda-Boyce material [47] were developed for modelling foamed or compressible elastomers. From among the biomaterials the Ogden material or its regularized versions ought to be mentioned, which are widely used in modelling soft tissues [35, 37].

The basic idea of alternatively modelling material degradation of hyperelastic materials within the framework of finite-strain elastoplasticity is to separate the elastic motion from the plastic motion by removing the plastic displacement field from the overall displacement field at a particular material point of the body; that is, expressing the constitutive equation of the material in the intermediate configuration of the body. It should be noted that the intermediate configuration is a configuration of the body at time $t > 0$ after the body has undergone elastic deformations only. Then the elastic part of the deformation gradient takes the following form [see also Eq. (18)]

$$\mathbf{F}^{el} = \mathbf{F} - \frac{\partial^0 \mathbf{u}^{pl}}{\partial \mathbf{X}}. \tag{28}$$

Considering von-Mises plasticity with isotropic hardening, where the reference definition of the yield surface ${}^{\tau}\Psi$ is given in the current configuration of the body in terms of Kirchhoff stress measures, Eq. (23) imply the following formulas for the spatial $\partial \dot{\mathbf{u}}^{pl}/\partial \mathbf{x}$ and material $\partial^0 \dot{\mathbf{u}}^{pl}/\partial \mathbf{X}$ gradients of the plastic velocity fields

$$\frac{\partial \dot{\mathbf{u}}^{pl}}{\partial \mathbf{x}} = \dot{\lambda} \cdot \frac{\partial^{\tau} \Psi}{\partial \boldsymbol{\tau}}, \tag{29}$$

$$\frac{\partial^0 \dot{\mathbf{u}}^{pl}}{\partial \mathbf{X}} = \frac{\partial \dot{\mathbf{u}}^{pl}}{\partial \mathbf{x}} \cdot \frac{\partial \mathbf{x}}{\partial \mathbf{X}} = \frac{\partial \dot{\mathbf{u}}^{pl}}{\partial \mathbf{x}} \cdot \mathbf{F}, \tag{30}$$

Then the material gradient of the Lagrangian plastic displacement field ${}^0\mathbf{u}^{pl}$ can be expressed in the following incremental form [48]

$$\frac{\partial^{n+1,0} \mathbf{u}^{pl}}{\partial \mathbf{X}} = \Delta t \cdot \dot{\lambda} \cdot \left(\frac{\partial^{\tau} \Psi}{\partial \boldsymbol{\tau}} \cdot {}^{n+1}\mathbf{F} \right) + \frac{\partial^{n,0} \mathbf{u}^{pl}}{\partial \mathbf{X}} = \Delta t \cdot \dot{\lambda} \cdot \sqrt{\frac{3}{2}} \cdot \left(\mathbf{N} \cdot {}^{n+1}\mathbf{F} \right) + \frac{\partial^{n,0} \mathbf{u}^{pl}}{\partial \mathbf{X}}. \tag{31}$$

In Eq. (31) the left superscripts ${}^{n+1}(\cdot)$, ${}^{n}(\cdot)$ denote the values of the variable $(\cdot)$ at time steps $n+1$ and n respectively, corresponding to discrete times ${}^{n+1}t$ and ${}^{n}t$, where $\Delta t = {}^{n+1}t - {}^{n}t$ is the time step size and $\mathbf{N}$ is the unit outward normal of the yield surface.

Equations (28)–(31) then have to be supplemented with the following loading/unloading discrete Kuhn-Tucker plastic optimization conditions

$$\dot{\lambda} \geq 0, \quad {}^{\tau}\Psi \leq 0, \quad \dot{\lambda} \cdot {}^{\tau}\Psi = 0. \tag{32}$$

In this paper natural rubber behavior is studied using the modified Mooney-Rivlin and the neo-Hookean hyperplastic material models. Both models are based on the multiplicative split of the Lagrangian elastic deformation gradient $\mathbf{F}^{el}$ into a Lagrangian volumetric part $\mathbf{F}^{el}_{v}$ and a Lagrangian isochoric part $\mathbf{F}^{el}_{iso}$

$$\mathbf{F}^{el} = \mathbf{F}^{el}_{iso} \cdot \mathbf{F}^{el}_{v}. \tag{33}$$

where

$$\mathbf{F}^{el}_{v} = \det\left(\mathbf{F}^{el}\right)^{\frac{1}{3}} \cdot \mathbf{I}, \tag{34}$$

$$\mathbf{F}^{el}_{iso} = \det\left(\mathbf{F}^{el}\right)^{-\frac{1}{3}} \cdot \mathbf{F}^{el}. \tag{35}$$

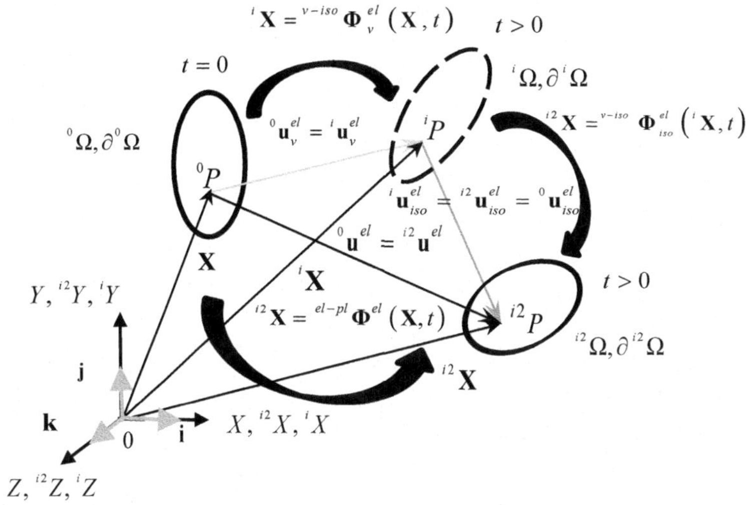

Fig. 2 The kinematics of elastic motion of a deformable body corresponding to the multiplicative split of the elastic deformation gradient into a volumetric part and an isochoric part

It should be noted that the multiplicative split, similarly as in Eq. (12), describes two pieces of motion, which are moreover constrained and where each part of the above gradient can be related directly to the corresponding displacement field depicted in Fig. 2. Then the proper Lagrangian multiplicative split of the elastic deformation gradient, consistent with the theory of nonlinear continuum mechanics, can be rewritten in the following form

$$\mathbf{F}^{el}(\mathbf{X},t) = \frac{\partial^{i2}\mathbf{X}}{\partial \mathbf{X}} = \mathbf{I} + \frac{\partial^{0}\mathbf{u}^{el}}{\partial \mathbf{X}} = \frac{\partial^{i2}\mathbf{X}}{\partial^{i}\mathbf{X}} \cdot \frac{\partial^{i}\mathbf{X}}{\partial \mathbf{X}}$$
$$= \mathbf{F}^{el}_{iso}(\mathbf{X},t) \cdot \mathbf{F}^{el}_{v}(\mathbf{X},t) = \mathbf{I} + \frac{\partial^{0}\mathbf{u}^{el}_{v}}{\partial \mathbf{X}} + \frac{\partial^{0}\mathbf{u}^{el}_{iso}}{\partial \mathbf{X}}, \tag{36}$$

where

$$\mathbf{F}^{el}_{iso}(\mathbf{X},t) = \frac{\partial^{i2}\mathbf{X}}{\partial^{i}\mathbf{X}} = \left(\mathbf{I} + \frac{\partial^{0}\mathbf{u}^{el}_{iso}}{\partial \mathbf{X}} \cdot \left[\mathbf{F}^{el}_{v}(\mathbf{X},t)\right]^{-1}\right) \text{ and}$$
$$\mathbf{F}^{el}_{v}(\mathbf{X},t) = \frac{\partial^{i}\mathbf{X}}{\partial \mathbf{X}} = \left(\mathbf{I} + \frac{\partial^{0}\mathbf{u}^{el}_{v}}{\partial \mathbf{X}}\right). \tag{37}$$

Equations (36) and (37) then have to be supplemented with the following constraint equations in order that the formulation of the material model be complete

$$F^{el}_{v,11} = F^{el}_{v,22} = F^{el}_{v,33} = \det\left(\mathbf{F}^{el}\right)^{\frac{1}{3}}, \tag{38}$$

$$F^{el}_{v,12} = F^{el}_{v,21} = F^{el}_{v,13} = F^{el}_{v,31} = F^{el}_{v,23} = F^{el}_{v,32} = 0, \tag{39}$$

$$\mathbf{F}^{el}_{iso} = \mathbf{F}^{el} \cdot \left(\mathbf{F}^{el}_{v}\right)^{-1}. \tag{40}$$

It should be emphasized here that, from the nonlinear continuum mechanics point of view, Eq. (33) is incomplete without Eq. (37) and the constraint equations Eqs. (34), (35) or Eqs. (38), (40) respectively. As a result, none of the Lagrangian elastic volumetric displacement field ${}^{0}\mathbf{u}^{el}_{v}$, the Lagrangian elastic isochoric displacement field ${}^{0}\mathbf{u}^{el}_{iso}$, the elastic volumetric motion ${}^{i}\mathbf{X} = {}^{v-iso}\boldsymbol{\Phi}^{el}_{v}(\mathbf{X}, t)$, the elastic isochoric motion ${}^{i2}\mathbf{X} = {}^{v-iso}\boldsymbol{\Phi}^{el}_{iso}\left({}^{i}\mathbf{X}, t\right)$, the overall elastic motion ${}^{i2}\mathbf{X} = {}^{el-pl}\boldsymbol{\Phi}^{el}(\mathbf{X}, t) = {}^{v-iso}\boldsymbol{\Phi}^{el}_{iso}\left[{}^{v-iso}\boldsymbol{\Phi}^{el}_{v}(\mathbf{X}, t), t\right]$, the corresponding Lagrangian and Eulerian elastic displacement fields ${}^{0}\mathbf{u}^{el} = {}^{i2}\mathbf{u}^{el}$ and the Lagrangian deformation gradients defined by Eqs. (36) and (37) respectively can be determined in accordance with their true physical meaning depicted in Fig. 2.

Taking into account the multiplicative split of the elastic deformation gradient (Eq. 33), the definition of the left elastic isochoric Cauchy-Green tensor can be expressed as

$$\mathbf{B}^{el}_{iso} = \mathbf{F}^{el}_{iso} \cdot \left(\mathbf{F}^{el}_{iso}\right)^{T} = \det\left(\mathbf{F}^{el}\right)^{-\frac{2}{3}} \cdot \mathbf{F}^{el} \cdot \left(\mathbf{F}^{el}\right)^{T} = \det\left(\mathbf{F}^{el}\right)^{-\frac{2}{3}} \cdot \mathbf{B}^{el}, \tag{41}$$

with the corresponding eigenvalue problem

$$\left(\mathbf{B}^{el}_{iso} - {}_{B}\lambda^{*} \cdot \mathbf{I}\right) \cdot \mathbf{v} = \mathbf{0}, \tag{42}$$

and its characteristic equation

$$\det\left(\mathbf{B}^{el}_{iso} - {}_{B}\lambda^{*} \cdot \mathbf{I}\right) = -{}_{B}\lambda^{*3} + I^{*}_{1} \cdot {}_{B}\lambda^{*2} - I^{*}_{2} \cdot {}_{B}\lambda^{*} + I^{*}_{3} = 0, \tag{43}$$

where

$$I^{*}_{1} = I^{*}_{1}\left(\mathbf{B}^{el}_{iso}\right) = tr\left(\mathbf{B}^{el}_{iso}\right), \tag{44}$$

$$\begin{aligned} I^{*}_{2} = I^{*}_{2}\left(\mathbf{B}^{el}_{iso}\right) &= \frac{1}{2} \cdot \left[\left[tr\left(\mathbf{B}^{el}_{iso}\right)\right]^{2} - tr\left[\left(\mathbf{B}^{el}_{iso}\right)^{2}\right]\right] \\ &= \frac{1}{2} \cdot \left[\left(I^{*}_{1}\right)^{2} - tr\left[\left(\mathbf{B}^{el}_{iso}\right)^{2}\right]\right], \end{aligned} \tag{45}$$

$$I^{*}_{3} = I^{*}_{3}\left(\mathbf{B}^{el}_{iso}\right) = \det\left(\mathbf{B}^{el}_{iso}\right) = 1. \tag{46}$$

Using the invariants in the above allows for the definition of the strain-energy density functions of the Mooney-Rivlin and neo-Hookean materials in the following forms [35, 37]:

$$\overline{\Psi}^{*}\left(I^{*}_{1}, I^{*}_{2}, J^{el}\right) = C_{10} \cdot \left(I^{*}_{1} - 3\right) + C_{01} \cdot \left(I^{*}_{2} - 3\right) + \frac{1}{d} \cdot \left(J^{el} - 1\right)^{2}, \tag{47}$$

$$\overline{\Psi}^*\left(I_1^*, J^{el}\right) = \frac{1}{2}\cdot G\cdot\left(I_1^* - 3\right) + \frac{1}{d}\cdot\left(J^{el} - 1\right)^2. \tag{48}$$

In Eqs. (47) and (48) $J^{el} = \det\left(\mathbf{F}^{el}\right)$ is the Jacobian of the elastic deformation gradient and $C_{10}, C_{01}, G = 2\cdot C_{10}, d$ are material parameters, where $K = 2/d$ is the bulk modulus, which relates the hydrostatic pressure to the purely volumetric component of the elastic deformation gradient $\mathbf{F}^{el}$ [35].

Then the stress constitutive functions of the materials in the intermediate configuration of the body, where only elastic deformations take place, can be determined as a Kirchhoff stress measure in accordance with the following formula [35, 37]

$$\boldsymbol{\tau}^{el} = \frac{\partial\overline{\Psi}^*}{\partial\mathbf{F}^{el}}\cdot\left(\mathbf{F}^{el}\right)^T. \tag{49}$$

Equation (49) then results in the following stress constitutive functions for the Mooney-Rivlin (Eq. 50) and the neo-Hookean materials (Eq. 51) in the intermediate configuration of the body

$$\boldsymbol{\tau}^{el} = 2\left(C_{10} + C_{01}\cdot I_1^*\right)\cdot dev\left[\mathbf{B}_{iso}^{el}\right] - 2\cdot C_{01}\cdot dev\left[\left(\mathbf{B}_{iso}^{el}\right)^2\right] + \frac{2}{d}\cdot J^{el}\cdot\left(J^{el} - 1\right)\cdot\mathbf{I}, \tag{50}$$

$$\boldsymbol{\tau}^{el} = G\cdot dev\left[\mathbf{B}_{iso}^{el}\right] + \frac{2}{d}\cdot J^{el}\cdot\left(J^{el} - 1\right)\cdot\mathbf{I}, \tag{51}$$

where

$$dev\left[\mathbf{B}_{iso}^{el}\right] = \mathbf{B}_{iso}^{el} - \frac{tr\left(\mathbf{B}_{iso}^{el}\right)}{3}\cdot\mathbf{I} \quad\text{and}$$
$$dev\left[\left(\mathbf{B}_{iso}^{el}\right)^2\right] = \left(\mathbf{B}_{iso}^{el}\right)^2 - \frac{tr\left[\left(\mathbf{B}_{iso}^{el}\right)^2\right]}{3}\cdot\mathbf{I}. \tag{52}$$

The stress constitutive functions of the materials in the current configuration of the body can then be calculated as

$$\boldsymbol{\tau} = \mathbf{F}\cdot\left[\left(\mathbf{F}^{el}\right)^{-1}\cdot\boldsymbol{\tau}^{el}\cdot\left(\mathbf{F}^{el}\right)^{-T}\right]\cdot\mathbf{F}^T = \mathbf{F}\cdot\left[\left(\mathbf{F}^{el}\right)^{-1}\cdot\left(\frac{\partial\overline{\Psi}^*}{\partial\mathbf{F}^{el}}\cdot\left(\mathbf{F}^{el}\right)^T\right)\cdot\left(\mathbf{F}^{el}\right)^{-T}\right]\cdot\mathbf{F}^T, \tag{53}$$

which result in the final formulas for the Mooney-Rivlin material [48]

$$\boldsymbol{\tau} = \mathbf{F}\cdot\left[\left(\mathbf{F}^{el}\right)^{-1}\cdot\left[2\left(C_{10} + C_{01}\cdot I_1^*\right)\cdot dev\left[\mathbf{B}_{iso}^{el}\right]\right]\cdot\left(\mathbf{F}^{el}\right)^{-T}\right]\cdot\mathbf{F}^T$$

$$-\mathbf{F}\cdot\left[\left(\mathbf{F}^{el}\right)^{-1}\cdot\left[2\cdot C_{01}\cdot dev\left[\left(\mathbf{B}_{iso}^{el}\right)^{2}\right]-\frac{2\cdot J^{el}}{d}\cdot\left(J^{el}-1\right)\cdot\mathbf{I}\right]\cdot\left(\mathbf{F}^{el}\right)^{-T}\right]\cdot\mathbf{F}^{T}, \quad (54)$$

and for the neo-Hookean material respectively

$$\boldsymbol{\tau}=\mathbf{F}\cdot\left[\left(\mathbf{F}^{el}\right)^{-1}\cdot\left[G\cdot dev\left[\mathbf{B}_{iso}^{el}\right]+\frac{2}{d}\cdot J^{el}\cdot\left(J^{el}-1\right)\cdot\mathbf{I}\right]\cdot\left(\mathbf{F}^{el}\right)^{-T}\right]\cdot\mathbf{F}^{T}. \quad (55)$$

The Kirchhoff stress measures described above can then be re-expressed as Cauchy's stress measures using the transformation $\boldsymbol{\sigma}=\boldsymbol{\tau}/J$, where $J=\det(\mathbf{F})$ is the Jacobian of the deformation.

2.3 The Reference Definition of the Yield Surface

Objective and thermodynamically consistent formulation of the plastic flow allows for the development of consistent material models. It will be shown later, that Eqs. (19)–(26) play a crucial role in developing an objective and thermodynamically consistent plastic flow which, for the first time, result in an elastoplastic material model whose analysis results are not affected by the description and particularities of the material model formulation. This also means that the constitutive equation and the yield surface of the material can be defined in whichever stress spaces without influencing the analysis results.

In this work the reference definition of the yield surface is given in the Kirchhoff stress space, because the theory requires uniaxial tensile testing of the modelled material, during which the deformation gradient is also determined. Since this is not the general case, defining the reference yield surface in the Kirchhoff stress space allows for the assessment of the deformation gradient from numerical uniaxial tensile testing at the least. Considering a Von Mises material model with isotropic hardening, the reference definition of the yield surface can be expressed in the following simple form [48] (Fig. 3):

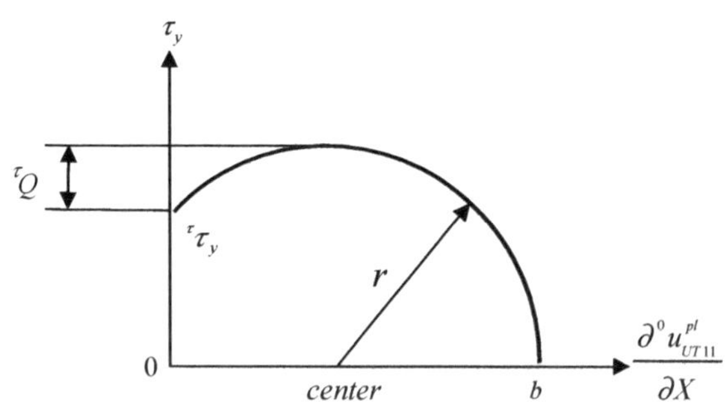

Fig. 3 Assumed uniaxial Kirchhoff stress vs. "plastic strain" curve used in the analysis

$$^{\tau}\Psi = {}^{\tau}\sigma_{eq}(\boldsymbol{\tau}) - \tau_y \le 0, \tag{56}$$

where

$$^{\tau}\sigma_{eq}(\boldsymbol{\tau}) = \sqrt{\frac{3}{2}} \cdot \boldsymbol{\Sigma\tau} : \boldsymbol{\Sigma\tau},\ \boldsymbol{\Sigma\tau} = \boldsymbol{\tau} - \frac{tr(\boldsymbol{\tau})}{3} \cdot \mathbf{I},\ \tau_y = \sqrt{r^2 - \left(a \cdot \frac{\partial^0 u^{pl}_{UT11}}{\partial X}\right)^2}, \tag{57}$$

$$r = {}^{\tau}\tau_y + {}^{\tau}Q,\quad center = \sqrt{r^2 - {}^{\tau}\tau_y^2},\quad \mathrm{a} = \frac{center + r}{b}, \tag{58}$$

$$\frac{\partial \dot{\mathbf{u}}^{pl}}{\partial \mathbf{x}} = \mathbf{d}^{pl} = \dot{\lambda} \cdot \frac{\partial^{\tau}\Psi}{\partial \boldsymbol{\tau}} = \dot{\lambda} \cdot \sqrt{\frac{3}{2}} \cdot \mathbf{N},\ \mathbf{N} = \frac{\boldsymbol{\Sigma\tau}}{\sqrt{\boldsymbol{\Sigma\tau} : \boldsymbol{\Sigma\tau}}},$$

$$\frac{\partial^0 \dot{\mathbf{u}}^{pl}}{\partial \mathbf{X}} = \dot{\lambda} \cdot \sqrt{\frac{3}{2}} \cdot \mathbf{N} \cdot \mathbf{F}, \tag{59}$$

$$\dot{e}^{pl} = \sqrt{\frac{2}{3} \cdot \mathbf{d}^{pl} : \mathbf{d}^{pl}} = \dot{\lambda},\ e^{pl} = \int_0^t \dot{e}^{pl} \cdot dt,\ \frac{\partial \dot{u}^{pl}_{UT11}}{\partial x} = \dot{\lambda} \cdot \sqrt{\frac{3}{2}},$$

$$\frac{\partial^0 \dot{u}^{pl}_{UT11}}{\partial X} = \frac{\partial \dot{u}^{pl}_{UT11}}{\partial x} \cdot F_{UT11} = \dot{\lambda} \cdot \sqrt{\frac{3}{2}} \cdot F_{UT11} \tag{60}$$

$$\frac{\partial^0 u^{pl}_{UT11}}{\partial X} = \int_0^t \frac{\partial^0 \dot{u}^{pl}_{UT11}}{\partial X} \cdot dt,\ F^{el}_{UT11} = F_{UT11} - \frac{\partial^0 u^{pl}_{UT11}}{\partial X},$$

$$\mathbf{F}^{el}_{UT} = diag\left[F^{el}_{UT11},\ 1,\ 1\right], \tag{61}$$

$$\mathbf{F}_{UT} = diag\left[F_{UT11},\ 1,\ 1\right],\ J^{el}_{UT} = \det\left(\mathbf{F}^{el}_{UT}\right). \tag{62}$$

Equations (56)–(58) define the yield surface used in the analysis, where the assumed uniaxial Kirchhoff yield stress vs. 'plastic strain' curve defined by Eq. $(57)_3$–(58) is depicted in Fig. 3. The material hardening/softening in the analysis is controlled by the parameter $\partial^0 u^{pl}_{UT11}/\partial X$, Eq. $(61)_1$, determined from the uniaxial tensile test of the material, where b is the maximum value of $\partial^0 u^{pl}_{UT11}/\partial X$ at which the material loses its integrity, i.e. where $\tau_y = 0$. Here ${}^{\tau}\tau_y$ is the constant yield stress of the material and ${}^{\tau}Q$ is the maximum hardening stress. The spatial and material gradients of the plastic velocity fields can then be calculated in accordance with Eq. (59), which have counterparts in uniaxial tensile testing (Eqs. $(60)_3$ and $(60)_4$). The equivalent plastic strain rate and the accumulated plastic strain, defined by the rest of Eq. (60), while informative, are not used in the analysis. The definitions of the deformation gradient, its elastic part and the Jacobian of the latter defined by Eqs. (61) and (62) respectively, which should have been determined during uniaxial tensile testing of the material, are needed in order to assess the values of these variables during numerical uniaxial tensile testing.

2.4 Calculation of the Plastic Multiplier and the Axial Component of the Deformation Gradient Determined from Numerical Uniaxial Tensile Test

When the plastic step occurs in the analysis, Eq. (56) is used directly to calculate the plastic multiplier value as well as the stress constitutive function values of the Mooney-Rivlin (Eq. 54) and neo-Hookean (Eq. 55) material within finite element analysis. The solution procedure requires the yield limit τ_y as an entry in these calculations, which normally is determined during material testing of the modelled material. Since the yield limit τ_y is a Kirchhoff stress measure, it is also required that the axial component F_{UT11} of the deformation gradient is determined during material testing. Moreover, when irreversible deformations take place in the specimen, the deformation gradient $F_{UT11} = F_{UT11}\left(\partial^0 u^{pl}_{UT11}/\partial X\right)$ becomes a function of the $\partial^0 u^{pl}_{UT11}/\partial X$ parameter only, due to the fact that τ_y is delimited by the uniaxial stress versus 'plastic strain' curve of the material, which together with the stress constitutive function of the material, as constraints, result in $\partial^0 u^{el}_{UT11}/\partial X = \partial^0 u^{el}_{UT11}/\partial X\left(\partial^0 u^{pl}_{UT11}/\partial X\right)$ (see also Fig. 3). Since F_{UT11} at present is not determined during material testing, in order to assess its value, numerical uniaxial testing of the material have been carried out by solving the following equation [48]

$$\tau_{UT11} - \tau_y = 0, \tag{63}$$

where τ_{UT11} is the axial component of the $\boldsymbol{\tau}_{UT} = \boldsymbol{\tau}\left(\mathbf{F} = \mathbf{F}_{UT}, \mathbf{F}^{el} = \mathbf{F}^{el}_{UT}\right)$ stress constitutive function, defined by Eqs. (54) and (55), in which the deformation gradient $\mathbf{F}$ was replaced by $\mathbf{F}_{UT}$ (Eq. $(62)_1$) and $\mathbf{F}^{el}$ by $\mathbf{F}^{el}_{UT}$ (Eq. $(61)_3$), while keeping the definition of the yield limit τ_y unchanged (Eq. $(57)_3$).

2.5 On the Objectivity and the Thermodynamic Consistency of the Formulation

There are a few requirements that a material model has to meet in order for the model and its analysis results to be independent of the description and particularities of the material model formulation. These are the requirements of objectivity and the requirements of thermodynamic consistency. Here, by description, we mean updated or total Lagrangian description and by the particularities of the material model formulation, the ability to formulate the problem in whatever stress spaces using internal mechanical power conjugate stress measures and strain rates. Meeting the requirements of objectivity ensures that the spatial tensors and the parts of the mixed-base tensors with spatial bases employed in the mathematical formulation of the stress constitutive function of the material undergo the same rotation $\mathbf{Q}$ with respect to the reference coordinate system, as the observer's coordinate system which moves arbitrarily at time $t > 0$ in a 3D Euclidian space, but was coinciding with the reference coordinate system at time $t = 0$. With respect to the above, it can be proven that the following transformations hold true under the change of the observer [32]

$$
{}^{+}\mathbf{F}^{el} = \mathbf{Q}\cdot\mathbf{F}^{el},\ {}^{+}\mathbf{F} = \mathbf{Q}\cdot\mathbf{F},\ \frac{\partial^{+0}\mathbf{u}^{pl}}{\partial\mathbf{X}} = \mathbf{Q}\cdot\frac{\partial^{0}\mathbf{u}^{pl}}{\partial\mathbf{X}},\ {}^{+}\mathbf{B}^{el} = \mathbf{Q}\cdot\mathbf{B}^{el}\cdot\mathbf{Q}^{T}
$$
$$
\text{and } {}^{+}\boldsymbol{\tau} = \mathbf{Q}\cdot\boldsymbol{\tau}\cdot\mathbf{Q}^{T}. \tag{64}
$$

In Eq. (64) the tensors ${}^{+}\mathbf{F}^{el}$, ${}^{+}\mathbf{F}$, $\partial^{+0}\mathbf{u}^{pl}/\partial\mathbf{X}$, ${}^{+}\mathbf{B}^{el}$, ${}^{+}\boldsymbol{\tau}$, from the observer's perspective connected to an arbitrarily moving coordinate system, denote identical tensors $\mathbf{F}^{el}$, $\mathbf{F}$, $\partial^{0}\mathbf{u}^{pl}/\partial\mathbf{X}$, $\mathbf{B}^{el}$, $\boldsymbol{\tau}$, when viewed by the observer connected to the reference coordinate system, where the rotation tensor $\mathbf{Q}$ defines the relative rotation of the moving coordinate system with respect to the reference coordinate system of the observers.

The requirements of thermodynamic consistency using Lagrangian formulation, which have recently been improved by Écsi and Éleszős to cover constitutive equations in rate forms [36], are actually based on the 1st principle of thermodynamics and its only aim is to ensure that the internal mechanical power at a material point of the modelled body and its arbitrary higher order time derivatives remain the same in all stress spaces throughout the analysis. The resulting formulas then allow for the transformation of the constitutive equation of the material from one stress space to another. When the constitutive equation of the material is defined in terms of finite strain or deformation measures, as in Eqs. (54) and (55) respectively, the transformations reduce to well know formulas defining the relationship between various stress measures and strain rates in continuum mechanics, which are conjugate with respect to the internal mechanical power

$$
dW = \dot{\mathbf{E}} : \mathbf{S}\cdot dV_0 = \frac{\partial^{0}\dot{\mathbf{u}}}{\partial\mathbf{X}} : \mathbf{P}\cdot dV_0 = \mathbf{d} : \boldsymbol{\tau}\cdot dV_0 = \mathbf{d} : \boldsymbol{\sigma}\cdot dv = dW^{el} + dW^{pl}. \tag{65}
$$

Considering Eqs. (3), (19)–(21), (26) and (27), it can be shown that the internal mechanical power at a material point of the body is the sum of the elastic mechanical power dW^{el} and the plastic mechanical power dW^{pl} (Eq. 27), where the latter can be expressed in terms of 'normality rules' (Eq. 26) as

$$
dW^{pl} = \dot{\lambda}\cdot\frac{\partial^{S}\Psi}{\partial\mathbf{S}} : \mathbf{S}\cdot dV_0 = \dot{\lambda}\cdot\frac{\partial^{P}\Psi}{\partial\mathbf{P}} : \mathbf{P}\cdot dV_0 = \dot{\lambda}\cdot\frac{\partial^{\tau}\Psi}{\partial\boldsymbol{\tau}} : \boldsymbol{\tau}\cdot dV_0
$$

$$= \dot{\lambda} \cdot \frac{\partial^{\sigma}\Psi}{\partial \boldsymbol{\sigma}} : \boldsymbol{\sigma} \cdot dv, \tag{66}$$

and where

$$dW^{el} = \dot{\mathbf{E}}^{el} : \mathbf{S} \cdot dV_0 = \frac{\partial^0 \dot{\mathbf{u}}^{el}}{\partial \mathbf{X}} : \mathbf{P} \cdot dV_0 = \mathbf{d}^{el} : \boldsymbol{\tau} \cdot dV_0 = \mathbf{d}^{el} : \boldsymbol{\sigma} \cdot dv. \tag{67}$$

Moreover, the equivalent stress in the definition of the yield surface (Eq. 56) can be manipulated into the following form

$$^{\tau}\sigma_{eq}(\boldsymbol{\tau}) = \sqrt{\frac{3}{2} \cdot \boldsymbol{\Sigma\tau} : \boldsymbol{\Sigma\tau}} = \sqrt{\frac{3}{2}} \cdot \frac{\boldsymbol{\Sigma\tau} : \boldsymbol{\Sigma\tau}}{\sqrt{\boldsymbol{\Sigma\tau} : \boldsymbol{\Sigma\tau}}} = \sqrt{\frac{3}{2}} \cdot \frac{\boldsymbol{\Sigma\tau}}{\sqrt{\boldsymbol{\Sigma\tau} : \boldsymbol{\Sigma\tau}}} : \boldsymbol{\tau} = \frac{\partial^{\tau}\Psi}{\partial \boldsymbol{\tau}} : \boldsymbol{\tau}. \tag{68}$$

Then after multiplying Eq. (56) with the plastic multiplier $\dot{\lambda}$ and the infinitesimal volume element dV_0 in the initial configuration of the body, the first term on the right hand side of Eq. (56) becomes the internal mechanical plastic power at a material point of the body

$$dW^{pl} = \dot{\lambda} \cdot {}^{\tau}\sigma_{eq}(\boldsymbol{\tau}) \cdot dV_0 = \dot{\lambda} \cdot \frac{\partial^{\tau}\Psi}{\partial \boldsymbol{\tau}} : \boldsymbol{\tau} \cdot dV_0 = \mathbf{d}^{pl} : \boldsymbol{\tau} \cdot dV_0, \tag{69}$$

and the second term on the right hand side of Eq. (56) becomes the multiple of the internal mechanical plastic power at a material point of the specimen coming from the tensile test of the modelled material (see also Eqs. $(60)_1$ and $(60)_3$)

$$\dot{\lambda} \cdot \tau_y \cdot dV_0 = \dot{e}^{pl} \cdot \tau_y \cdot dV_0 = \sqrt{\frac{2}{3}} \cdot \frac{\partial \dot{u}^{pl}_{UT11}}{\partial x} \cdot \tau_y \cdot S_0 \cdot dX = \sqrt{\frac{2}{3}} \cdot dW^{pl}_{UT}, \tag{70}$$

where $S_0 = S_0(X)$ is the constant cross sectional area of the specimen in uniaxial tensile testing of the material in the initial configuration of the body of the specimen and dW^{pl}_{UT} is the plastic power of the specimen at its arbitrary material point. This means that the solution of Eq. (56) for the plastic multiplier value actually ensures that the internal mechanical plastic power at a material point of the modelled body dW^{pl} be the $\sqrt{2/3}$ multiple of the internal mechanical plastic power at a material point of the specimen coming from the tensile test of the modelled material,$dW^{pl}_{UT} = \partial \dot{u}^{pl}_{UT11}/\partial x \cdot \tau_y \cdot S_0 \cdot dX$ i.e. $dW^{pl} = \sqrt{2/3} \cdot dW^{pl}_{UT}$. Additionally, Eqs. (27) and (66) also hold true, we have proved that the formulation of the plastic flow and the normality rules (Eq. 26) are also thermodynamically consistent.

3 Numerical Example

As a numerical example, tensile testing of a cuboid shaped specimen, with dimensions 1 m × 0.1 m × 0.1 m, manufactured from natural rubber, have been studied using the presented Mooney-Rivlin and neo-Hookean material models. One end of the specimen was fixed and the second end underwent a prescribed $v = 0.02\ \mathrm{m\,s^{-1}}$ velocity. The transverse degrees of freedoms at all nodes at the specimen's ends were constrained.

The body was initially at rest except for the nodes at the moving end of the specimen, where initial axial velocity was prescribed $v = 0.02\ \mathrm{m\,s^{-1}}$. The analysis was run as dynamic until failure to converge, using implicit time integration and $\Delta t = 1$ s time step size. In the analysis material parameters of natural rubber experimentally determined by the collective Shahzad et al. [49] were used. Table 1 outlines the material properties used in the numerical experiment. Figure 4 depicts the spatially discretized body of the specimen.

Figure 5 depicts a few selected results in the current configuration of the body at the end of the numerical analysis using the Mooney-Rivlin material model. These are the axial displacement field, the axial velocity field, the axial acceleration field, the von Mises stress as a Cauchy's stress measure, the accumulated plastic strain and the parameter $\partial^0 u^{pl}_{UT11}/\partial X$ which controls the material hardening. In Fig. 6 a few time history curves are shown using the Mooney-Rivlin and neo-Hookean material models in the analyses. These are the accumulated plastic strain time history curve, the $\partial^0 u^{pl}_{UT11}/\partial X$ parameter time history curve and the axial stress time history curve at node 29 of element 77 of the spatially discretized specimen depicted in Fig. 4.

Table 1 Material properties of the natural rubber specimen

ρ_0 [kg m^{-3}]	1520
C_{10} [Pa]	333,900
C_{01} [Pa]	−337
d [−]	0.0015828
${}^{\tau}\tau_y$ [Pa]	3,109,540.0
${}^{\tau}Q$ [Pa]	621,908.0
b [−]	2.0

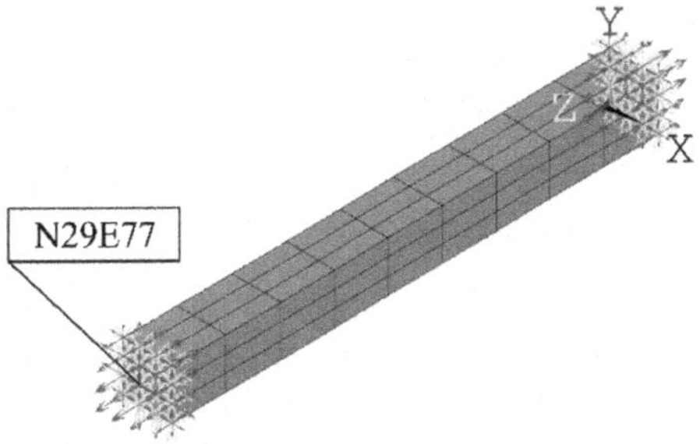

Fig. 4 Spatially discretized specimen in the analysis

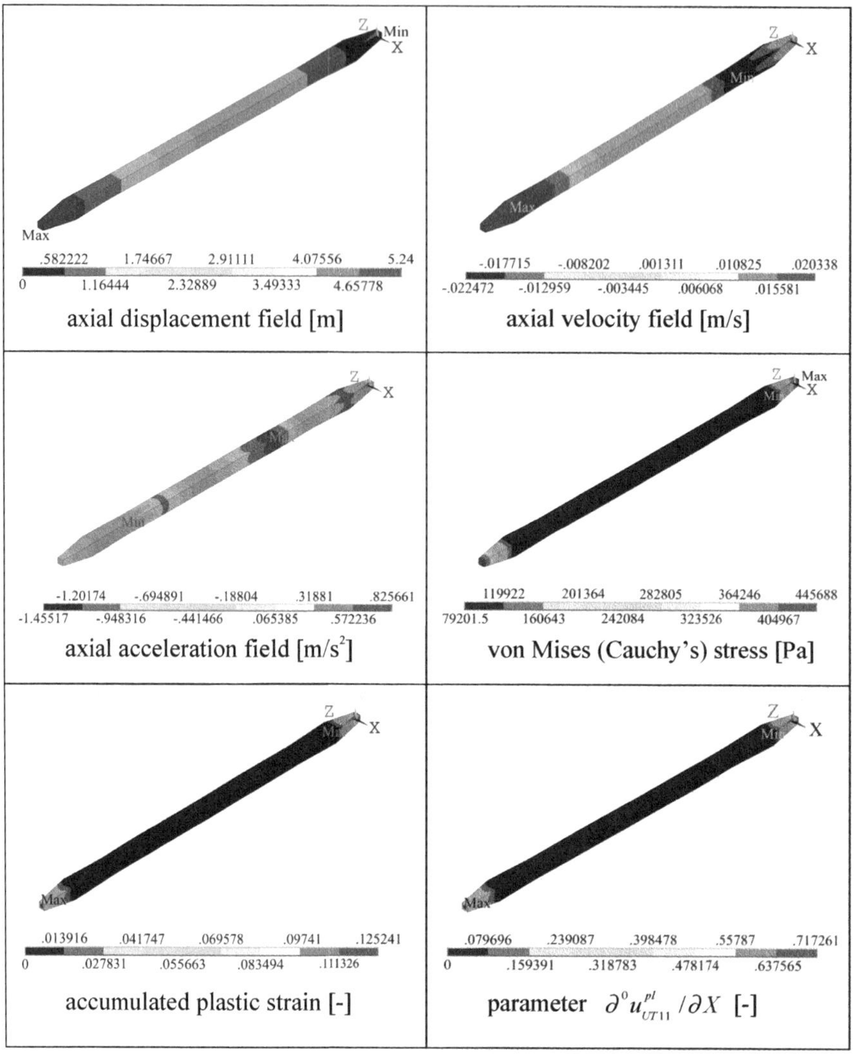

Fig. 5 A few selected results at the end of the analysis using Mooney-Rivlin material

The time in the curves have been replaced by the estimated value of the stretch parameter at the moving end of the specimen, calculated as one plus the axial displacement while utilizing the fact that the initial length of the specimen was also a meter. As can be seen in the figures, material degradation takes place at the specimen ends, where the cross sectional area of the specimen remains the smallest. The material degradation takes place in just about 0.5 s starting at 5.5 s, when the Von Mises stress value reaches the yield limit of the material and after a short period of time the maximum load bearing capacity of the cuboid shaped specimen is reached,

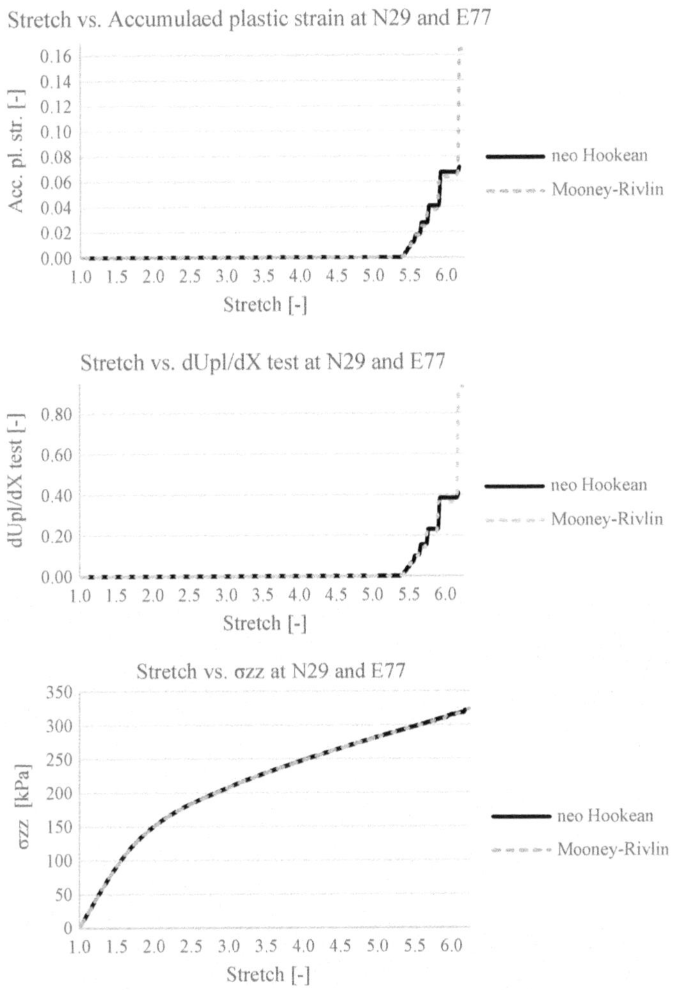

Fig. 6 A few selected time-history curves

where the analysis stops to converge. Although they are limited by the accuracy of the material models in elastic loading, the analysis results seem to be reasonable.

4 Conclusions

In this paper an alternative method for modelling material degradation of hyperelasic materials within the framework of finite-strain plasticity have been presented using the modified Mooney-Rivlin and the neo-Hookean material models. The models are based on the first nonlinear continuum theory of finite deformations of elastoplastic media, which allows for the development of objective and thermodynamically consistent material models. Therefore, the material models and their results when used in numerical analyses are no longer affected by the description and the particularities of model formulation. Moreover, the models can be used to study the behavior of the deforming body even considering thermodynamic aspects, as it relates the model's internal mechanical power density to the internal mechanical power density of the specimen from the tensile test of the modelled material. In this paper natural rubber behavior was studied. Although, the analysis results seem to be reasonable, proper verification of the models require a few modifications in contemporary material testing, as the deformation gradient has to be determined during uniaxial tensile testing of the material in addition.

Acknowledgements Funding from the VEGA grant 1/0740/16 resources is greatly appreciated.

References

1. Lee EH, Liu DT (1967) Finite strain elastic-plastic theory particularly for plane wave analysis. J Appl Phys 38:19–27
2. Lee EH (1969) Elasto-plastic deformation at finite strains. J Appl Mech 36:1–6
3. Kroner E, Teodosiu C (1972) Lattice defect approach to plasticity and viscoplasticity. In: Sawczuk A (ed) Problems of plasticity. Noordhoff, Warsaw
4. Mande J (1964) Contribution theorique a l´etude de l´ecrouissage et des lois de l´ecoulement plastique, In: Proceedings of the 11th international congress on applied mechanics, pp 502–509
5. Mandel J (1973) Equations constitutives et directeurs dans les milieu plastiques et viscoplastiques. Int J Solids Structs 9:725–740
6. Mandel J (1974) Thermodynamics and plasticity. In: Delgado JJ et al (ed) Macmillan, NY
7. Kratchovil J (1973) On a finite strain theory of elastic-inelastic materials. Acta Mech 16:127–142
8. Sidoroff F (1974) Un modele viscoelastique non linearie avec configuration. IJMec 13:679–713
9. Nemat-Nasser S (1979) Decomposition of strain measures and their rates in finite deformation elastoplasticity. Int J Solids Structs 15:155–166
10. Nemat-Nasser S (1982) On finite deformation elasto-plasticity. Int J Solids Structs 18(10):857–872
11. Dafalias YF (1984) The plastic spin concept and a simple illustration of its role in finite plastic transformations. Mech Mater 3:223–233
12. Dafalias YF (1985) The plastic spin. J Appl Mech 52:865–871
13. Lubarda VA, Lee EH (1981) A correct definition of elastic and plastic deformation and its computational significance. Appl Mech 48:35–40
14. Agah-Tehrani R, Lee EH, Mallet RI, Oden RT (1986) The theory of elastic-plastic deformation at finite strain with induced anisotrophy modeled as combined isotropic-kinematic hardening. Metal Forming Report, Renselaer Polytechnic Institute, Troy, NY, June
15. Lubliner J (1984) A maximum-dissipation principle in generalized plasticity. Acta Mech 52:225–237
16. Lubliner J (1986) Normality rules in large-deformation plasticity. MOM 5:29–34
17. Simo JC, Ortiz M (1985) A unified approach to finite deformation elastoplasticity based on the use of hyperelastic constitutive equations. Comput Meth Appl Mech Eng 58:79–116

18. Simo JC (1988) A framework for finite-strain elastoplasticity based on maximum plastic dissipation and the multiplicative decomposition: part I. Continuum formulation. Comput Meth Appl Mech Eng 66:199–219
19. Simo JC (1988) A framework for finite-strain elastoplasticity based on maximum plastic dissipation and the multiplicative decomposition: part II. Computational aspects. Comput Meth Appl Mech Eng 68:1–31
20. Taylo GI, Elam CF (1923) The distortion of aluminum crystals during a tensile test. Proc Royal Soc Lon A102:643–667. (Bakerian lecture)
21. Taylo GI, Elam CF (1925) The plastic extension and fracture of aluminum crystals. Proc Royal Soc Lond A108:28–51
22. Taylo GI (1938) Analysis of plastic strain in a cubic crystal. In: Lessels JM (ed) Stephen Timoshenko 60th anniversary volume. Macmillan, NY
23. Hill R (1966) Generalized constitutive relations for incremental deformation of metal crystals by multislip. J Mech Phys Solids 14:95–102
24. Rice JR (1971) Inelastic constitutive relations for solids: an internal-variable theory and its application to metal plasticity. J Mech Phys Solids 19:433–455
25. Hill R, Rice JR (1972) Constitutive analysis of elastic-plastic crystals at arbitrary strain. J Mech Phys Solids 20:401–413
26. Asaro RJ, Rice JR (1977) Strain localization in ductile single crystals. J Mech Phys Solids 25:309–338
27. Asaro RJ (1979) Geometrical effects in the inhomogeneous deformation of ductile single crystals. Acta Metall 27:445–453
28. Asaro RJ (1983) Micromechanics of crystals and polycrystals. In: Advances in applied mechanics, vol 23, Academic Press, NY
29. Peirce D, Asaro RJ, Needlemann A (1982) An analysis of nonuniform and localized deformation in ductile single crystals. Acta Metall 30:1087–1119
30. Peirce D (1983) Shear band bifurcation in ductile single crystals. J Mech Phys Solids 31:133–153
31. Spencer AJM (1980) Continuum mechanics. Longman London
32. Holzapfel GA (2000) Nonlinear solid mechanics, a continuum approach for engineering. Wiley, Chichester
33. Borisenko AJ, Tarapov IE (1968) Vector and tensor analysis with applications. Translated from Russian by Silverman RA. Dover Publications Inc, NY
34. Simo JC, Hughes TJR (1998) Computational inelasticity. Springer, NY
35. De Souza Neto EA, Perić D, Owen DRJ (2008) Computational methods for plasticity, theory and applications. Wiley, Singapore
36. Écsi L, Élesztős P (2018) An alternative material model using a generalized J2 finite-strain flow plasticity theory with isotropic hardening. Int J Appl Mech Engrg 23(2):351–365
37. Hackett RM (2016) Hyperelasticity primer. Springer, NY
38. Ogden RW (1984) Non-linear elastic deformations. Ellis Horwood, Chichester
39. Ogden RW (1972) Large deformation isotropic elasticity—on the correlation of theory and experiment for incompressible rubberlike solids. Proc R Soc Lond A326:565–584
40. Hencky H (1933) The elastic behavior of vulcanized Rubber. J Appl Mech 1:45–53
41. Green AE, Zerna W (1954) Theoretical elasticity. Oxford University Press, Oxford
42. Yeoh OH (1990) Characterization of elastic properties of carbon-black-filled rubber vulcanizates. Rubber Chem Technol 63:792–785
43. Gent AN (1996) A new constitutive relation for rubber. Rubber Chem Technol 69:59–61
44. Dill EH (2007) Continuum mechanics: elasticity, plasticity, viscoelasticity. CRC, Boca Raton
45. Bower AF (2010) Applied mechanics of solids. CRC, Boca Raton
46. Blatz PJ, Ko WL (1962) Application of finite elasticity to the deformation of rubbery materials. Trans Soc Rheol 6:223–251

47. Arruda EM, Boyce MC (1993) A three-dimensional constitutive model for the large stretch behavior of rubber elastic materials. J Mech Phys Solids 41:389–412
48. Jerábek R, Écsi L (2019) Numerical study of material degradation of a silicone cross-shaped specimen using a thermodynamically consistent Mooney-Rivlin material model. Mater Sci Forum 952:258–266
49. Shahzad M, Kamran A, Siddiqui MZ, Farhan M (2015) Mechanical characterization and FE modelling of a hyperelastic material. Mater Res 18(5):918–924

Geometric Nonlinear Numerical Analysis of Shells in the Form of a Hyperboloid of Revolution

Ismael Taha Farhan Farhan, Gil-oulbé Mathieu and Timur Soibnazarovich Imomnazarov

Abstract This paper provides information on hyperboloids of revolution surfaces and their classification, using methods of differential geometry and computer design. All surfaces are plotted by means of computer graphics. The geometric nonlinear numerical analysis of thin shells in the form of hyperboloid of revolution surfaces on buckling resistance and stress-strain state is given in this paper, using the finite elements method in a computer software. Comparison is done with the result of the finite elements linear analysis of their buckling resistance.

Keywords Hyperboloids of revolution · Differential geometry · Geometric nonlinear numerical analysis · Finite elements linear analysis · Linear analysis · Buckling resistance

1 Introduction

In geometry, a hyperboloid of revolution, sometimes called circular hyperboloid, is a surface that may be generated by rotating a hyperbola around one of its principal axes. A hyperboloid is a surface that may be obtained from a hyperboloid of revolution by deforming it by means of directional scaling, or more generally, of an affine transformation.

A hyperboloid is a quadric surface, which is a surface that may be defined as the zero set of a polynomial of degree two in three variables. Among quadric surfaces, a hyperboloid is characterized by not being a cone or a cylinder, having a centre of symmetry, and intersecting many planes into hyperbolas. A hyperboloid has also three pairwise perpendicular axes of symmetry, and three pairwise perpendicular planes of symmetry.

Let's focus, in this paper, on hyperboloid of one sheet as presented in Fig. 1 (Figs. 2 and 3).

I. T. F. Farhan (✉) · G. Mathieu · T. S. Imomnazarov
Department of Civil Engineering, Engineering Academy, Peoples' Friendship, University of Russia, (RUDN University), 6 Miklukho-Maklaya Street, 117198 Moscow, Russian Federation
e-mail: ismael.civilengineer@gmail.com

A. Öchsner and H. Altenbach (eds.), *Engineering Design Applications II*,
Advanced Structured Materials 113, https://doi.org/10.1007/978-3-030-20801-1_6

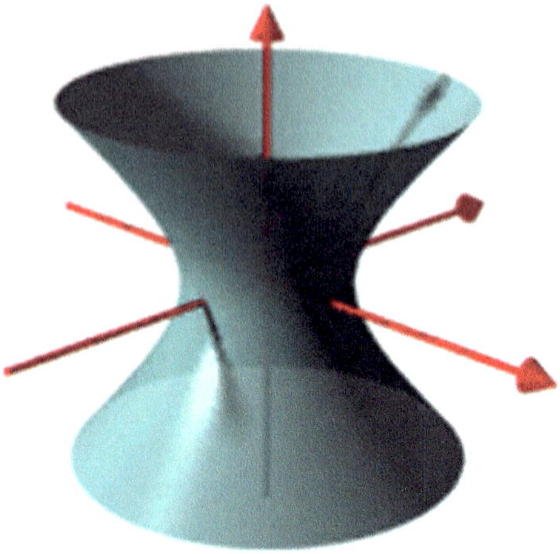

Fig. 1 Hyperboloid of one sheet

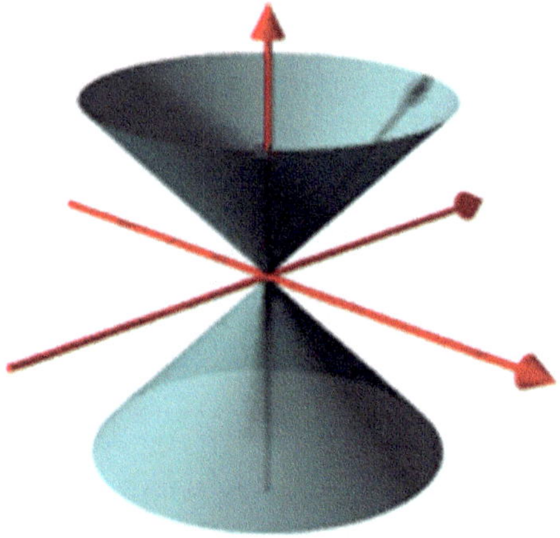

Fig. 2 Conical surface in between

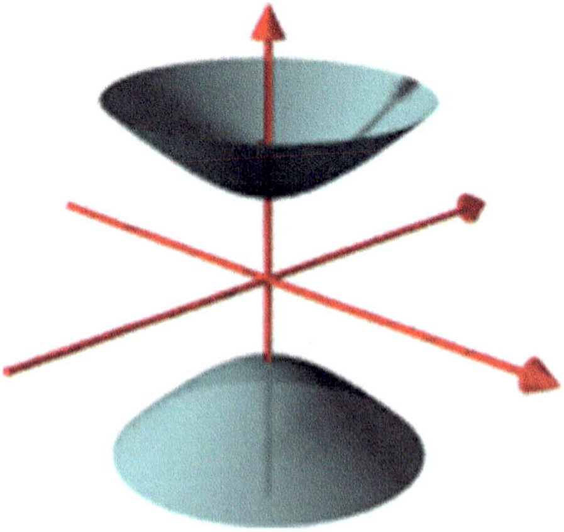

Fig. 3 Hyperboloid of two sheets

2 Methods of Setting and Shaping the Surface

The one-sheeted hyperboloid of revolution is the only ruled surface of revolution of general form, which can degenerate into a cylinder, cone or plane. The one-sheeted hyperboloid of revolution can also be obtained by rotating the hyperbola

$$\frac{x^2}{a^2} - \frac{z^2}{c^2} = 1 \tag{1}$$

With the convex side facing the rotation axis Oz (Fig. 4) in the xOz plane. Hyperbola will be the meridian.

Of this surface. The single-sheeted hyperboloid of rotation is a double ruled surface. Through any point of the surface there pass two straight lines lying entirely on this hyperboloid (Fig. 5). The possibility of linear construction is of great practical importance, since it allows to optimally place the reinforcement in monolithic reinforced concrete shells.

The one-sheeted hyperboloid of revolution can be specified in the implicit form by:

$$\frac{x^2 + y^2}{a^2} - \frac{z^2}{c^2} = 1 \tag{2}$$

Or in parametric form by:

$$\begin{aligned} x(u, v) &= -a \cdot \sin u \pm a \cdot v \cdot \cos u; \; y(u, v) \\ &= -a \cdot \cos u \pm a \cdot v \cdot \sin u; \; z(v) = \pm c \cdot v \end{aligned} \tag{3}$$

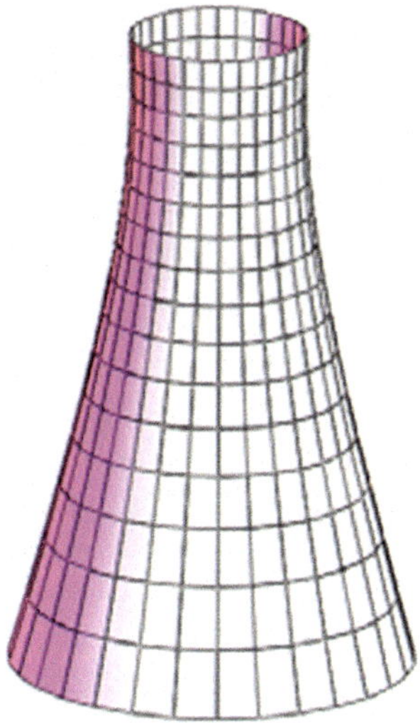

Fig. 4 Varieties of hyperboliods of revolution

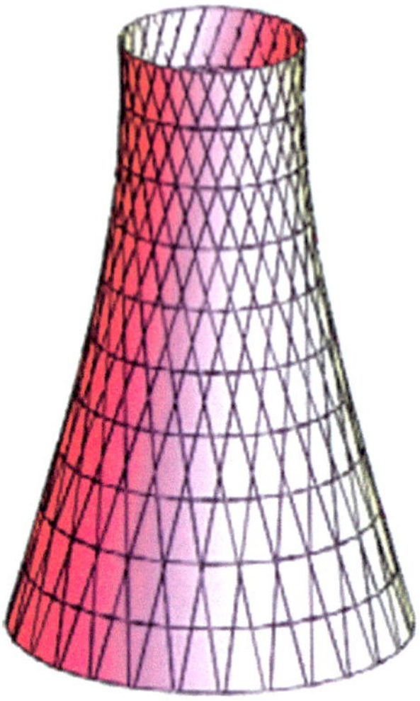

Fig. 5 Varieties of hyperboliods of revolution

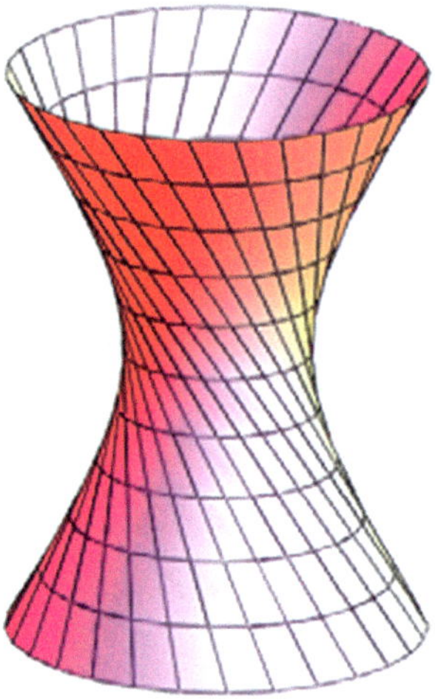

Fig. 6 Varieties of hyperboliods of revolution

If $a = c$, then the hyperboloid is said to be regular.

The coordinate lines v ($u = const$) coincide with one family of straight lines, and the lines u are parallels of a one-sheeted hyperboloid of revolution.

In Fig. 6, the hyperboloid is shown with allowance for the upper signs in the parametric equations of the surface, and in Fig. 7—considering the lower signs. There are other forms of specifying the surface in question.

The one-sheeted hyperboloid of revolution is a surface of negative Gaussian curvature, so it will lie on both sides of its tangent plane.

The hyperbolic shape of thin-walled structures creates optimal conditions for aerodynamics, strength and stability.

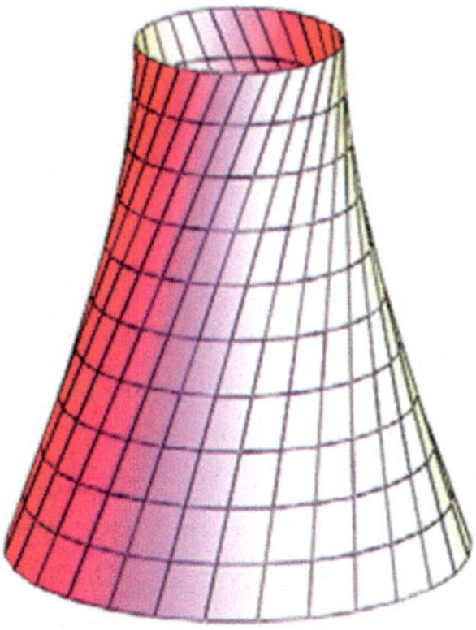

Fig. 7 Varieties of hyperboliods of revolution

3 Overview of Literature

The most complete work on the history of the development of hyperboloid structures in domestic construction is the monograph of IA. Petropavlovskaya [1] with 684 titles of the bibliography.

The article [2] gives a comparative analysis of the stability of the initial form of equilibrium of reticulated shells in the form of single-sheeted hyperboloids of revolution. The analyses are performed both considering only geometric and double (geometric and physical) nonlinearity. The influence of the shape of the generatrix of a single-sheeted hyperboloid of revolution and the physical nonlinearity of the material on its stability in the indicated formulations of the problem is considered. The curves of the equilibrium states of the shells under load acting on the upper base are given.

In the review article [3] are summarized principal achievements of science and engineering in the sphere of design, construction, and static, vibrational, and buckling analysis of thin-walled constructions and buildings in the shape of hyperbolic surfaces of revolution. These shells are useful as hyperbolic cooling towers, TV towers, reinforced concrete water tanks, and arch dams. They are also used as supports for electric power transmission lines and as high chimneys. Several public and industrial buildings having the hyperbolic form are described in the review. The basic results of theoretical and experimental investigations of stress-strain state, buckling, and vibration are summarized. The influence of temperature and moisture on the stress-strain state of the shells in question is also analysed.

In book [4] Reddy J.N. presents the theory and computer implementations of the finite element method as applied to nonlinear problems of heat transfer and similar field problems, fluid mechanics (flows of incompressible fluids), and solid mechanics (elasticity, beams and plates). Both geometric as well as material nonlinearities are considered, and static and transient (i.e.: time-dependent) responses are studied.

Trushin S.I. and Petrenko Ph.I. In paper [5] investigated the influence of form of hyperboloid's generatrix on its stress-strain state, stability and fundamental frequency. The results of numerical analysis of reticulated shells under different loads are given.

The article [6] studies the impact of the shape of the generatrix of a discrete reticulated shell in the form of a one-sheet hyperboloid of revolution on its stability. It studies the stability of a reticulated hyperboloid with five types of frames formed from the generatrices of the hyperbolae with one asymptote. For comparison, the study examines an additional frame with rectilinear generatrices—the asymptotes of hyperbolae generatrices, the properties of which are different from those of shells with a curvilinear rod network. The work presents the results of the numerical analysis of the reticulated shell performed with the help of finite element method in a geometrically non-linear setting. The article also presents the curves of the equilibrium states of the shell under horizontal and vertical concentrated loads. The data obtained has made it possible to conclude on the impact of the used generatrix on the stability of the frame of reticulated shells in the form of hyperboloid of revolution.

4 Linear and Geometric Nonlinear Analysis of Thin Shells in the Form of Hyperboloid of Revolution

Geometric nonlinearity, in general, represent the cases when the relations among kinematic quantities (i.e., displacement, rotation, and strains) are nonlinear.

Such nonlinearities often occur when deformation is large. Due to the large rotation, linear analysis cannot be used to accurately represent the deformation. For the linear problems, the relation between strain and displacement is linear. For example, in the case of a one-dimensional bar element, this relation can be written as

$$\varepsilon(x) = \frac{du(x)}{dx} \tag{4}$$

If the displacement is doubled, the strain will also be doubled, which is a fundamental property of linear problems. Note that the above relation is valid only when displacement and its gradient are infinitesimal. The above relation is not accurate, and the following definition of strain needs to be introduced:

$$E(x) = \frac{du}{dx} + \frac{1}{2}\left(\frac{du}{dx}\right)^2 \tag{5}$$

Table of comparison of the linear and nonlinear results

Dimension of the design model:
top diameter = 2.5 m
bottom diameter = 6.5 m
height = 4 m
the concrete thickness = 15 cm

THE DESIGN MODEL

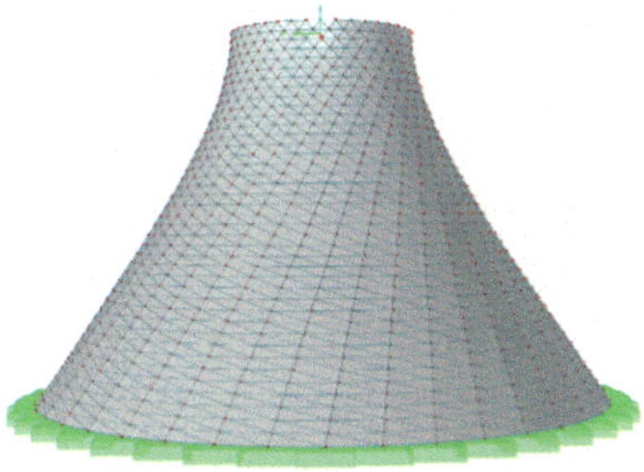

LINEAR ANALYSIS

NONLINEAR ANALYSIS

Isofield of internal forces
Isofield of n_x [kN/m]

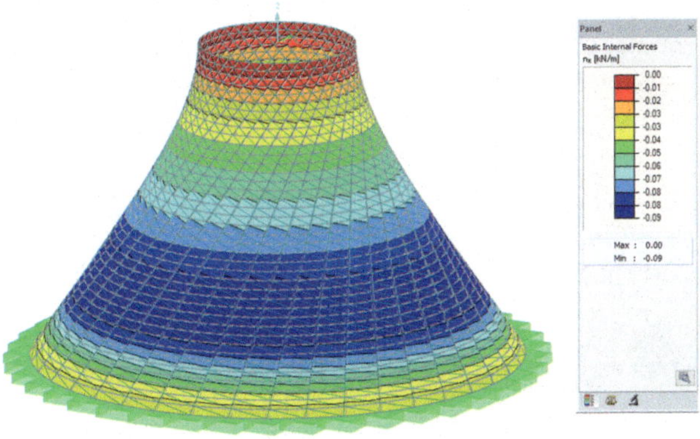

Internal force n_x varies from the minimum value ($n_x = -0.09$ kN/m) to the maximum one ($n_y = 0.00$ kN/m) for both linear and geometric nonlinear analyses.

(continued)

(continued)

Isofields of internal forces
Isofield of n_y [kN/m]

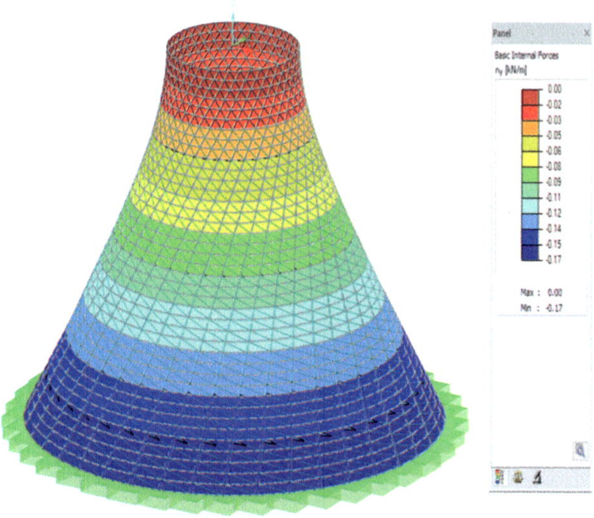

Internal force n_y varies from the minimum value ($n_y = -0.17$ kN/m) to the maximum one ($n_y = 0.00$ kN/m) for both linear and geometric nonlinear analyses.

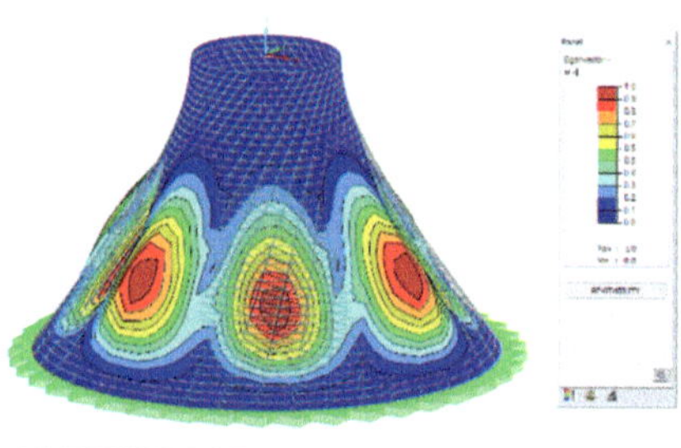

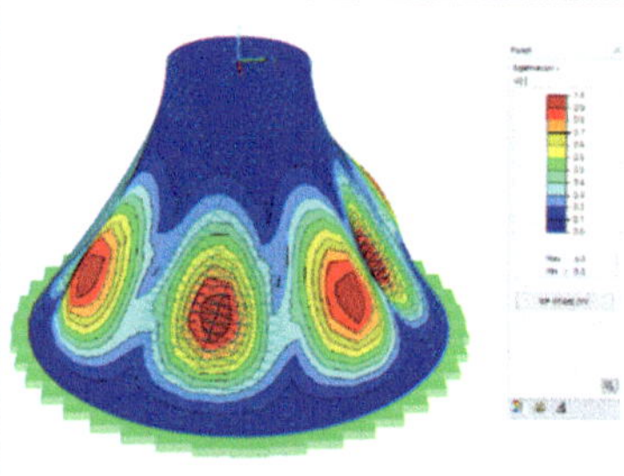

The value of Eigenvector U from the maximum is (U = 1.0 mm) and the minimum is (U = 0.0 mm) for both linear and geometric nonlinear analyses

Isofields of U_x—displacement along the x-axis

(continued)

(continued)

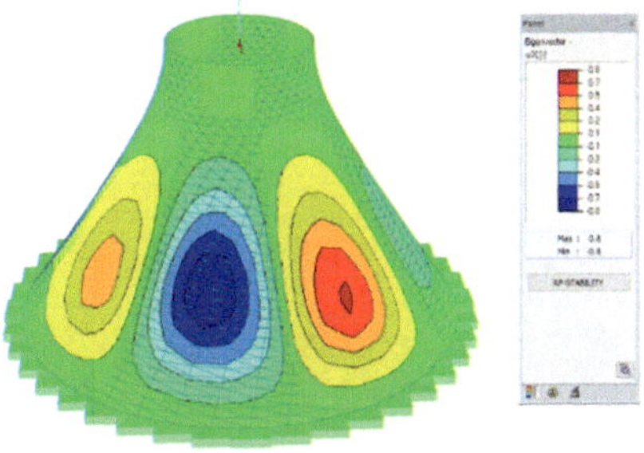	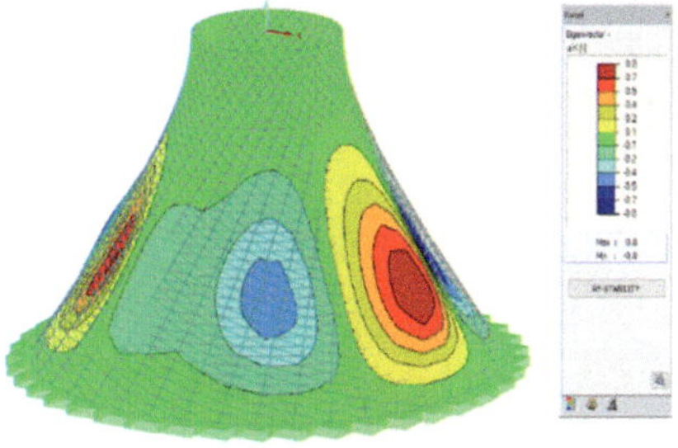
The Eigenvector U_x varies from the minimum value ($U_x = -0.8$ mm) to the maximum one ($U_x = 0.8$ mm) for both linear and geometric nonlinear analyses	
Isofields of U_y—displacement along the y-axis	
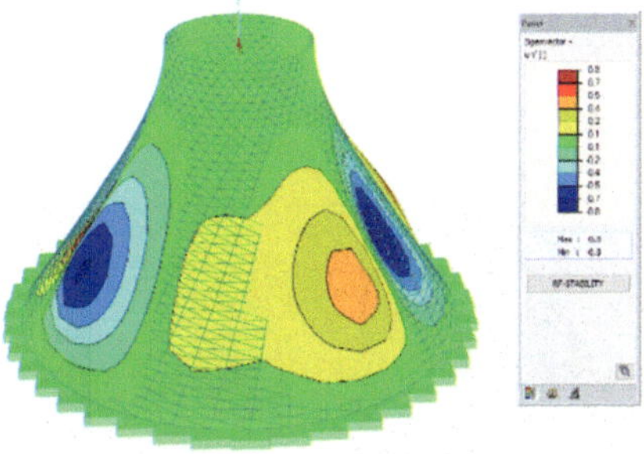	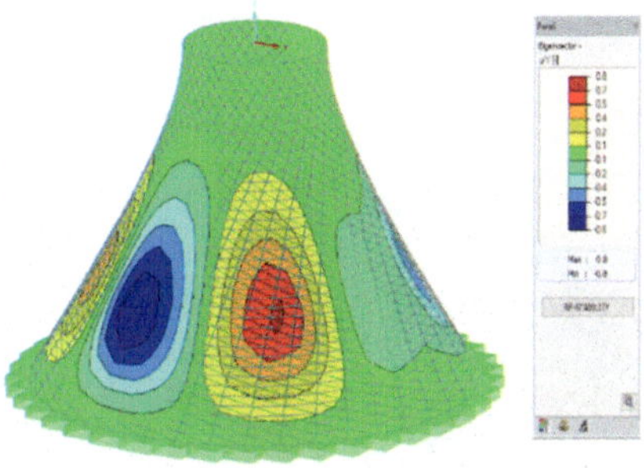
The Eigenvector U_y varies from the minimum value ($U_y = -0.8$ mm) to the maximum one ($U_y = 0.8$ mm) for both linear and geometric nonlinear analyses	
Isofields of U_z—displacement along the z-axis	
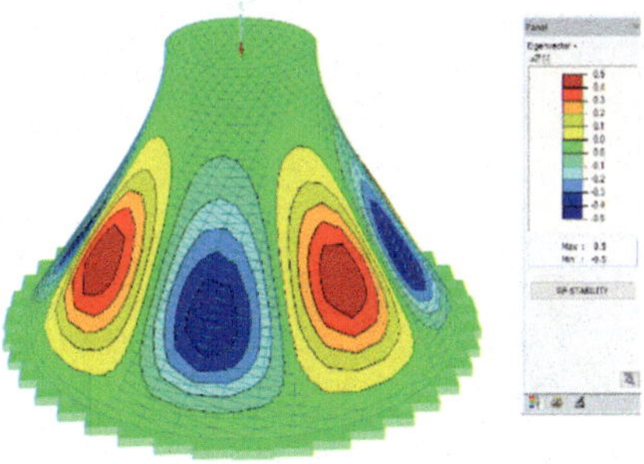	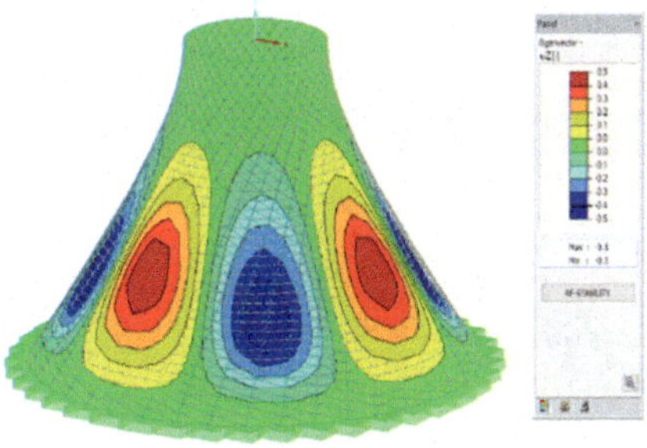
The Eigenvector U_z varies from the minimum value ($U_z = -0.5$ mm) to the maximum one ($U_z = 0.5$ mm) for both linear and geometric nonlinear analyses	

(continued)

(continued)

Isofields of φ_x—angular displacement around the x-axis

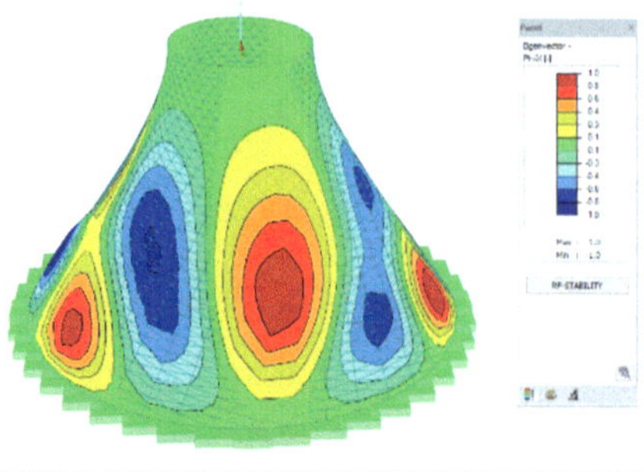

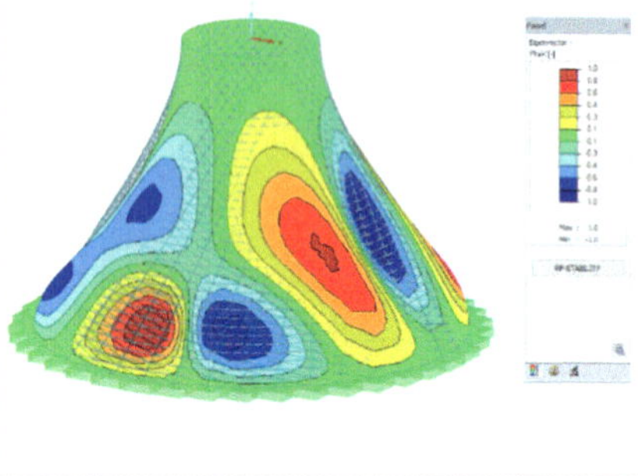

The Eigenvector φ_x varies from the minimum value ($\varphi_x = -1.0$ rad) to the maximum one ($\varphi_x = 1.0$ rad) for both linear and geometric nonlinear analyses

Isofields of φ_y—angular displacement around the y-axis

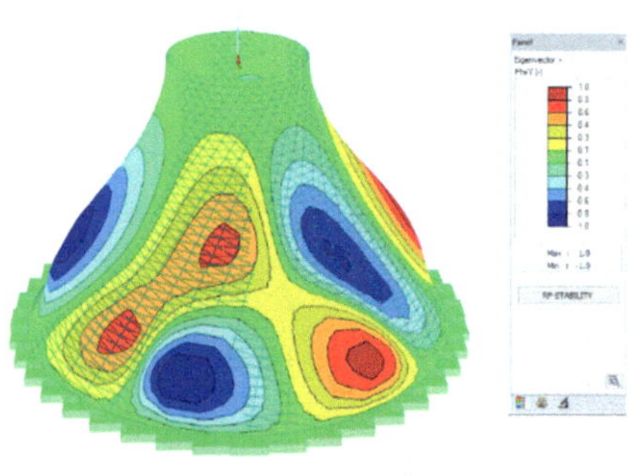

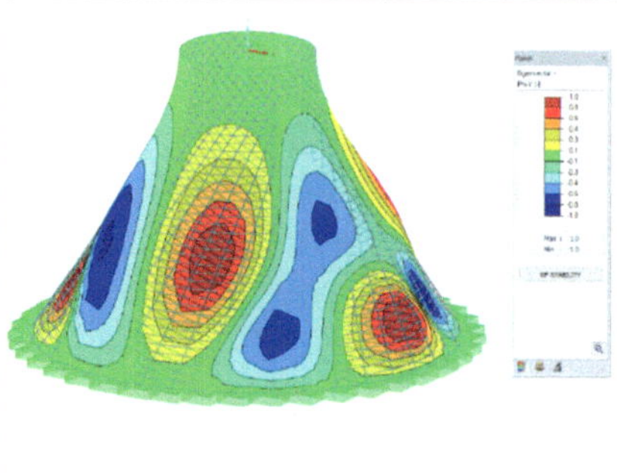

The Eigenvector φ_y varies from the minimum value ($\varphi_y = -1.0$ rad) to the maximum one ($\varphi_y = 1.0$ rad) for both linear and geometric nonlinear analyses

Isofields of φ_z—angular displacement around the z-axis

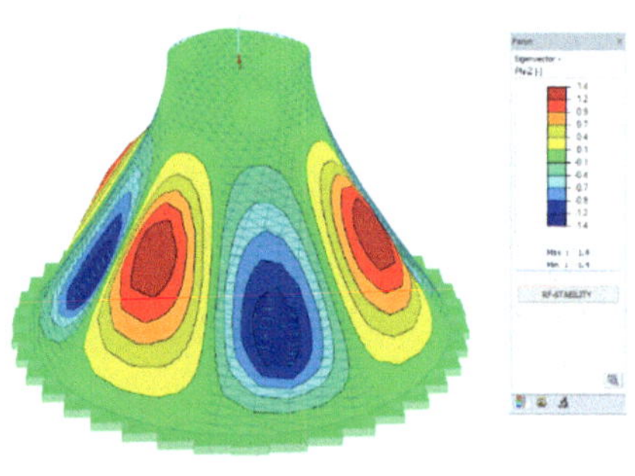

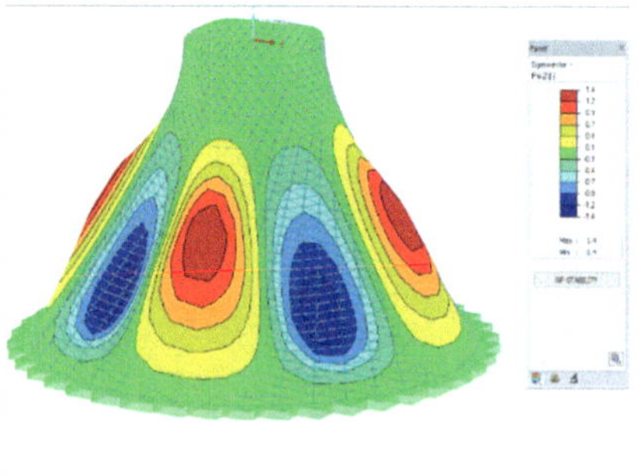

The Eigenvector φ_z varies from the minimum value ($\varphi_z = -1.4$ rad) to the maximum one ($\varphi_x = 1.4$ rad) for both linear and geometric nonlinear analyses

5 Conclusion

The linear and the geometric nonlinear investigations of the stability behaviour of thin shells in the form of a hyperboloid of revolution by the software program R-FEM show that for both linear and geometric nonlinear analyses:

1. internal force n_x varies from the minimum value ($n_x = -0.09$ kN/m) to the maximum one ($n_x = 0.00$ kN/m);
2. internal force n_y varies from the minimum value ($n_y = -0.17$ kN/m) to the maximum one ($n_y = 0.00$ kN/m);
3. the value of Eigenvector U from the maximum is ($U = 1.0$ mm) and the minimum is ($U = 0.0$ mm);
4. the Eigenvector U_x varies from the minimum value ($U_x = -0.8$ mm) to the maximum one ($U_x = 0.8$ mm);
5. the Eigenvector U_y varies from the minimum value ($U_y = -0.8$ mm) to the maximum one ($U_y = 0.8$ mm);
6. the Eigenvector U_z varies from the minimum value ($U_z = -0.5$ mm) to the maximum one ($U_z = 0.5$ mm);
7. the Eigenvector φ_x varies from the minimum value ($\varphi_x = -1.0$ rad) to the maximum one ($\varphi_x = 1.0$ rad);
8. the Eigenvector φ_y varies from the minimum value ($\varphi_y = -1.0$ rad) to the maximum one ($\varphi_y = 1.0$ rad); and
9. The Eigenvector φ_z varies from the minimum value ($\varphi_z = -1.4$ rad) to the maximum one ($\varphi_x = 1.4$ rad);
10. The view from the top of these graphical results shows the dispositions of the isofields of each value of the linear and angular displacements are axisymmetric for both linear and geometric nonlinear analyses;
11. Negligible differences are noticed in the axisymetrical dispositions of the isofields of each value of the linear and angular displacements;
12. The similarity between the result of linear and geometric nonlinear analyses for this investigated thin shell subject to selfweight may not be generalized to other loading cases and heating;
13. The similarity between the linear and the geometric nonlinear displacements in the concerned shells subject to selfweight shows that these displacements are considered not large.

References

1. Petropavlovskaya IA (1988) Hyperboloid constructions in structural mechanics—Nauka, Moscow. p. 230
2. Trushin SI, Petrenko PhI (2017) Influence of physical nonlinearity on the calculated indicators of stability of reticulated one-sheet hyperboloid of revolution with different forms of generatrixes. Struct Mech Eng Constr Build 4:50–56. https://doi.org/10.22363/1815-5235-2017-4-50-56

3. Krivoshapko SN (2002) Static, vibration and buckling analysis and applications to one-sheet hyperboloidal shells of revolution. Appl Mech Rev 55(3):241—270, https://doi.org/10.1115/1.1470479
4. Reddy JN (2005) An introduction to nonlinear finite element analysis. Oxford University Press, New York P 463
5. Trushin SI, Petrenko PhI (2014) The influence of the morphology of reticulated hyperboloid on its stress strain state, stability and fundamental frequencies. Struct Mech Anal Struct 4:59–64
6. Trushin SI, Petrenko PhI (2016) Analysis of the stability of flexible reticulated shells in the form of a hyperboloid of revolution. Sci Rev 6:95–99

Numerical Experiments on Helicoids Stress-Strain State

Marina Rynkovskaya, A. S. Markovich, M. Gil-Oulbe and S. V. Strashnov

Abstract Helicoidal surfaces are widely used in various structures and their elements, especially in mechanical engineering, as well as in civil engineering and architecture. However there are still different classifications in both mathematical and engineering fields, and the general types of these surfaces are not completely investigated. In this paper, there are numerical experiments on two types of ruled helicoids stress-strain state: oblique and convolute. The results shows the differences in stress behavior of these two types, and proves that the preliminary design according to the specific requirements can be effective in the case when an engineer chooses the proper type of a helicoids for a particular structure. It is also shown how the presented investigation can be continued for a future research and used in practical engineering tasks.

Keywords Helical surface · Oblique helicoid · Convolute helicoid · Numerical analysis

1 Introduction

Generally there are five types of ruled helicoids [1]: Right; Oblique; Developable; Pseudo-developable; Convolute; but the most general of them are oblique and convolute (or pseudo-developable helicoid of a general type). There are also various classifications of ruled helicoids, for example, some classifications can be found in [2, 3]. However the ruled helicoids can also be divided into just two types: "oblique" and "convolute" (or closed and open) since they are the most general types which can be used in practice while they can be designed with different slopes (without and with inner hole respectively). Since this classification is reasonable in the case of the specific requirements for the structure, further we will consider these two types only.

M. Rynkovskaya (✉) · A. S. Markovich · M. Gil-Oulbe · S. V. Strashnov
Department of Civil Engineering, Peoples' Friendship University of Russia (RUDN University), 6 Miklukho-Maklaya St, 117198 Moscow, Russian Federation
e-mail: marine_step@mail.ru

A. Öchsner and H. Altenbach (eds.), *Engineering Design Applications II*, Advanced Structured Materials 113, https://doi.org/10.1007/978-3-030-20801-1_7

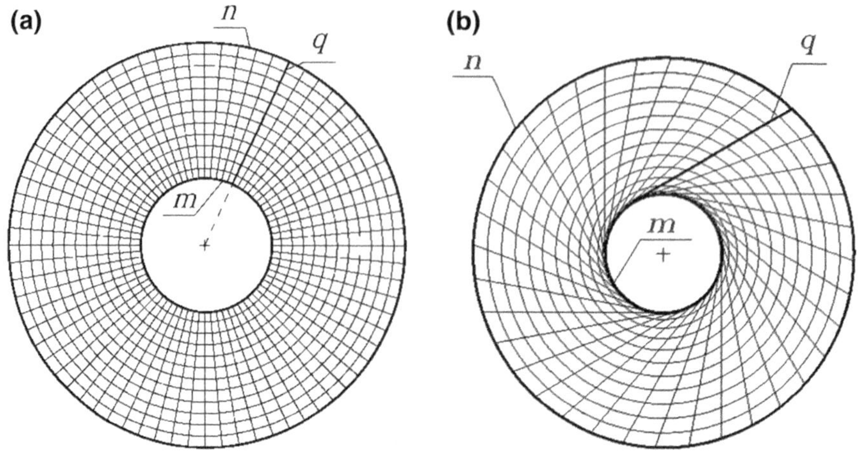

Fig. 1 Projections of helicoids: **a** oblique; **b** convolute

Table 1 Oblique (model 1) and convolute (model 2) helicoids

Type of helicoids	Geometrical characteristics		Differences in forming process	Practical characteristics
	Gaussian curvature	Middle curvature		
Oblique	$K < 0$	$H \neq 0$	The generatrix intersects the axis of the helicoid	Any slope available Well-known structure Can be built without inner hole
Convolute	$K < 0$	$H \neq 0$	The generatrix never intersects the axis of the helicoids (the generatrix is tangent to the interior cylinder)	Any slope available Not available without inner hole

The main characteristics and the differences between oblique and convolute helicoids are shown in Fig. 1 and Table 1.

Concerning the description from geometrical point of view, oblique and convolute helicoids have many similarities except the fact that the generatrix of an oblique helicoid intersects the axis of the helicoid, while the generatrix of a convolute helicoid is tangent to the inner cylinder (or to the inner circle in the projection).

According to the Encyclopedia of Analytical Surfaces [2] an oblique helicoid is "a helical ruled surface formed by a director straight line that intersects the axis of the helicoids under a constant angle α not equal to 900 and rotates with constant angular speed around the same axis".

The parametric equations of an oblique helicoid are the following:

$$x = x(u, v) = u \sin \alpha \cos v$$
$$y = y(u, v) = u \sin \alpha \sin v$$
$$z = z(u, v) = cv + u \cos \alpha$$

where *u, v* are coordinate lines; *c* is the displacement of a generatrix straight line after its rotation at one radian; α is an acute angle of the axis of the helicoids (an axis z) with the straight generatrix.

According to the same Encyclopedia [2] a convolute helicoid is "formed by a straight line that moves in the space all the time intersecting a helix and touching the side surface of a right circular cylinder with a radius *a*. The axis of the helix coincides with the axis of the cylinder but the angle between these skew lines is not equal to π/2".

The parametric equations of a convolute helicoid are the following:

$$x = x(t, v) = a \cos v - t \sin \gamma \sin v$$
$$y = y(t, v) = a \sin v + t \sin \gamma \cos v$$
$$z = z(t, v) = pv + t \cos \gamma$$

where *a* is the shortest distance in a straight line from an axis *Oz*; γ is the angle of the generatrix straight line with the axis of helicoid *Oz*; parameter *t* determines the disposition of a point *M* lying on the straight generatrix. The equations give both the spaces of a convolute helicoid (when $t > 0$ and $t < 0$).

Since both surfaces seem to be similar at the first glance of an engineer, except the fact that oblique helicoids can be used when the closed structure (without inner hole) is required, while the convolute helicoid can be used only for structures with inner holes, it looks reasonable to investigate the stress-strain state of the structures in the shape of both these helicoids to find out which type can represent more effective behavior.

There have been found several papers about stress-strain state of the oblique helicoids [4], as well as for the right [5], developable [6, 7] and pseudo-developable [8], but there have not been found any papers about the calculations of convolute helicoids, except the paper on geometry investigations [9]. The purpose of this paper is to analyze the stress and strain response of two most general types of helicoids (oblique and convolute) and to show which type is more effective for the practical tasks.

2 Methods and Materials

Let us analyze the structures for a ramp (Fig. 2) in the shape of two types of helicoids: oblique (model 1) and convolute (model 2) with similar geometric parameters, and calculate them by the means of commercial software (LIRA SAPR 2013 R3, based on the finite element method), considering stresses and strains for the strength analysis and also analyzing both these structures for the stability. The geometric characteristics of the structures are determined in accordance with the civil engineering codes for the ramps in parking zones and are shown in Table 2. The physical characteristics of the structures are shown in Table 3. We apply two loads: the dead load (which is calculated by the software and is equal to 4 kN/m^2) and active load which is determined equal to 10 kN/m^2.

There were conducted two numerical experiments with different boundary conditions.

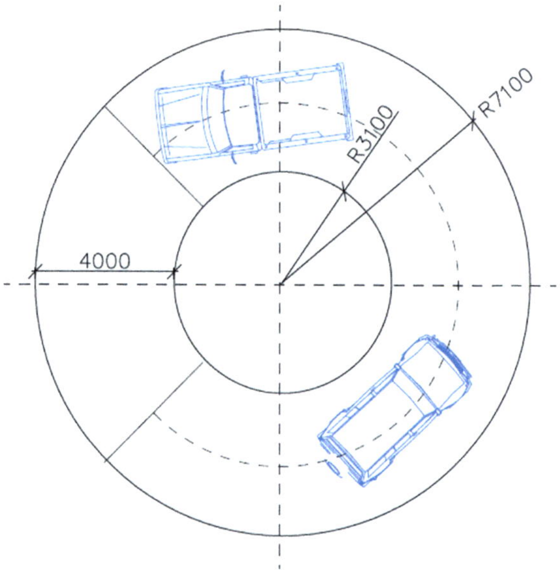

Fig. 2 Projection of a ramp structure

Table 2 Geometric characteristics for calculated models

Parameter	Units of measurement	Value
Interior radius	m	3.1
Exterior radius	m	7.1
Hitch height	m	3
Depth of the section	m	0.16
Inclination angle	0	3
Transverse slope	%	6

Table 3 Physical characteristics for calculated models

Parameter	Symbol	Units of measurement	Value
Poison's ratio	ν	–	0.18
Young's modulus	E	MPa	2.55×10^6
Specific weight	ρ	N/m^3	25.5

Table 4 Results for stability analysis of two models (test 1)

Buckling form	Stability safety factor		Divergence, %
	Model 1	Model 2	
I	1.66	2.40	−44.7
II	2.50	2.61	−4.4
III	3.52	3.57	−1.42
IV	4.98	4.68	6.02
V	6.22	5.73	7.87

Table 5 Results for stability analysis of two models (test 2)

Buckling form	Stability safety factor		Divergence, %
	Model 1	Model 2	
I	6.2806	5.2064	17.1
II	13.5837	8.1898	39.7
III	18.0671	10.1712	43.7
IV	19.2523	11.5469	40.0
V	21.8809	13.0984	40.1

The boundary conditions for the first numerical experiment (Test 1) are defined as the following: clamped linear edges at the beginning and at the end of a helicoid; free inner (interior) and outer (exterior) helical curves.

The boundary conditions for the second numerical experiment (Test 2) are defined as the following: clamped linear edges at the beginning and at the end of a helicoid; clamped inner (interior) helical curve; free outer (exterior) helical curve.

The results for the stability analysis of both structures are shown in Tables 4, 5.

The results of the static analysis for linear and angular displacements, bending and twisting moments, normal and shear forces are shown in Tables 6, 7, 8 and 9.

3 Results and Discussion

The results of the stability analysis for the first numerical experiment (Test 1) show that for the first buckling form stability safety factor for convolute helicoid is 44%

Table 6 Results for displacements in two models (test 1)

Parameter		Units of measurement	Value		Divergence, %
			Model 1	Model 2	
Displacement along *Z*	Minimum	mm	−665	−400	40
	Maximum		0	0	0
Displacement along *Y*	Minimum	mm	−3.56	−1.34	62
	Maximum		37.3	29.7	20
Displacement along *X*	Minimum	mm	−0.02	−35.1	>100
	Maximum		74.1	0.62	>100

Table 7 Results for stresses and moments of two models (test 1)

Parameter		Units of measurement	Value		Divergence, %
			Model 1	Model 2	
Normal stress N_x per m^2	Minimum	Ton/m^2	−4860	−2380	51
	Maximum		1650	5110	>100
Normal stress N_y per m^2	Minimum	Ton/m^2	−6920	−2350	66
	Maximum		2990	1220	59
Bending moment M_x per m	Minimum	(Ton*m)/m	−95.6	−87.1	9
	Maximum		17.9	2.4	86
Bending moment M_y per m	Minimum	(Ton*m)/m	−17.7	−50.8	>100
	Maximum		22.1	12.1	45
Twisting moment M_{xy} per m	Minimum	(Ton*m)/m	−13.7	−7.7	43
	Maximum		22.3	23.7	−6
Shear force Q_x per m	Minimum	Ton/m	−78.7	−833	>100
	Maximum		195	1000	>100
Shear force Q_y per m	Minimum	Ton/m	−17.6	−480	>100
	Maximum		490	235	52

Table 8 Results for displacements in two models (test 2)

Parameter		Units of measurement	Value		Divergence, %
			Model 1	Model 2	
Displacement along Z	Minimum	mm	−65.2	−62	5
	Maximum		0	0	0
Displacement along Y	Minimum	mm	−8.4	−7.5	11
	Maximum		8.7	7.2	17
Displacement along X	Minimum	mm	−9.76	−4.36	55
	Maximum		3.93	7.91	>100

Table 9 Results for stresses and moments of two models (test 2)

Parameter		Units of measurement	Value		Divergence, %
			Model 1	Model 2	
Normal stress N_x per m^2	Minimum	Ton/m^2	−935	−856	8
	Maximum		1680	1340	20
Normal stress N_y per m^2	Minimum	Ton/m^2	−783	−815	−4
	Maximum		1130	964	15
Bending moment M_x per m	Minimum	(Ton*m)/m	−4.6	−7.1	−53
	Maximum		0.5	0.67	33
Bending moment M_y per m	Minimum	(Ton*m)/m	−4.4	−5.9	34
	Maximum		2.3	4.3	−88
Twisting moment M_{xy} per m	Minimum	(Ton*m)/m	−2.3	−3.3	−42
	Maximum		4.4	2.5	42
Shear force Q_x per m	Minimum	Ton/m	−6.6	−175	>100
	Maximum		6.3	47.4	>100
Shear force Q_y per m	Minimum	Ton/m	−7.2	−32.7	>100
	Maximum		6.9	−5.7	>100

higher than the safety factor for oblique helicoid, while the divergence for second and third buckling forms would have been 4 and 1% accordingly (Table 4).

The results of the stability analysis for the second numerical experiments (Test 2) show that the stability safety factor for oblique helicoid is higher than for convolute helicoid. The divergence in the stability safety factors for the first buckling form is around 17%, while for other buckling forms the divergence is around 40% (Table 5).

The strength analysis shows more complicated results for both numerical experiments (Test 1 and Test 2). The results for Test 1 and Test 2 are shown in Tables 6, 7, 8 and 9 respectively.

According to the results of Test 1, the schemes for normal displacements (along axis *Z*) distribution for both models are similar, but the extreme values of normal displacements differ up to 40% (Table 6), while for tangential displacements (along axes *X* and *Y*) the zones with extreme values are not similar, and the zones with positive and negative values of displacements differ greatly (Table 6).

The schemes of normal stresses N_x and N_y distribution are similar with some differences in positive and negative signs. Thus, the zone of extreme values of N_x for convolute helicoid is near the quarter of a screw ($\pi/2$), and these extreme values are positive, while for oblique helicoid the zone with extreme values of N_x is close to the upper support edge, and the extreme values there are negative. For the values of N_y, the schemes of forces distribution are the same, but the signs of values are opposite.

According to results for moments M_x and M_y the schemes of distribution are similar but zones are different, as well as the signs of values. For twisting moments

M_{xy}, the schemes are rather different, while the values are similar except one zone near the support of convolute helicoids which shows extreme values in two finite elements.

Shear forces Q_x and Q_y vary considerably for model 1 and model 2. The highest values in both models appear near linear fixed supports, however convolute helicoid has additional problems along the inner helical curve (with extreme values), while oblique helicoid has smoother distribution and lower values of shear forces in these zones. In other zones the values of shear forces seem to be rather similar for both models.

According to the results of Test 2, in both structures the maximum normal displacements (along the axis Z) appear at the free curved edges (exterior helical curves) but the values differ by about 5% (Table 7).

The tangential displacements (along the axes X and Y, Table 7) show the difference in maximum values up to 17% for the displacement along axis Y, while the difference in maximum values for the displacements along axis X is much bigger (about 100%). Great divergences may be caused by some problems with triangular finite elements mesh and require further analysis, for example by analytical methods or alternative numerical methods.

The normal stresses N_x and N_y differ by up to 20 and 15% respectively; the bending moments M_x and M_y differ by up to 53 and 88% respectively; the twisting moments M_{xy} differ by about 42%, while the shear forces Q_x and Q_y differ by more than 100%.

The analysis proves that the helical structure in the shape of a convolute and oblique helicoids have different stress and strain response which vary for different parameters (stresses and strain) and have to be analyzed more by other calculating methods.

According to the stability analysis an oblique helicoid shows better stability behavior in comparison with convolute helicoids when it has fixed inner helical edge, but oblique helicoid has worse stability safety factor in comparison with convolute helicoid when it has free inner helical edge. It can be caused by the differences in geometrical formation, because the generatrix of oblique helicoid leads to the axis of helicoids, while the generatrix of convolute helicoid never leads to its axis. However, the boundary conditions also influence stability behavior of these two types of helicoids.

For the future research, more investigations are required to prove the links between geometry, stress-strain and stability behavior. Two other types of helicoids may also be compared: right and pseudo-developable helicoids, since they are the particular cases of oblique and convolute helicoids respectively (when the generatrix lays in the plane which is perpendicular to the axis of helicoid), and both of them can be used for staircases. However only right helicoid is widely used for staircases while pseudo-developable helicoid seems to be unknown by designers. Besides, more investigations can be done for the fifth type of helicoids, the developable helicoid, which differs from other types by its developable properties and can be used for mechanical engineering products and machine elements. Some numerical experiments can be done for other types of structures and materials, for example, for laminated skew plates (Kumar and

Chakrabarti [10] in the shape of helicoids, as well as some experiments in laboratory can be conducted to verify all obtained results.

4 Conclusion

The helical ramp structures in the shapes of two types of helicoids are investigated for the strength and stability. The obtained strength analysis results for both tests (with different boundary conditions) show that these structures have different stresses and strains responses, and the divergences are mostly bigger for Test 1 (with free helical edges) than for Test 2 (with fixed inner helical edge). Stresses and strains in ramps in the shape of convolute helicoid are mostly less than in ramps in the shape of oblique helicoids, except a number of special zones, which occur near the inner helical edge, especially for shear forces distribution. There are also differences in signs of values and zones of stresses and strains distribution which require further investigation to discover the links between the geometry (the types of helicoids) and stress-strain behavior for different boundary conditions or special requirements.

On the contrary, the stability analysis shows better response of convolute helicoids in comparison with oblique helicoids only for Test 1 (free boundary conditions for helical edges), while for Test 2 (fixed inner helical edges) the stability analysis shows the opposite results in stability behavior: stability safety factors are higher for ramps in the shape of oblique helicoids than for ramps in the shape of convolute helicoids, which can be explained taking into account the generatrix positions relatively to the axis of these two types of helicoids. The analysis shows that the stability behavior for these two different types of helicoids may depend on boundary conditions, thus, it can be useful to conduct deeper research on stability behavior of all five types of helicoids with various boundary conditions.

Generally, the differences in results are considerable (especially for shear forces) and have to be taken into account while designing the structures. It proves that for the preliminary design it can be reasonable to choose the proper geometry (proper type of helicoid) beforehand to reduce the stresses, strains or material consumption. Since shear forces vary greatly near the linear supports, the zones around linear supports require more investigations in order to prevent these zones by special arrangements like smooth transition from the helicoids part to flat slab of the storey. Special attention must be paid to the flexible structures which require stability analysis.

Besides, it would be useful to develop the recommendations for designers and engineers, how to optimize the structure according to the exploitation requirements using different ruled helicoids types.

Acknowledgements The publication was prepared with the support of the RUDN University Program "5–100".

References

1. Krivoshapko SN, Rynkovskaya M (2017) Five types of ruled helical surfaces for helical conveyers, support anchors and screws, MATEC Web Conf **95**, 06002. http://dx.doi.org/10.1051/matecconf/20179506002
2. Krivoshapko SN, Ivanov VN (2015) Encyclopedia of analytical surfaces, Springer, p 752
3. Jean Paul V (2018) A review of geometry investigations of helicoids, IOP Conf Series Mater Sci Eng **371**(012029):1–8 https://doi.org/10.1088/1757-899x/371/1/012029
4. Tupikova EM (2016) Investigation of VG Rekatch's method of stress—strain analysis of the shell of long shallow oblique helicoid form. Struct Mech Analysis Constr 1:14–19
5. Rynkovskaya M, Ivanov V (2019) Analytical method to analyze right helicoid stress-strain. In: Öchsner A, Altenbach H (eds) Engineering design applications. Adv Struct Mater 92:157–171. Springer
6. Krivoshapko SN (2009) Geometry of linear surfaces with the return rib and linear theory for torso-helicoid calculation, RUDN, p 357
7. Rynkovskaya MI (2014) Application of Runge-Kutte method and running tri-diagonal matrix for open helicoidal shell calculation. Struct Mech Eng Constr Build 3:77–80
8. Halabi SM (2001) Moment theory of thin helicoidal pseudo-developable shells. Struct Mech Eng Constr Build 10:61–67
9. Pylypaka SF (2002) Control of bending of ruled surfaces on an example of a screw conoid. Prikladna Geometriya ta Inzhenerna Grafika 70:180–186
10. Kumar A, Chakrabarti A (2017) Failure analysis of laminated composite skew laminates. Procedia Engineering 173:1560–1566

Transversely Cracked Beams with Quadratic Function's Variation of Height

Matjaž Skrinar

Abstract This paper considers the modelling of transversely cracked slender beams by a new one-dimensional finite element where the cross-section's height variation is given as a general quadratic function. The derivations of a closed-form stiffness matrix and a load vector are based on a simplified computational model where the cracks are represented by means of internal hinges endowed with rotational springs that take into account the cross-section's residual stiffness. These derived at expressions efficiently upgrade the stiffness matrix as well as the load vector coefficients which were previously presented for the non-cracked beams. Although the derivations were rather straightforward, all the derived terms are written entirely in closed-symbolic forms. These solutions, together with the solutions of second order governing differential equation of bending thus define an 'exact' finite element for the implemented simplified computational model. The derivations are complemented by two comparative case studies. They demonstrated that elaborated solutions may be effectively implemented for structural analyses as the presented expressions produced excellent results that were confirmed independently by more thorough 3D models.

Keywords Beam finite element · Stiffness matrix · Load vector · Quadratic function's variation of height · Transverse cracks

1 Introduction

Although beams are rather simple structural elements they also represent a very common type of load-bearing structural component. Generally, a beam is a straight bar of an arbitrary cross-section, but the majority of research is focused onto beams with constant cross-sections which are also the most frequent ones. The derivations

M. Skrinar (✉)
Faculty of Civil Engineering, Transportation Engineering and Architecture, University of Maribor, Maribor, Slovenia
e-mail: matjaz.skrinar@um.si

A. Öchsner and H. Altenbach (eds.), *Engineering Design Applications II*,
Advanced Structured Materials 113, https://doi.org/10.1007/978-3-030-20801-1_8

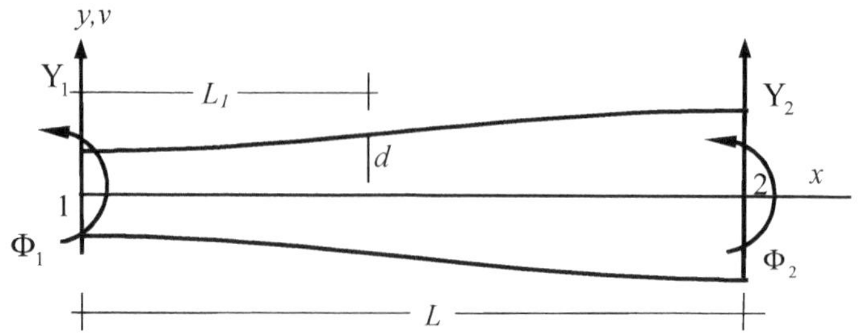

Fig. 1 Beam finite element with nodes and degrees of freedom

of stiffness matrix and the corresponding load vector for the prismatic beams can thus be found in numerous references (e.g. [5]).

Nevertheless, some papers consider beams with non-uniform cross-sections. Cracked stepped beams are considered in the works by Biondi and Caddemi [1], as well as by Skrinar [3], where the beams with a linear variation of heights are also considered.

Recently, the solutions for non-cracked slender beams where the height variation is given as a general quadratic function were also presented [4]. This paper effectively upgrades these solutions as it presents the derivation of a stiffness matrix and a load vector of the beam element with an arbitrarily located transverse crack.

2 Mathematical Model Formulation

The discussed solutions are appropriate for small deflections' evaluations of slender elastic homogeneous Euler-Bernoulli beams with a uniform modulus of elasticity E and width b. The quadratic variation of the considered element's height is given in general form as:

$$\mathrm{h}(x) = 2 \cdot \chi(x) \tag{2.1}$$

with

$$\chi(x) = 2 \cdot \left(A_1 + A_2 \cdot x + A_3 \cdot x^2\right). \tag{2.2}$$

The finite element of total length L is assumed to have four degrees of freedom altogether: transverse displacement Y_1 and rotation Φ_1 at the left-end (node 1), as well as transverse displacement Y_2 and rotation Φ_2 at the right-end (node 2), seen in Fig. 1. Upward displacements and anticlockwise rotations are taken as positive.

The effect of the transverse crack located at the distance L_1 from the left end is introduced through the simplified computational model as given by Okamura [2]. In this model, each crack is represented by means of an internal hinge endowed with a rotational spring that takes into account the cross-section's residual stiffness. The rotational spring' stiffness is denoted as K_r and for its evaluation, several definitions can be found in the references.

3 Derivation of Stiffness Matrix

The derivation of the stiffness matrix follows the essential steps of the non-cracked beam as already given by Skrinar and Imamović. The coefficients of the stiffness matrix are thus again obtained from two separate derivations on two statically determinate structures, however by taking also the crack into the account. All the stiffness matrix coefficients are obtained from two cantilever substructures: clamped at both the right and left-ends, respectively.

3.1 Cantilever Clamped at the Right-End

A cantilever clamped at the right-end is considered while deriving the first two columns of the stiffness matrix. It was shown by Skrinar and Imamović that the coefficients of the first row of the stiffness matrix of beam finite element with quadratic height variation (FEQVH) are given as:

$$\frac{\{\Phi_1(\delta x_2), -Y_1(\delta x_2), -\Phi_1(\delta x_2), \Phi_1(\delta x_2)\cdot L + Y_1(\delta x_2)\}}{\Phi_1(\delta x_2)\cdot Y_1(\delta x_1) - Y_1(\delta x_2)^2} \tag{3.1}$$

while the coefficients of the second row and the column are:

$$\frac{\{-Y_1(\delta x_2), Y_1(\delta x_1), Y_1(\delta x_2), -Y_1(\delta x_1) - Y_1(\delta x_2)\cdot L\}}{\Phi_1(\delta x_2)\cdot Y_1(\delta x_1) - Y_1(\delta x_2)^2}. \tag{3.2}$$

In Eqs. (3.1)–(3.2), $Y_1(\delta x_1)$ and $Y_1(\delta x_2)$ represent free end's transverse displacements due to a virtual vertical upward force $\delta x_1 = 1$ and a virtual anticlockwise bending moment $\delta x_2 = 1$ at the free end, respectively. In the same way, $\Phi_1(\delta x_1)$ and $\Phi_1(\delta x_2)$ represent free end's rotations due to the same virtual loads. Both displacements and rotations can be presented as combinations of contributions from non-cracked parts as well as from the crack:

$$Y_1(\delta x_1) = \int_{x=0}^{L} \frac{x^2}{\mathrm{EI}(x)}\cdot dx + \frac{L_1^2}{K_r} = \bar{Y}_1(\delta x_1) + \frac{L_1^2}{K_r} \tag{3.3}$$

$$Y_1(\delta x_2) = \int_{x=0}^{L} \frac{-x}{\mathrm{EI}(x)} \cdot dx - \frac{L_1}{K_r} = \bar{Y}_1(\delta x_2) - \frac{L_1}{K_r} \tag{3.4}$$

$$\Phi_1(\delta x_1) = \int_{x=0}^{L} \frac{-x}{\mathrm{EI}(x)} \cdot dx - \frac{L_1}{K_r} = \bar{\Phi}_1(\delta x_1) - \frac{L_1}{K_r} \tag{3.5}$$

$$\Phi_1(\delta x_2) = \int_{x=0}^{L} \frac{1}{\mathrm{EI}(x)} \cdot dx + \frac{1}{K_r} = \bar{\Phi}_1(\delta x_2) + \frac{1}{K_r}, \tag{3.6}$$

In Eqs. (3.3)–(3.6), $\bar{Y}_1(\delta x_1)$, $\bar{Y}_1(\delta x_2)$, $\bar{\Phi}_1(\delta x_1)$ and $\bar{\Phi}_1(\delta x_2)$ represent values of the displacements and rotations for the non-cracked cantilever clamped at the right-end, already completely analytically presented by Skrinar and Imamović.

3.2 Cantilever Clamped at the Left-End

To complete the derivation of the stiffness matrix the process of the free-end's vertical displacement and rotation calculation is now repeated for a cantilever, clamped at the left-end. The third row column and column coefficients are further evaluated as:

$$\frac{\{-\Phi_2(\delta x_4), Y_2(\delta x_4) - \Phi_2(\delta x_4) \cdot L, \Phi_2(\delta x_4), -Y_2(\delta x_4)\}}{\Phi_2(\delta x_4) \cdot Y_2(\delta x_3) - Y_2(\delta x_4)^2} \tag{3.7}$$

while the coefficients of the fourth row and the column are:

$$\frac{\{Y_2(\delta x_4), -Y_2(\delta x_3) + Y_2(\delta x_4) \cdot L, -Y_2(\delta x_4), Y_2(\delta x_3)\}}{\Phi_2(\delta x_4) \cdot Y_2(\delta x_3) - Y_2(\delta x_4)^2}. \tag{3.8}$$

where $Y_2(\delta x_3)$ and $Y_2(\delta x_4)$ represent right end's transverse displacements due to a virtual vertical upward force $\delta x_3 = 1$ and a virtual anticlockwise bending moment $\delta x_4 = 1$ at the free end, respectively. Similarly, $\Phi_2(\delta x_3)$ and $\Phi_2(\delta x_4)$ represent free end's rotations due to a virtual vertical upward force $\delta x_3 = 1$ and a virtual anticlockwise bending moment $\delta x_4 = 1$ at the free end, respectively.

Also these displacements and rotations can be easily presented as combinations of contributions from non-cracked parts as well as from the crack:

$$Y_2(\delta x_3) = \bar{Y}_2(\delta x_3) + \frac{L_1^2}{K_r} - 2 \cdot \frac{L \cdot L_1}{K_r} + \frac{L^2}{K_r} \tag{3.9}$$

$$Y_2(\delta x_4) = \bar{Y}_2(\delta x_4) + \frac{L}{K_r} - \frac{L_1}{K_r} \tag{3.10}$$

$$\Phi_2(\delta x_3) = \bar{\Phi}_2(\delta x_3) + \frac{L}{K_r} - \frac{L_1}{K_r} \tag{3.11}$$

$$\Phi_2(\delta x_4) = \bar{\Phi}_2(\delta x_4) + \frac{1}{K_r}. \tag{3.12}$$

In Eqs. (3.9)–(3.12), $\bar{Y}_2(\delta x_2)$, $\bar{Y}_2(\delta x_4)$, $\bar{\Phi}_2(\delta x_3)$ and $\bar{\Phi}_2(\delta x_4)$ represent values of the displacements and rotations for the non-cracked cantilever clamped at the left-end. Also these expressions were already presented in analytical forms.

It is clearly evident from Eqs. (3.3)–(3.6) and Eqs. (3.9)–(3.12) that the influence of a crack can be effectively introduced in a rather elegant way.

4 Derivation of the Load Vector Due to the Uniform Load q Over the Whole Element

Similarly to the coefficients of the stiffness matrix, the load vector's coefficients are also presented as combinations of the already known terms for the non-cracked element, with the terms that take into account the location and severeness of the crack.

The first two coefficients of the load vector are thus given as:

$$F_{n_{1,q}} = \frac{\Phi_1(\delta x_2) \cdot Y_{1,q} - Y_1(\delta x_2) \cdot \Phi_{1,q}}{\Phi_1(\delta x_2) \cdot Y_1(\delta x_1) - \Phi_1(\delta x_1) \cdot Y_1(\delta x_2)} \tag{4.1}$$

$$M_{n_{1,q}} = \frac{-\Phi_1(\delta x_1) \cdot Y_{1,q} + Y_1(\delta x_1) \cdot \Phi_{1,q}}{\Phi_1(\delta x_2) \cdot Y_1(\delta x_1) - \Phi_1(\delta x_1) \cdot Y_1(\delta x_2)} \tag{4.2}$$

while the remaining two coefficients can be easily derived from the first two as:

$$F_{n2,q} = L \cdot q - F_{n_{1,q}} \tag{4.3}$$

$$M_{n2,q} = F_{n_{1,q}} \cdot L - M_{n1,q} - \frac{L^2 \cdot q}{2}. \tag{4.4}$$

where $Y_{1,q}$ and $\Phi_{1,q}$ represent nodal transverse displacement and rotation of the free-end due to the applied distributed load q, respectively. They can be expressed as:

$$Y_{1,q} = \bar{Y}_{1,q} + \frac{q \cdot L_1^3}{2 \cdot K_r} \tag{4.5}$$

$$\Phi_{1,q} = \bar{\Phi}_{1,q} - \frac{q \cdot L_1^2}{2 \cdot K_r}. \tag{4.6}$$

In Eqs. (4.5)–(4.6), $\bar{Y}_{1,q}$ and $\bar{\Phi}_{1,q}$ represent values of the displacement and rotation for the non-cracked cantilever clamped at the right-end due to the uniform load q, as already previously presented in analytical forms.

5 Calculation of Exact Transverse Displacements Along the Finite Element

To obtain transverse displacements outside finite element's nodes, shape or interpolation functions are generally implemented. However, in the presented derivations the interpolation functions were neither needed nor derived. Therefore, the only option to obtain the transverse displacements is to solve the governing differential equations.

The vector of element's secondary variables *{Q}* obtained by implementing element's stiffness matrix and element's vector of nodal displacements and rotations allows for nodal bending moments (and shear forces) to be obtained. Both discrete nodal's moments values, M_1 and M_2, in combination with transverse load q, allow the function of bending moments to be constructed. Further, the known differential relation between bending moments and transverse displacements allows the function of transverse displacements to be obtained. This differential equation can always be solved for any combination of constants A_1, A_2 and A_3.

However, a single analytical solution cannot be obtained as various values and combinations of constants A_1, A_2 and A_3 produce different mathematical forms of the functions. Therefore, just one of these solutions will be presented below.

For $4{\cdot}A_1{\cdot}A_3 > A_2^2$ the solutions for transverse displacement to the left and to the right of the crack are given as

$$\mathrm{v}_1(x) = \mathrm{C}_2 + \mathrm{C}_1 \cdot \mathrm{x} - \frac{\mathrm{f}_6 + \mathrm{f}_7 \cdot \mathrm{x}}{\chi(\mathrm{x})} + (\mathrm{f}_5 \cdot \mathrm{x} - \mathrm{f}_9) \cdot \mathrm{ArcTan}\left(\frac{\chi'(\mathrm{x})}{\mathrm{k}}\right) + \mathrm{f}_{10} \cdot \mathrm{Ln}(\chi(\mathrm{x})) \quad 0 \le x \le L_1 \tag{5.1}$$

$$\mathrm{v}_2(x) = \mathrm{C}_4 + \mathrm{C}_3 \cdot \mathrm{x} - \frac{\mathrm{f}_6 + \mathrm{f}_7 \cdot \mathrm{x}}{\chi(\mathrm{x})} + (\mathrm{f}_5 \cdot \mathrm{x} - \mathrm{f}_9) \cdot \mathrm{ArcTan}\left(\frac{\chi'(\mathrm{x})}{\mathrm{k}}\right) + \mathrm{f}_{10} \cdot \mathrm{Ln}(\chi(\mathrm{x})) \quad L_1 \le x \le L \tag{5.2}$$

with the following abbreviations:

$$\mathrm{k} = \sqrt{4 \cdot \mathrm{A}_1 \cdot \mathrm{A}_3 - \mathrm{A}_2^2}$$

$$\mathrm{f}_1 = \frac{3 \cdot \left(2 \cdot \mathrm{A}_2 \cdot \mathrm{A}_3 \cdot \mathrm{L} \cdot \mathrm{M}_1 + \mathrm{A}_1 \cdot \left(\mathrm{A}_2 \cdot L \cdot q + 2 \cdot \mathrm{A}_3 \cdot \left(2 \cdot \mathrm{M}_1 - 2 \cdot \mathrm{M}_2 + L^2 \cdot q\right)\right)\right)}{8 \cdot \mathrm{A}_3 \cdot b \cdot E \cdot k^2 \cdot L}$$

$$\mathrm{f}_2 = \frac{3 \cdot \left(\mathrm{A}_2^2 \cdot L \cdot q + 2 \cdot \mathrm{A}_3 \cdot L \cdot (2 \cdot \mathrm{A}_3 \cdot \mathrm{M}_1 - \mathrm{A}_1 \cdot q) + \mathrm{A}_2 \cdot \mathrm{A}_3 \cdot \left(2 \cdot \mathrm{M}_1 - 2 \cdot \mathrm{M}_2 + L^2 \cdot q\right)\right)}{8 \cdot \mathrm{A}_3 \cdot b \cdot E \cdot k^2 \cdot L}$$

$$\mathrm{f}_3 = \frac{3 \cdot \mathrm{A}_2}{8 \cdot \mathrm{A}_3 \cdot b \cdot E \cdot k^4 \cdot L} \cdot \left(\mathrm{A}_2^2 \cdot L \cdot q + 2 \cdot \mathrm{A}_3 \cdot L \cdot (6 \cdot \mathrm{A}_3 \cdot \mathrm{M}_1 + \mathrm{A}_1 \cdot q) + 3 \cdot \mathrm{A}_2 \cdot \mathrm{A}_3 \cdot \left(2 \cdot \mathrm{M}_1 - 2 \cdot \mathrm{M}_2 + L^2 \cdot q\right)\right)$$

$$\mathrm{f}_4 = \frac{2 \cdot \mathrm{A}_3 \cdot \mathrm{f}_3}{\mathrm{A}_2}$$

$$f_5 = \frac{4 \cdot A_3 \cdot f_3}{A_2 \cdot k}$$

$$f_6 = \frac{2 \cdot A_1 \cdot f_2 - A_2 \cdot f_1}{k^2}$$

$$f_6 = \frac{2 \cdot A_1 \cdot f_2 - A_2 \cdot f_1}{k^2}$$

$$f_7 = \frac{A_2 \cdot f_2 - 2 \cdot A_3 \cdot f_1}{k^2}$$

$$f_8 = -4 \cdot A_3^2 \cdot (f_1 + 2 \cdot A_1 \cdot f_3) - A_2^3 \cdot f_4 + 2 \cdot A_2 \cdot A_3 \cdot (f_2 + A_2 \cdot f_3 + 2 \cdot A_1 \cdot f_4)$$

$$f_9 = \frac{1}{2 \cdot A_3 \cdot k^3} \cdot \left(A_2^3 \cdot (f_5 \cdot k - 2 \cdot f_4) - 8 \cdot A_3^2 \cdot (f_1 + 2 \cdot A_1 \cdot f_3) \right.$$
$$\left. +4 \cdot A_2 \cdot A_3 \cdot (f_2 + A_2 \cdot f_3 + 2 \cdot A_1 \cdot f_4 - A_1 \cdot f_5 \cdot k)\right)$$

$$f_{10} = \frac{2 \cdot f_4 - k \cdot f_5}{4 \cdot A_3}.$$

The constants of integration C_1, C_2, C_3 and C_4 are further obtained from the general kinematic continuity conditions at the crack location as well as from actual kinematic boundary conditions of each considered situation.

6 Numerical Examples

The implementation of presented expressions is shown in two examples differing in geometry, boundary conditions and applied loads. The relative crack depth of 0.5 was considered in both examples for all the cracks just to minimize its influence on the results. Among all known definitions, Okamura's definition was selected for the rotational spring stiffness evaluation.

The quality of results obtained by the newly presented expressions is confirmed by comparing the obtained values for transverse displacements and the reactions at the supports to the results from alternative 3D finite element models where a detailed realistic description of the cracks was established.

6.1 *Example 1—Cantilever Beam with a Concentrated Load at the Free End*

In the first example, an 8 m long multi-cracked cantilever clamped at the left end and loaded with a vertical concentrated load of 10 kN at the right was analysed, as shown in Fig. 2. For the considered cantilever the height *h* followed a parabolic distribution given by a polynomial of second degree.

$$h(x) = 0.5 - 0.075 \cdot x + 0.0046875 \cdot x^2$$

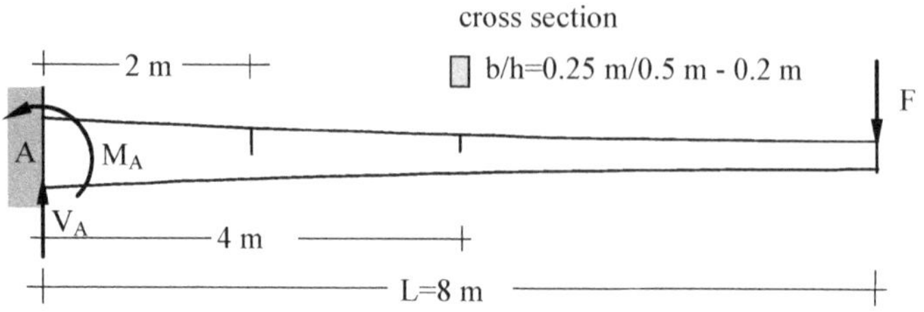

Fig. 2 First example structural setup

The Young modulus of the material was $E = 30$ GPa, with Poisson ratio 0.3, whilst the constant width b of the cantilever was 0.25 m. Two cracks were introduced at the locations of 2 and 4 m from the clamped end, respectively.

Due to the presence of two cracks, two newly derived CFEQVH beam finite-elements were required to model the beam. Their lengths were 3 and 5 m, respectively. The computational model consisted of three nodes and it had altogether six degrees of freedom. However, by taking into account the known zero kinematic conditions at the left end of the element (node 1) the discrete unknowns (vertical displacements and rotations at the node 2 as well as at the free end i.e. node 3) were evaluated from a simple system of four linear equations:

$$\begin{bmatrix} 21.9249 & -21.0779 & -0.8509 & 1.8289 \\ -21.0779 & 44.8185 & -2.4257 & 3.5544 \\ -0.8509 & -2.4257 & 0.8509 & -1.8289 \\ 1.8289 & 3.5544 & -1.8289 & 5.5901 \end{bmatrix} \times 10^6 \cdot \begin{Bmatrix} Y_2 \\ \Phi_2 \\ Y_3 \\ \Phi_3 \end{Bmatrix} = \begin{Bmatrix} 0 \\ 0 \\ -10000 \\ 0 \end{Bmatrix}$$

From these values, nodal vertical reaction force and bending moment of the clamped-end were evaluated over the element's stiffness matrix. Since the structure is statically determinate these values were effortlessly verified by using basic static equilibrium.

To further obtain the distribution of transverse displacements between the nodes the constants of integration C_1, C_2, C_3 and C_4 were evaluated. For the considered case these constants are given as:

$$\begin{aligned}
\mathrm{C}_1 &= \Phi_1 - \frac{\mathrm{f}_1}{\mathrm{A}_1^2} - \frac{\mathrm{f}_3}{\mathrm{A}_1} - \mathrm{f}_5 \cdot \mathrm{ArcTan}\left(\frac{\mathrm{A}_2}{k}\right) \\
\mathrm{C}_2 &= Y_1 + \frac{\mathrm{A}_2 \cdot \mathrm{f}_1 - 2 \cdot \mathrm{A}_1 \cdot \mathrm{f}_2}{\mathrm{A}_1 \cdot \mathrm{A}_2^2 - 4 \cdot \mathrm{A}_1^2 \cdot \mathrm{A}_3} + \frac{\mathrm{f}_8 \cdot \mathrm{ArcTan}\left(\frac{\mathrm{A}_2}{\mathrm{k}}\right)}{\mathrm{A}_3 \cdot k^3} + \frac{\mathrm{A}_2 \cdot \mathrm{f}_5 \cdot \mathrm{ArcTan}\left(\frac{k}{\mathrm{A}_2}\right)}{2 \cdot \mathrm{A}_3} \\
&\quad - \frac{\mathrm{f}_4 \cdot \mathrm{Ln}(\mathrm{A}_1)}{2 \cdot \mathrm{A}_3} + \frac{k \cdot \mathrm{f}_5 \cdot \mathrm{Ln}(4 \cdot \mathrm{A}_1 \cdot \mathrm{A}_3)}{4 \cdot \mathrm{A}_3} \\
\mathrm{C}_3 &= \frac{2 \cdot \mathrm{C}_1 \cdot K_r \cdot L + 2 \cdot L \cdot \mathrm{M}_1 - 2 \cdot L_1 \cdot \mathrm{M}_1 + 2 \cdot L_1 \cdot \mathrm{M}_2 - L^2 \cdot L_1 \cdot \mathrm{q} + L \cdot L_1^2 \cdot \mathrm{q}}{2 \cdot K_r \cdot L} \\
\mathrm{C}_4 &= \mathrm{C}_2 + (\mathrm{C}_1 - \mathrm{C}_3) \cdot L_1
\end{aligned} \tag{6.1}$$

In Eqs. (6.1) Y_1 and Φ_1 represent the element's left node transverse displacement and rotation, respectively.

By implementing Eqs. (5.1), (5.2) and (6.1) the obtained approximate transverse displacement functions were:

$$
\begin{aligned}
v_1(x) &= -7.98918 \times 10^{-2} + 2.84444 \times 10^{-3} \cdot \mathrm{x} + \frac{3.03407 - 0.37926 \cdot \mathrm{x}}{106.6667 - 16 \cdot \mathrm{x} + \mathrm{x}^2} \\
&\quad + 5.80620 \times 10^{-2} \cdot \mathrm{ArcTan}(1.22475 - 0.153093 \cdot \mathrm{x}) \qquad 0 \le x \le 2m \\
v_2(x) &= -7.61468 \times 10^{-2} + 9.71939 \times 10^{-4} \cdot \mathrm{x} + \frac{3.03407 - 0.37926 \cdot \mathrm{x}}{106.6667 - 16 \cdot \mathrm{x} + \mathrm{x}^2} \\
&\quad + 5.80620 \times 10^{-2} \cdot \mathrm{ArcTan}(1.22475 - 0.153093 \cdot \mathrm{x}) \qquad 2m \le x \le 4m \\
v_3(x) &= -6.71686 \times 10^{-2} - 1.27262 \times 10^{-3} \cdot \mathrm{x} + \frac{3.03407 - 0.37926 \cdot \mathrm{x}}{106.6667 - 16 \cdot \mathrm{x} + \mathrm{x}^2} \\
&\quad + 0.058062 \cdot \mathrm{ArcTan}(1.22475 - 0.153093 \cdot \mathrm{x}) \qquad 4m \le x \le 8m
\end{aligned}
$$

The results obtained from the simplified model were further compared to the results obtained from the more detailed computational model of the considered structure. These results were obtained through a 3D solid finite elements computational model by implementing the finite element program COSMOS/M. The transverse displacements and reactions were obtained from a computational model consisting of 48,800 3D solid elements with 74,583 nodes. Vertical and horizontal displacements of discrete nodal points were obtained by solving approximately 223,461 linear equations.

The results for several significant discrete parameters (transverse displacements and reactions) from all the applied computational models were compared and are summarised in Table 1. The table confirms that the presented finite element offers excellent results that are from the engineering point of view basically identical to the solutions from the more detailed (and also more computational effort demanding) 3D finite model. For example, the discrepancy at the free end where maximum displacement occurs is approximately 0.47%. The table also demonstrates that the new FE's values for reactions in the left support are identical to the values from the basic static analysis.

A comparison between transverse displacements along the axis obtained by two implemented models is given in Fig. 3, where no discrepancies are noticeable.

6.2 *Example 2—Propped Cantilever with a Uniform Load*

In the second example, a multi-cracked propped cantilever was analysed, seen in Fig. 4. It was loaded by a vertical uniform load of 10 kN/m over the whole length of 8 m.

Table 1 Comparisons of the results for the cantilever from two different computational approaches

Parameter/model	2 CFEQVH FE	COSMOS 48,000 3D FE
Vertical reaction [N] at the left support	10,000	10,000
Bending moment [Nm] at the left support	80,000	80,000
Displacement [mm] at first crack (x = 2 m)	−2.1378	−2.1256
Displacement [mm] at node (x = 3 m)	−7.2735	−7.3151
Displacement [mm] at second crack (x = 4 m)	−14.4973	−14.5830
Displacement [mm] at the free end (x = 8 m)	−77.3495	−77.7179

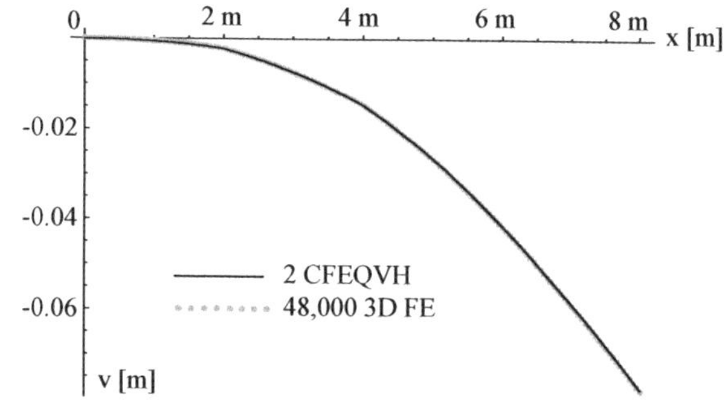

Fig. 3 Comparison of transverse displacements of cantilever from two applied models

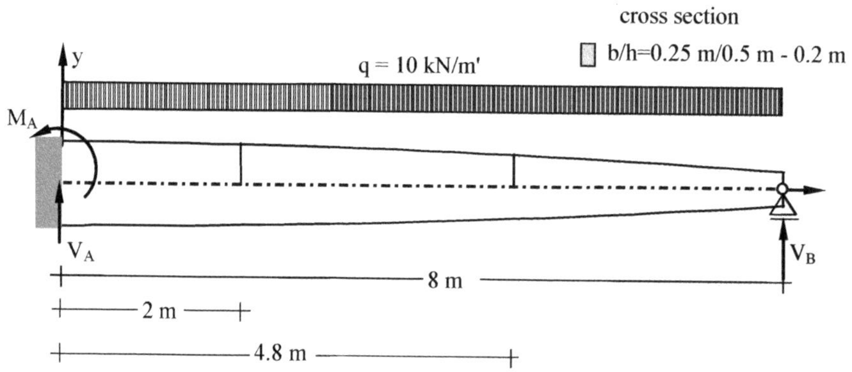

Fig. 4 Second example structural setup

For this structure the height h was varied from 0.5 m at left-end to 0.2 m at right-end following the distribution:

$$h(x) = 0.5 - 1.25 \times 10^{-2} \cdot \mathrm{x} - 3.125 \times 10^{-3} \cdot \mathrm{x}^2$$

The Young modulus of the material was $E = 30$ GPa with a Poisson ratio 0.3, whilst the width b of the cantilever was 0.25 m. Two cracks were introduced at

Table 2 Comparisons of results for the second example using two different computational models

Parameter/model	2 CFEQVH FE	COSMOS/M 4000 3D FE
Vertical reaction [N] at the left support	53,610.4	53,500.7
Bending moment [Nm] at the left support	108,883.3	107,999.8
Vertical reaction [N] at the right support	26,389.6	26,512.0
Displacement [mm] at first crack (x = 2 m)	−2.0588	−2.0847
Displacement [mm] at node (x = 4 m)	−6.5672	−6.5617
Displacement [mm] at second crack (x = 4.8 m)	−7.8787	−7.8405

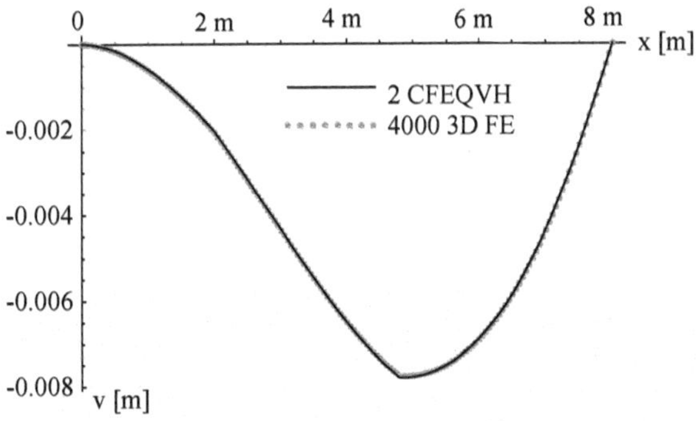

Fig. 5 Comparison of transverse displacements of propped cantilever from both applied models

the locations of 2 and 4.8 m from the clamped end, respectively. The computational model consisted of two CFEQVH beam finite-elements of equal lengths. The discrete unknowns (vertical displacement and rotations at the node as well as at the free end) were evaluated from a simple system of just three linear equations.

This example was additionally analysed by implementing a computational model with 3D solid finite elements in the finite element program COSMOS/M, exclusively to verify the results. The discrete values of some representative quantities are compared in Table 2. Furthermore, Fig. 5 represents the distribution of transverse displacements from both models. Both, Fig. 5 and Table 2 clearly confirm that there is an apparent agreement from both models.

7 Conclusions

The stiffness matrix, as well as the load vector due to a uniform continuous load of a beam finite element with a quadratic function's variation of height, were upgraded to include a transverse crack into account. A straightforward modification of already presented expressions significantly expanded the utilisation of the model as it allows transverse cracks to be included in a very efficient way. However, all the obtained terms are yet again written entirely in closed-symbolic forms. The CFEQVH finite element is fully compatible with standard beam finite elements and it can be easily incorporated into existing software for FEM structural analysis.

Two numerical examples demonstrated that elaborated solutions can be effectively implemented for structural analyses as the presented expressions produced excellent results that were confirmed independently by more thorough 3D models. The derived at expressions thus offer—although rather short and compact—a reliable and efficient computational model.

Acknowledgements The author acknowledges the financial support from the Slovenian Research Agency (research core funding No. P2-0129 (A) "Development, modelling and optimization of structures and processes in civil engineering and traffic").

References

1. Biondi B, Caddemi S (2007) Euler-Bernoulli beams with multiple singularities in the flexural stiffness. Eur J Mech A/Solids 26(5):789–809
2. Okamura H, Liu HW, Chorng-Shin C (1969) A cracked column under compression. Eng Fract Mech 1(3):547–564
3. Skrinar M (2013) Computational analysis of multi-stepped beams and beams with linearly-varying heights implementing closed-form finite element formulation for multi-cracked beam elements. Int J Solids Struct 50(14–15):2527–2541
4. Skrinar M, Imamović D (2019) Beams with quadratic function's variation of height. In: Öchsner A, Altenbach H (eds) Engineering design applications. Springer International Publishing
5. Smith IM, Griffiths DV (1997) Programming the finite element method. John Wiley & Sons, Chichester

Effect of Vapour-Grown Carbon Nanofibres on Thermo-Mechanical Properties of High-Functionality Based Resin Used in CFRP Strengthening System Subjected Severe Service Conditions

R. Al-Safy, R. Al-Mahaidi, G. P. Simon and J. Habsuda

Abstract Carbon-based nanomaterials are commonly used to produce nanocomposites with polymeric matrices due to their unique properties. Vapour-grown carbon nanofibres (VGCF) are employed in this work to modify highly cross-linked epoxy-based resin (DGOA) used as a bonding agent in CFRP/concrete system. The influence of such modification on adhesive properties as well as on bond performance between CF-fabrics and concrete at various services conditions is addressed. Transmission electronic microscopy (TEM) technique was employed to characterize the structure of VGCF. The latter was identified as stacked-cup and bamboo-like fibres with various thickness of grapheme walls as well as of the hollow cores. The inclusion of (1 and 2 wt%) of VGCF into DGOA was found to reduces the glass transition temperature (T_g) measured by Differential Scanning Calorimetry (DSC) technique. Results of tests at elevated temperatures showed improvement in bond strength in tension of about 7 °C for samples with 2 wt% of VGCF-modified DGOA when compared with unmodified DGOA. The exposure of loaded CFRP/concrete specimens to different temperature and humidity cycles for different exposure periods (21 and 44 days) results in achieving sufficient bond performance using VGCF-modified DGOA. This bond performance was conducted by means of single-lap shear test. The strain along the bonded length of survived CFRP/concrete specimens was measured as well using 3D image correlation photogrammetry.

Keywords VGCF · DGOA · CFRP · Concrete · Durability

R. Al-Safy (✉)
Materials Engineering Department, Mustansiriyah University, Baghdad, Iraq
e-mail: edu.ralsafy@uomustansiriyah.edu.iq

R. Al-Mahaidi
Swinburne University of Technology, Melbourne, VIC 3122, Australia

G. P. Simon · J. Habsuda
Department of Materials Engineering, Monash University, Melbourne, VIC 3800, Australia

A. Öchsner and H. Altenbach (eds.), *Engineering Design Applications II*,
Advanced Structured Materials 113, https://doi.org/10.1007/978-3-030-20801-1_9

1 Introduction

Carbon-based nanomaterials are commonly used to produce nanocomposites with polymeric matrices [1–4] due to their unique properties [5, 6]. Vapour-grown carbon nanofibre (VGCF) is a class of these carbon nanofibres which have been received great attention by researchers in materials and aerospace applications [7, 8] due to their special structure, high aspect ratio (high length/diameter ratio) and their low cost in comparison with other carbon-based nanomaterials [9–11]. Therefore, VGCF is considered as a promising candidate to work as nanofiller in various polymeric composites as reported in literatures.

In one of the reported investigations [12], two types of carbon-based nanomaterials (fluorinated single wall carbon nanotubes (FSWCNT) and vapor grown carbon fibres (VGCF)) were used to reinforced epoxy matrix (DGEBF) in order to the study the effect of such reinforcement on the thermo-physical properties of epoxy nanocomposites. In terms of the VGCF/epoxy nanocomposites, various concentration levels (up to 4 wt%) of a commercially available VGCF (PR-19-PS) were dispersed in the epoxy matrix using sonication technique with the assistance of the acetone to functionalize the dispersion. An increment of 0.48 GPa was obtained for the storage modulus at room temperature (measured by DMTA) for samples with 1.5 wt% of VGCF. The glass transition temperature (T_g) was found to be reduced slightly as the percentage of VGCF was increased. This reduction was attributed to the reduction of the cross-linked density which might be related to the chemical reaction between hydroxyl group of VGCF surface and the curing agent (hardener).

The effect of inclusion various percentages of VGCF (1, 2 and 3 wt%) on mechanical and thermal properties of SC-15 epoxy matrix was explored by many group of researchers [13]. The diameter of the VGCF was 200 nm while its length was in the range (20–100 μm). The VGCF was added to the epoxy and mixed using ultrasonic cavitation method. The latter was found to be suitable for inclusion VGCF into epoxy when VGCF $\leq$ 2 wt%. Also, it was reported that the storage modulus of the epoxy, measured by DMTA, increases as the addition of VGCF increases. An improvement of 65% in the storage modulus at room temperature was achieved when VGCF was 3 wt% at 30 °C. Additionally, the introduction of VGCF into epoxy matrix was found to increase the T_g. As the VGCF percentages increases, the T_g was increases as well. This is due to the polymer molecule segments that close to the filler surface being less mobile. Highest enhancement in the tensile strength, fracture toughness and fatigue performance of nanocomposites was obtained when 2 wt% of VGCF were added to the epoxy.

Another investigation [14] was conducted to address the influence of different addition of VGCF (5,10 and 20 wt%) to epoxy which was commercially available as Epikote 825. The latter was used: either after diluted with acetone to produce low viscosity resin; or used without acetone to produce (high viscosity resin). The properties of VGCF was 150 nm in diameter and the fibre length was (10–20 μm). The T_g of the nanocomposites was found to increase significantly with the increase in the concentrations of VGCF for low viscosity epoxy and the highest increment

(26 °C) in T_g was conducted for 20 wt% of VGCF. This increment may attributes to the restriction of VGCF to the molecular mobility. However, the T_g for high viscosity epoxy resins was reduced when 20 wt% of VGCF was added. This might be related to less interaction between the fillers and epoxy due to availability of voids as well as the poor dispersion of VGCF in the epoxy matrix. In order to study mechanical properties of the nanocomposites, different percentages of VGCF (3, 5, 7 and 10 wt%) were added to two types of epoxy (low and high viscosity). It was found that the mechanical properties (tensile strength and Young's modulus) were enhanced when VGCF was added up to 5 wt% and this is related to the good dispersion of VGCF with less voids in the nanocomposites. A reduction in the mechanical properties was reported when VGCF was added with concentration level higher than 5 wt%. This reduction was due to cracks which showed only in high viscosity epoxy nanocomposite for samples with 10 and 20 wt% of VGCF content at 10 and 20 wt%.

The work in the present paper reports an experimental investigation that carried out to modify highly cross-linked epoxy resin with VGCF to be used in civil engineering applications; particularly in strengthening concrete with CFRP. The VGCF modified adhesive will acts as a saturant to CF-fabrics and as bonding agent between CF-fabrics and concrete. The influence of such modification on adhesive properties as well as on bond performance between CF-fabrics and concrete at various services conditions is addressed. Additionally, the structure of VGCF and thermal properties of VGCF-reinforced adhesive were explored as well using various characterization techniques (TEM and DSC).

2 Experimental Work

2.1 *Materials*

A two-part epoxy system of room temperature curing is used as bonding agent in CFRP/concrete system. High functionality based-resin N,N-diglycidyl-4-glycidyloxyaniline (DGOA) is utilized as Part A. This low viscosity resin was purchased from Sigma-Aldrich, Australia. The hardener was Part B of MBrace® Saturant epoxy adhesive which is provided by BASF, Australia. Based on our previous work [15], the monomer and the hardener were mixed using mixing ratio of (100:50) by weight.

A type of carbon-based nanofiber known as VGCF was used in this work to reinforce DGOA. The VGCF was commercially-available carbon nanofibre, Grade PR-24 XT-LHT, which was added with two concentrations (1 and 2 wt%) to DGOA. The PR-24 XT-LHT is provided by Pyrograf®-III, Applied Sciences, Inc., USA. The carbon nanofiner is a black, fluffy agglomerate with an average diameter of 100 nm and surface area of 43 m^2/g as provided by the manufacturer [16].

The CFRP/concrete system that used in this investigation is composed of two main elements: concrete specimens and CF-fabrics. Concrete of normal strength was used to fabricate concrete specimens with dimensions of 75 × 75 × 250 mm. These samples were strengthened externally with CF-fabrics (S & P C-Sheets 240) which is supplied by BASF, Chemical Construction Pty. Ltd., Melbourne, Australia with fibre modulus of 240 GPa and tensile strength of 3800 MPa.

2.2 *Preparation of VGCF-Reinforced Epoxy*

The dispersion of VGCF into highly cross-linked resin (DGOA) was carried out using Speed Mixer^{TM} DAC150 FV to produce a mixture of (DGOA and VGCF). Details of mixing procedure are reported in our previous work [17]. This was followed by the addition of the hardener (Part B) to the mixture of (DGOA and VGCF) and mixed manually until homogenous mixture was achieved.

2.3 *Fabrication of CFRP Strengthening System*

Concrete specimens were prepared for bonding with CF-fabric using sand-blasting technique to prepare the surface of the concrete prior to the application of CFRP. A brush was then used to remove the dust from the surface. In order to enhance the properties of adhesion of concrete, a thin layer of primer of room temperature curing [15] was applied, using a brush, on the area of concrete surface (75 × 150 mm) which will be bonded later with CF-fabrics. The technique that followed for bonding CF-fabrics to concrete was wet-lay up technique [17]. The two-part epoxy system that reinforced with VGCF was used as saturant to CF-fabrics and as bonding agent between CF-fabrics and concrete substrates.

3 Characterization of VGCF

Transmission Electronic Microscopy (TEM) was utilized to characterize the structure of carbon nanofibres (PR-24 XT-LHT). Low concentrations of them were diluted with water followed by sonication for 1 min using an Uultrasonicator prob (model Vibracell 750 W and 20 kHz). The sonication was applied in order to achieve well dispersion of VGCF without causing any damage to the fibres. After drying at ambient conditions, the sonicated mixture (PR-24 XT-LHT+ diluted water) was place on TEM grids. Various magnifications up to 100,000 (100 K) were utilized to capture bright field TEM images which were later analyzed manually.

4 Thermal Properties of VGCF-Reinforced Adhesives

The change in the heat capacity (ΔC_p) and the glass transition temperature (T_g) of adhesive that reinforced with VGCF were conducted by Differential Scanning Calorimetry (DSC) (Perkin–Elmer Pyris1). The testing program involves the exposure of small solid sample of the VGCF-reinforced adhesive to three thermal cycles (heating-cooling-heating). The temperature for the heating was from 20 to 120 °C with 10 °C/min heating rate. At the end of each cycle, the temperature was held for 1 min prior to the start of the other cycle.

5 Tests on CFRP/Strengthening System

5.1 Bond Strength on Tension at Elevated Temperatures

The bond performance of CFRP/concrete samples was evaluated by means of adhesion (pull-off) test under elevated temperatures. Cubic CFRP/concrete specimens with dimensions of 75 × 75 × 75 mm were used for this test. These samples were obtained from cutting CFRP/concrete samples that prepared previously (refer to Sect. 2.3) to the desired size in order to be placed in a steel frame. In order to isolate the tested area from the rest of the material, each cubic sample was cored on CFRP/concrete surface. A disc was attached to the isolated area using high-temperature resistance adhesive. As illustrated in Fig. 1 the cubic sample is placed in a steel frame which was place inside Instron Chamber. The latter was connected to Instron loading machine which permits the load application with a displacement of 0.5 mm/min. The procedure of test involves the application of 20% of the ultimate load to pull-off the disc that attached to the cubic sample. This was associated with the gradual increase in the temperature until the pulling-off of the attached disc. In order to record the temperature at which the bond was lost, a dummy CFRP/concrete sample that contains a thermocouple (Type K) in the adhesive layer was placed inside the chamber.

5.2 Accelerated Durability Tests

A total of eight CFRP/concrete specimens were exposed to two different accelerated durability tests (A1 and A2) which based on cycles of temperature and humidity. The first test (A1), which was conducted for 504 h, involves the application of constant temperature of 50 °C and ramped humidity from 30 to 90%. The second durability test (A2), which was performed for 1060 h, includes the application of constant humidity at 90% and cyclic temperature (20–50 °C). The exposure regime was carried out inside an environmental chamber using steel frames to apply constant

Fig. 1 Arrangements of adhesion (pull-off) tests for CFRP/concrete samples at elevated temperatures

load (40% of the ultimate) during the whole period of exposure. The application of load was maintained by provide the steel frame with springs with a displacement of 4 mm for each 1 kN of load application. At the end of the exposure period, single-lap shear test at ambient conditions was performed for the survived samples in order to measure the residual strength using Instron loading machine. This test was associated with the usage of 3D image correlation photogrammetry (ARAMIS) to record the strain along the bonded length (150 mm).

6 Results and Discussion

6.1 TEM Micrographs for Vapour-Grown Carbon Nanofibres (VGCFs)

TEM images for PR-24 XT-LHT are shown in Figs. 2, 3 and 4 using different levels of magnifications. The carbon nanofibres are observed to have different lengths and shapes, as illustrated in Fig. 2 to form a stacked-cup carbon nanofibre structure. In such structure, fibres are formed fro open or closed individual nanocones stacked together [18, 19]. Some fibres are relatively straight tubes with different thicknesses for both the walls and hollow cores, while other fibres rolled around their lengths. Table (1) shows these variations in the thickness of the walls and hollow cores of VGCF that used in this work. The listed measurements in Table (1) were conducted using ImageJ software.

A bamboo-like structure was recognized from TEM images in Figs. 3 and 4 for PR-24 XT-LHT. The bamboo segments are formed from the straight tubes of the fibres are connected via sections with structural defects. Within the same segment, an inter-layer (arrow I) and outer layer (arrow O) was observed (Fig. 4 (left)). Along

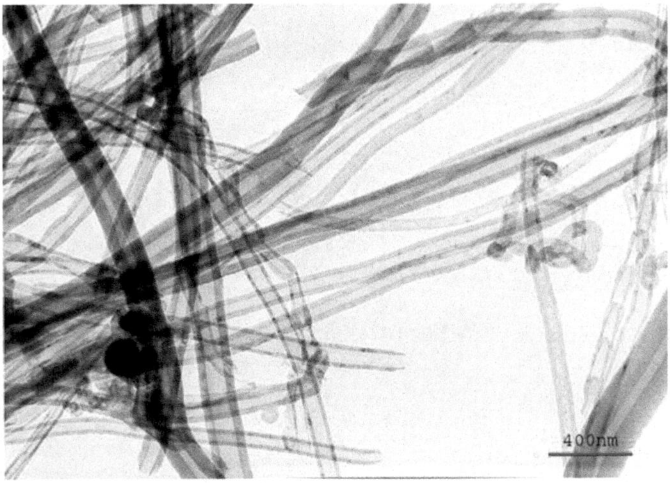

Fig. 2 TEM micrograph for VGCF-PR24-LHT at low magnification level

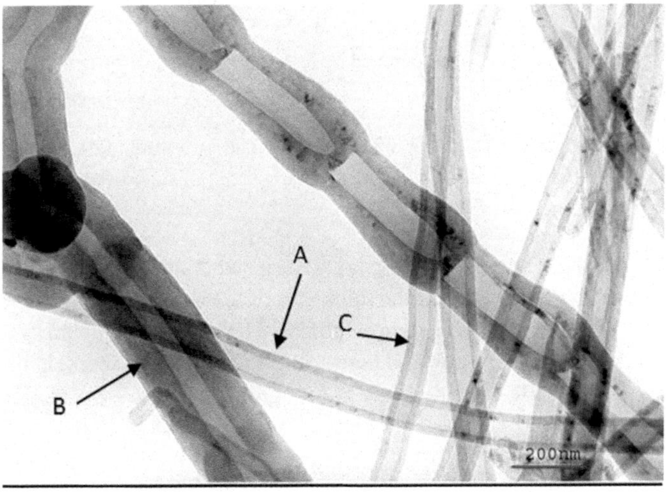

Fig. 3 TEM micrograph for VGCF-PR24-LHT at medium magnification level

Table 1 Measured thickness of carbon nanofibres (PR-24 XT-LHT) using ImageJ software

Level of magnification	Symbol of fibres	Measured thickness of hollow core (nm)	Measured thickness of walls (nm)
Medium	A	15	19
	B	55	80
	C	37	20
High	D	47	27
	E	57	35

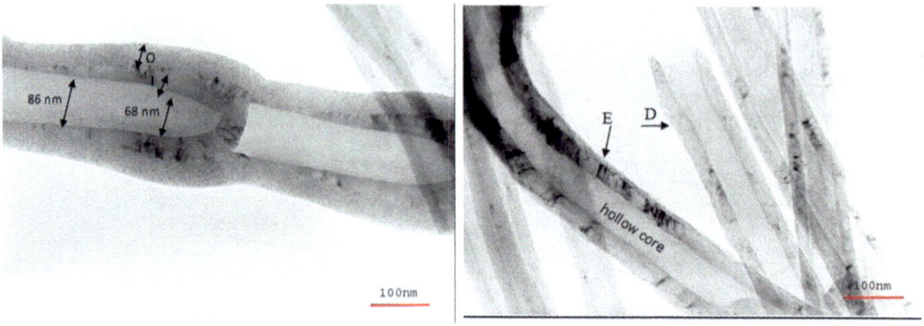

Fig. 4 TEM micrograph for VGCF-PR24-LHT at high magnification level

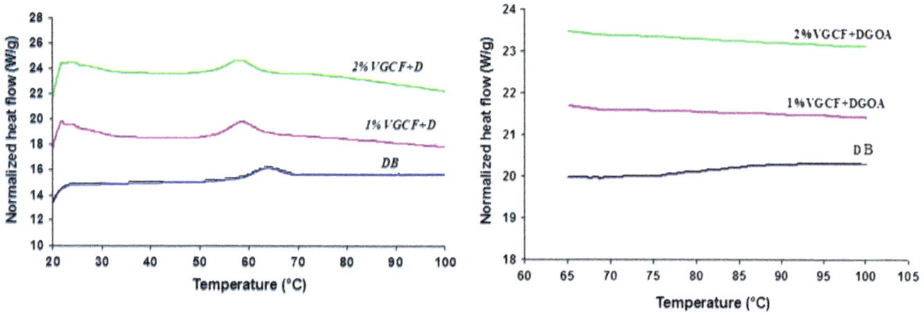

Fig. 5 DSC thermograms for VGCF-modified DGOA-based resin: from first heating scan (*left*), and from second heating scan (*right*)

the segment, the thickness of the inter-layer varies and almost disappeared towards the bottom of segment. In terms of the thickness of the hollow core of the segment, it was noted that it varies along the same segment. The measured thickness was 68 nm and 86 nm in the near the end cup and in the middle of the same segment respectively as illustrated in Fig. 4 (left) at high-level of magnification.

6.2 DSC Measurements

Figure 5 shows the DSC traces for VGCF-reinforced adhesive. These traces are related to the fist heating scan (Fig. 5 left) and the second heating scan (Fig. 5 right) for adhesive that reinforced with 1 and 2 wt% of VGCF. For comparison purposes, DSC traces for control sample (without VGCF) was obtained from our previous reported work [20] and presented by the curve DB in Fig. 5.

A single peak was observed in DSC traces in Fig. 5 (left) due to aging of the material with time under ambient condition. However, a transition step in DSC traces (Fig. 5 (right)) was noted for all VGCF-reinforced DGOA. The shape of this step

Table 2 DSC measurements of DGOA-based resin modified with VGCF

VGCF (wt%)	Onset T_g (°C)	T_g: half C_p extrapolated (°C)	Comments
0	58	60	1st heating scan
	75	82	2nd heating scan
1	52	53	1st heating scan
	52	53	2nd heating scan
2	51	52	1st heating scan
	65	71	2nd heating scan

was observed to be similar within the adhesive that reinforced with VGCF but it was stepped for control sample.

In terms of the DSC measurements, they were conducted from the two heating scan and were listed in Table (2). The measurements of the control sample were obtained from our previous work [20]. It was found that, the addition of VGCF results into no significant change in ΔC_p. This is due to less mobility of molecules of the highly cross-linked resin that used to be reinforced with VGCF. Also, the reinforcement of the high-functionality based resin with VGCF was found to reduce the T_g of the adhesive in comparison with the control sample (without VGCF).

6.3 Bond Performance of CFRP/Concrete System at Elevated Temperatures

Figure 6 illustrates the obtained results from pull-off test that performed at elevated temperatures. The reinforcement of 1 wt% of VGCF was observed to cause slight improvements in the bond-loss temperature. However, an enhancement of 7 °C was recoded for specimens with 2 wt% of VGCF in comparison with control sample. The visual characterization of failure pattern showed pure failure in the primer/concrete interface for all CFRP/concrete samples when VGCF was used to reinforce DGOA (Fig. 7). This may related to the adhesion strength between the primer and concrete which was assumed to be lower than the cohesive strength of such samples.

6.4 Bond Performance of CFRP/Concrete System at Harsh Environmental Conditions

6.4.1 Exposure to A1 Accelerated Durability Test

None of the loaded concrete samples that strengthened with CF-fabric using VGCF-reinforced DGOA failed in the environmental chamber for durability test (A1) that

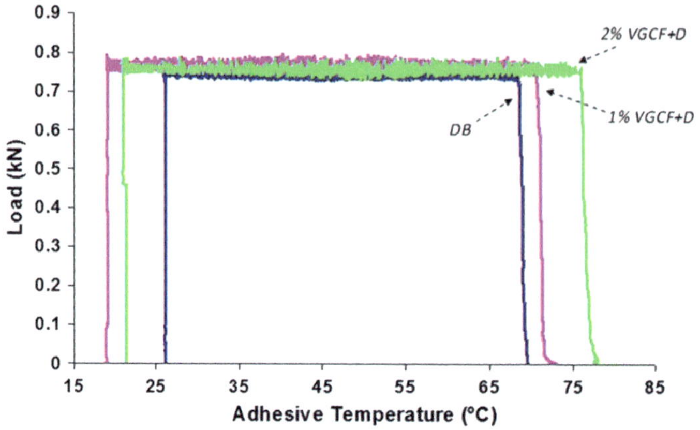

Fig. 6 Adhesion (pull-off) test results at elevated temperatures of CFRP/concrete system using VGCF-modified high functionality resin (DGOA)

Fig. 7 Failure pattern of adhesion pull-off tests of CFRP/concrete samples at elevated temperatures using high functionality-based resin (DGOA) modified with: 1 wt% VGCF (*left*), and 2 wt% VGCF (*right*)

based on constant temperature of 50 °C and ramped humidity from 30 to 90% for 21 days. The residual strengths of these samples are listed in Table (3). The value of the control sample was obtained from our previous work [20] for comparison purposes. As a comparison with the reported results of the unconditioned-unloaded specimens of our previous work [18] there was no significant reduction in the residual strength of all samples regardless of some minor experimental variations. Same failure pattern (concrete rupture) (Fig. 8) was observed for all tested single-lap shear samples at ambient conditions.

In terms of strain measurements along the bonded length, Fig. 9 illustrates the strain distribution. A maximum strain was recoded to be within the range of 6000–8500 microstrain using the photogrammtery, depending on the failure load. For load levels up to 10 kN, the effective bonded length for tested samples was found to be up to 45 mm from the loaded edge. In terms of recoded residual strain readings,

Table 3 Single-lap shear test results for CFRP/concrete specimens subjected to A1 accelerated durability test

Sample code	Sustained load (%)	Samples survived inside the chamber	No. of tested samples	Failure load (kN)	Average failure load (kN)
D:B[a]	40	3	3	18.42,23,18.77	20.06
1% VGCF+D[b]	40	2	2	19.99,18.94	19.47
2% VGCF+D	40	2	2	18.92,18.31	18.62

[a]High functionality based resin (DGOA): Part B
[b]DGOA: Part B

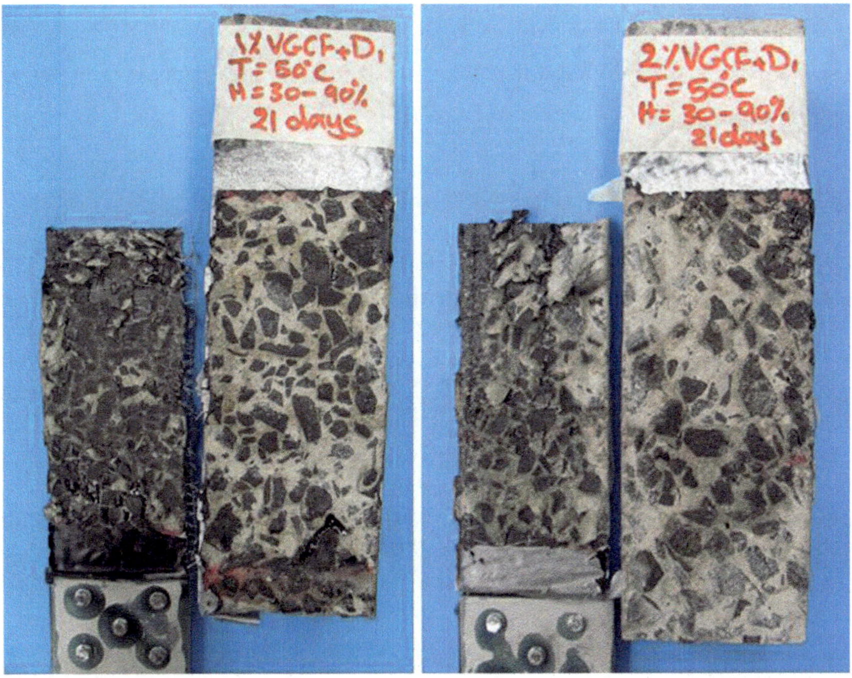

Fig. 8 Failure mode for tested CFRP/concrete samples after the end of A1 accelerated durability test for 21 days exposure period using VGCF-modified DGOA-based resin: 1 wt% VGCF (*left*), and 2 wt% VGCF (*right*)

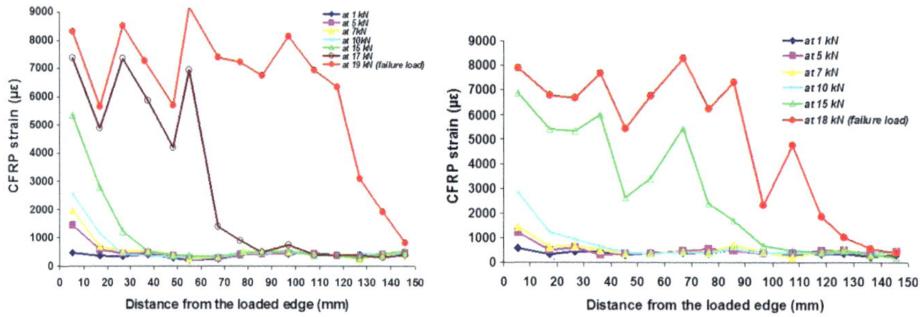

Fig. 9 Strain measurements along the bonded length of tested single-lap shear CFRP/concrete specimens at ambient temperature after A1 accelerated durability test: 1 wt% VGCF-reinforced DGOA (*left*), and 2 wt% VGCF- reinforced DGOA (*right*)

it was observed to be slightly lower than those of the ultimate values for all tested samples.

6.4.2 Exposure to A2 Accelerated Durability Test

All four loaded concrete specimens that strengthened with CF-fabrics using VGCF-reinforced adhesive were survived inside the environmental chamber when expose to durability test (A2) for 44 days. Table (4) lists the residual strength which was conducted from testing the survived samples using single-alp shear test at ambient conditions. A slight increase in the failure load was recorded for samples with VGCF-reinforced adhesive in comparison with failure loads of control samples [20]. Figure 10 illustrates the failure mode for all tested specimens and it was concrete rupture. The readings of photogrammtery for the strain along the bonded length showed lower values than the ultimate values for 40% load level as illustrates in Fig. 11 for concrete samples that strengthened with CF-fabrics using VGCF-reinforced DGOA. This is due to the effect of harsh environmental conditions on the interfacial bond for these specimens.

Table 4 Single-lap shear test results for CFRP/concrete specimens subjected to A2 accelerated durability test

Sample code	Sustained load (%)	No. of exposed samples	No. of tested samples	Failure load (kN)	Average failure load (kN)
D:B[a]	40	4	4	16.13,15.98, 18.6,17.06	16.94
1% VGCF+D[b]	40	2	2	19.79,20.75	20.27
2% VGCF+D	40	2	2	20.65,17.9	19.28

[a]High functionality based resin (DGOA): Part B
[b]DGOA: Part B

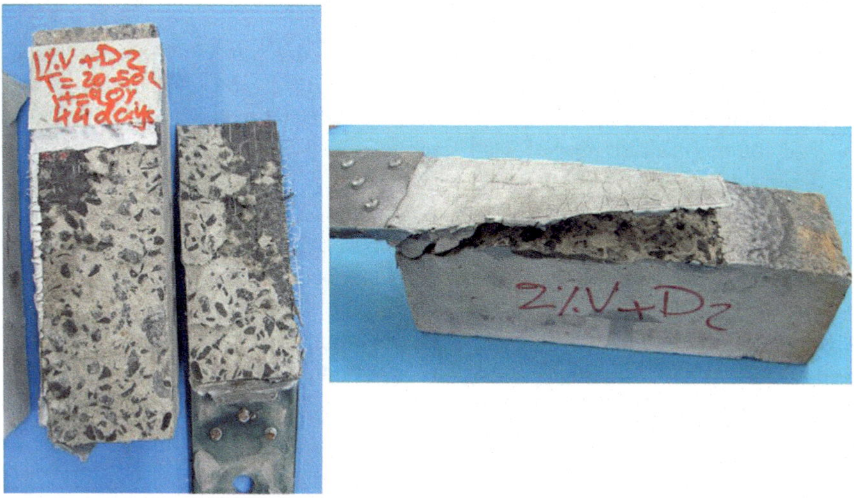

Fig. 10 Failure mode for tested CFRP/concrete samples after 44 days exposure using: **a** 1 wt% VGCF modified DGOA (*left*), and **b** 2 wt% VGCF-modified DGOA (*right*)

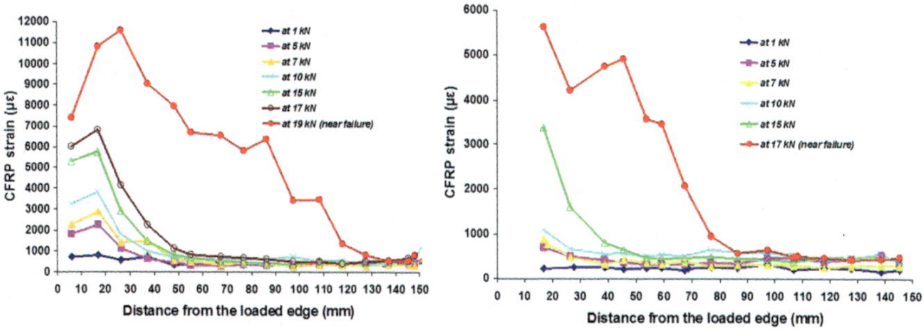

Fig. 11 Strain measurements along the bonded length of single-lap shear CFRP/concrete specimens at ambient temperature after 44 days using: 1 wt% VGCF-modified DGOA (*left*), and 2% VGCF-modified DGOA (*right*)

7 Conclusions

The influence of inclusion VGCF to highly-cross-linked epoxy-based resin on thermo-mechanical properties of the adhesive and on the bond efficiency of CFRP/concrete system under different environmental conditions was investigated and reported in this paper. The following observations can be drawn:

1. Different types and structures for the commercially-available VGCF, PR-24 XT-LHT, including stacked-cup and bamboo-like fibres with different thicknesses for the hollow core and graphene walls was observed from TEM characterization.
2. The addition of VGCF to DGOA was found to cause no significant change in ΔC_p since the molecular of the resin has less mobility due to the highly cross-linked. For the same reason that related to the nature of the resin, the T_g was reduced with the addition VGCF.
3. Enhancement by 7 °C in the bond-loss temperature was found by reinforcing the highly cross-linked based resin (DGOA) with 2% VGCF, compared with unmodified DGOA-based resin from testing the bond strength via pull-off test. The common failure mode for all samples with VGCF-reinforced DGOA was observed in the primer/concrete interface.
4. Appropriate bond behaviour was observed for loaded CFRP/concrete specimens using VGCF-reinforced high-functionality based resin when these samples were exposed to harsh environmental condition that based on cyclic temperature and humidity under sustainable load. The residual strength was found not to be affected negatively for all tested concrete samples that strengthened externally with CF-fabrics using highly-cross linked resin that reinforced with VGCF.

References

1. Al-Safy R, Al-Mahaidi R, Simon GP et al (2014) Thermo-mechanical characterization of VGCF-modified adhesive for bond between CFRP and concrete subjected to combined effect of temperature and humidity. Adv Struct Eng 17:1718–1823
2. Al-Safy R, Al-Mahaidi R, Simon GP (2011) Thermal and mechanical characterizations of nanomaterial-modified adhesive used in bonding CFRP to concrete. J Adhes 87:842–857
3. Kumar S, Rath T, Mahaling RN et al (2007) Study on mechanical, morphological and electrical properties of carbon nanofiber/Polyetherimide composites. Mater Sci Eng, B 141:61–70
4. Xu LR, Bhamidipati V, Zhong WH et al (2004) Mechanical property characterization of a polymeric nanocomposite reinforced by graphitic nanofibers with reactive linkers. Compos Mater 38:1563–1582
5. Hammel E, Tang X, Trampert M et al (2004) Carbon nanofibers for composite applications. Carbon 42:1153–1158
6. Zeng QH, Yu AB, Lu GO et al (2005) Clay-based polymer nanocomposites: research and commercial development. J. Nanosci and Nanotechnol 5:1574–1592
7. Ting JM, Lake ML (1995) Vapor-grown carbon-fiber reinforced carbon composites. Carbon 33:663–667
8. Patton RD, Pittman CUJ, Wang L et al (2001) Vapor grown carbon fiber/phenolic matrix composites for rocket nozzles and heat shield. In: Proceedings of the 4th conference on aerospace materials, processes, and environmental technology, NASA/CP-2001-210427

9. Sherman LM (2007) Carbon nanotubes lots of potential—if the price is right. Plast Technol 83:68–73
10. Breuer O, Sundararaj U (2004) Big returns from small fibers: a review of polymer/carbon nanotube composites. Polymer Compos 25:630–645
11. Tibbetts GG, Lake ML, Strong KL et al (2007) A review of the fabrication and properties of vapor-grown carbon nanofiber/polymer composites. Compos Sci and Technol 67:1709–1718
12. Miyagawa H, Rich MJ, Drzal LT (2006) Thermophysical properties of epoxy nanocomposites reinforced by carbon nanotubes and vapor grown carbon fibers. Thermochim Acta 442:67–73
13. Zhou Y, Pervin F, Jeelani Sh (2007) Effect vapour grown carbon nanofiber on thermal and mechanical properties of epoxy. J Mater Sci 42:7544–7553
14. Choi Y-K, Sugimoto K-I, Song S-M et al (2005) Mechanical and physical properties of epoxy composites reinforced by vapor grown carbon nanofibers. Carbon 43:2199–2208
15. Al-Safy R, Al-Mahaidi R, Simon GP (2010) CFRP strengthening of concrete structures using modified adhesives. In: The 5th civil engineering conference in the Asian Region and Australasian structural engineering conference (CECAR 5 and ASEC 2010), Sydney, New South Wales, Australia, 09–12 Aug 2010, (USB copy)
16. Material properties for PR-24 XT-LHT obtained from (Pyrograf®-III, Applied Sciences, Inc., USA) web site: (http://pyrografproducts.com/Merchant5/merchant.mvc?Screen=cp_nanofiber)
17. Al-Safy R, Al-Mahaidi R, Simon GP (2010) Thermal and mechanical characterization of nanomaterials modified bonding adhesive used in CFRP/concrete system. In: The 4th international conference on advanced computational engineering and experimenting (ACE-X 2010), Paris, France, (Abstract Book-CD copy)
18. Terrones H, Hayashi T, Muoz-Navia M et al (2001) Graphitic cones in palladium catalysed carbon nanofibres. Chem Phys Lett 343:241–250
19. Endo M, Kim YA, Hayashi T et al (2002) Structural characterization of cup-stacked-type nanofibers with an entirely hollow core. Appl Phys Lett 80:1267–1269
20. Al-Safy R, Al-Mahaidi R, Simon GP (2013) A study of using high functionality-based resin for bond between CFRP and concrete under harsh environmental conditions. J Compos Struct 95:295–306

Study on Closed-Die Forging in Stainless Steel as Standard ASTM F138 in Grain Size Function

Givanildo Alves dos Santos, Wagner Figueiredo, Ylich Peter Schmitt, Maurício Silva Nascimento, Fabio Miranda, Gilmar Ferreira Batalha and Antonio Augusto Couto

Abstract Forging is a bulk deformation process in metal working commonly employed in the manufacture of metallic materials prostheses. Depending on the compressive loads applied to the material, structures (grain sizes) are formed which increase its strength. By forging an ASTM F138 stainless steel, the heating of material in a suitable temperature is of fundamental importance to be formed without the presence of folds arising from the process and as a result we can obtain a proper microstructure. This work aims to define the accurate temperature in relation to the strain rate applied in the hot die forging process. Besides, it is used a closed-die forging in which cavity has different thickness that stimulates the profile of an orthopedic plate. Four ranges of temperature used in the process vary between 1000 and 1200 °C, with a strain rate ranging from 400 to 600 s^{-1}. Results demonstrate that there is a transitional temperature close to 1050 °C, where temperatures of 1000–1040 °C provide

G. A. dos Santos · W. Figueiredo · Y. P. Schmitt · M. S. Nascimento
Department of Mechanics, Federal Institute of São Paulo, IFSP, São Paulo, SP, Brazil
e-mail: gvalvs@gmail.com

W. Figueiredo
e-mail: wgnrfig@gmail.com

Y. P. Schmitt
e-mail: ylichps@gmail.com

M. S. Nascimento
e-mail: mauriciofisico@gmail.com

F. Miranda
Department of Mechanics, Paulista University, UNIP, São Paulo, SP, Brazil
e-mail: fabio.miranda.1976@gmail.com

G. F. Batalha
Department of Mechatronics and Mechanical Engineering, University of São Paulo, USP, São Paulo, SP, Brazil
e-mail: gilmar.epusp@gmail.com

A. A. Couto (✉)
Department of Materials Science and Engineering, Department of Materials Engineering, Nuclear and Energy Research Institute, IPEN, Mackenzie Presbyterian University, UPM, São Paulo, SP, Brazil
e-mail: acouto@ipen.br

A. Öchsner and H. Altenbach (eds.), *Engineering Design Applications II*,
Advanced Structured Materials 113, https://doi.org/10.1007/978-3-030-20801-1_10

satisfactory grain size and temperatures of 1100 and 1200 °C, showed limitations in grain size, greater than those recommended by ASTM F 621.

Keywords ASTM F138 · Closed-die forging · Strain rate · Grain size

1 Introduction

Austenitic stainless steels are used on a large scale in products of technical applications which require specific properties, which can be adapted according to design needs depending on conformation and specific heat treatments [1, 2]. These metals, because they are constituted with high percentages of Cr and Ni, present greater resistance to the conformation, necessitating forces that are up to 50% larger than those necessary for the conformation of medium carbon steel (C1020, C1025, C1030).

Studies related to manufacturing technology show the influence of manufacturing processes and their respective variables on the microstructure and properties of engineering materials [3–7]. Austenitic stainless steels have excellent cold workability [2] with the deformation applied, increase the hardness, improve the mechanical strength and, as a disadvantage, reduce the corrosion resistance. In the hot forming (also called hot working) of authentic stainless steels, it is possible to obtain complex geometries, in which the mechanisms of recovery, recrystallization and grain growth coexist and are dependent on the temperature, time, deformation and rate of applied deformation, making it possible to change the microstructure and the properties of the material [2, 8–10].

Another aspect of austenitic stainless steels is the medium to low stacking fault energy [11]. Depending on the desired mechanical properties, the thermal treatment of solubilization after forming can be recommended. In the case of hot forming, the temperature, time and conformation rate controls are necessary to achieve a microstructure that meets the specific technical standards of each product.

In general, metallic materials can be considered the most important engineering materials; they are used as biomaterials because of their excellent thermal conductivity and mechanical properties. The main property required of a metal as biomaterial is that it does not illicit an adverse reaction when placed in services, which means to be a biocompatible material [12].

Austenitic stainless steel is commonly used as an engineering material for the manufacture of medical equipment and devices, as well as prostheses and surgical tools. Due to the importance of the use of the material in these contexts, the present work aims to present the result of a study that relates the variables that influence the bulk deformation process, and to analyze the metallographic characteristics based on the current standard specification.

In Fe–Cr–Ni alloys, C and N have a solid phase influence on the stabilization of austenite. However increased levels of C contribute to saturate the austenite and cause the removal of chromium, favoring the formation of carbides [13].

Austenite is a solid solution of carbon or another solute in the gamma-iron crystallizing in the face-centered cubic (FCC) crystal structure. These steels are not magnetic, however, in some cold hardened alloys light magnetism appears. Due to the retention of a metastable form of austenite at room temperature, the steels of this group are referred to as austenitic. The FCC structure at room temperature imparts a faceted granular microstructure in these steels. Note that the ferrite is α iron, stable at room temperature and has a body-centered cubic (BCC) crystal structure [14].

Many properties of the materials are dependent on their microstructure. A typical example is the influence of grain size on the yield strength of the polycrystalline solids. In general, properties that are dependent on the microstructure are determined by the amount, size, shape and distribution of the phases and the crystalline defects. The mechanical properties, hardness, ductility, toughness, corrosion resistance, wear resistance, among others, are dependent on the microstructure [15].

Austenitic stainless steel presents excellent weldability and higher corrosion resistance in relation to the steels of the other groups of stainless steels. It combines the yield strength with the high tensile strength and good elongation, offering the best properties for cold work.

These steels have good mechanical properties combined with excellent corrosion resistance. They have high resistance to attack by corrosive agents and have good toughness at low temperatures. The thermal conductivity corresponds to a quarter of the carbon steel. In the machining process, the heat concentrates in the cutting edge is not transferred to the material or to the chips [16].

Such materials cannot be hardened by heat treatment, however, they can be hardened by cold forming, which worsens machinability and reduces their potential for corrosion resistance. The hot behavior of austenitic stainless steels is similar, with no significant changes in the heating and bulk deformation process, therefore, some factors must be observed [12].

- The volume to be processed;
- Variations of cross-sections in product format;
- The applied deformation;
- The initial grain size;
- The initial and final working temperature;
- The speed of conformation; and
- The form of heating.

Stainless steels are more difficult to forge than carbon steels and low alloy steels due to their higher mechanical strength at high temperatures and the possible limitation of the maximum temperature applied in order to avoid microstructural damages.

They can be forged in a wide temperature range of 980–1250 °C. Care must be taken with austenitic steels which in their composition are susceptible to δ iron formation, which may occur above 1100 °C. At temperatures below recrystallization, grains can be stretched perpendicular to the applied force.

Another equally important constraint on the conformation of the austenitic stainless steels applies to the final shape temperature. The larger the cross-sectional shape, the lower the cooling rate of the material, the higher the microstructure exposure at

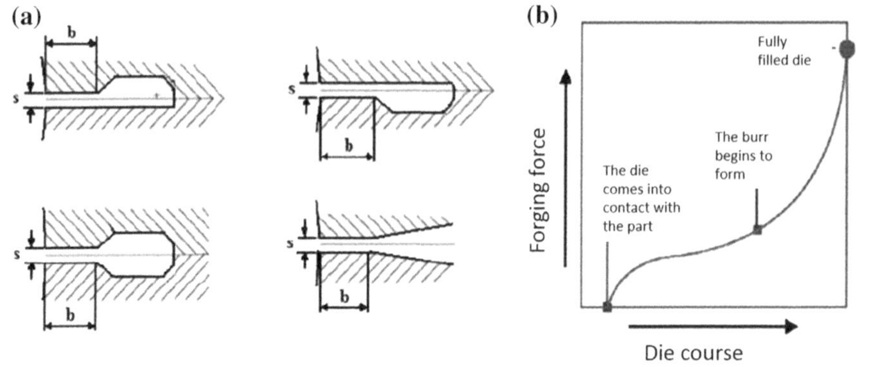

Fig. 1 **a** Types of flash rails and **b** Typical force of die forging. Adapted from Schaeffer [17]

high temperatures and grain growth may occur. Short times of exposure to heat are recommended, to inhibit the formation of oxides, avoiding intergranular oxidation on the surface. It is recommended to provide oversize to ensure complete removal of those imperfections that can be generated in the forming process.

Therefore, in the closed-die forging, the material flow, determined by the die closure line together with the b/s ratio, according to Fig. 1, where "b" is the thickness of the flash and "s" the neck length, should be established, where possible, in order to facilitate filling of the cavity and the texture of the material with the smallest volume of material possible [17].

Excess material should never fill the flash basin, and the b/s ratio should be worked in order to adjust the forces involved. Note that the larger "b" and the smaller "s", the greater the pressure inside the forging cavity.

The softening is a phenomenon of thermal activation that occurs by recovery and/or recrystallization, being these responsible for the grain refinement. When a mechanical energy is applied to the material together with a thermal energy recovery and dynamic recrystallization occurs, when the thermal energy is applied subsequent to a mechanical application, the static softening occurs. The phenomena of recovery, recrystallization and grain growth occur sequentially, but may coexist in a hot forming process. Materials with high stacking fault energy are prone to dynamic recovery while low stacking fault materials, such as austenitic steels, tend to have a higher occurrence of dynamic recrystallization [18].

When deforming a material increases the concentration of dislocations, with the generated thermal energy, a greater rearrangement and reduction of the dislocations by atomic diffusion occurs as a function of the high temperatures. In the hot forming (generally above 1000 °C), immediately after the applied deformation, an internal energy and a condensation of dislocations are generated which are then annihilated or redistributed by diffusion due to the present thermal energy (dynamic recovery). New configurations of dislocations with low stacking fault energy are generated and can form sub grains from the tangles of dislocations. In a process of isothermal annealing, the static recovery has a rapid onset and is slowing with the reduction

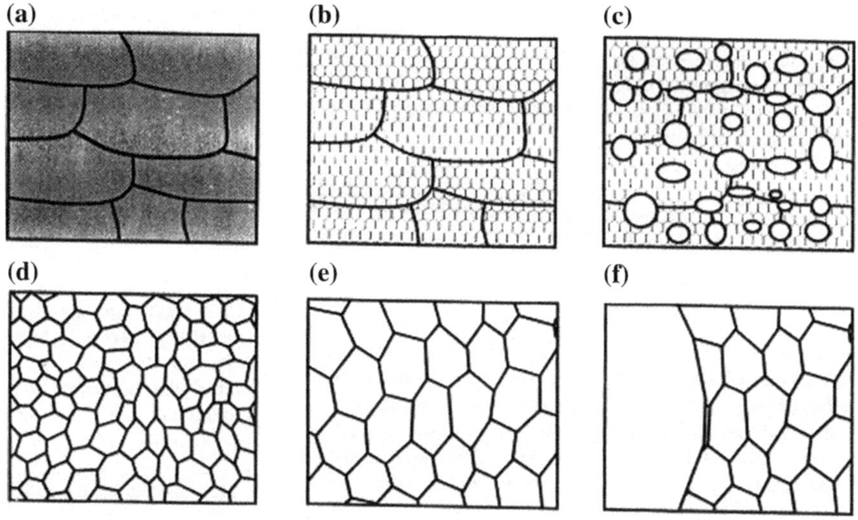

Fig. 2 Schematic diagram of the main processes with thermal treatment. **a** Deformed state, **b** recovered, **c** partially recrystallized, **d** completely recrystallized, **e** grain growth, **f** abnormal grain growth [21]

of the motive power, since the kinetics of recrystallization occurs by nucleation and growth processes, with a slow onset, increasing to a maximum speed of reaction, and returns to be slow [14, 19, 20].

According to Dieter [21], six important variables influence the amount of pre-strain, temperature, time, initial grain size, composition and degree of recovery or polygonization prior to the start of recrystallization (Fig. 2).

After the recrystallization, grain growth depends on the temperature, time and chemical composition of the material. However, large grain size is detrimental to the mechanical properties of the material which implies grain size control during processing involving high temperatures.

Since grain contours are deformed regions of the material, there is a mechanical energy associated with them. The grain growth occurs because in this way the total area of contours is reduced, reducing the associated mechanical energy.

Knowing the mechanisms of recovery, recrystallization and grain growth, work can be performed on the metals with hot or cold deformation, with or without subsequent heat treatment, in order to achieve the desired properties in the material.

2 Materials and Methods

The ASTM F138 steel used consists of austenitic stainless steel based on extra-low carbon chromium-nickel-molybdenum. Manufactured with ESR/VAR process,

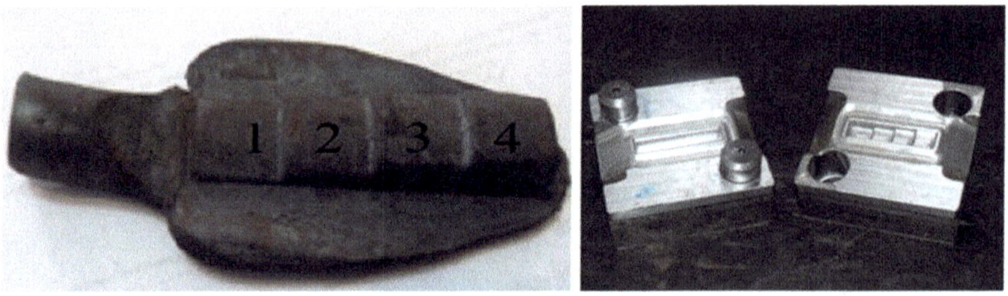

Fig. 3 Left: forged (piece). Right: Forging die

which ensures the product, besides the perfect control of the chemical composition, a high homogeneity of the structure and a high degree of cleaning [22].

The most complete international standard specifying the use of 316 LVM stainless steel for implants is ASTM F138 "*Standard Specification For Wrought 18 Chromium 14 Nickel 2.5 Molybdenum Stainless Steel Bar and Wire for Surgical Implants*" (UNS S31673). This standard details the standards to be followed to preserve the quality of stainless steel F138. The material used complies with this standard (Table 1).

The methodology for grain size analysis was performed according to ASTM E112, using the intercept method, which involves the actual counting of the number of grains by the intersections of grain boundaries. A metallographic microscope with coupled camera and AxioVision SE64 Rel. 4.9.1 software were used as essential equipment for accurate sizing, since hardware and software integrate automatically, producing images and graphical readings generating tables and records of the obtained results. Because ASTM F138 steel is a non-equiaxed grain austenitic material, to minimize dimensional distortions the use of circles, horizontal, vertical and cross lines has proven to be the best option since all orientations are tested equally (ASTM E 112).

The sample (left in Fig. 3) was designed in the CIMATRON CAD-CAM software, which provided the projected area (A_p) data. Four different degrees of deformation were defined, starting with a 3/4 diameter material. Four different temperatures were used: 1000, 1040, 1100 and 1200 °C, four samples being manufactured for each temperature. The permitted average grain size should be thinner than 4 and no grain larger than 3, according to ASTM E 1112. This definition is found in the standard for forged implants: ASTM F621. The forging die (right in Fig. 3) was made of H12 tool steel.

Heating of the material was carried out in a gas furnace in which the temperature was controlled with type K thermocouple. The temperature of the piece with controlled optical pyrometer RAYTEK, model RAYMR15BSF. The heating time was set to 10 min in order to homogenize the material temperature. The equipment used for forging was a gravity drop hammer with a falling weight of 541.4 kg of a heavy ram. The cutting height of the tool was defined as $H = 2.415$ m.

The specimens were prepared for micrographic analysis using a water-cooled disc-cutting machine and sanded with sandpapers of 220, 320, 400 and 600 mesh. The polishing was done on felt with 1 μm alumina paste. The etchant used to reveal

Table 1 Chemical composition ASTM F138 (% by weight)

C	Mn	Si	Cr	Ni	Mo	P	S	Cu	N	Fe
0.030 max.	2.00 max.	0.75 max.	17.0–19.0	13.0–15.0	2.25–3.00	0.025 max.	0.003 max.	0.50 max.	0.10 max.	Bal.

Table 2 Demonstration of deformations and strain rates generated

	Sample 4	Sample 3	Sample 2	Sample 1
hf (mm)	14.5	12.0	9.5	7.0
$A_p(mm^2)$	280.25	235.57	189.25	141.23
ε	0.28	0.46	0.70	1.00
$\dot{\varepsilon}(s^{-1})$	411.05	449.25	499.27	569.35

the microstructure was water, 50% hydrochloric acid (HCl) and 50% nitric acid (HNO_3) for a time of 20 s. Computational optical microscopy was used using the circular intercepts method described in ASTM E112. Each sample was analyzed in five different regions.

3 Results and Discussions

The strain (ε) was calculated using Eq. (1). The deformation rate ($\dot{\varepsilon}$) was calculated using Eq. (2) where h_i is the initial height of the material, h_f the final deformed height and v is the velocity. Adopting $h_i = 19.05$ mm and $v = 6852.85$ mm/s, the calculated values are presented in Table 2.

$$\varepsilon = \frac{d_h}{hi} = \int \frac{d_h}{h} = ln\frac{hi}{hf} \tag{1}$$

$$\dot{\varepsilon} = \frac{ln\frac{hi}{hf}}{hi - hf} * v \tag{2}$$

The regions indicated in Fig. 4 define where the grain size transition occurs. The red region (upper region) defines the occurrence of larger grains. In the blue region [central region: die division (parting line)] recrystallization occurs and partial recovery occurs in the purple region (lower region).

3.1 *Micrographics of the Different Heated Regions Without Forging*

Figure 5 shows the influence of temperature on grain growth. In the material heated to 1000 °C grain is verified according to ASTM of 6.0–8.0, average size of 7.5. In the material heated to 1040 °C, grain according to ASTM of 6.0–7.0, average size of 6.9 is verified. In the material heated to 1100 °C grain is verified according to ASTM of 3.0–4.5, average size of 3.9. In the material heated to 1200 °C, ASTM

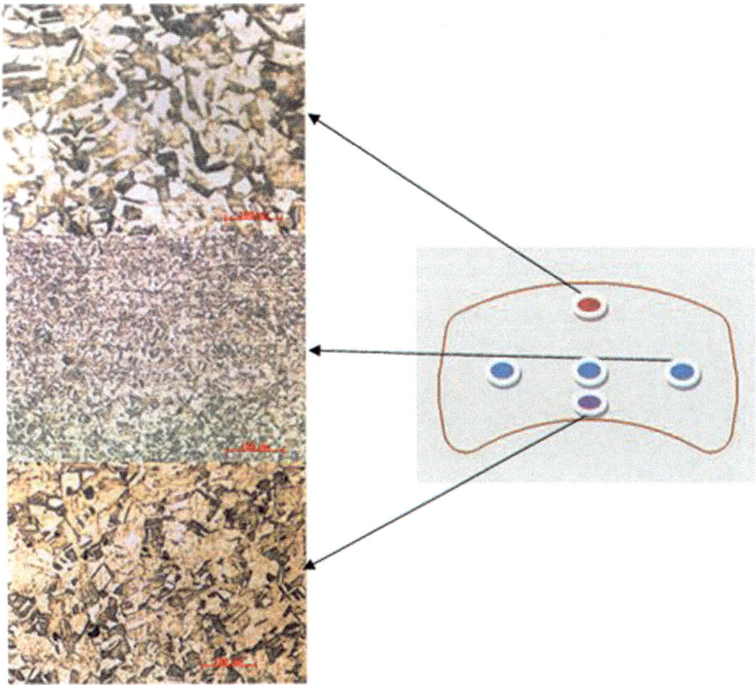

Fig. 4 Granular structure and regions analyzed

grain of 1.0–2.0 is verified. Average size of 1.7. All materials were heated in a time of 10 min.

It is observed in the heating that a great grain growth takes place in the temperatures above 1040 °C. This phenomenon confers on the work of Geronimo [23], which detected this transition between 1000 and 1100 °C.

3.2 Micrographics of Different Regions Forged at 1000 °C

A grain growth is observed in the upper region in relation to the material heated at 1000 °C and undeformed. The deformation of the grain with respect to the undeformed structure can be seen (Fig. 6).

It is observed in the central region (die division) the occurrence of recrystallization in all the deformations, noting that there was partial recrystallization for $\varepsilon = 0.28$ (Fig. 7).

The heterogeneity of the grain size with evidence of deformation is observed in the lower region, generating partial recrystallization in the samples (Fig. 8).

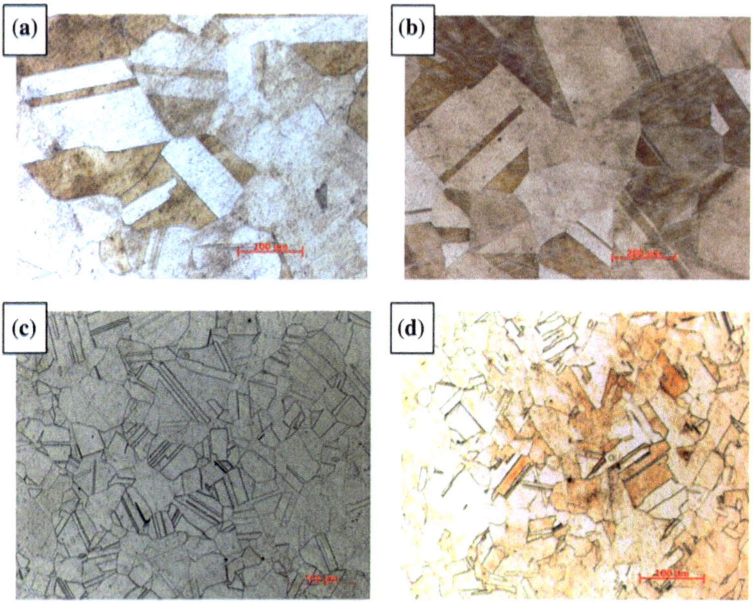

Fig. 5 Micrograph of the structure of the material heated at different temperatures. **a** 1000 °C. **b** 1050 °C. **c** 1100 °C. **d** 1200 °C

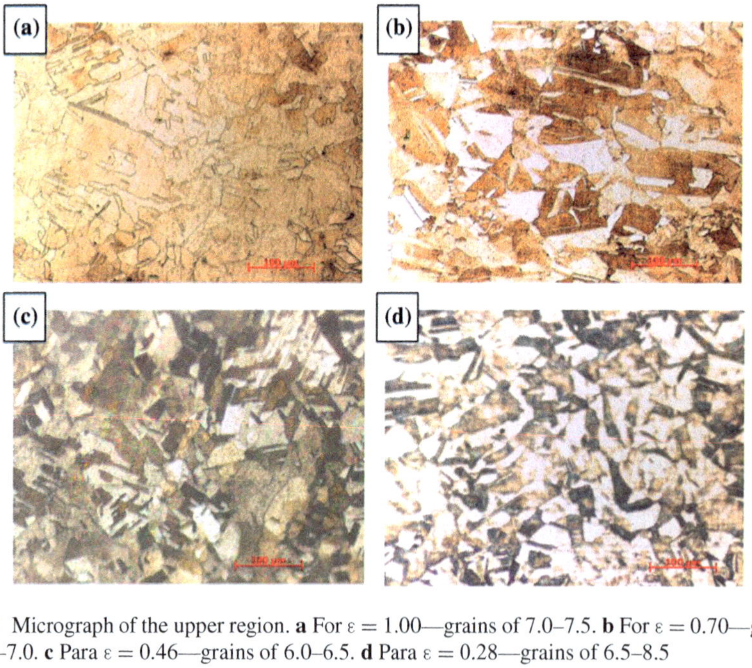

Fig. 6 Micrograph of the upper region. **a** For $\varepsilon = 1.00$—grains of 7.0–7.5. **b** For $\varepsilon = 0.70$—grains of 6.5–7.0. **c** Para $\varepsilon = 0.46$—grains of 6.0–6.5. **d** Para $\varepsilon = 0.28$—grains of 6.5–8.5

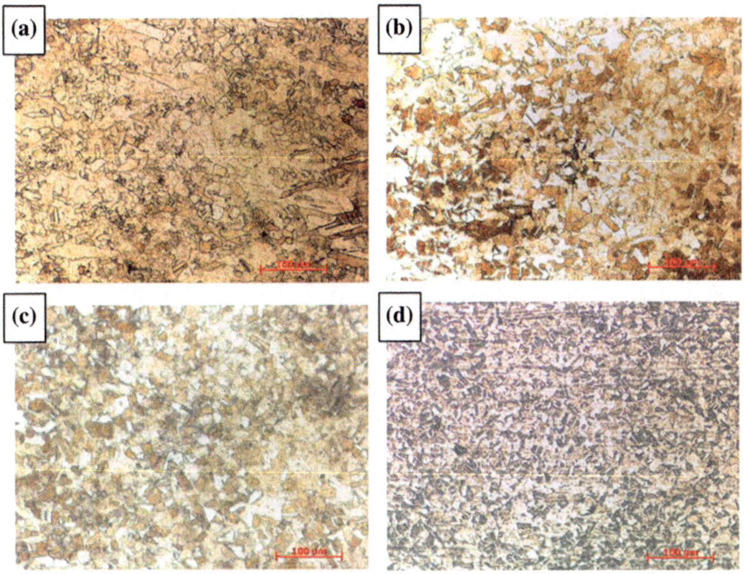

Fig. 7 Micrograph of the division region of the die. **a** For ε = 1.00—grains of 9.0–9.5. **b** For ε = 0.70—grains of 8.5–9.0. **c** For ε = 0.46—grains of 8.5–9.5. **d** For ε = 0.28—grains of 6.5–8.5

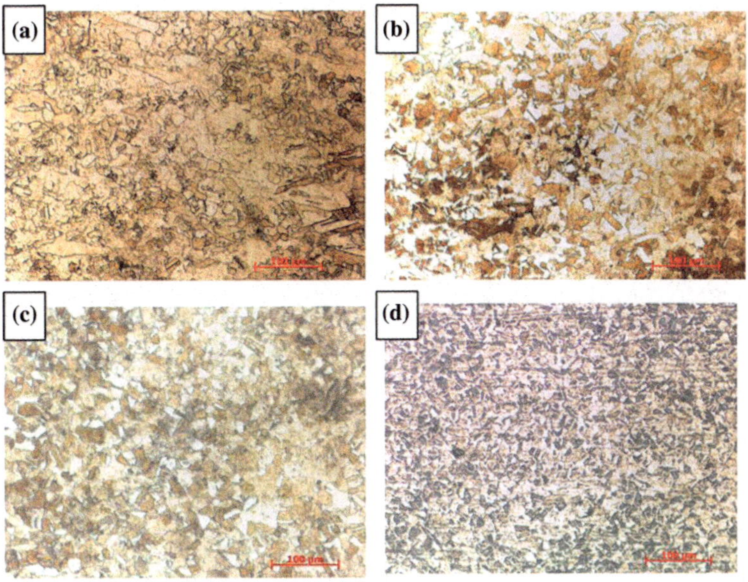

Fig. 8 Micrographs of the lower region. **a** For ε = 1.00—grains of 7.0–7.5. **b** For ε = 0.70—grains of 7.0–7.5. **c** For ε = 0.46—grains of 6.5–7.0. **d** Para ε = 0.28—grains of 7.0–7.5

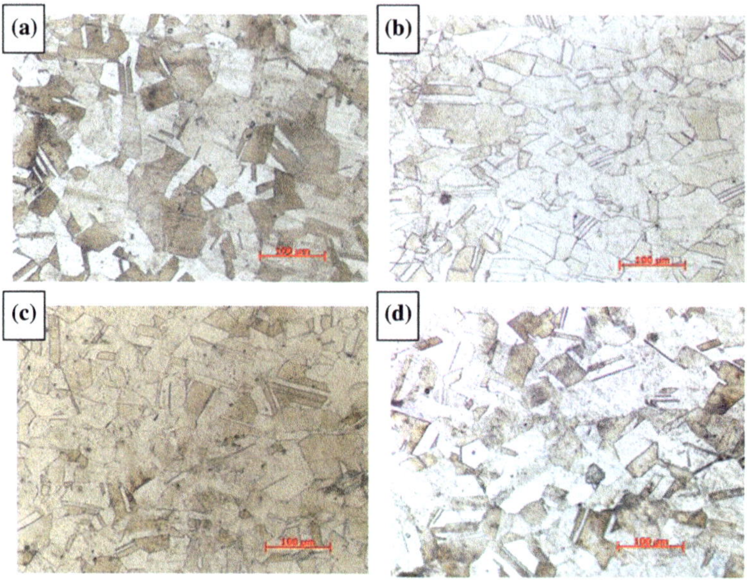

Fig. 9 Micrograph of the upper region. **a** For ε = 1.00—grains of 6.0–6.55. **b** For ε = 0.70—grains of 5.5–6.0. **c** For ε = 0.46—grains of 6.0–7.0. **d** For ε = 0.28—grains of 6.0–7.0

3.3 Micrographics of Different Regions Forged at 1040 °C

A similarity to that occurring at 1000 °C is observed in the upper region: a grain growth in relation to the material heated at 1040 °C and undeformed. The grains presented in a more homogeneous form (Fig. 9).

It is observed in the central region the occurrence of recrystallization in the deformations of 0.46; 0.70 and 1.00. For deformation of 0.28, there was no significant change in structure (Fig. 10).

The lower region had little variation in structure at all deformations. The grains, in shape, are altered but not reduced in size. Comparing with the deformations at 1000 °C it is verified that there was no partial recrystallization. Note that the material flow is concentrated in the region of the die division (Fig. 11).

3.4 Micrographs of Different Regions Forged at 1100 °C

It is observed in the upper region that, for deformation of 1.00, there was a great reduction in grain size. For the other deformations grain growth is observed (Fig. 12).

It is observed in the region of division of the die the occurrence of recrystallization in all the deformations, evidencing a greater plasticity of the material (Fig. 13).

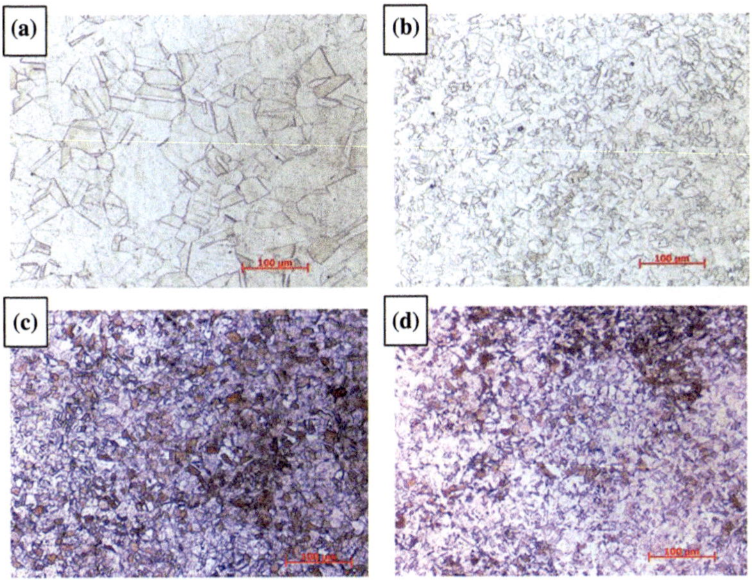

Fig. 10 Micrograph of the region of the die division. **a** For ε = 1.00—grains of 9.5–10.5. **b** For ε = 0.70—grains of 9.5–10.5. **c** For ε = 0.46—grains of 9–10.0. **d** For ε = 0.28—grains of 6.0–7.0

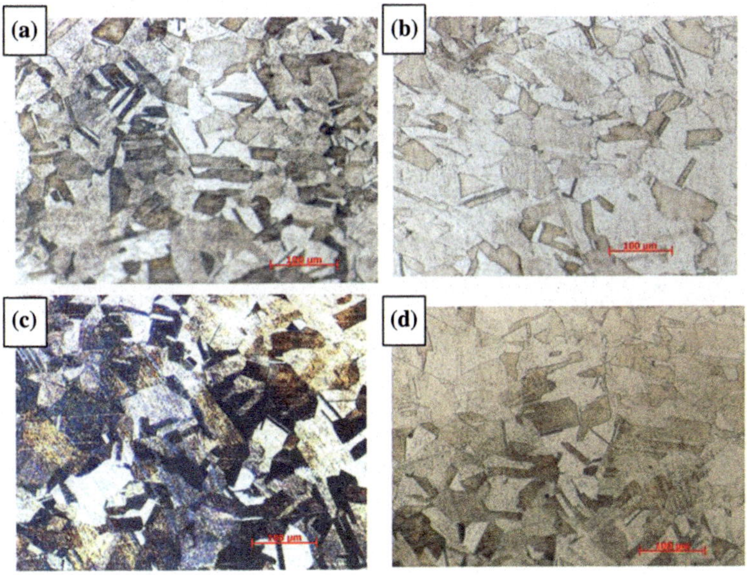

Fig. 11 Micrographs of the lower region. **a** For ε = 1.00—grains of 6.5–7.5. **b** For ε = 0.70—grains of 6.0–7.0. **c** For ε = 0.46—grains of 6–7.0. **d** For ε = 0.28—grains of 6.0–7.0

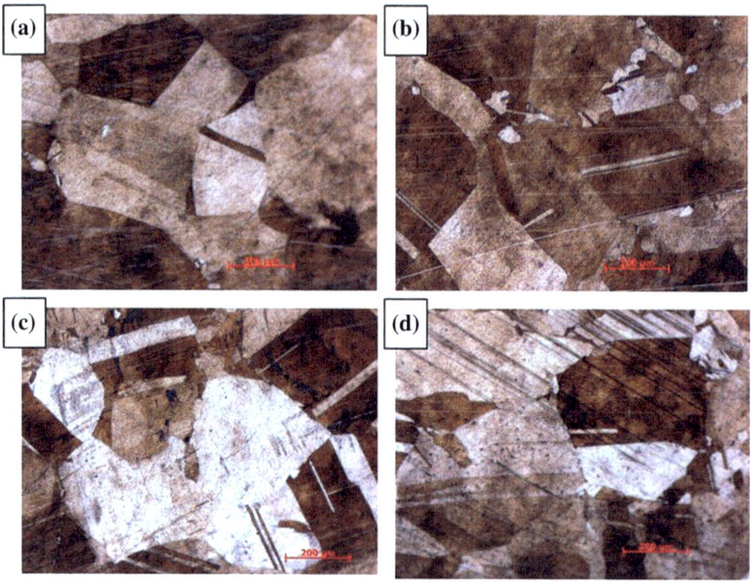

Fig. 12 Micrograph of the upper region forged at 1100 °C. **a** For $\varepsilon = 1.00$—grains of 6.5–7.0. **b** For $\varepsilon = 0.70$—grains of 3.0–5.0. **c** For $\varepsilon = 0.46$—grains of 3.0–4.5. **d** For $\varepsilon = 0.28$—grains of 3.0–4.5

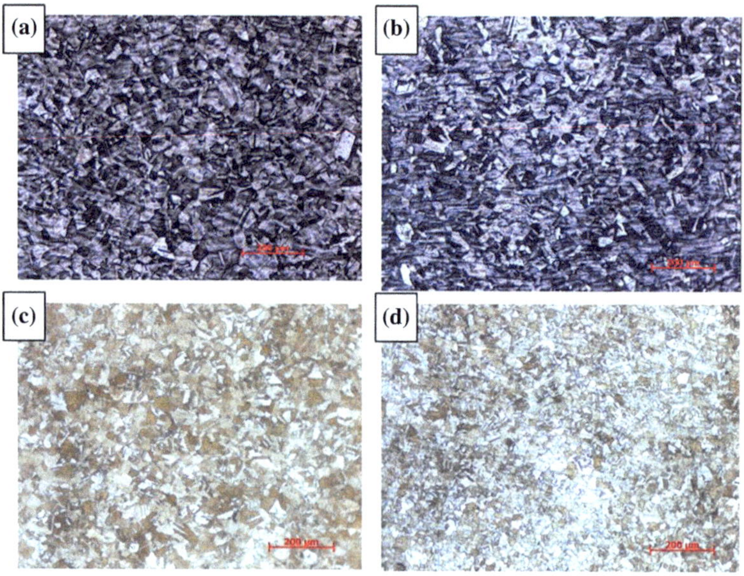

Fig. 13 Micrograph of the division region of the die. **a** For $\varepsilon = 1.00$—grains of 8.0–8.5. **b** For $\varepsilon = 0.70$—grains of 7.5–8.0. **c** For $\varepsilon = 0.46$—grains from 8.0 to 9.0. **d** For $\varepsilon = 0.28$—grains of 8.0–9.0

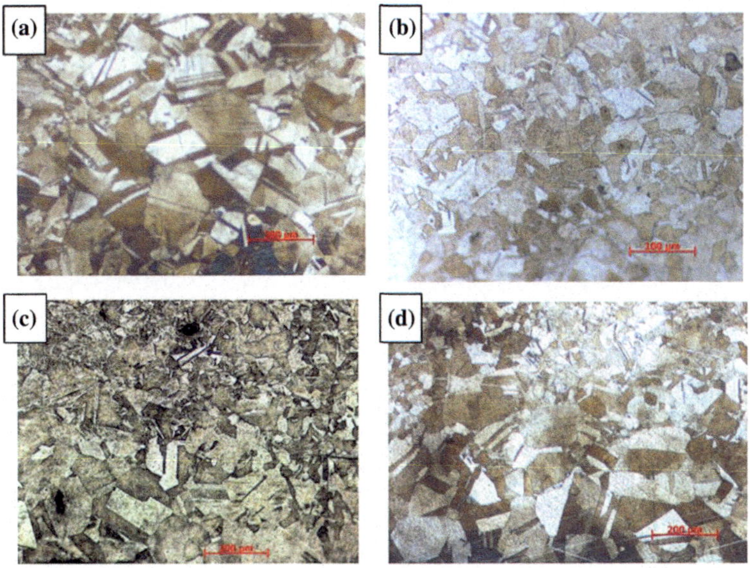

Fig. 14 Micrographs of the lower region. **a** For ε = 1.00—grains of 7.0–8.0. **b** For ε = 0.70—grains of 3.5–7.5. **c** For ε = 0.46—grains of 3.5–5.0. **d** For ε = 0.28—grains of 3.5–4.5

In the lower region, for ε = 1.00, it is observed that there was total recrystallization. The same does not occur with the other deformations, however the area of observation was restricted to 2 mm of height by 5 mm of width. Following the standard of the previous temperatures, grain growth is verified in relation to the heated material (Fig. 14).

3.5 Micrographics of Different Regions Forged at 1200 °C

It is observed in the upper region the reduction of grain size, but of little significance. It is observed that the higher the working temperature, the lower the heat absorption by the material in the deformation (Fig. 15).

It is observed in the region of division of the die the occurrence of recrystallization in all the deformations, showing great plasticity of the material (Fig. 16).

It is observed in the lower region that there was recovery followed by partial recrystallization. Unlike the other temperatures, grain size reduction is observed (Fig. 17).

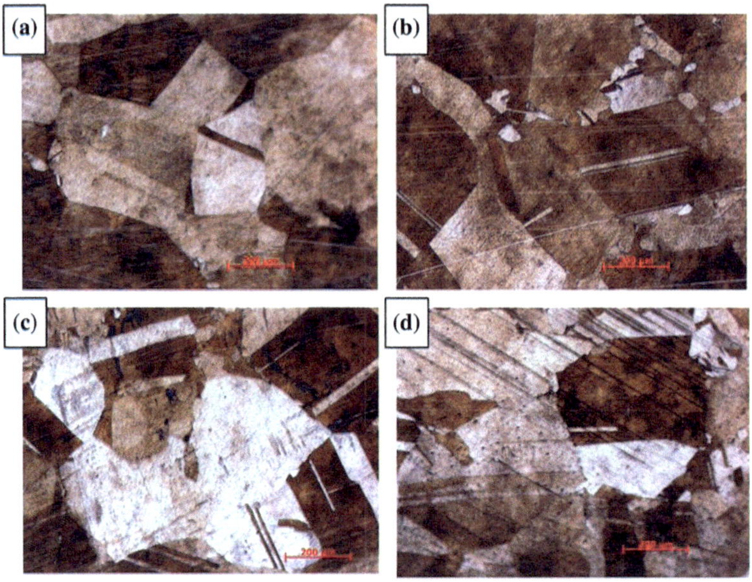

Fig. 15 Upper region micrograph. **a** For ε = 1.00—grains of 1.5–2.0. **b** For ε = 0.70—grains 1.0–2.0. **c** For ε = 0.46—grains of 1.0–1.5. **d** For ε = 0.28—grains of 1.0–1.5

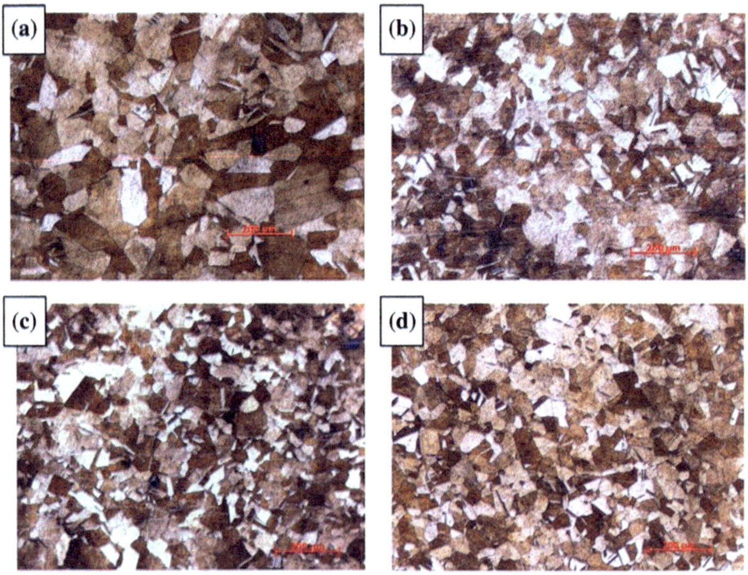

Fig. 16 Micrograph of the region of the die division. **a** For ε = 1.00—grains of 5.5–6.0. **b** For ε = 0.70—grains of 5.5–6.0. **c** For = 0.46—grains of 4.5–5.0. **d** For ε = 0.28—grains of 4.0–4.5

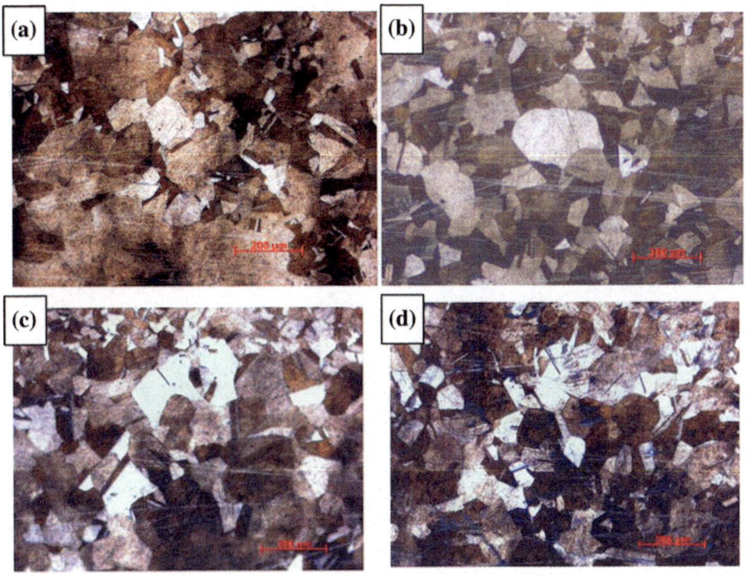

Fig. 17 Micrograph of the lower region. **a** For ε = 1.00—grains of 3.5–4.0. **b** For ε = 0.70—grain from 3.0 to 4.0. **c** For ε = 0.46—grain from 3.0 to 4.0. **d** For ε = 0.28—grains of 2.0–4.0

3.6 Grain Size

There are, therefore, two very distinct groups. The first is forged with temperatures up to 1040 °C and the second with temperatures above 1040 °C. Although the average grain of sample 1 at 1100 °C, according to Table 3, is within the norm specified, grain 2 was detected in some samples, which disqualifies the forged sample.

It is observed that the material before forging has an abrupt growth at temperatures of 1100 and 1200 °C. In the region, where the die dividing line is located at about 4 mm, there was total recrystallization of the grains and it was observed that the recrystallization of the grain was proportional to the initial size. All the grains in this region are within the normative parameters.

In the samples forged with temperature up to 1040 °C, a small increase of the grain size is observed in the upper and lower regions. This is due to the increase in temperature during the forging process due to the high rate of deformation applied. Therefore, it is possible to verify the sensitivity of the material to grain growth when at elevated temperatures. All samples forged up to 1040 °C met the standard.

In the samples forged with temperatures of 1100 °C, a small increase in grain size is observed in the upper and lower regions, however, since the initial grain is already above the 89.8 μm (grain size 4) specified in the standard, no sample meets the specifications. At 1200 °C it is observed that the grains in the upper and lower regions had reduction indicating excellent plasticity of the material, but not enough to recrystallize grains within the normative parameters.

Table 3 Experimental results in their respective regions in the sample

UPPER REGION OF THE DIE

Temperature	1000	1040	1100	1200
Material	31.02	35.14	96.81	241.55
Sample 1	30.41	47.42	33.13	210.65
Sample 2	37.07	51.53	124.94	222.06
Sample 3	42.65	36.52	108.67	233.15
Sample 4	44.9	35.61	100.95	234.35

DIE DIVISION REGION

Temperature	1000	1040	1100	1200
Material	31.02	35.14	96.81	241.55
Sample 1	14.22	12.80	20.13	51.40
Sample 2	17.01	11.91	26.56	48.97
Sample 3	15.34	13.71	23.99	67.83
Sample 4	18.54	24.75	24.63	77.36

LOWER REGION OF THE DIE

Temperature	1000	1040	1100	1200
Material	31.02	35.14	96.81	241.55
Sample 1	28.40	36.44	36.95	103.08
Sample 2	32.09	45.65	108.67	103.48
Sample 3	34.34	38.65	96.35	113.75
Sample 4	33.80	38.22	105.28	103.08

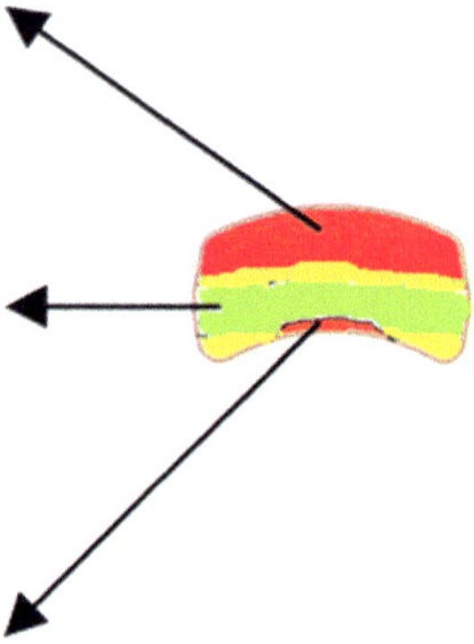

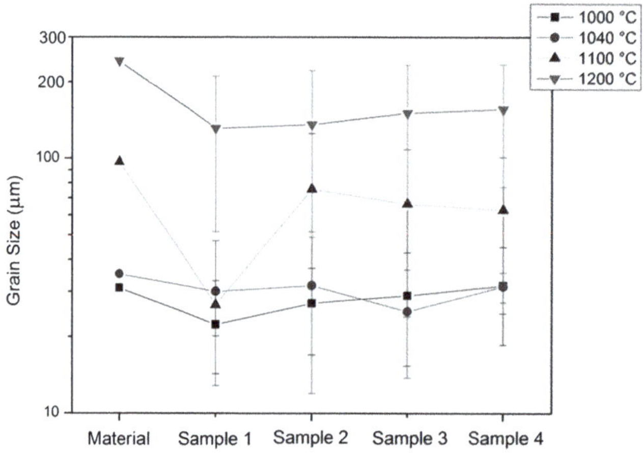

Fig. 18 Representative graph of grain size obtained as a function of temperature

The influence of the die shape on the material flow during forging is clear. Figure 18 shows the grain size variation in the same cross section with its respective applied temperatures and samples. In the lower region, where the material initially comes in contact with the die, a similar process occurs, however the die format allows the material to flow and consequently grain reduction. In the yellow region, as in Table 3, the partial recrystallization of the grains, and grains more coarse than the region of the division of the die, is observed.

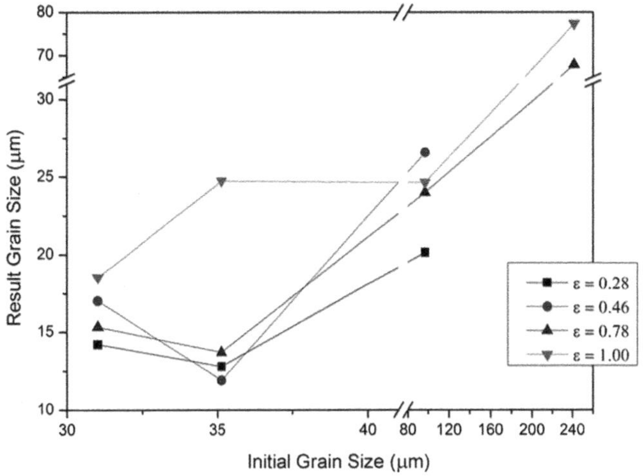

Fig. 19 Grain size resulting as a function of deformation applied at temperatures of 1000, 1040, 1100 and 1200 °C

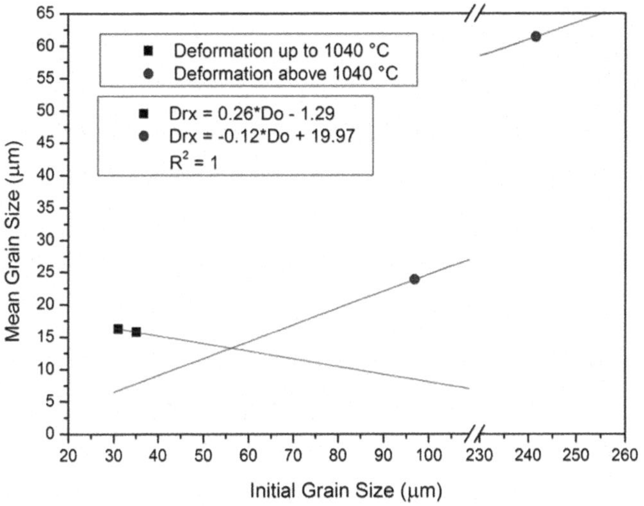

Fig. 20 Mean grain size by applied deformation

Therefore, in this study, the influence of temperature and deformation on the recrystallization and recovery of the grain is verified, reporting that the rate of deformation has a greater influence on the generation of thermal energy.

From the data obtained and demonstrated in Fig. 18, the data can be rearranged to obtain the resulting grain size as a function of the applied deformation, initial shape temperature and initial grain size, and this is demonstrated in Fig. 19.

Two straight line equations can be fitted, one with the mean of the data obtained at the temperatures of 1000 and 1040 °C and the second with the average of the data obtained at the temperatures of 1100 and 1200 °C, as indicated in Fig. 20.

For temperatures above 1040 °C, Drx = 0.26D0-1.29 is obtained, and for temperatures up to 1040 °C it obtains Drx = −0.12D0 + 19.97, where Drx is the dynamically recrystallized grain size and D_0 the grain size initial. Equationing the two equations we have $D_0 = 55.95$ µm.

Substituting the value of D_0 in one of the equations we can predict a recrystallized grain Drx = 13.26 µm. Interpolating D_0 between the temperatures of 1040 and 1100 °C, the transition temperature is 1021.08 °C. In fact the transition temperature according to Villares Metals [22] is 1050 °C. There is thus an error of approximately 3%.

4 Conclusions

The following conclusions are possible from the analysis of the results obtained.

- ASTM F138 Austenitic Stainless Steel exhibits moderate grain growth to a temperature close to 1050 °C.
- Grain growth is directly proportional to the increase in temperature.
- High deformation rate broadens the initial forging temperature, which directly influences grain growth.
- The shape of the die directly interferes with the flow of the material and the result of the granular structure.
- With hot axial forging, homogeneity cannot be obtained in the granular structure of ASTM F138.
- The temperature and the deformation applied are the two variables that influence the final result of the resulting grain size.

Acknowledgements The authors would like to thank IFSP, IPEN, Wieland Forjados and CAPES for their partial support to this work.

References

1. Ishida MA (2009) Avaliação Comparativa de Barras Laminadas do Aço AISI 316L com e sem Tratamento Térmico de Solubilização. [Masters thesis (Engineering)] Porto Alegre, Universidade Federal do Rio Grande do Sul
2. Padilha AF, Guedes LC (1994) Aços Inoxidáveis Austeníticos: Microestrutura e Propriedades. Hemus, São Paulo
3. Miranda F, Rodrigues D, Nakamoto FY, Frajuca C, Santos GA, Couto AA (2016) Microstructural evolution of composite 8 WC-(Co, Ni): effect of the addition of SiC. Defect Diffus Forum 371:78–85. https://doi.org/10.4028/www.scientific.net/DDF.371.78

4. Miranda F, Rodrigues D, Nakamoto FY, Frajuca C, Santos GA (2017) The influence of the sintering temperature on the grain growth of tungsten carbide in the composite WC-8Ni. Mater Sci Forum 899:424–430. https://doi.org/10.4028/www.scientific.net/MSF.899.424
5. Nascimento MS, Frajuca C, Nakamoto FY, Santos GA, Couto AA (2017) Correlação entre variáveis térmicas de solidificação, microestrutura e resistência mecânica da liga Al-10%Si-2%Cu. Materia, Rio de Janeiro 22:e11774. https://doi.org/10.1590/s1517-707620170001.0106
6. Nascimento MS, Franco ATR, Frajuca C, Nakamoto FY, Santos GA, Couto AA (2018) an experimental study of the solidification thermal parameters influence upon microstructure and mechanical properties of Al-Si-Cu alloys. Mater Res 21(5):e20170864. https://doi.org/10.1590/1980-5373-mr-2017-0864
7. Santos GA, Goulart PR, Couto AA, Garcia A (2017) Primary dendrite arm spacing effects upon mechanical properties of an Al 3 wt%Cu 1 wt%Li Alloy. In: Andreas O, Holm A (eds) Advanced structured materials, vol 33. 1edn. Singapore: Springer Singapore, pp 215–229, . https://doi.org/10.1007/978-981-10-1602-8_19
8. Hendler BR (2010) Avaliação do Tamanho de Grão de um Forjado a Quente Via Simulação Numérica. Universidade Federal do Rio Grande do Sul, Porto Alegre
9. Tavares LN, Rhary GVO, Gama JLL (2007) Estudo da Cinética de Crescimento de Grão em Metal. In: II CONNEPI—Congresso de Pesquisa e Inovação da Rede Norte Nordeste de Educação Tecnológica, 2007, João Pessoa. II CONNEPI
10. Nascimento LA (2010) Estudo da Recristalização Dinâmica durante a deformação a quente de um aço ISO 5832-9. [Masters thesis (Physics)] São Luís, Universidade Federal do Maranhão
11. Balacin O (2018) Simulação física do processamento a quente de materiais metálicos—Universidade Federal de São Carlos. [Cited from: 16 Jun 2018]. Available from: https://www.researchgate.net/profile/Oscar_Balancin/post/Tips_for_performing_a_hot_torsion_test/attachment/59d6297cc49f478072e9c44c/AS%3A272470379040799%401441973408545/download/Simula%C3%A7%C3%A3o+F%C3%ADsica.pdf
12. Santos GA (2017) The importance of metallic materials as biomaterials. Adv Tissue Eng Regenerative Med 3(I)
13. Padilha AF, Siciliano Jr F (1996) Encruamento, recristalização, crescimento de grão e textura. Associação Brasileira de Metalurgia e Materiais, 2 ed. São Paulo
14. Callister Jr. WD (2016) Ciência e Engenharia de Materiais: Uma Introdução, LTC, 9,ed, Rio de Janeiro
15. Santos GA (2015) Tecnologia dos materiais metálicos—propriedades, estruturas e processos de obtenção, São Paulo, Érica
16. Camargo R (2008) Verificação da Usinabilidade dos Aços Inoxidáveis Austeníticos através do Processo de Furação. [Masters thesis (Engineering)] Campinas, UNICAMP
17. Schaeffer L (2006) Introdução ao projeto de ferramentas para forjamento a quente em matriz fechada. Ferramental. N. 9, pp 19–28
18. Geronimo FHC (2006) Estudo da recristalização dinâmica do aço inoxidável austenítico F138, utilizado em próteses ortopédicas. [Masters thesis (Engineering)] São Carlos UFSCAR
19. Reed-Hill RE (1982) Princípios de Metalurgia Física. Guanabara Dois, 2 edn, Rio de Janeiro, 1982
20. Bresciani Filho E, Zavaglia CAC, Button ST, Gomes E, Nery FAC (1997) Conformação Plástica dos Metais. Campinas, Editora da Unicamp, 5 ed
21. Dieter EG (1976) Mechanical metallurgy. McGraw-Hill, New York
22. Villares Metals (2003) Aços inoxidáveis para implantes ortopédicos, v. 138
23. Geronimo FHC, Balancin O (2013) Competição entre recuperação e recristalização dinâmica do aço inoxidável austenítico ASTM F138 utilizado em implantes ortopédicos. Tecnologia Metalurgia, Materiais e Mineração. São Paulo, 10(2), pp 162–169

Contribution to an Electrical Transport in Montmorillonite/Polyaniline Composite

S. Rusnáková, K. Karvanis, P. Koštial, Z. Koštialová-Jančíková and A. Zimula

Abstract The paper deals with the electrical properties of Montmorillonite (MMT)/Polyaniline (PANI) composites. These materials show specific electrical properties as relatively high anisotropic electrical conductivity and the thermally induced electrical transport shows relatively large spectrum of charge transport mechanisms. The paper presents measurements of dc conductivity versus temperature as well as the measurements of ac. conductivity at ambient temperature for the investigated samples.

Keywords PANI/MMT structure · dc. and ac. · Electric conductivity

1 Introduction

Clay minerals are widely represented not only in the older sedimentary rocks, but also in the form of the basis of recent sediments and soils. Major component of the clay minerals and sheet silicates form the minor component is a mineral of allophane, also hydroxides, oxy-hydroxides, or oxides of aluminium and iron [1].

All phyllosilicates are generally made up of SiO_4 pyramids, of which three oxygen atoms shared with the neighbouring tetrahedra with which creates an infinite two-dimensional grid [2]. The structural influence on the PANI and PANI/MMT electrical properties has been investigated in [3].

The work [4] is focused on the effect of MMT content on the thickness, morphology and electrical conductivity of PANI/MMT nanocomposite films. Authors in [5] fabricated, through various conditions, pressed tablets from polyani-

S. Rusnáková (✉) · K. Karvanis
Department of Production Engineering, Faculty of Technology, Tomas Bata University in Zlín, Vavrečkova 275, 760 01 Zlin, Czech Republic
e-mail: rusnakova@utb.cz

P. Koštial · Z. Koštialová-Jančíková · A. Zimula
Department of Material Engineering, Faculty of Metallurgy and Materials Engineering, VŠB – Technical University of Ostrava, 17. Listopadu 15/2172, 708 33 Ostrava, Czech Republic
e-mail: pavel.kostial@vsb.cz

A. Öchsner and H. Altenbach (eds.), *Engineering Design Applications II*,
Advanced Structured Materials 113, https://doi.org/10.1007/978-3-030-20801-1_11

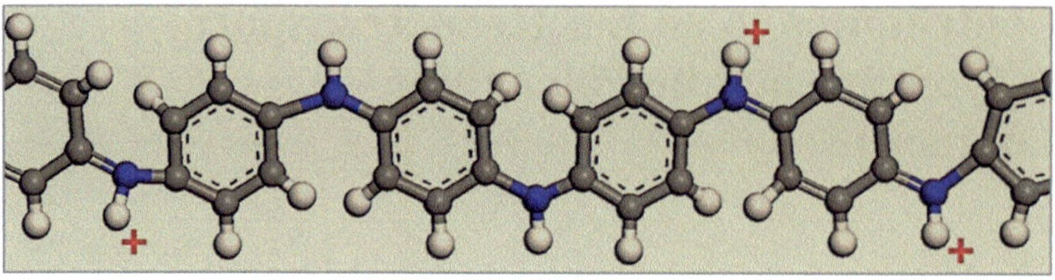

Fig. 1 Structure of conductive PANI [11]

line/phyllosilicate nanocomposites in order to optimize the anisotropic conductivity of the composites by ordering of fiat phyllosilicate particles intercalated with polyaniline (PANI).

In the work [6] has been proved that the nanostructure of PANI affects the conductivity. Sapurina and Stejskal [7] studied the mechanism for the oxidation of aniline and the formation of supramolecular polyaniline structures. In the work [8] authors studied the production of polyaniline films on glass surface immersed in the reaction mixture during the oxidation of the aniline.

Clay minerals with the layered structure (phyllosilicates) have been often used as matrices for such type of organo-inorganic composites. Interaction of guest species with negatively charged structure can offer new applications of these composites [9, 10]. Different technological aspects of PANI/MMT preparation are in details described in [11–14]. Electrical transport in polymer clay nanocomposites is described in [15].

In this paper, the electrical properties of Montmorillonite (MMT)/Polyaniline (PANI) composites are studied. The obtained results classify the studied material as an interesting piezoresistor or sensor material for specific applications.

2 Theory

Polyaniline exists in various forms depending on a degree of oxidation or reduction. Figure 1 depicts a conductive form of polyaniline called emeraldine salt. Polyaniline chain is composed of a system of conjugated double bonds (in the structure are regularly alternating single and double bonds). For electrical conductivity, in addition to conjugation, it is important the presence of the charge carriers which mediate their transfer along the chain. This is achieved by doping or protonation of the nitrogen in the polyaniline chain.

According to X-ray analysis which was presented in the past [3] shows that the morphology of pure PANI particles as strongly anisotropic.

In Fig. 2 the part (a) is a model view of PANI/MMT containing two chains of polyaniline with 4.4 wt% H_2O and 8Na + cations. The lower part of the figure (b) shows a side view of the model. Water molecules and Na + cations are not shown for clarity. In the side view (b) the "molecular capacitor" created by Hydrogen and Oxygen atoms can be seen [16].

Specific features of PANI/MMT structure which were obtained by X-ray diffraction have been presented in [3]. The platy silicate particles in pellets are preferentially oriented with their 001 plane parallel to the flat pellet surface (IP). First basal reflection 001 at 26-6.5° corresponds to the basal spacing 1.31 nm and indicates the change of interlayer structure of pristine MMT after one step composite preparation. The relatively sharp 001 profile in comparison with other PANI/clay intercalates indicates that MMT was fully intercalated by PANI chains. Intercalation of PANI has significant effect on the in-plane conductivity of pellets, as we will see. Different views on the structure of intercalated PANI/MMT structure are shown in Fig. 2.

In the point of view of the electrical transport, MMT bears the negative charge, which is compensated by Na cations. Π-electrons from PANI chains are other source of charges cooperating on electrical conductivity. Nitrogen in PANI bound protons and create cations. Such structure serves as a natural capacitor.

In the electrical point of view the capacitor in Fig. 3 can be described by the following schema and the well-known relations 1.1–1.3.

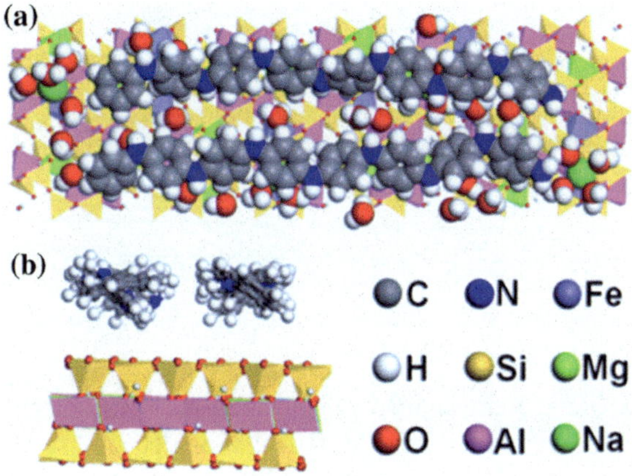

Fig. 2 Model of structure PANI/MMT. **a** Two chains of polyaniline, **b** Side view of the model [16]

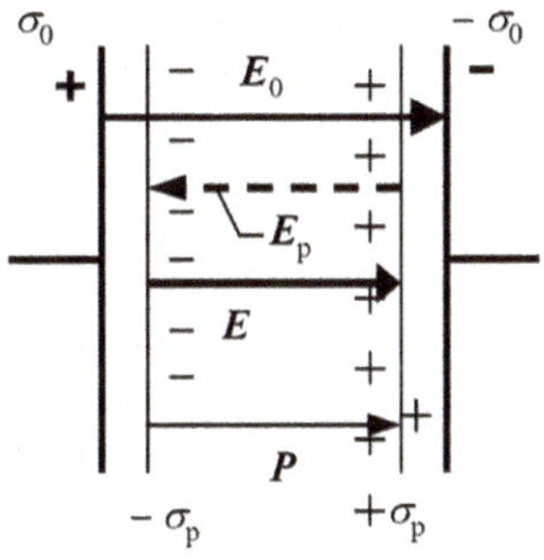

Fig. 3 Schema of the capacitor

Table 1 Dielectric parameters for PD and IP surfaces

Surface	ε'	ε''	C [F]	σ_p [C/m^2]	σ_0 [C/m^2]	$\varepsilon'' = \sigma_{ac}/\omega \cdot \varepsilon_0$
PD	1.5×10^8	2.8×10^{11}	1.6×10^{-5}	0.32	2.2×10^{-9}	2.7×10^{11}
IP	8.7×10^6	3.3×10^{11}	6.1×10^{-7}	0.015	1.77×10^{-9}	3.26×10^{11}

For distinguished intensities (in relation to Fig. 1) of electric field the following types can be written:

$$E_0 = \frac{\sigma_0}{\varepsilon_0} \tag{1.1}$$

$$E_p = \frac{\sigma_p}{\varepsilon_0} = \frac{P}{\varepsilon_0} \tag{1.2}$$

and for final intensity the equation is valid

$$E = \frac{\sigma_0 - \sigma_p}{\varepsilon_0} < E_0 \tag{1.3}$$

where σ_0, σ_p are corresponding surface charge densities, P is the polarization vector and ε_0 is vacuum dielectric constant [17–19].

Further known relations used in the final numerical analysis (Table 1) are the following:

$$\sigma_p = \varepsilon_0(\varepsilon_r - 1)U/d, \tag{1.4}$$

$$\sigma_0 = \varepsilon_0 U/d, \tag{1.5}$$

where U is the applied voltage and d is the distance between the electrodes.

In the point of view of temperature dependence of specific electric conductivity it is described by the relation [20]

$$\sigma = \sigma_0 * \mathrm{e}^{-\frac{W}{kT}}, \tag{1.6}$$

where σ is the dc. specific electric conductivity, k is the Boltzmann's constant, T is the absolute temperature and W is an energy. In the impurity mode of specific conductivity W plays the role of half value of impurity activation energy W_a. At higher temperatures intrinsic electrical conductivity takes place and in this case W plays the role of the forbidden gap width.

Description of ac. electric field we start with equation for the complex dielectric conductivity.

$$\varepsilon^* = \varepsilon' - i\varepsilon'', \tag{1.7}$$

where ε' is the real part of the permittivity describing polarization effect in dielectric, ε'' is the imaginary part of the dielectric permittivity which describes losses caused by conductive charges and i is the imaginary unit.

Equation 1.5 can be then rewritten as

$$\varepsilon'' = \varepsilon''_{pol} + \varepsilon''_{\text{vod}} = \varepsilon''_{pol} + \frac{\sigma}{\omega\varepsilon_0}, \tag{1.8}$$

σ is the ac. specific conductivity and ω is the angular frequency [20].

The following relations are valid for parallel connection of resistor and capacitor in ac. field [15]:

$$\varepsilon' = \frac{C_p}{C_0}, \tag{1.9}$$

$$\varepsilon'' = \frac{1}{C_0\omega R_p}, \tag{1.10}$$

$$\varepsilon'' = \frac{\sigma}{\omega\varepsilon_0} \rightarrow \sigma = \varepsilon'' 2\pi f \varepsilon_0, \tag{1.11}$$

where C_p and C_0 are the capacities of parallel and vacuum capacitor, f is the frequency.

3 Experimental Condition

The nanocomposites PANI/MMT were prepared by intercalation (into the interlayer space of montmorillonite) by product of oxidative polymerization of aniliniumsulfate (using ammonium peroxydisulfate). In practice, 200 cm^3 solution of aniliniumsulphate was mixed with 200 cm^3 solution of ammonium peroxydisulfate and 8 g of montmorillonite Portaclay. The mixture was stirred for 6 h at room temperature then filtered under vacuum and washed with dilute hydrochloric acid and finally dried in an oven at 40 °C for 24 h [3]. The nanocomposite PANI/MMT plates were compressed using a hand press LECO (applied pressure 28 MPa), without any binders and lubricants and samples with dimensions 10 mm × 10 mm × 4 mm were prepared.

4 Results and Discussion

Further, the electrical properties of studied samples are presented. We have focused our attention on the dc. and ac. electric conductivity.

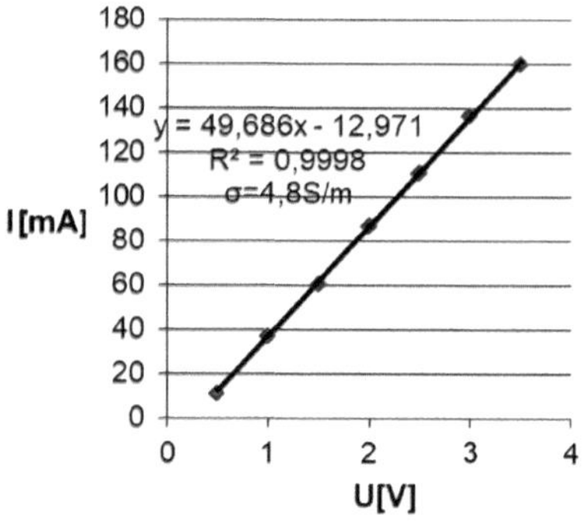

Fig. 4 V-A characteristic for IP surface

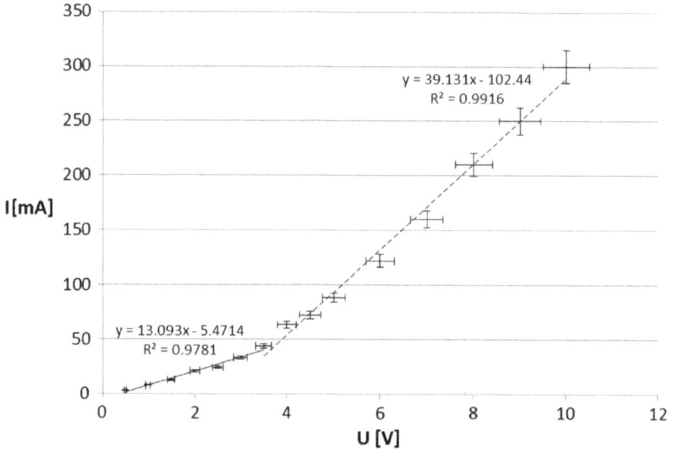

Fig. 5 V-A characteristic of IP surface for Au contact with 5 percentage segments. Dashed line σ = 3.25 S/m, full line σ = 1.09 S/m

Figure 4 shows the Volt-Ampere characteristics (V-A) for in-plane (IP) direction surface, results for surface with direction perpendicular to IP (PD) are similar. Main differences are in the values of the specific conductivity. It has been found that $\sigma = 4.8$ [S/m] for IP surface and $\sigma = 3.9$ [S/m] for PD one. V-A characteristics are clearly linear with very good value of R^2 presented in the plot (that means the contact is ohmic). Both values are plotted in bar diagram in Fig. 5.

On the other hand it is necessary to underline the fact that usage of contact with presence of liquid is not possible because of sample porosity and subsequent capillary effect changes the electrical properties of the samples. From the potential practical applications of these materials it is not a good message.

It has been a reason why we tested also evaporated (dry) Au contacts. Nonlinearity (at smaller voltage) of V-A characteristic is visible on Fig. 5. The same situation is in the case of PD surface. Also, the absolute values of the specific conductivity differ

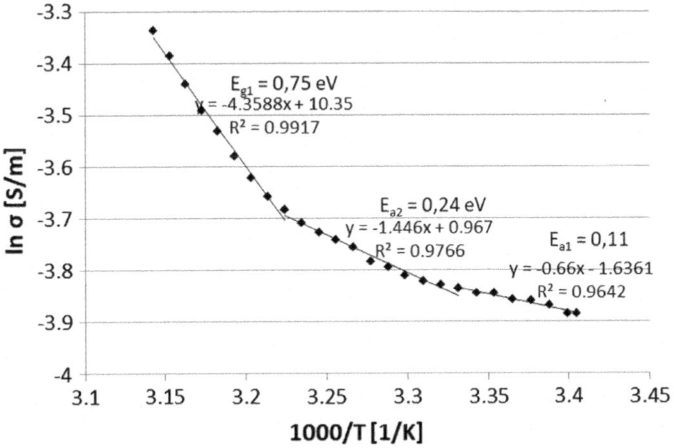

Fig. 6 Temperature dependence of specific conductivity for virgin sample and IP surface

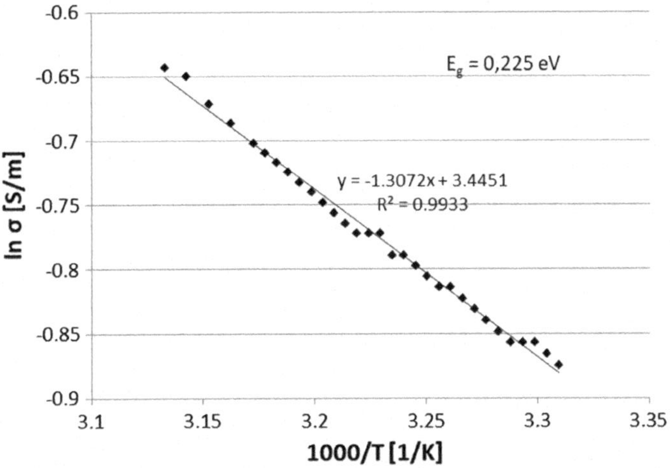

Fig. 7 Temperature dependence of specific conductivity for virgin sample and PD surface

from that in Fig. 5 which is probably caused by interaction of Au with structure of the composite mainly above 4 V.

In Figs. 6 and 7, we present the temperature dependences of specific conductivity for virgin samples and on both studied surfaces (see Eq. 1.4). For IP surface, we can see relatively complicated electrical transport with three breaks on the plot. First two breaks from left side of temperature scale corresponds to impurity conductivity with activation energies $E_{a1} = 0.11$ eV (probably π-electrons from PANI structure) and $E_{a2} = 0.24$ eV (probably Na cations) and the final break represents intrinsic conductivity where $E_g = 0.75$ eV is equal to forbidden gap width (Fig. 6).

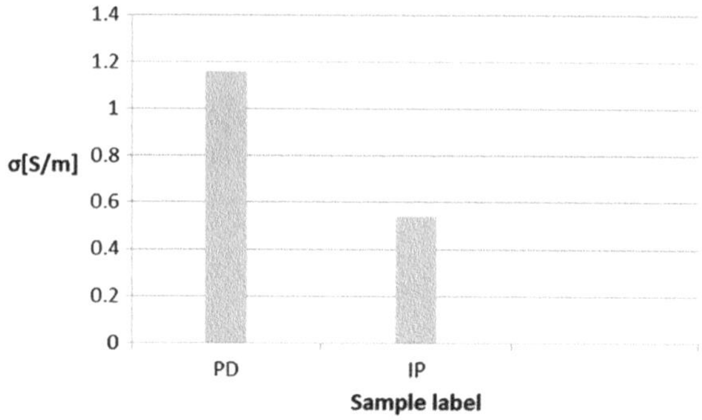

Fig. 8 Changes in specific conductivity after sample overheating up to 50 °C

On the other hand, the temperature dependence of the specific conductivity for PD surface is presented in Fig. 7. As it can be seen in this graph, the plot is relatively flat without breaks with unique activation energy equal to 0.225 eV which corresponds to the forbidden gap width.

Figure 8 depicts the differences between the conductivity of PD and IP surface after overheating. It is necessary to underline the sensitivity of the intercalated PANI to overheating up to 50 °C. As it can be seen, the usual form for the specific conductivity of virgin samples $\sigma_{pd} < \sigma_{ip}$ is now inversed. Such behavior is probably as a result of irreversible changes in PANI intercalates caused by overheating.

5 The Final Analysis Concerns Ac. Conductivity Measurements

The samples for measurements of electrical and dielectric properties were prepared without coating of standardly used conductive film.

For the AC measurements, an LCR Hi-tester Hioki 3522-50 was used, in the frequency range of 0.2 Hz–100 kHz. The isothermal measurement was performed at 20 °C, for 10 min. Values of electrical conductivity (direct and alternate), complex permittivity.

The measurement was carried out in the frequency range of 0.2 Hz–100 kHz. From the values of ac. electrical capacity and electrical resistivity the values of ε', ε'' were calculated (1.9, 1.10). Dependences of ε', ε'' versus frequency for PD surface are plotted in Figs. 9 and 10. The low frequency data are as inserts in relevant Figures. Shape of Figures for IP surface was resembled. The numerical values of dielectric parameters are presented in Table 1. From the data analysis we can see relatively high values of ε'' which reflects probably influence of moving π-electron charges and corresponding losses. This value is greater for IP surface, where is also higher the ac. and dc. electrical conductivity.

On the other hand polarization charge is substantially smaller for IP surface (see also values of σ_p calculated from Eqs. 1.4 and 1.5 for frequency 0.2 Hz). Notably, very

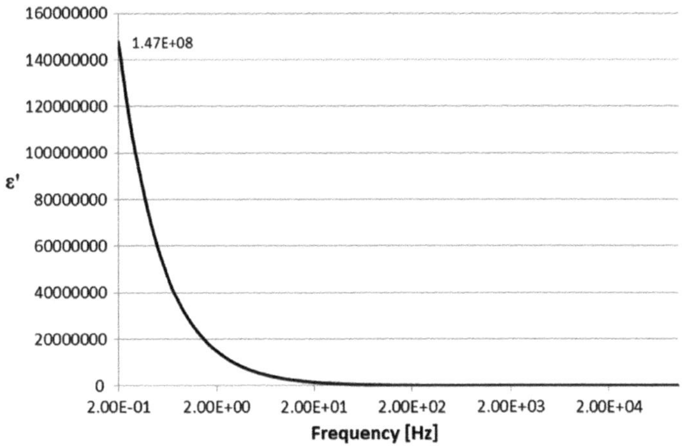

Fig. 9 ε′ versus frequency for PD surface

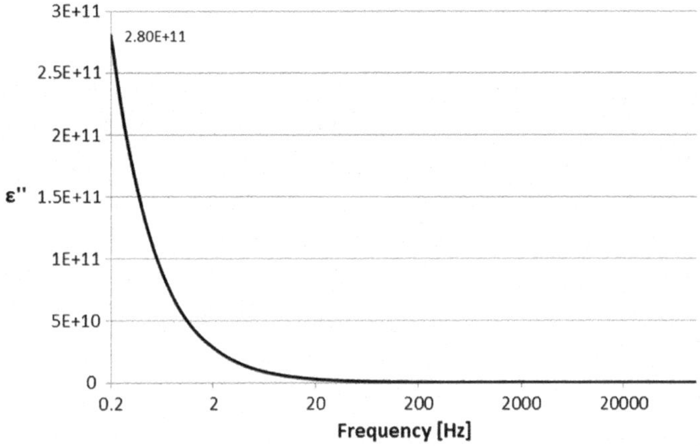

Fig. 10 ε″ versus frequency for PD surface

good agreement between both measured and calculated values of ε'' (first and last column) was obtained. Finally, in Fig. 11 the measurement of σ_{ac} versus frequency is presented. Its value for frequency range of 0.2–1000 Hz corresponds very good to the σ_{dc} reported above ($\sigma_{ac} = 3.12$ S/m at 0.2 Hz and $\sigma_{dc} = 3.9$ S/m). It must be underlined the fact, that both values were obtained by independent methods.

6 Conclusions

On the base of presented results it can be concluded:

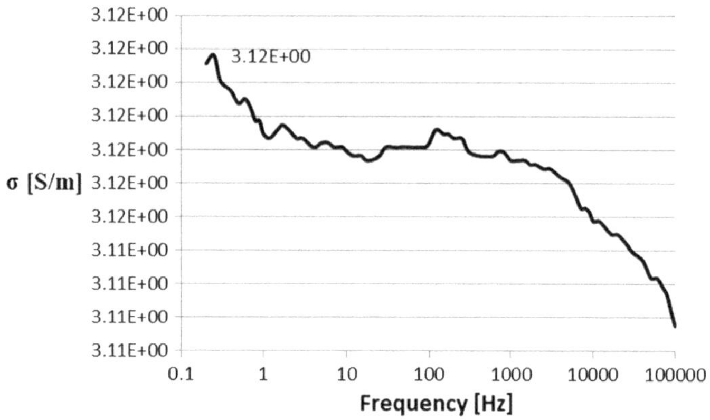

Fig. 11 σ_{ac} versus frequency

- Samples under investigation show anisotropy of *dc.* and *ac.* specific electric conductivity and mainly their relatively high values.
- *V-A* characteristics are linear.
- Temperature dependence of the σ_{dc} has relatively complicated shape for the IP surface.
- The studied samples show piezoresistive effect.

Acknowledgements This paper was created with the financial support of the internal grant of TBU in Zlín No. IGA/FT/2018/004 funded from the resources of specific university research.

References

1. Weiss Z, Kužvart M (2005) Jílové minerály: jejich nanostruktura a využití. Vyd. 1. Praha: Karolinum, ISBN 80-246-0868-5
2. Silicate Structures. Tulane University [online]. USA, [cit. 2016-02-27], http://www.tulane.edu/~sanelson/eens211/silicate_structures08.htm
3. Tokarský J, Mamulová-Kutláková K, Neuwirthová L, Kulhánková L, Stýskala V, Matějka V, Čapková P (2013) Texture and electrical conductivity of pellets pressed from PANI and PANI/montmorillonite intercalate. Acta Geodynamica et Geomaterialia 10(3):371–377. https://doi.org/10.13168/AGG.2013.0036
4. Kulhánková L, Tokarský J, Matějka V, Peikertová P, Vallová S, Mamulová-Kutláková K, Stýskala V, Čapková P (2014) Electrically conductive and optically transparent polyaniline/montmorillonite nanocomposite thin films. Thin Solid Films 562:319–325. https://doi.org/10.1016/j.tsf.2014.05.006
5. Tokarský J, Kulhánková L, Stýskala V, Mamulová-Kutláková K, Neuwirthová L, Matějka V, Čapková P (2013) High electrical anisotropy in hydrochloric acid doped polyaniline/phyllosilicate nanocomposites: effect of phyllosilicate matrix, synthesis pathway and pressure. Appl Clay Sci 80–81:126–132. https://doi.org/10.1016/j.clay.2013.06.029
6. Gregory RV, Kimbrell WC, Kuhn HH (1989) Conductive textiles. Synth Met 28:823–835

7. Sapurina I, Stejskal J (2008) The mechanism of the oxidative polymerization of aniline and the formation of supramolecular polyaniline structures. Polym Int 57:1295–1325. https://doi.org/10.1002/pi.2476
8. Sapurina I, Riede A, Stejskal J (2001) In-situ polymerized polyaniline films: 3. Film formation. Synth Metals 123(3):503–507. https://doi.org/10.1016/S0379-6779(01)00349-6
9. Lu J, Zhao X (2002) Electrorheological properties of a polyaniline-montmorillonite clay nanocomposite suspension. J Mater Chem 12(9):2603–2605. https://doi.org/10.1039/B203921D
10. Lerf A, Čapková P: Nalva H. S. (Ed.) (2004) Encyclopedia of nanoscience and nanotechnology, vol 2, pp 639–694, American Scientific Publishers, Stevenson Ranch, CA, USA
11. Zheng H, Feng X, Zhou L, Ye Y, Chen J (2016) Intercalated polyaniline-kaolinite nanocomposite prepared via in situ mechanochemical synthesis. J Appl Polym Sci 133(32)
12. do Nascimento GM, Pradie NA (2016) Deprotonation, Raman dispersion and thermal behavior of polyaniline-montmorillonite nanocomposites. Synthetic Metals 217:109–116. https://doi.org/10.1016/j.synthmet.2016.03.016
13. Abd El-Ghaffar MA, Youssef AM, Abd El-Hakim AA (2015) Polyaniline nanocomposites via in situ emulsion polymerization based on montmorillonite: preparation and characterization. Arab J Chem 8(6):771–779. https://doi.org/10.1016/j.arabjc.2014.01.001
14. Lagaly G, Ogawa M, Dékány I (2006). In: Bergaya F, Theng BKG, Lagaly G (eds) Handbook of clay science, developments in clay science, vol 1, pp 309–377. Elsevier, Amsterdam
15. Piao SH, Kwon SH, Choi HJ (2016) Stimuli-responsive polymer-clay nanocomposites under electric fields. Materials 9(1):52. https://doi.org/10.3390/ma9010052
16. Tokarský (2016) Habilitation thesis, Technical University of Ostrava
17. Hamann C, Burghardt H, Frauenheim T (1988) Electrical conduction mechanisms in solids. VEB, Berlín
18. Owen AE (1963) Electric conduction and dielectric relaxation in glass. In: Progress in ceramic science, vol 3 (Burke J.). Pergamon Press, Paris
19. Daniel VV (1967) Dielectric relaxation. Academic press, New York
20. Kubliha M (2007) Exploitation of electric dielectric methods for material research non-metallic materials. Alumni Press, Trnava

Chosen Physical Properties of Menzolit BMC 3100

S. Rusnáková, K. Karvanis, P. Koštial, Z. Koštialová Jančíková and A. Zimula

Abstract This work deals with the study of physical properties of the composite Menzolit BMC 3100 (further BMC), which has been developed especially for use in the front headlamps reflectors in automotive industry. The IR spectra of the BMC were measured and described together with Charpy impact strength and Brinell hardness.

Keywords Thermogravimetric analysis · Hardness

1 Introduction

Polymer matrix composites play a very important role in modern, mainly automotive industry [1–4]. Nowadays, composite materials, in most cases fiber reinforced polymers, are used in many applications where light weight, high specific modulus and high strength are critical issues. In this paper, the constituents of these materials and their special advantages compared to traditional materials are described. Further details are outlined regarding the present markets of polymer composites in Europe and their special applications in the automotive industry. In particular, the manufacturing of parts from thermoplastic as well as thermosetting, short and continuous fiber reinforced composites is emphasized in [5].

The question about the future use of this material remains as modern lighting systems are already working with "cold" light emitting diodes (LEDs) and laser, where thermoplastic materials such as polycarbonate have sufficient temperature resistance.

S. Rusnáková (✉) · K. Karvanis
Department of Production Engineering, Faculty of Technology, Tomas Bata University in Zlín, Vavrečkova 275, 760 01 Zlín, Czech Republic
e-mail: rusnakova@utb.cz

P. Koštial · Z. Koštialová Jančíková · A. Zimula
Department of Material Engineering, Faculty of Metallurgy and Materials Engineering, VŠB—Technical University of Ostrava, 17. listopadu 15/2172, 708 33 Ostrava, Czech Republic
e-mail: pavel.kostial@vsb.cz

A. Öchsner and H. Altenbach (eds.), *Engineering Design Applications II*, Advanced Structured Materials 113, https://doi.org/10.1007/978-3-030-20801-1_12

However, the properties of an advanced composite depend not only on the type of the matrix and the reinforcing materials but also on a factor which is distinct from the composition: the geometry of the reinforcement. A particular type of the composites is the injection moldable short fiber reinforced thermoplastics in which can be controlled only the concentration of the strengthening fibers; not their exact dimensions or orientation.

The Menzolit BMC 3100 [6] is a special material developed for the production of spotlight headlamps in the automotive industry. This polymer matrix composite replaces the traditional metal reflectors due to its simpler and less expensive usage in serial production.

2 Experimental Conditions

For the qualitative analysis of an organic matter the effect of the infrared absorption and its detection by an IR spectrophotometer can be used. Since the BMC is a composite material consisting of polymer, mineral filler and glass fiber, its resulting IR spectrum is somewhat complex (Fig. 1), and software evaluation may give misleading results in the wrong procedure. Therefore, it is necessary to separate the components from the spectrum and to obtain only the bands corresponding to the polymer matrix.

Inorganic filler analysis was performed on a sample through a thermogravimetric analysis of the BMC material. This is the rest after burning in the AIR atmosphere. The inorganic residue was identified by FTIR as CaO (characteristic wavelength from graph, peak is marked by arrow). Calcium oxide (Fig. 2) was formed by the thermal decomposition of the original mineral filler $CaCO_3$ (Fig. 3).

By separating the bands corresponding to the $CaCO_3$ filler, a polymer matrix spectrum was obtained which was subjected to further identification. The resulting spectrum was identified as a styrene-acrylate copolymer. The polymer component in uncured BMC material is an unsaturated resin containing acrylate groups and styrenes. Then, their reaction creates a thermoset (Fig. 4).

Also, in order to complete the data concerning about the BMC, its Brinell hardness at room temperature was measured, by ZWICK/ROELL instrument which was recorded as (369.4 ± 9.2) HB.

Determination of the Charpy impact strength was performed on a fully automated device 5.5 Ceast. The principle of this measurement is the determination of the amount of the impact work consumed during the initiation and spread of the crack into the quarry. The stroke is perpendicular to the longitudinal axis of the test body. The measured value found to be (11.9 ± 2.4) kJ/m^2.

The BMC uncured material contains a reactive polyester resin, filler and glass fibers in the form of a plastic mixture, suitable for processing in screw-type injectors. Glass fibers are by far the most predominant type of fibers used in the reinforced polymer industry and they are also among the most versatile. Although melting glass and drawing it into fibers is an ancient technique, long continuous fibers drawn from

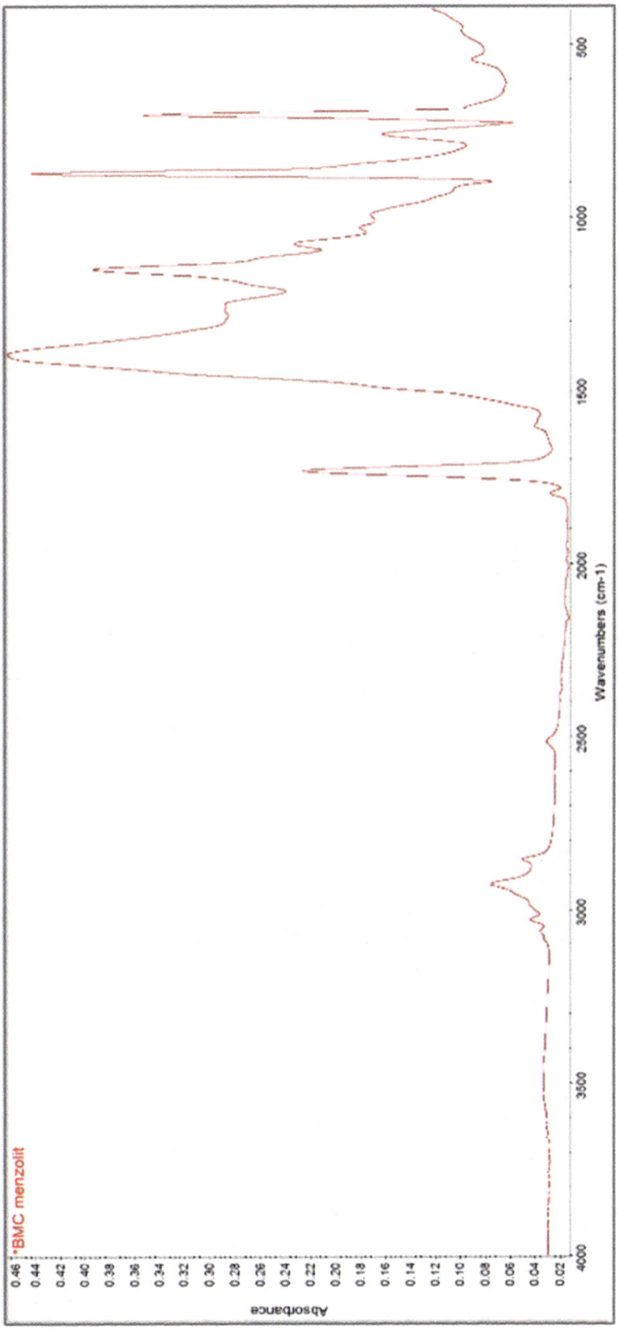

Fig. 1 IR spectrum of BMC—absorbance versus wavelength

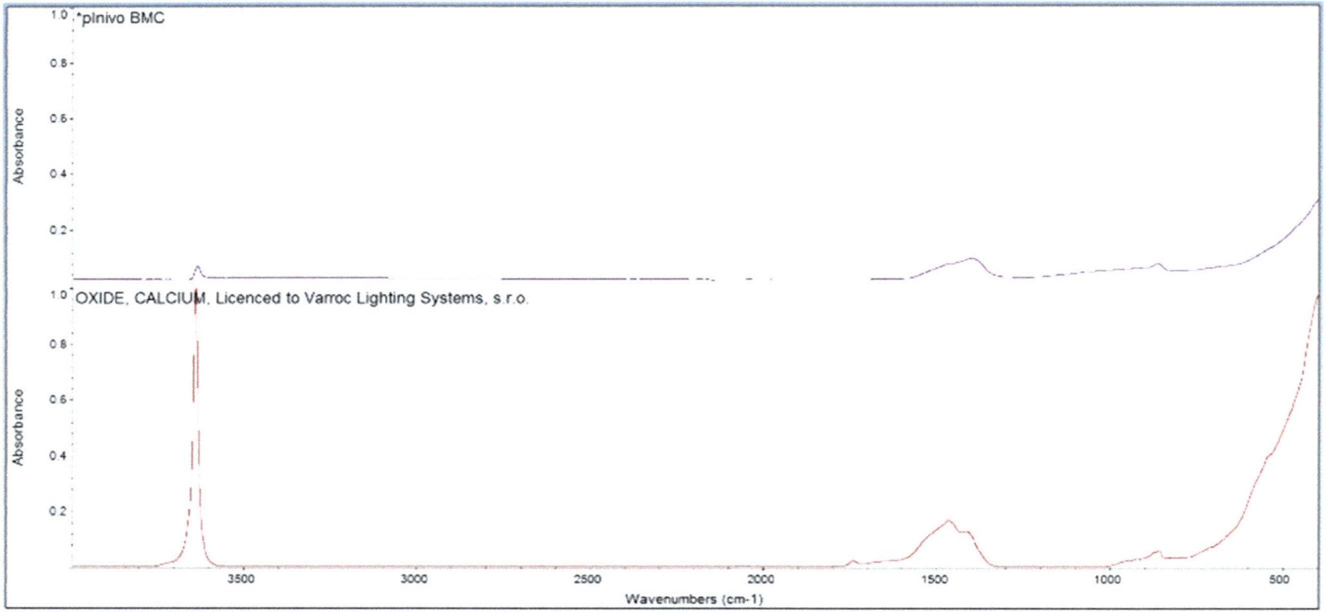

Fig. 2 IR spektrum—absorbance versus wavelength. Upper curve—for the sample obtained from thermogravimetric analysis of BMC material. Down curve-comparative spectrum from the library of inorganic spectra

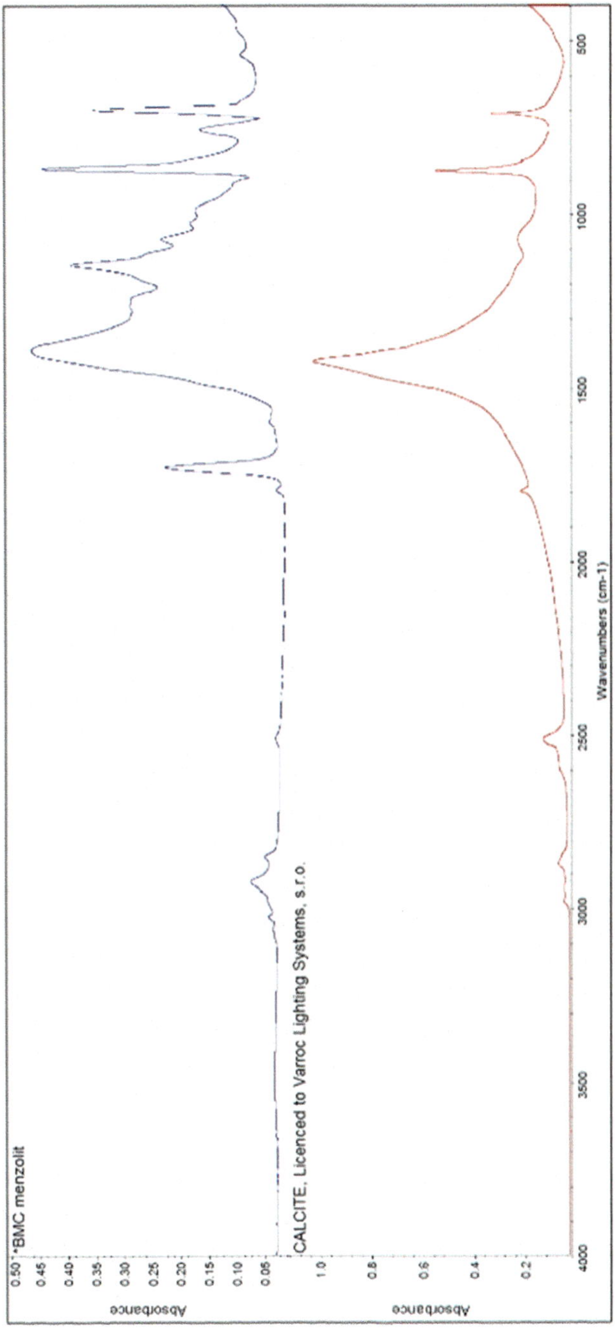

Fig. 3 IR spectrum-absorbance versus wavelength. Upper curve—BMC spectrum. Down curve-comparative sample of limestone $CaCO_3$ from the library of inorganic spectra

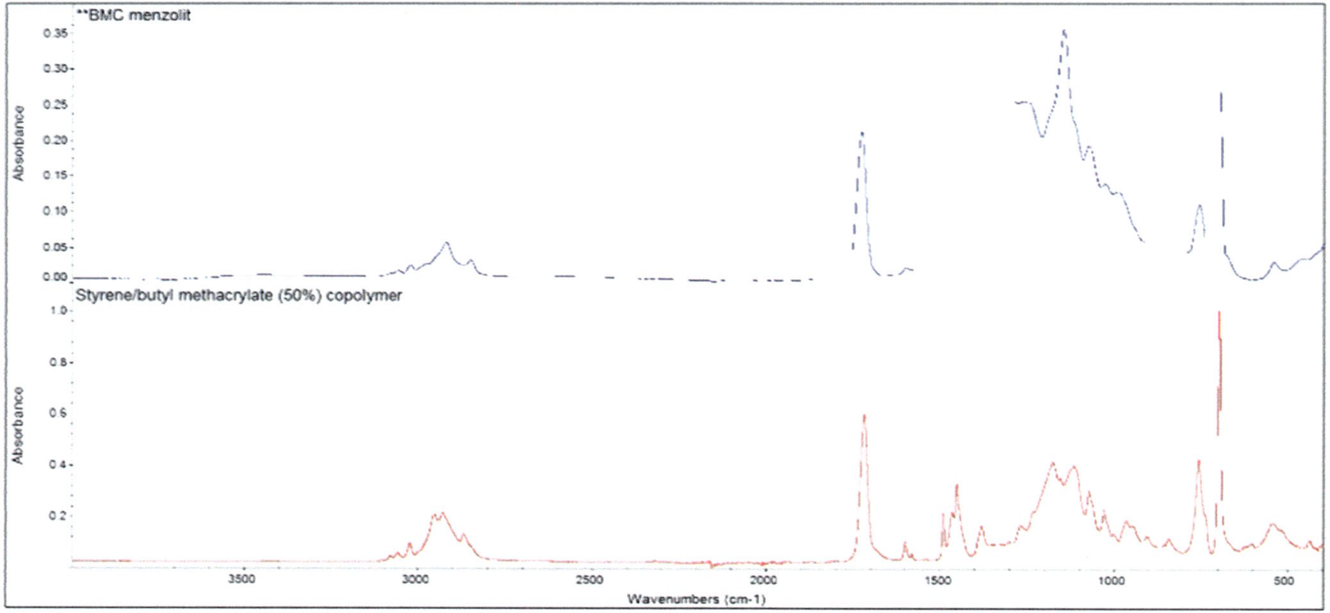

Fig. 4 IR spectrum absorbance versus wavelength. Upper curve—BMC after separation of $CaCO_3$ peak. Down curve—a comparative sample of the styrene-acrylate copolymer from the polymer spectra library

glass were introduced in the 1930s by Owens-Corning as glass wool and to these were given the name fiberglass.

The distribution of the filler (glass fiber) is not preferentially oriented, but they are randomly arranged or form the bonded bundles connected by the mineral filler and the resin.

The length of the glass fibers is about 4.5 mm while their diameter is about 14 μm. It should be noted that during the material processing, especially in the press, breaking and shortening can occur.

This composite material has a relatively high temperature resistance; it is sufficiently heat-resistant to withstand long-term temperatures around 200 °C generated by halogen light sources. In addition, thanks to the mold injection technology, parts of precisely defined shapes and sizes can be manufactured, characteristics which are absolutely essential for the applications of the optical components. Subsequent coatings in the form of varnishing and vacuum coating enable the production of perfect parabolic systems for the reflection and routing of light rays of all kinds. Moreover, in the case of combustion, products from this material do not melt, drip or excessively smoke. This material is recommended to be injected into heated steel molds, preferably with chrome-plated tools.

3 Conclusions

The material Menzolit BMC 3100 is very few described in the scientific literature excluding the material list of the producer. The material list offers some values of its physical properties characterizing it but details about the materials' composition are not known. This paper presents details about the composition of the BMC obtained by the study of the IR spectra and finally the material components are described. Also, the results of measurements of Brinell hardness and Charpy impact strength are presented.

Acknowledgements This paper was created with the financial support of the internal grant of TBU in Zlín No. IGA/FT/2018/004 funded from the resources of specific university research.

References

1. Ehrenstein GW (2009) Polymerní kompozitní materiály. Scientia, Praha. ISBN 978-80-86960-29-6
2. Ferry JD (1980) Viscoelastic properties of polymers. Wiley, New York
3. Brinson HF, Brinson LC (2014) Polymer engineering science and viscoelasticity. Springer, Berlin
4. Friedrich K, Almajid AA (2013) Manufacturing aspects of advanced polymer composites for automotive applications. Appl Compos Mater 20:107–128. https://doi.org/10.1007/s10443-012-9258-7
5. http://www.build-on-prince.com/glass-fiber.html
6. Product Information: menzolit® BMC 3100, 21 March 2014

Structural Analysis by Finite Element Method in Ball Valves to Improve Their Mechanical Properties

M. Egure-Hidalgo, J. M. Aburto-Barrera, C. R. Torres-San Miguel, J. Martinez-Reyes and B. Romero-Ángeles

Abstract Thermoelectric power plants have steam lines that are mainly structured by pipes, y-pattern globe valves and ball valves. According to international databases (Rogers in The Guardian 18, 2011 [7]), the most common problems in valves are erosion and material thermal expansion that produce pressure drop, heat loss, material leakage and crack nucleation. For these reasons, while engineering process is applied, structural analyses are generated to show the material behavior under a linear external agent (Stolarski et al. in Engineering analysis with ANSYS software. Butterworth-Heinemann, 2018 [8]). However, the steam tends to behave in a saturated way before it becomes in a superheated steam to improve thermal efficiency and work capability. So, through Mollier diagrams, ASME and ASTM standards, mechanical properties of materials and the boundary conditions that supply a fixed value constrain, and contact constrains presented in ball valves. It's proposed a variable structural elastic-linear static analysis with an isotropic material using the Finite Element Method. Where a homogeneous behavior in the variation of pressure and temperature represent the structural changes of the valve while is working with saturated and superheated steam with the aim to consider and understand the critical components of the assembly, which in this case of study are; the body, the ball and the seats. In such a way

M. Egure-Hidalgo · J. M. Aburto-Barrera · C. R. Torres-San Miguel · B. Romero-Ángeles (✉)
Escuela Superior de Ingeniería Mecánica y Eléctrica, Sección de Estudios de Posgrado e Investigación, Unidad Profesional Adolfo López Mateos "Zacatenco", Instituto Politécnico Nacional, Edificio 5, 2° Piso Col. Lindavista, C.P. 07360 Ciudad De México, Mexico
e-mail: romerobeatriz97@hotmail.com

M. Egure-Hidalgo
e-mail: eguremil@gmail.com

J. M. Aburto-Barrera
e-mail: josemaria620@hotmail.com

C. R. Torres-San Miguel
e-mail: ctorress@ipn.mx

J. Martinez-Reyes
Escuela Superior de Ingeniería Mecánica y Eléctrica, ISISA, Unidad Profesional Adolfo López Mateos "Zacatenco, Instituto Politécnico Nacional, Edificio 2, Col. Lindavista, C.P. 07360 Ciudad De México, Mexico
e-mail: jacmartinez@esimez.mx

A. Öchsner and H. Altenbach (eds.), *Engineering Design Applications II*,
Advanced Structured Materials 113, https://doi.org/10.1007/978-3-030-20801-1_13

that, applying a load similar to the fluid characteristics and unloading the system, total displacements and principal stresses can be obtained during each step. And through Von Mises failure theory. It can be obtained the more prone areas to failure by elongations/contractions that could affect the correct function of the valve during service.

Keywords Finite element method · Ball valve · Structural analysis · 3D modeling

1 Introduction

Valves are mechanical elements used for fluid control and may be considered important and a reference for nuclear plants since they are engaged not only for the transferring of fluids, but also about the protection of main elements like the turbine and steam generator. During operation, the body and other fluid contact elements are subjected to repetitive mechanical and thermal loads hence, either plastic deformations nor cracks nucleation can be reached by these elements. So, it's important to develop during their design process an analysis concerning fluid and heat transference to improve the comprehension on the behavior of the valve under real working conditions to obtain an ideal flow model to optimize the internal structure of the mechanism.

The following study simulates the structural behavior of the internal components of a ball valve under two different loading conditions to foresee the total deformations, as well as the maximum stresses to establish which part is the most prone to failure. The first condition is a numerical analysis for a single load corresponding to saturated steam. For a second condition, it is depicted the change of phase, going from superheated to saturated steam, ending with a transition from the linear load of saturated steam to a total discharge of the system, represented as the depressurization due to a complete shut of the line. These conditions are present in the conduction of steam and are common in a thermoelectric or a nuclear plant, so for a better approach to the behavior of a valve, simulations must thoroughly consider the actual states of the fluid.

2 Methods

The valve used for the analysis was of floating ball type, class 4500#, with SW (socket weld) endings, ¾″ NPS diameter, quarter turn operation, and full port. The 3D model was developed with Solidworks® software to understand the geometry of the components and establish the working conditions that are applied to each part once it is assembled, as shown in Fig. 1.

Since the body and the mounting plate comprise a single piece, it can be understood that W equals the mass of the handle, A is the mounting area of the handle, P is

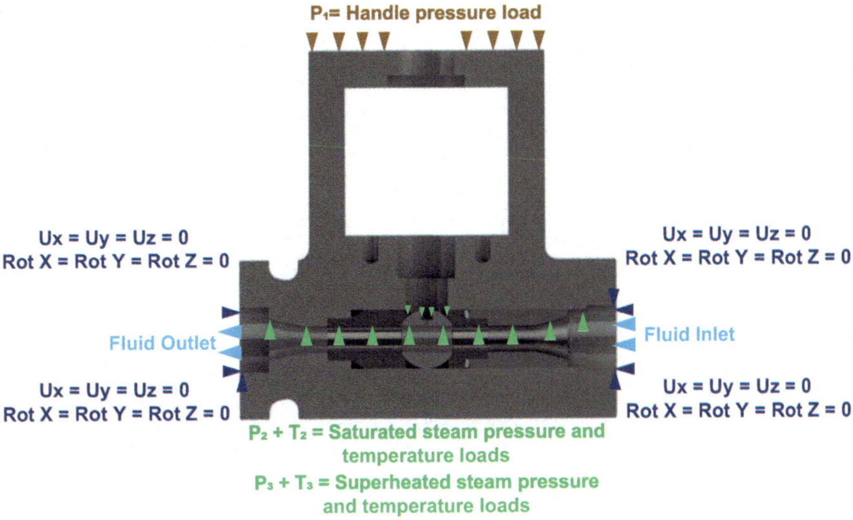

Fig. 1 Boundary conditions and external agents affecting the internal parts of the valve

pressure, T represents the temperature, and specifically P_1 is considered the pressure coming from the handle.

Therefore, by applying Newton's second law [2].

$$F = m\frac{dv}{dt} = ma \quad (1)$$

It can be obtained that pressure P_1 corresponds to the normal force component F on surface A [6].

$$P_1 = \frac{F}{A} \quad (2)$$

At the same time, the ideal wall thickness is calculated on base to cylindrical pressure vessels theory [4] and A.S.M.E. code section VIII [5] to verify that the valve wall thickness employed in the design matches or exceeds the obtained measurement. In addition, it is performed to preserve a factor safety that meets the standard requirements.

For the first analysis, two internal loads are applied to each element, corresponding to the values of saturated steam (P_2 and T_2). On the other hand, for the second analysis, different loads are applied, that corresponds to a change between superheated steam (P_3 and T_3) to saturated steam (P_2 and T_2). To perform the analysis, the input data is as follows:

$$W_{handle} = 5.37\,\text{kg}$$

Table 1 Material mechanical properties [3]

Piece	Material	Young modulus (MPa) [E]	Poisson ratio [ν]	Density (g/mm^3) [ρ]
Body	ASTM A182 F316 (UNS S31600)	193,000	0.275	0.008
Ball and seat	AISI 410SS (UNS S41000)	210,000	0.28	0.0078
T-sleeve and pusher seat	AISI 416SS (UNS S41600)	210,000	0.28	0.0077

$$A_{handle} = 8880.61\ \text{mm}^2$$

$$P_1 = \frac{5.37 * 9.81}{8880.61} = 0.00593\ \text{N/mm}^2$$

$$P_2 = 2611\ psig = 18.002\ \text{N/mm}^2$$

$$T_{\perp} = 357\,^{\circ}\text{C}\,[674.6\,\text{F}]$$

$$P_3 = 3910\,\text{psig} = 27\ \text{N/mm}^2$$

$$T_3 = 538\,^{\circ}\text{C}\,[1000.4\,\text{F}]$$

After confirming that the fluid pressure mainly acts on the body, seat, ball, pusher seat and t-sleeve, each part was analyzed by means of the Finite Element Method (FEM) in the ANSYS® APDL software, using high order element type of brick and tetrahedral shapes, with 20 nodes per element and three degrees of freedom per node (Solid 20 nodes 186) for a static structural analysis [1] of linear-elastic and isotropic type to determine the stress field at which contacting parts are subjected under the working loads corresponding to each of the conditions previously described. Prior to discretization, mechanical properties of the material need to be set. Values mentioned in Table 1, accomplish the characteristics of the physical device.

The pre-process, due to the characteristics of the body and ball geometries, they were initially generated in the Solidworks® software and later exported as IGES (Initial Graphics Exchange Specification) to ANSYS® APDL application. The t-sleeves, pusher seat and seat were generated by a controlled mesh in the second program, obtaining the results of Table 2.

Table 2 Meshing results

Case study	Part name	# of elements	# of nodes
1	T-sleeve	163,80	75,360
2	Pusher seat	8100	37,740
3	Ball	72,033	111,588
4	Seat	14,040	62,580
5	Body	493,597	697,239

2.1 Boundary Conditions and External Agents

An important part to be considered for the numerical analysis is the correct application of boundary conditions and external agents because an erroneous application will result in a simulation differing with the actual behavior of the part and it won't be possible to understand the contact conditions that exist in the assemblies.

As shown in Fig. 1, the first study case, is for the t-sleeve, located inside the right hand of the body of the valve. Restricted on the cylindrical surface. Within assembly, it keeps contact with the body of the valve, so a boundary restriction is applied in relation to the planes and rotations *y* and *z*. The planar surface that contacts the spring and the pusher seat has a restriction on *x*. While the red area represented in Fig. 2 corresponds to pressures and temperatures 2 and 3 belonging for saturated and superheated steam values.

The pusher seat is the element located on the right side next to the ball as shown on Fig. 1 and is restricted by the cylindrical surface, and contact is maintained with the valve body with respect to the planes *y* and *z*, in addition the spherical surface that contacts the ball has a restriction on *x* (constraints represented with arrows aquamarine). The red area shown in Fig. 3 corresponds to the pressures and temperatures 2 and 3.

The ball or valve plug is considered as the third study case. As a floating ball type, it is restricted in the axes *x*, *y* and *z* on the spherical surface, where the housing groove of the rod restricts rotations. Specifically, planar faces restrict rotation in *y* and *z*, while the curved portion prevents rotation in *x*. Therefore, the green arrows represent constraints with respect to the *x*, *y* and *z* axes, the orange arrows represent the constraints of the rotations in *x*, *y* and *z*, and the red area shown in Fig. 4 are the pressures and temperatures 2 and 3.

The fourth study case is for the seat located next to the ball on the left side in the assembly of Fig. 1 and it is restricted in the cylindrical surface which during assembly keeps contact with the valve body at the planes *y* and *z*. In addition, there is

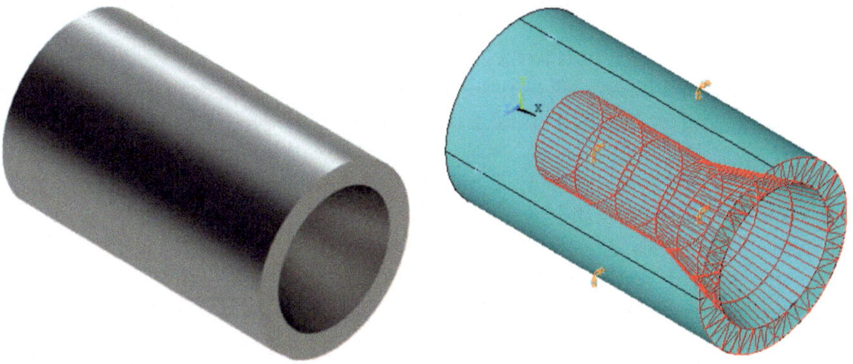

Fig. 2 Boundary conditions and external agents on t-sleeve

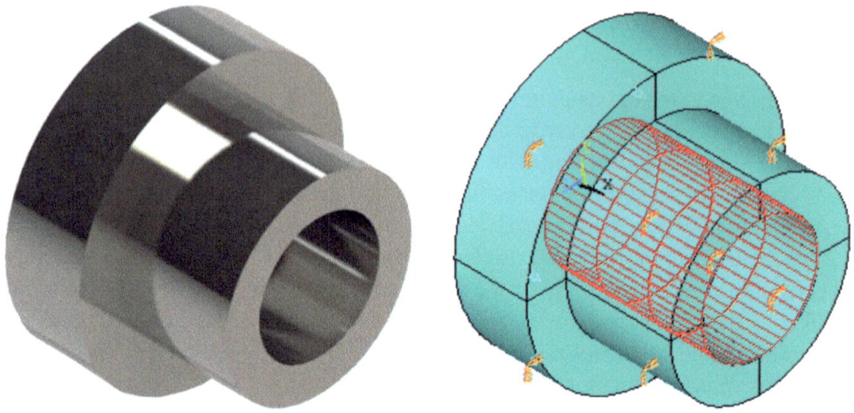

Fig. 3 Boundary conditions and external agents on pusher seat

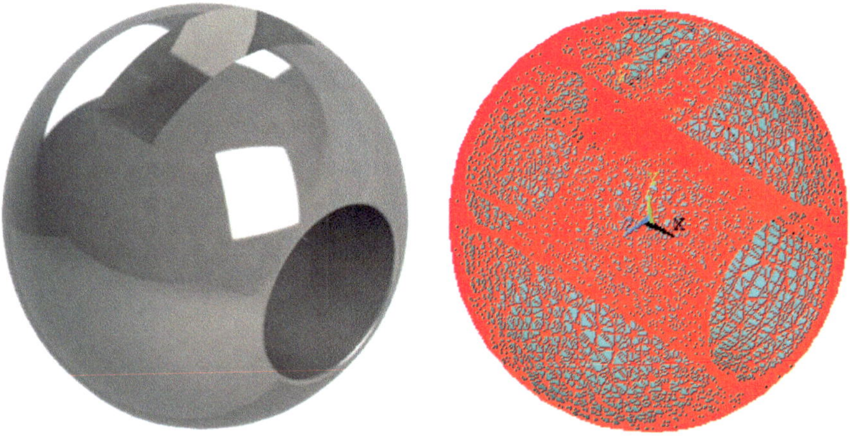

Fig. 4 Boundary conditions and external agents on ball

a restriction in x with respect to the planar surface of the body of the valve which is always in contact with as well as to the spherical part that is kept in contact with the ball. The red area represented in Fig. 5 corresponds to the pressures and temperatures 2 and 3.

Finally, the valve body is the fifth study case. It has a free mesh due to its geometry and is totally restricted by the Socket Weld housing of both inputs. The red area is the pressure and temperature of the fluid, applied over the body's area when passing by, although if the ball did lack the equalizer bore only the conical area would receive the applied loads of the external agents, where the pressures and temperatures 2 and 3 (Fig. 6). The upper part is the area in which the pressure 1 was applied, as shown in Fig. 1, and corresponds to the weight of the handle.

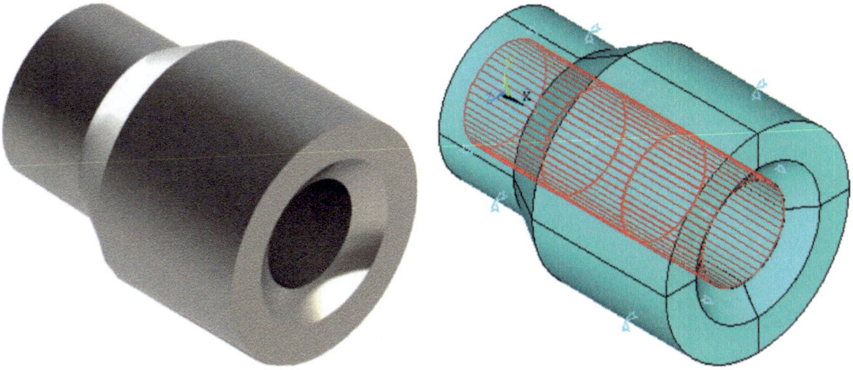

Fig. 5 Boundary conditions and external agents on seat

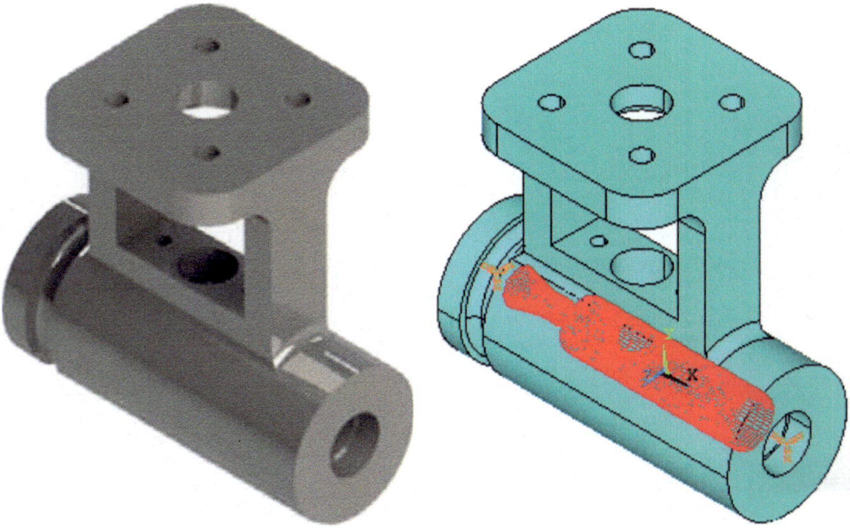

Fig. 6 Boundary conditions and external agents on body

3 Results

Once the pre-processor is solved, the ANSYS® APDL program automatically displays a graphic solution obtained by the post-processor in color gradients that provide the resulting displacements (mm) and stresses (MPa). Thus, as shown in Figs. 7 and 8, the total displacements (mm) and maximum stresses, based on the Von Mises (MPa) failure theory corresponding only to the first analysis related to saturated steam transport are as follows:

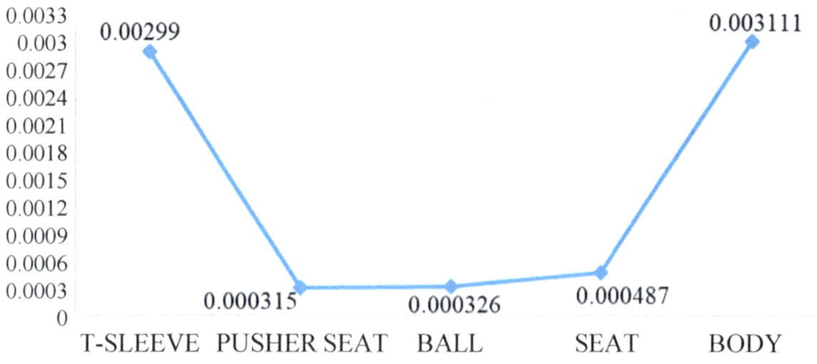

Fig. 7 Total displacements in mm related to saturated steam

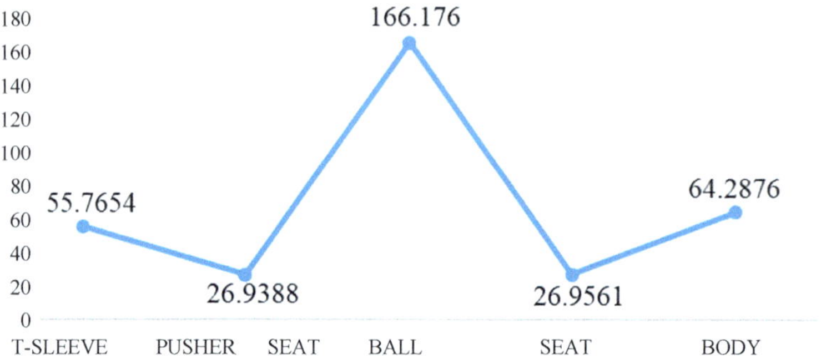

Fig. 8 Maximum stresses by von Mises in MPa for saturated steam

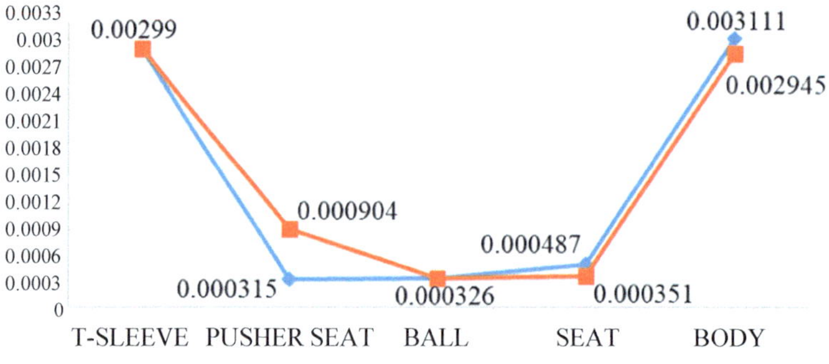

Fig. 9 Total displacements in mm between first and second analyses

Once applied the loads that correspond to superheated steam with a temperature change like saturated steam, the results are depicted in orange color, meaning total displacements (mm) and the maximum stresses (MPa) (Figs. 9 and 10).

The obtained results make evident that the greatest change in geometry is suffered by the body and the pusher seat (Figs. 11, 12, 13 and 14) due to pressure and thermal expansion, hence it is shown that applying a single load in the structural analysis is

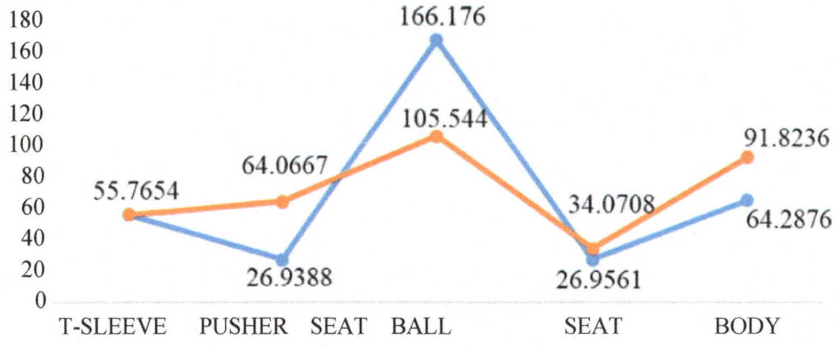

Fig. 10 Maximum stresses by von Mises in MPa between first and second analyses

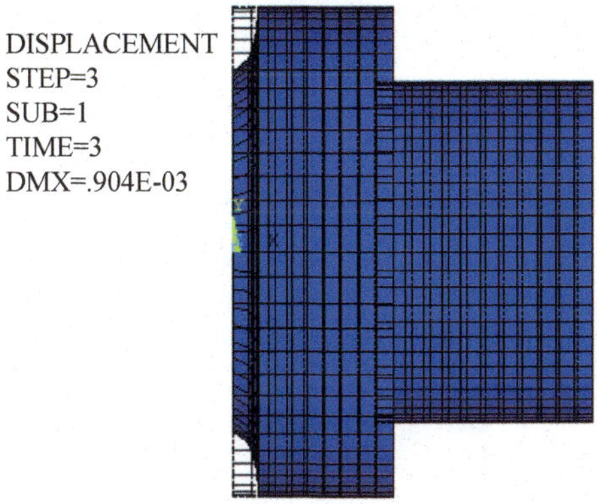

Fig. 11 Total displacements related to pusher seat in the second analysis

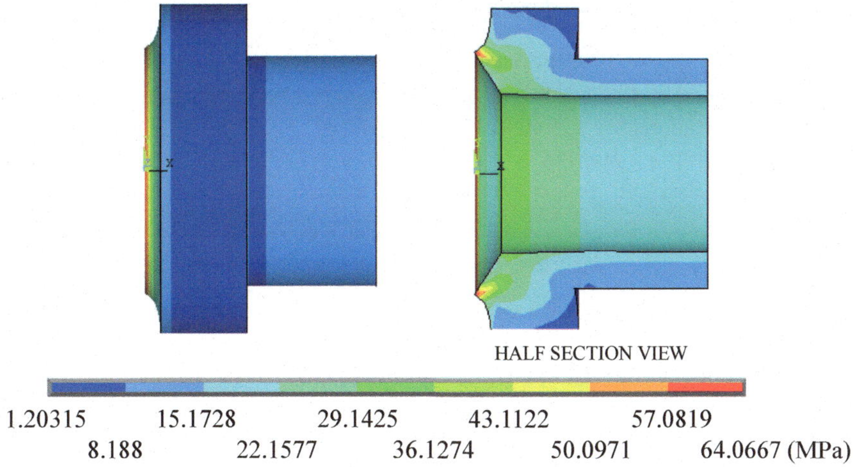

Fig. 12 Maximum stresses by von Mises failure related to pusher seat in the second analysis

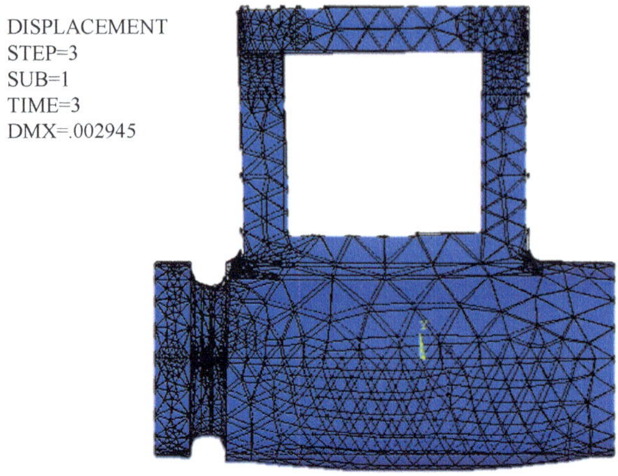

Fig. 13 Total displacements related to the body in the second analysis

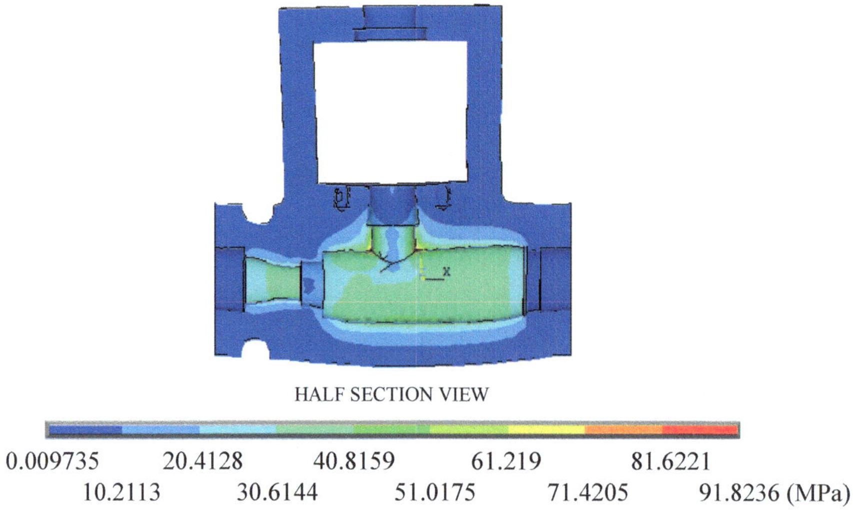

Fig. 14 Maximum stresses by von Mises failure related to the body in the second analysis

not always the best option since a fluid rarely behave in linear and continuous way through time. Therefore, these variations in the analysis can help to estimate the life span of the product under extreme conditions.

In sum, this analysis allows the Design Engineer to predict possible structural failures that can be solved through variations of shape or material. The use of FEM allows to broaden the perspective relating the kind of issues and analysis that must be performed (prior to design and manufacturing) in any type of product, since it correlates the behavior of the parts during service, as well to increase the service time. In this specific case, as a product experiencing a growing demand in the market due to its liability as replacement part or to work in association with globe and Y-pattern valves for venting and draining service in high pressure and temperature systems, it is necessary to consider this type of analyses to achieve better efficiency in the finished product.

Acknowledgements The authors thank to the Instituto Politécnico Nacional and the Consejo Nacional de Ciencia y Tecnología f or the support provided in the elaboration of this work.

References

1. ANSYS Inc. (2011) Ansys mechanical APDL element reference, pp 61,62,975,1066
2. Bedford A, Fowler W (2000) Dinámica: Mecánica para ingeniería, vol 1. Pearson Educación
3. Flow-Tek Inc. (2016) High performance severe service: Metal Seated Ball Valves, Ed Bray International Inc. https://www.bray.com/ball-valves/severe-service/series-m4-severe-service-ball-valve. Accessed Dec 2017
4. Kinsky R (1986) Engineering mechanics and strength of materials. McGraw-Hill
5. Miller U, Pastor T (2009) Section VIII—division 1: rules for construction of pressure vessels. Companion guide to the ASME boiler and pressure vessel code, vol 2, 3rd edn. ASME Press
6. Mott R (1996) Mecánica de fluidos aplicada. Pearson Educación, pp 75–94
7. Rogers S (2011) Nuclear power plant accidents: listed and ranked since 1952. The Guardian, 18
8. Stolarski T, Nakasone Y, Yoshimoto S (2018) Engineering analysis with ANSYS software. Butterworth-Heinemann

Modelling of Cavitation Around Hydrofoils with Included Bubble Dynamics and Phase Changes

Galina Ilieva and Christo Pirovsky

Abstract Cavitation is an important problem in hydraulic machines as it affects their performance and cause damages. After the flow velocity and forces acting over blades are increased there are regions with presence of pressure drop well below the vaporization pressure of the liquid and cavitation is formed. Cavitation is associated with erosion, noise and vibrations, leading to efficiency decrease and machinery damages. That is why cavitation, its development and effects must be taken into account during the design phase of rotating machines, propellers and others. The present paper deals with a numerical modeling and analysis of cavitation around a hydrofoil. It aims at creation of a relevant model to simulate the cavitating flow around a hydrofoil. User Defined Code, presenting bubble growth and two-phase interaction effects, is written and implemented to the main mathematical model in Fluent. Current computational methodology is applicable to study cavitation in various exploitation regimes.

Keywords Bubble growth and implosion · Cavitation · Numerical simulation · Two-phase mixture · Schnerr-Sauer model

1 Introduction

Hydrodynamic cavitation is a process of vaporization, followed by bubble generation and bubble implosion, as a result of pressure drop.

Hydrodynamic cavitation is observed in flows passing through channels (nozzles, injectors) at a specific flow velocity or over rotating elements in the flowing fluid,

G. Ilieva (✉)
Research and Development Laboratory "Heat Turbomashines", Technical University of Varna, Varna, Bulgaria
e-mail: galinaili@yahoo.com

C. Pirovsky
Research and Development Laboratory "Vibration Control and Diagnostics of Machinery and Structures", Technical University of Varna, Varna, Bulgaria
e-mail: christo.pirovsky@tu-varna.bg

A. Öchsner and H. Altenbach (eds.), *Engineering Design Applications II*,
Advanced Structured Materials 113, https://doi.org/10.1007/978-3-030-20801-1_14

such as marine propellers, hydro turbines, pumps, nozzles, injectors, and underwater bodies, [1]. Cavitation appears in hydro turbines due to specific combination of design shapes and operating conditions.

What is behind the cavitation phenomenon? Cavitation occurs when the static pressure of the fluid under consideration decreases below its vapor pressure, then the fluid boils and, in the flow volume, a large number of small vapor bubbles are formed. Next, these small bubbles can travel with the main stream to higher pressure zones where the vapors condense and the bubbles suddenly collapse, because the vapor is condensed again. Another observed process is that bubbles formed over the streamed surfaces due to the higher pressure collapse and hit the surfaces. This results in the formation of pitting and cavities on surfaces that are being impacted. The process of cavities formation and high pressures, acting over the surfaces, takes place in several nanoseconds. It results in the formation of large amplitude shock-waves, as demonstrated in [2] and leading to damages of the material, it fails by fatigue, added by corrosion, [3].

Also, the process of bubble generation, growth and collapse, results in very high energy levels and local temperatures over the bubble surfaces, for a short time. Over the years, it is proved that uncontrolled cavitation is very damaging; however by controlling the flow of the cavitation, the power can be non-destructive. The cavitation phenomenon must stay under control for the purposes of enhanced performance of high-speed marine vessels, turbines, projectiles, as well as in material processing technologies and etc.

Cavitation, as phenomena, is one of the major problems in hydraulic machines and decreases significantly their efficiency. For instance, when flow velocity increases and pressure loading over streamed blades in hydro-turbines decreases, small regions where the pressure drop goes below the vaporization pressure of the liquid are observed and cavitation is developed. The cavitation phenomenon is accompanied not only with erosion effects, but with noise and vibrations that significantly decrease machinery performance. Also, cavitation in hydraulic machinery is a 3D complex and unsteady process and requires strong studies, both theoretically and experimentally, to be revealed in details and controlled to enhance the performance of machines and others where it appears.

Numerical simulation, research and analysis of cavitation and its effects flows is still an ongoing research effort, having the ambitious goal to model 3D complex, two-phase and unsteady flow with included specifics of phase changes and bubble dynamics.

Research, analysis and subsequent control of the cavitation processes can be achieved by advanced mathematical models and experimental research approaches. The first exact classical solution on this problem is the well- known solution by Helmholtz in 1868, [4].

Early studies of cavitation phenomenon are based on the potential flow theory. This approach is found to be able to describe partially cavitating two-dimensional hydrofoils, including the re-entrant jet cavity closure model, see [5]. However, these models are not able to predict the behavior in 3D complex cavitating flows.

The earliest studies on the theory of cavitating flows with free boundaries and supercavitation were published in [6–8].

Another research works that describe the applied calculation techniques based on the principle of cavity expansion independence, theory of pulsations and stability of elongated axisymmetric cavities, are described in [1, 9].

In [10] are discussed the advances in the area for the last three decades and the combination of classical mathematical approaches with contemporary computer methods. Also, elaboration of nonlinear numerical methods to solve 3D cavitation problems; improve known theories; develop theories to solve for axisymmetric and nearly axisymmetric flows, etc., are presented in [10].

Last years, have been studied and developed many cavitation models, aiming to solve a definite set of equations. In 1999–2011, several researchers, [11, 12] suggested for application the Rayleigh model. This model has a limitation—describes the case of inertia-controlled growth of a spherical bubble in fluid, under variable pressure of the main flow. This model can't describe and predict the process of bubble's collapse and neglects numerous effects that determine the behavior of the cavitation bubbles.

Many models that have been developed, describe the flow as single fluid or multi fluid, using the the Eulerian—Eulerian or Eulerian—Lagrangian approaches, but they are empirical to a certain level.

For the purposes of correct modeling, analysis and understanding the cavitating flow must be treated as a 3D unsteady multi fluid, as it is, but this is a complex problem which can't provide satisfying results, due to the existing not so advanced models. In the engineering practice cavitating flow is considered as a single fluid, the cavitation area is modeled as an area having pressure lower than the vapour value, at the same conditions.

The progress, demonstrated by scientists in [13], in the development of single fluid models for the purposes of modeling of 3D unsteady cavitating flows is shown in [12, 14–19]. Models, described in the aforementioned research works, allow single fluid solvers to be applied to the conservation equations, which describe the behavior of multi fluid, without increase in the computational time, due to the increase in the number of conservation equations describing the multi flow behaviour.

As discussed in [20], a widely used approach is to simulate cavitating flows via transport equation models, solving an additional transport equation for mass or volume fraction of vapor. An appropriate source or sink terms are included to model the mass transfer between the liquid and vapor phases, [18, 21–23].

In [16] the Rayleigh-Plesset equation is coupled to the main set of flow equations to obtain solution for the local void fraction.

Mainly efforts in cavitation modelling are focused on the correct calculation of mass fraction. One approach has been proposed in [22], by modelling the phase transition process very similar to the chemically reacting flows. The model is applied in [18] for modeling of various flows with observed cavitation. Also, in [17, 22] artificial compressibility with a special preconditioning formulation is included to solve the main set of multiphase equations and the implemented volume fraction equations.

In [24] is discussed the elaboration and implementation of a model in which the evaporation/condensation parameters (introduced by Merkle) are avoided.

A different approach is proposed by Schnerr and Sauer in [25]. They considered the vapour-liquid mixture as containing a large number of spherical bubbles.

In [22] is shown a "full-cavitation model" with included in it terms related to the phase-change rate, which is derived from the Rayleigh-Plesset equation for bubble dynamics and local flow conditions.

In [26, 27] various models are compared to establish a cavitation model based on the interfacial dynamics. This model gives opportunity to direct interpretation of the empirical parameters in existing transport equation-based models. The mixture density is calculated in a function of the transport processes.

One of the most important issues in the cavitating flow simulations is the turbulence model. Cavitation is an unsteady phenomena, strong interactions between the cavity and the boundary layer during cavity development are observed, also there are processes of phase changes. The Navier–Stokes equations set, averaged by Reynolds—(RANS), has been widely used to model turbulent flows. However, as stressed above, the turbulence model is crucial to obtain physically real picture of cavitation and its effects. As discussed in [28] the RANS models with eddy viscosity turbulence models have limited application and capability to simulate unsteady cavitating flows.

In [29], an idea to account for the local turbulence intensity by additional term—pturb', is presented, following [19]:

$$\mathrm{pv} = \mathrm{psat} + \mathrm{pturb}' \tag{1}$$

where pv is the vaporation pressure, $psat$ is the saturation pressure in function of liquid temperature, $pturb' = 0.39.\ \rho.k$ is related to the turbulent kinetic energy k.

This modification shows that phase-change is more intensive in regions where vorticity generation enhances the turbulence.

In [20] for the case of 2D modelling RANS set of equations is applied with the use of k-ω SST turbulence model. The k-ω SST was found as a model giving enough good prediction of boundary layer detachment characteristics for flow around hydrofoil NACA66, [30].

Based on the literature survey, it is obvious that 3D, unsteady and with included bubble dynamics simulations have to be performed and addressed to: fluid physical properties, working conditions and effects of their variation on flow dynamic; cavitation and how it affects the overall efficiency performance and materials strength; bubble growth and implosion; noise and vibrations, pressure pulsation effects, due to the rapid pressure changes; phase changes and mass transfer, etc.

This will contribute for detailed research of cavitation and its specifics under various conditions, also to consider specific criteria and approaches to surpass the cavitation and increase the efficiency of hydro machines, nozzles, marine propellers, etc.

This research presents a part of an ongoing research aiming to develop numerical methodology for complex cavitating flows with included User Defined Code with developed model related to the two-phase change and mass transfer. This code is implemented to the main model of flow equations and bubble dynamics to model the cavitation with its all specifics and characteristics over a hydrofoil NACA66. The obtained results are verified, current methodology is applicable to study cavitation in various rotating machines, propellers, projectiles and others that could be affected by cavitation in various exploitation regimes.

2 Problem Formulation

2.1 Brief Description of the Applied Mathematical Model

Together with flow equations set in ANSYS-FLUENT, a two-phase cavitation model is implemented, [31].

For the purposes of cavitation modelling, the equation that governs the liquid-vapor mass transfer (evaporation and condensation) is a vapor transport equation, [31]:

$$\frac{\partial}{\partial t}(\alpha.\rho v) + \nabla\left(\alpha.\rho v.\overrightarrow{Vv}\right) = Re - Rc \tag{2}$$

where, v—vapor phase, α—vapor volume fraction, $\overrightarrow{Vv}$—vapor phase velocity, *Re* and *Rc*—mass transfer source terms connected to the growth and collapse of vapor bubbles.

In Eq. (2), terms *Re* and *Rc* describe the processes of mass transfer between the vapor and liquid phases in cavitation. In ANSYS FLUENT, they are modeled based on the Rayleigh-Plesset equation describing the growth of a single vapor bubble in a liquid, [31].

The phase changes—generation and condensation of vapor bubbles, is described of transport equation for a single spherical bubble, the well-known Rayleigh–Plesset equation, Eq. (1):

$$\begin{aligned} R_{bubble}\frac{d^2 R_{bubble}}{dt^2} + \frac{3}{2}.\left(\frac{dR_{bubble}}{dt}\right)^2 &= \frac{P_{bubble}-P}{\rho_{liquid}} \\ &+ \frac{4vliquid\dot{R}_{bubble}}{R_{bubble}}.\frac{2.\sigma}{\rho_{liquid}.R_{bubble}} \end{aligned} \tag{3}$$

where first term $\left(R_{bubble}\frac{d^2R_{bubble}}{dt^2}\right)$ presents the driving term, the second term $\left(\frac{3}{2}.\left(\frac{dR_{bubble}}{dt}\right)^2\right)$ is the viscous term, on the right hand side are placed both the inertia and surface tension terms.

In the homogenous two-phase flow model formulations there is no slip between the vapor and liquid phase, i.e. the phases share the same velocity field. The distribution of the vapor phase is calculated by solving an additional transport equation for vapor phase and the influence of the vapor phase on the motion of the mixture is accounted by applying mixture density and mixture viscosity as physical properties [31], as follows:

$$\rho = (1-\alpha).\rho_{liquid} + \alpha.\rho v \; and \; \mu = (1-\alpha).\mu_{liquid} + \alpha.\mu v \tag{4}$$

2.2 Schnerr-Sauer Mass Transfer Model [31]

Schnerr-Sauer's model is based on bubble dynamics and the amount of vapor in a control volume is calculated from the number of nesting vapor bubbles and the average radius of these bubbles.

According to the Schnerr-Sauer model, the mass transfer is calculated applying Eq. (5):

$$\begin{aligned} \dot{m} = -3_{\rho v}\left[\mathrm{n0}.\frac{4}{3}.\pi.\left(\alpha 2 - \alpha 3\left(1 - \frac{\rho v}{\rho_{liquid}}\right)\right)\right]^{1/3} \\ .sign(Pv - P).\left[2/3.(Pv - P)/\rho_{liquid}\right]^{0.5} \end{aligned} \tag{5}$$

Depending on the pressure value $P < Pv$ or $P > Pv$, value of $\dot{m}$ is:

$$\begin{aligned} \dot{m}(+) = Cv.\left(\rho v.\rho_{liquid}/\rho\right).\alpha(1-\alpha).3/R_{bubble}.\left[2/3.(Pv - P)/\rho_{liquid}\right]^{0.5} \\ \mathrm{if\,P} < \mathrm{Pv} \end{aligned} \tag{6}$$

$$\begin{aligned} \dot{m}(-) = Cc.\left(\rho v.\rho_{liquid}/\rho\right).(1-\alpha).3/R_{bubble}.\left[2/3.(P - Pv)/\rho_{liquid}\right]^{0.5} \\ \mathrm{if\,P} > \mathrm{Pv} \end{aligned} \tag{7}$$

Cv and *Cc* are empirical coefficients. In the above equation $n0$ is the number density of micro bubbles and α is the vapor volume fraction, equal to the volume vapor divided by the total volume, Eq. (8):

$$\frac{d\alpha}{dt} = (1-\alpha).\frac{4.\pi.n0.R_{bubble}2.\dot{R}_{bubble}}{1 + n0.4.\pi.R_{bubble}3/3} \tag{8}$$

$$if\; \alpha* = 4.\pi.\mathrm{n_0}.\mathrm{R_{bubble}}^3/3, \; \mathrm{then}\; \alpha = \alpha*/(1+\alpha*) \tag{9}$$

For flows without mass transfer $d\alpha/dt$ is equal to zero. In case of mass transfer, the continuity equation could be written in the form of Eq. 10:

$$\frac{\partial\alpha}{\partial t} + \nabla(\rho.\vec{c}) = \left(\frac{(1-\alpha).n0.\rho_{liquid}}{(1-\alpha).\rho_{liquid} + \alpha.\rho v}\right).\left(\frac{n0}{1 + n0.4.\pi.R_{bubble}3/3}\right) \\ .\frac{d}{dt}\left(4/3.\pi.\mathrm{R_{bubble}}^{3}\right) \quad (10)$$

There are several factors greatly influencing the numerical stability and affecting the solution convergence—high pressure differences between the inlet and exit, large ratio of liquid to vapor density, and large phase change rates between the liquid and vapor. Another very import concern is that poorly defined boundary and initial conditions lead to an unrealistic pressure field distribution and unexpected cavitating zones. That is why, one must keep in mind that both the segregated (SIMPLE, SIMPLEC, and PISO) and coupled pressure-based solvers can be used in cavitation. However, the coupled solver is more robust and converges faster for cavitating flows in rotating machinery.

The model of Schnerr and Sauer is enough robust and there is no need, in most cases, for specific initial conditions. However, in some very complicated cases, it may be beneficial to obtain a realistic pressure field before substantial cavities are formed. The relaxation factor for vapor is 0.5 or higher unless the solution diverges or all the residuals oscillate excessively. For the segregated solver, the relaxation factor for pressure should be no less than the value for the momentum equations and etc., [31].

2.3 Problem Set-Up

This paper is targeted to numerical modeling and simulation of cavitating flow applying the mixture model implemented in Fluent. The two-phase flow is described via the Rayleigh equation applied to bubbles, uniformly distributed in the computational domain. User Defined Code is written and introduces a developed model related to modeling of the two-phase change and mass transfer processes.

The object under consideration is a 3D model of hydrofoil NACA66 [32], positioned at angle of attack 6°, [32], see Fig. 1.

The computational domain is shown in Fig. 2. The spanwise length is chosen as ~$0.5C$ in present paper, due to the significant need of computational time, as the spanwise domain is usually set as at least twice the thickness of the hydrofoil to resolve streamwise vortex, [33].

Geometry modeling is performed in GAMBIT. The flow domain is discretized with tetrahedral 3D elements with resolved boundary layer of quadrilateral elements around the profile walls, Fig. 3. The created mesh has relatively coarse structure at distance from the hydrofoil and it density increases as the distance to the foil decreases. The resulted flow domain mesh is of around 1.2 million nodes. First row of elements is with height of 0.01, the boundary layer consists of 18 rows of elements and has thickness calculated based on the theory of Schlighting and flow conditions.

Boundary zones are velocity_inlet, pressure_outlet, symmetry and wall. Defined boundary conditions are as follows: inlet—velocity 5 m/s; outlet-static pressure 101,200 Pa; top wall-wall; bottom wall-wall; front and back symmetry; profile walls-wall.

As working fluid is used water with following physical characteristics: density and dynamic viscosities of liquid water are $\rho_{\text{liquid}} = 1000$ [kg/m^3] and $\mu_{\text{liquid}} = 9.97 \times 10^{-4}$ [Pa s] at temperature of 293 K. According to the case under consideration, the vapor density of water is $\rho\text{v} = 0.0175$ [kg/m^3], the vapor viscosity is $\mu\text{v} = 9.725^{-6}$ [Pa s], saturated pressure at 293 K is $P\text{v} = 2330$ Pa.

Reynolds number is calculated as $\text{Re} = V\infty.c/\nu l = 7{,}522{,}166$ and the cavitation number is

$$\sigma = (p\infty - pv)/\left(0.5\rho l V^2\infty\right) \tag{11}$$

Cavitation number is phase change pressure drop to kinetic energy per unit volume and describes the potential of the fluid to start the cavitation phenomena. Lower cavitation number means more intense cavitation. Unfortunately, as stated in [27] σ "is not strictly a parameter of similitude for cavitating flows". In fact, it is a necessary but not a sufficient condition.

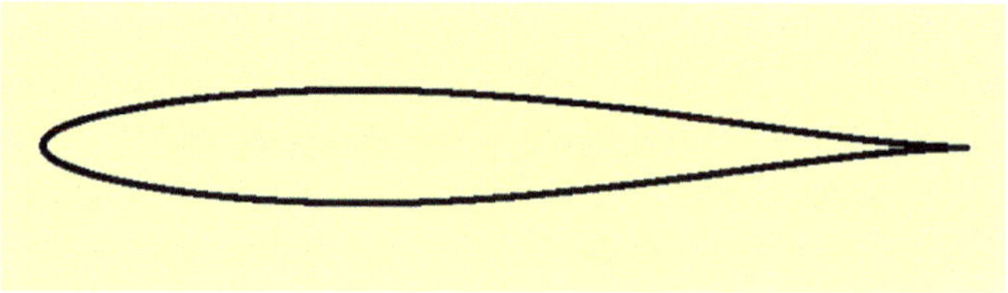

Fig. 1 Profile NACA66

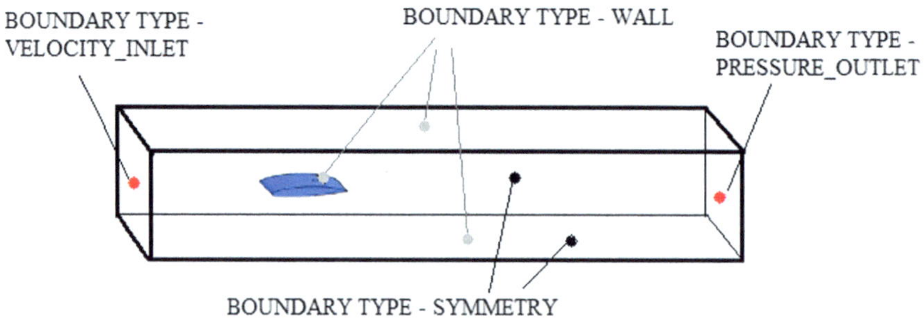

Fig. 2 Computational domain and imposed boundary zones

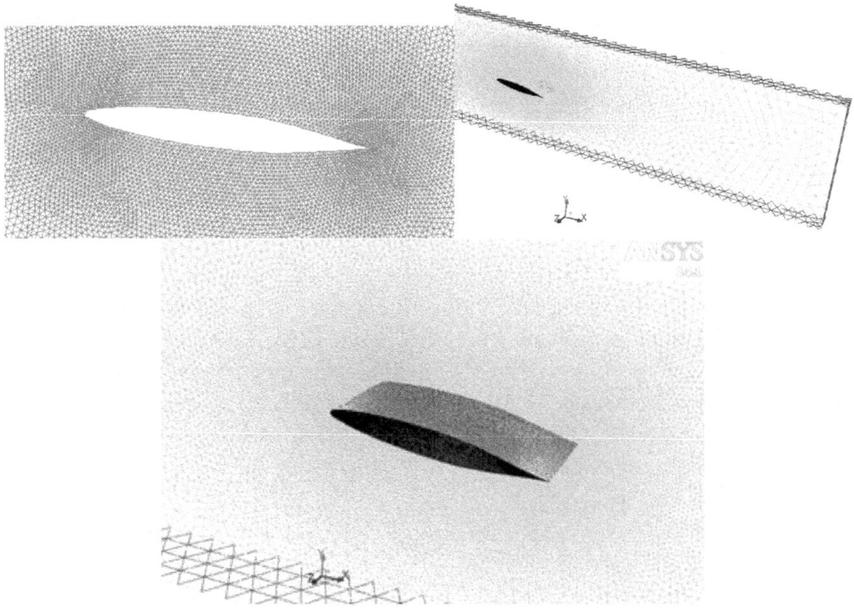

Fig. 3 Part of the elements' discretization around the hydrofoil

A particular value of cavitation number that corresponds to the particular inlet pressure P, at which cavitation is initially observed, is called cavitation inception number—σi, [34]. There exists three modes of cavitation, as it is well known:

- incipient cavitation: $\sigma = \sigma i$;
- cloud cavitation: $\sigma < \sigma i$;
- supercavitation: $\sigma \ll \sigma i$.

Values of the surface pressure coefficient is provided by Eq. (12)

$$C_P = (P - P\infty)/\left(0.5\,\rho V^2\infty\right) \tag{12}$$

Initially, NACA 66 hydrofoil is at a fixed angle of attack of $\alpha = 6°$ in steady flow; after that unsteady flow is under consideration.

A temporal convergence study is carried out in the case of partial sheet cavitation, at an angle of attack of $\alpha = 6°$ with $\sigma = 1.42$.

A computationally-intensive unsteady calculations are necessary to accurately simulate the process of bubble formation, growth and break-off. The numerical simulation starts at time 0 s, time step is 10^{-4} s, (*Tref*/200, where *Tref* = *C/Vinlet* and *Vinlet* is the inflow velocity at the domain inlet, [35, 36].

Another important dimensionless numbers used to describe the flow field are the lift and drag coefficients. Lift and drag coefficients are defined as follows, Eqs. (13) and (14):

$$Cl = Flift/0.5\,\rho.Vinlet^2.A \tag{13}$$

$$C_D = Fdrag/0.5\,\rho.Vinlet^2.A \tag{14}$$

where *Flift*—lift force acting on the profile, N; *Fdrag*—drag force formed around the streamed profile, N; *A* is cross section area in m^2; *Vinlet* is inlet velocity of the incoming flow, in m/s.

Standard k-ε turbulence model, which includes two extra transport equations to represent the turbulence of the fluid, is chosen to model the turbulence effects in the flow. This two-equation model allows for the accounting of effects such as convection and the diffusion of turbulent energy, it provides very good overall turbulence modelling without requiring an excess of additional computational time.

Discretization schemes are SIMPLE for pressure-velocity coupling, PRESTO for pressure equation and for the rest equations QUICK. Convergence for unknown parameters is 10^{-4}, convergence criterion for continuity and volume fraction is set to 10^{-5} for improved accuracy.

3 Results

Research on cavitation phenomena is performed in the case of a profile inclined at angle of 2°, 4°, 6°, 8° and 10°. Flow speed is 5, 10 and 15 m/s.

In Fig. 4 values of lift coefficient in function of angles of attack for 3 flow speeds are shown.

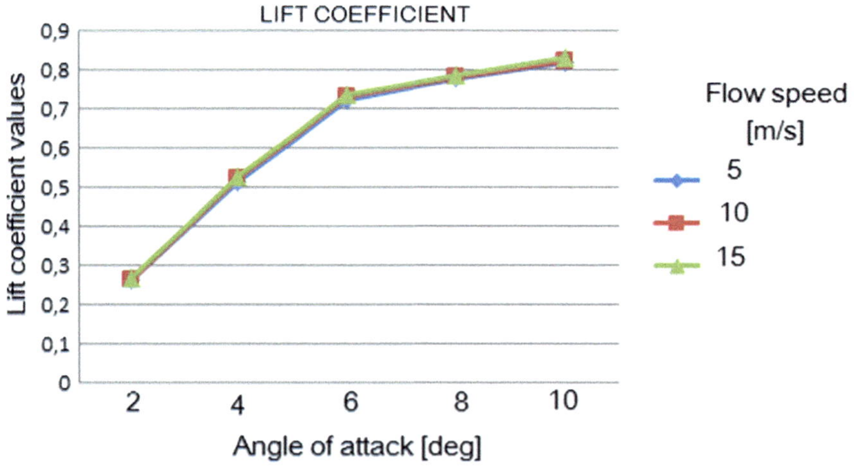

Fig. 4 Lift coefficient values in function of angles of attack for three different speeds of the incoming speed

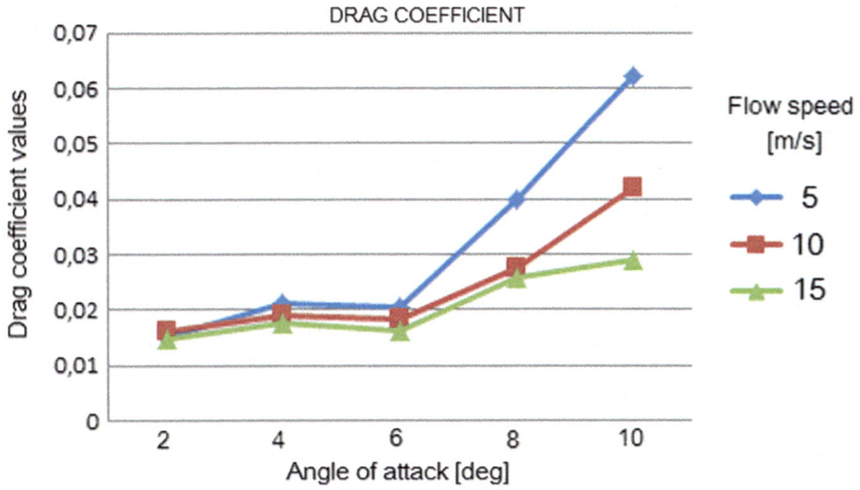

Fig. 5 Drag coefficient values in function of angles of attack for three different speeds of the incoming speed

In Fig. 4 the predicted lift coefficient values for different speeds of the incoming flow are compared. The lift coefficient increases in the range $\sigma = 1.28$–1.54, and with development of partial sheet cavitation after the leading edge. The increase in lift is attributed to the suction effect created by the small cavitation bubble caused by high pressure gradient and observed just after the leading edge.

As the cavity, formed over the suction surface, grows with decrease of the cavitation number (i.e., $\sigma < 1.2$), the lift coefficient decreases. Both the slight increase in the lift coefficient for $1.28 < \sigma < 1.54$ and subsequent drop in lift coefficient for $\sigma < 1.2$, are observed. The same situation is discussed in [20].

As its shown in Fig. 5 drag force, acting on the streamed profile, decreases with increase in speed of the incoming flow (Re number, respectively). Harsh decrease in values is observed for angles of attack in the range from 6° to 10°.

The obtained results show that cavitation phenomena work against lower drag coefficients. In case of increase in drag and decrease (even insignificant) of lift coefficient leads to efficiency decrease, Fig. 6.

Cavitation inception is observed in case of cavitation number $\sigma \approx 2.32$ in the phase of computational research. For the range of $1.45 < \sigma <$ around 2.0, a phenomena of stable cavitation is noticed. Both, the proposed methodology and the UDF code give acceptable results for stable cavitation and its inception.

For values of cavitation number less than 1.45, the cavity becomes unstable, as noticed in results, obtained by the proposed methodology. However, experimental observations predict partial sheet cavitation for the cavitation numbers in the range of 1.28–1.54, [20]. The partial sheet cavitation is characterized by a partial break of the cavity with a stable vapor region at the region of the leading edge, as discussed and shown in [20, 37].

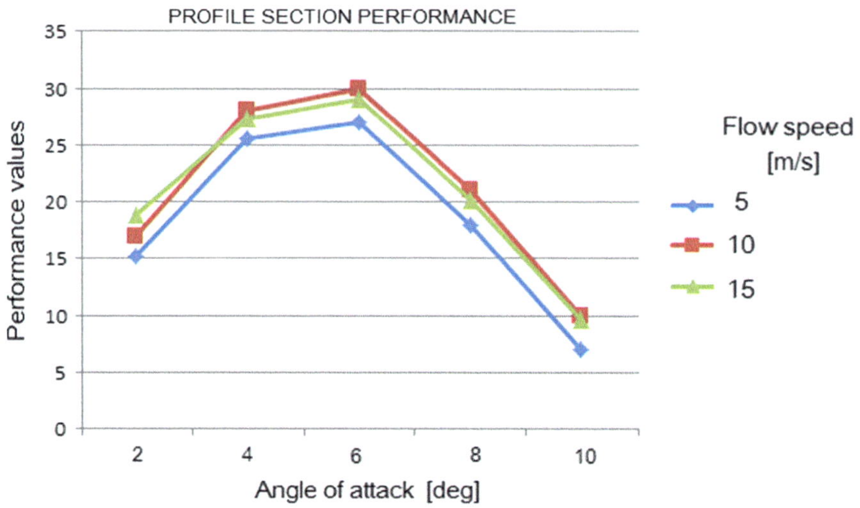

Fig. 6 Profile performance for three different flow speeds, in function of angle of attack

In case of decreased cavitation number, the break point of the cavity, as found in our research works, moves closer and closer to the leading edge of the profile under consideration. When the cavitation number arrives to a value of around 1.02 or equal to 1.00, an unsteady sheet cavitation occurs.

In Fig. 4 predicted lift coefficients (see Eq. 13) for different angles of attack/cavitation numbers are compared. The lift coefficient increases slightly for cavitation number in the range between 1.28 and 1.54.

The slight increase in lift values could be related to the suction effect created by cavitation bubble formed at the leading edge and the high pressure gradient observed between the pressure and suction sides of the profile under consideration.

As the cavitation number decreases its value below 1.2, lift coefficient decreases.

As it is related to the drag coefficient, comparisons of the numerically predicted drag coefficient values, in function of angle of attack and various speeds for the incoming flow, are shown in Fig. 5.

It should be noted that the observation of over- or under-prediction of lift and drag values may be due to the attributed 3D effects, boundary layer effects, presence of angle of attack, etc. Current results show that the increase in the cavitation intensity leads to increase of drag coefficient values. Last mentioned, together with specific variations in lift coefficient values, is a prerequisite to decrease in efficiency, for angles of attack of 8° and 10°, Fig. 6.

4 Conclusions

Our main task aimed at creation of a relevant approach and code to simulate cavitation phenomena around a hydrofoil. A User Defined Code with included equations for purposes of modelling of bubble growth, two-phase interaction effects, etc., is written and implemented to the main mathematical model in Fluent.

As a result of the proposed methodology, developed User Defined Code and the fulfilled research, cavitation phenomena, variation in Cl, Cd, efficiency and other basic parameters, as well as flow structures, could be observed.

This research provides insights on the impact of the blade design, profile section, exploitation regime and conditions on the flow parameters distribution and formation of cavitation conditions around streamed surfaces. This study could contribute to better understanding of the mechanisms of blade streaming, formation and development of cavitation, mass and heat transfer in cavitation conditions and provide approaches to overcome the phenomena of cavitation.

The numerical scheme with its logical sequence and the developed User Defined Code can be applied as a basic methodology to attain all parameters needed for complex research of factors and conditions leading to cavitation to appear and its relation to energy efficiency decrease, noise, stresses and vibrations and others; research on exploitation in nominal and variable operating regimes and could provide more light on the approaches to escape from that undesirable phenomena.

References

1. Knapp RT, Daily JW, Hammitt FG (1970) Cavitation. McGraw-Hill, New York
2. Avellan F, Dupont P, Ryhming IL (1989) Generation mechanisms and dynamics of cavitation vortices downstream of a fixed leading edge cavity. In: 17th symposium on naval hydrodynamics. National Academy Press, Washington, DC, pp 317–329
3. Modi PN, Seth SM (2002) Hydraulics and fluid mechanics
4. Helmholtz H (1868) Uber diskontinuierliche Flüssigkeitsbewegungen. Monatsber. Akad. Wiss., Berlin, 23, 215–228
5. Krishnaswamy P (2000) Flow modelling for partially cavitating hydrofoils, Ph.D. thesis, Technical University of Denmark
6. Birkhoff EH, Zarantonello G (1957) Jets, wakes and cavities. Academic Press, New York
7. Gurevich MI (1978) Theory of jets of ideal fluid. Nauka, Moscow (in Russian)
8. Logvinovich GV (1969) Hydrodynamics of flows with free boundaries. Naukova dumka, Kiev (In Russian)
9. Epshtein LA (1970) Dimensionality and similarity methods in the problems of the hydromechanics of vessels. Sudostroyenie, Leningrad (in Russian)
10. Terentiev A, Kirschner I, Uhlman J (2011) The hydrodynamics of cavitating flows. Backbone Publishing Company
11. Alajbegovic A, Grogger HA, Philipp H (1999) Calculation of transient cavitation in nozzle using the two-fluid model. In: Proceedings of the ILASS-Americas'99 annual conference, pp 373–377
12. Yuan W, Sauer J, Schnerr HG (2001) Modeling and computation of unsteady cavitating flows in injection nozzles. Mec Ind 2(5):383–394

13. Kanfoudi H, Lamloumi H, Zgolli R (2012) Numerical investigation for steady and unsteady cavitating flows. “Advances in Modeling of Fluid Dynamics”, INTECH
14. Chen Y, Heister SD (1996) Modelling hydrodynamic nonequilibrium in cavitating flows, Tr. ASME J Fluids Eng 118:172–178
15. Chen Y, Heister D (1995) Two-phase modelling of cavitated flows. ASME Cavitation Multiph Forum 24:799–806
16. Kunz RF, Boger DA, Stinebring DR (2000) A preconditioned Navier-Stokes method for two-phase flows with application to cavitation prediction. Comput Fluids 29:849–875
17. Ahuja V, Hosangadi A, Arunajatesan S (2001) Simulations of caviting flows using hybrid unstructured meshes. J Fluids Engng 123:331–339
18. Singhal AK, Athavale MM, Li H, Jiang Y (2002) Mathematical basis and validation of the full cavitation model. 124:617–624
19. Kubota A, Kato H, Yamaguchi H (1992) A new modelling of cavitating flows: a numerical study of unsteady cavitation on a hydrofoil section. J Fluid Mech 240:59–96
20. Ducoin A, Huang B, Young YL (2012) Numerical modeling of unsteady cavitating flows around a stationary hydrofoil. Int J Rotating Mach 2012, Article ID 215678
21. Menter FR (1992) Improved two-equation k-turbulence models for aerodynamic flows. NASA Technical Memorandum
22. Merkle CL, Feng JZ, Buelow PEO (1998) Computational modeling of the dynamics of sheet cavitation. In: Proceedings of the 3rd international symposium on cavitation, Grenoble, France (1998) vol 2
23. Kunz RF, Boger DA, Chyczewski TS, Stinebring DR, Gibeling HJ (1999) Multiphase CFD analysis of natural and ventilated cavitation about submerged bodies. In: Proceedings of the3rd ASME/JSME joint fluid engineering conference paper FEDSM99-7364
24. Senocak I, Shyy W (2002) Evaluation of cavitation models for Navier-Stokes computations, Proceedings of the ASME Fluids Engineering Division Summer Meeting, Paper FEDSM 2002-31011
25. Schnerr GH, Sauer J (2001) Physical and numerical modeling of unsteady cavitation dynamics. In: Proceedings of the 4th international conference on multiphase flow, New Orleans, USA
26. Tran TD, Nennemann B, Vu TC, Guibault F (2014) Numerical simulation of unsteady sheet/cloud cavitation. IOP Conf Ser: Earth Environ Sci 22:052012
27. Şenocak I (2002) Computational methodology for the simulation of turbulent cavitating flows. Dissertation thesis, University of Florida, USA
28. Ji B, Luo XW, Arndt REA, Peng X, Wu Y (2015) Large Eddy Simulation and theoretical investigations of the transient cavitating vortical flow structure around a NACA66 hydrofoil. Int J Multiph Flow 68:121–134
29. Bernard S, Muntean S, Susan-Resiga RF, Anton I (2004) Numerical simulation of two-phase cavitating flow in turbomachines. In: 6th international conference on hydraulic machinery and hydrodynamics Timisoara, Romania, October 21–22, 2004
30. Ducoin A, Astolfi JA, Deniset F, Sigrist J-F (2009) Computational and experimental investigation of flow over a transient pitching hydrofoil. Eur J Mech B 28(6):728–743
31. Ansys- Fluent 12.0/12.1 documentation, information on https://www.afs.enea.it/project/neptunius/docs/fluent
32. Information on www.airfoiltools.com
33. Sagaut P (2002) Large Eddy simulation for incompressible flows. Springer
34. Brennen C (1994) Hydrodynamics of pumps, Concepts NREC
35. Coutier-Delgosha O, Reboud JL, Fortes-Patella R (2003) Evaluation of the turbulence model influence on the numerical simulations of unsteady cavitation. J Fluids Eng 125(1):38–45
36. Coutier-Delgosha O, Hofmann M, Stoffel B, Fortes PR, Reboud JL (2003) Experimental and numerical studies in a centrifugal pump with two-dimensional curved blades in cavitating condition. J Fluids Eng 125(6):970–978
37. Leroux JB, Astolfi JA, Billard JY (2004) An experimental study of unsteady partial cavitation. J Fluids Eng 126(1):94–101

Implementation of Beamforming Codes in 3D CFD Simulations for the Localization and Visualization of Rotating Sound Sources

Christian Maier and Wolfram Pannert

Abstract The reduction of acoustic emission is one very important topic in product development of rotating sound sources and should be considered in early stages of development. Rotating airfoils of helicopter blades (Farassat in Theory of noise generation from moving bodies with an application to helicopter rotors. Langley Research Center, Harnpton, 1975 [7]), rotating wind turbines (Ramachandran et al. J Sound Vib 14:3058–3080, 2014 [19]) or fan blades (Dougherty in Locating and quantifying broadband fan sources using in-duct microphones. In: 16th AIAA/CEAS aeroacoustics conference, 2004 [5]), (Minck et al. in Fan noise analysis using a microphone array. In: Fan 2012—international conference on fan noise, technology, and numerical methods, 2012 [15]), (Herold and Sarradj Noise Control Eng J 63(6):546–551, 2015 [10]) cause unwanted noise. In this paper, a method is presented that allows the localization of stationary and rotating sound sources in the CFD simulation process. This enables a more detailed view on the geometry and offers a basic approach for an optimization concerning the localization and visualization of sound sources using a microphone array with suitable beamforming algorithm. A virtual microphone array (acoustic camera) is built in the CFD simulation and coupled to various beamforming codes. The method is presented on a rotating dipole (cylinder) and the results are compared to measurements with a real microphone array. To resolve the acoustical scales large eddy simulations (LES) are performed. The pressure data of the microphones is exported and the localization of the sound sources can be done in a post-processing procedure in an early stage of development with numerical software like Matlab or GNU Octave. Both time domain algorithms with retarded time (Maier in Practical and theoretic aero-acoustics, 2013 [14]) and frequency domain algorithms (Pannert and Maier in J Sound Vib 333(7):1899–1912, 2014 [17]) can applied to the CFD simulation data to visualize sound sources. Here the time domain version is used. With this approach, a method is presented that examined a rotating cylinder from a LES Fluent simulation with a rotational speed

C. Maier (✉) · W. Pannert
Department of Mechanical Engineering, University of Applied Sciences Aalen, Beethovenstrasse 1, 73430 Aalen, Germany
e-mail: christian.maier@onlinehome.de

W. Pannert
e-mail: wolfram.pannert@hs-aalen.de

A. Öchsner and H. Altenbach (eds.), *Engineering Design Applications II*,
Advanced Structured Materials 113, https://doi.org/10.1007/978-3-030-20801-1_15

of 702 rpm anti clockwise that rotates at a radius of 0.48 m. This rotating dipole emits sound that the microphone array can visualize and assigning to the cylinder sound. By post processing these data it is possible to visualize the sound emitted by the cylinder. The novelty of this work is the visualization of rotating sound sources of a CFD Large Eddy Simulation.

Keywords Rotating beamforming · Large eddy simulation · Moving sound sources · Acoustic dipole

Nomenclature

$\vec{x}$	Position at observer
$\vec{y}$	Position at sound source
Q	Moving monopole
c	Speed of sound
n	Time step
p'	Pressure distribution
q	Source distribution
r	Distance between observer and sound source
t	Observer time at position $\vec{x}$
δ	Dirac-delta function
τ	Retarded time at position $\vec{y}$

Abbreviations

CAA	Computational aero-acoustics
CFD	Computational fluid dynamics
CFL	Courant-Friedrich-Levy number
D + S	Delay and sum
dp	Double precision
FW-H	Ffowcs-Williams & Hawkins
LES	Large eddy simulation
M	Mach number
NS	Navier-Stokes
pbns	Pressure based Navier-Stoke
PSD	Power spectral density
Re	Reynolds number
Sr	Strouhal number

1 Introduction

The challenge in the prediction of aero-acoustic effects with Computational Fluid Dynamics (*CFD*) is the fact that fluid flows contain much more energy than the sound waves that use the flow as a transport medium. Aero-acoustic simulations for near field sound prediction is easier compared to far field problems with the direct method Computational Aero-Acoustics (*CAA*). Because it have mesh up completely resolved up to the far field where the microphone array is placed. This leads to very big meshes and is often not feasible nowadays with computers to get a result in convenient time.

For far field predictions hybrid approaches were done, where the flow part (*CFD*) is decoupled from the acoustic part (*CAA*). In this approach, an analogy called *Ffowcs-Williams and Hawking* (*FW-H*) (Sound generated by turbulence and surfaces in arbitrary motion [8], is used to calculate the sound in the far field [2].

The analogy following FW-H is also called extended Lighthill Equation [6] and can be used to estimate the noise level and distribution of flowed objects concerning the far-field in computer simulations using a microphone array. Because the direct method (*CAA*) is too costly and time consuming it is not efficient to mesh up to the physical far-field nowadays to place points in the computer simulation that read out the pressure fluctuations. The microphone array that captures the pressure fluctuations using FW-H can be outside the meshed domain. This saves a lot of mesh points and therefore simulation time as well. This is why it is convenient to use the extended Lighthill Equation (*FW-H*) to determine the noise by the numerically calculated source terms with respect to an integral following a simplified version of the *Greens Function* [12, 13, 16]. However, it is still an approximation.

Pressure fluctuations on the surface of the body of interest are transported with the analogy to the receiver using *Green's Function* and do not use the fine discretized mesh. For a correct calculation of the pressure fluctuations large eddy simulations (*LES*) are performed to resolve the acoustical scales completely at the sound origin walls [9, 18]. The resolved delamination of the flow on the surface of the body is necessary to capture the smallest eddies that cause sound at these walls. To control the resolution of the resolved eddies at the wall the simulations were carried out with a $y+$ value smaller than one. This guarantees that the smallest eddies are catched up by the filter used in *LES* for the NS equations. With the analogy of *FW-H* it is possible to rebuild a microphone array (in the flow simulation using virtual microphones/pressure points) in the far field without meshing up to this point. This microphone array has the same position in the simulation as the real microphone array in the experiments and measurements. For each microphone, a time signal with its pressure fluctuations can be signal processed, exported and used for the beamforming algorithms in frequency or time domain to localize the sound sources after a Fourier transformation with GNU Octave e.g. for example.

In the examined case of a rotating sound source, the pressure data is transformed to a rotating frame of reference and the Doppler shift of the source frequency, caused by the movement, is compensated. The standard *Delay and Sum beamforming* (D + S) method can be applied in time and frequency domain [4].

2 Theory for the Rotating Beamforming in Time Domain

The movement of a point source can be treated with the Greens function approach for solving the inhomogeneous wave equation. Taking the inhomogeneous wave equation for pressure fluctuations of a stationary sound source located at $\vec{x}$

$$\frac{1}{c^2}\frac{\partial^2 p'}{\partial t^2} - \Delta p' = q\left(\vec{x}, t\right) \tag{1}$$

where $q\left(\vec{x}, t\right)$ is a the source distribution. The solution for free space conditions without boundaries for p' can be calculated from an integral formulation

$$p\left(\vec{x}, t\right) = \frac{1}{4\pi}\int_{\mathbb{R}^3} \frac{q\left(\vec{y}, t - \left|\vec{x} - \vec{y}\right|/c\right)}{\left|\vec{x} - \vec{y}\right|} d^3\vec{y} \tag{2}$$

with

$$\left|\vec{x} - \vec{y}\right| = r \tag{3}$$

and the retarded time τ

$$\tau = t - \left|\vec{x} - \vec{y}\right|/c \tag{4}$$

The signal which was emitted at time τ at a position $\vec{y}$ and is observed at time t at the point $\vec{x}$. For the general source distribution, a concrete source distribution can be inserted. The simplest model for a moving monopole is the distribution

$$q\left(\vec{x}, t\right) = Q(t)\delta\left(\vec{x} - \vec{x_s}(t)\right) \tag{5}$$

with $\left(\vec{x} - \vec{x_s}(t)\right)$ as the actual time dependent position and $Q(t)$ as the amplitude of the monopole sound source. Figure 1 shows the situation for a moving sound source and a fixed observer position. The observer points are the microphone positions at the microphone array. It is necessary to calculate the distance between sound source and the microphone position for every time step τ_n to calculate the time delay to the observer position [6, 2]. In the retarded time approach, the retarded emission time τ is calculated back from the receiving time t via

$$\tau = t - r(\tau)/c \tag{6}$$

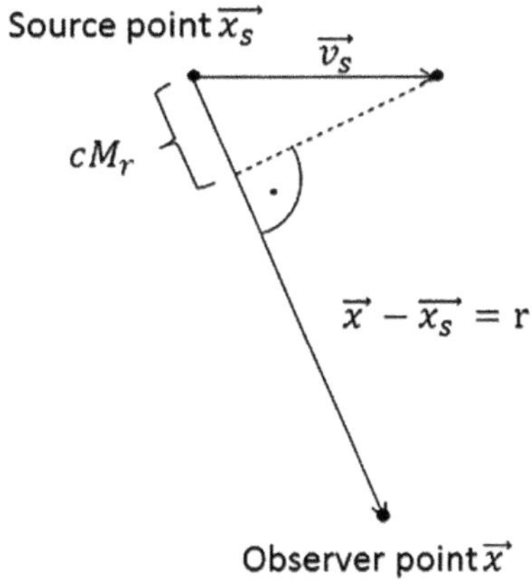

Fig. 1 Movement of the source term to a fix observation point

and cannot be calculated analytically in general cases due to the complicated dependence of $r(\tau)$ from τ. It can numerically found as a root of Eq. (7) [2]. Algorithms that treat that problem can be found in [1] or [11].

In the advanced time approach, which is applied in this work, the receiver time t can be calculated via

$$t = \tau + r(\tau)/c \tag{7}$$

This is much easier, but results in unequally spaced time samples t_n when using equally spaced time samples τ_n. In Fig. 2 the situation is shown for a moving source. An emitted signal at the time τ arrives at the observer position $\vec{x}$ at the time t. The speed of sound is c. In the case of a stationary source, the retarded time only depends on the position of the source $\vec{x_s}$. In the case of a moving source it depends on $\vec{x_s}(\tau)$. This time delay is calculated for every time step n.

$$\Delta t_n(\tau_n) = t_n - \tau_n = \frac{\left|\vec{x} - \vec{x_s}(\tau_n)\right|}{c} \tag{8}$$

The condition $\tau_n > t_n$ is always fulfilled for the case that the source term moves with subsonic speed and $\Delta t_n(\tau_n)$ is always positive. Working with these time delays the motion of the source can be compensated in the received microphone signals and the moving source is imaged at a fixed position which corresponds to the position at time $\tau = 0$. This is the start position of the calculation. For example the start position of one rotation.

In Fig. 3, the typical set up for investigating a fan with an acoustic camera is shown. On the left side, the microphone array is positioned in front of the rotating sound source. It is necessary to compensate the movement of the sound source. To compensate this movement, in this case the rotation of the cylinder, it is necessary to shift the time signals of every microphone for every time step at an amount,

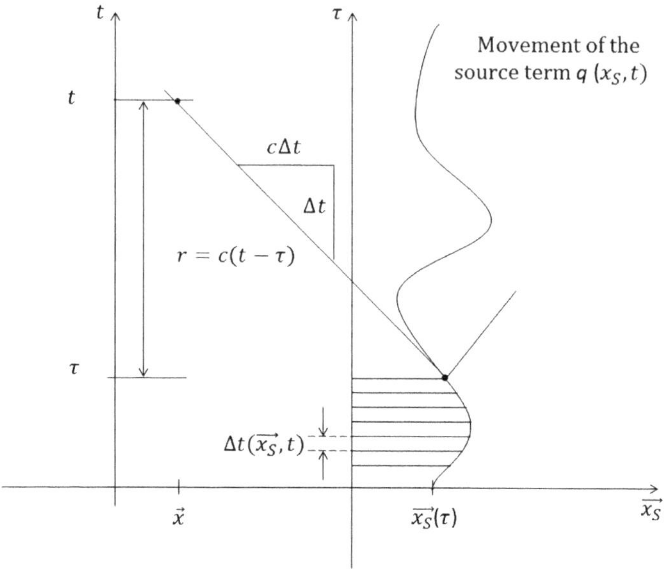

Fig. 2 Retarded time emitted from a moving sound source in space-time layer

Fig. 3 Microphone array with moving sound source in front of it

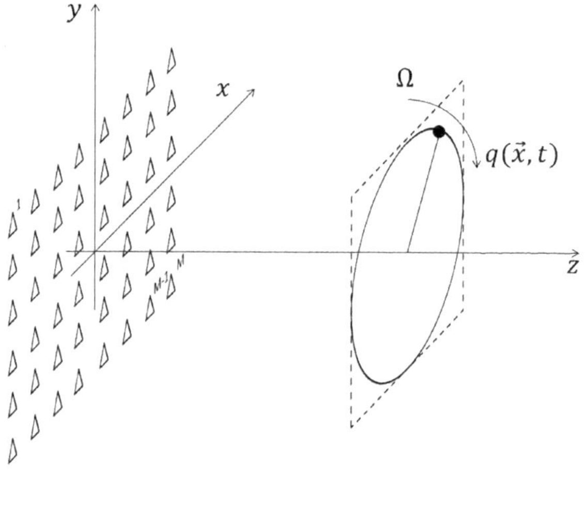

which is due to the change in distance between the moving source and the selected microphone. These shifted signals are used then to calculate the beam pattern.

In Fig. 4 simulated signals for a rotating source are shown. The pressure signal shows the varying frequency due to the Doppler shift (Fig. 4a). The radial motion of the source is subsonic Fig. 4b shows the spectrum of the microphone signal in (a). The radius at which the sound source rotates is 0.65 m and the frequency of rotation is 100 Hz. The frequency spectrum (Fig. 4b) shows the peak no longer at the position of 1500 Hz. This is the effect due to the Doppler shift; the peak is now shifted in

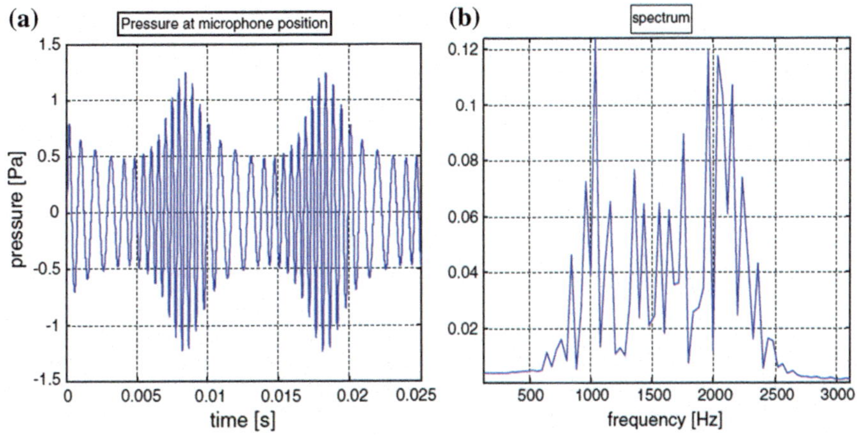

Fig. 4 Received signal from a rotating source. Distance microphone—plane of rotation in distance D = 1 m; rotation speed n = 6000/min; frequency of the source f = 1500 Hz. **a** Shows the microphone signal and **b** the spectrum of the pressure signal

positive and negative frequency away from the emitted 1500 Hz. Beside this, this frequency spectrum is not symmetric, because the motion between the source and the receiver also has an influence to the amplitude of the signal.

3 To the CFD

Acoustic waves have to be resolved completely in time and space. With 20–30 points per wavelength it is resolved in time an can be calculated with

$$\Delta t \approx \frac{1}{f\text{max}}/30 \tag{9}$$

where Δt is the time step for the maximal frequency required to be analyzed, divided by at least 20–30 supporting points. This is the time step for CAA and can be found in the ANSYS/FLUENT® "Aeroacoustics Simulation with ANSYS CFD" teaching script [16].

To yield a good frequency resolution it is necessary to calculate/estimate the simulation time with

$$f = 1/T \tag{10}$$

When setting the acoustic simulation time to 0.083 s for one rotation of the cylinder a frequency resolution of 12 Hz is given. This is coarse but allows to detect the peak and assign it to the theoretical peak following

$$f = \frac{Sr \cdot u}{l} \tag{11}$$

With Sr as the Strouhal Number. For a wide range of range and for many engineering problems and also the flowed/rotating cylinder the Strouhal Number can be set to 0.2.

4 To the Directivity Pattern of a Flowed Cylinder

When a cylinder is flowed by a medium, e.g. air, it propagates sound at the cylinder wall. This sound emitted from the cylinder wall has a directivity pattern and radiates the sound orthogonal to the flow. Figure 5 shows the flowed cylinder rotating anti clockwise in the velocity plot with the directivity of the sound emitted (dashed line) exemplarily. Investigations from Maier [14] shows that the dynamic of the sound from the flow direction to the orthogonal direction are at order of 8 db.

5 Measurements and Simulations

For the measurements and experiments, the same parameters are set as in the simulations. Therefore, the distance from cylinder to microphone array is 1.15 m and the array is rectangular shaped and has a length of 0.80 m with 48 microphones for

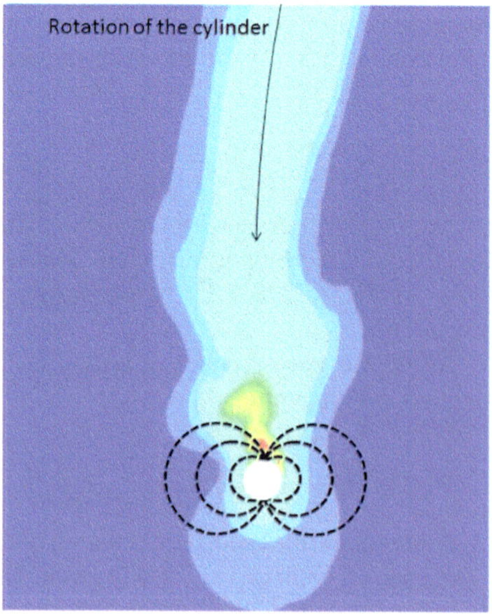

Fig. 5 Directivity pattern of the flowed cylinder with a rotational speed of 702 rpm

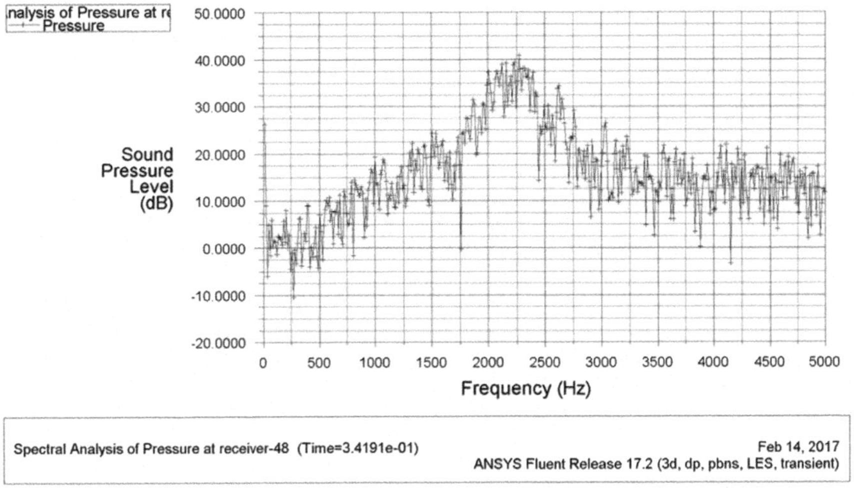

Fig. 6 Sound pressure level (SPL) of one microphone distanced 1.15 m away from the cylinder

the measurements. A diameter of 3 mm for the cylinder is chosen because the peak frequency is shifted to a higher frequency that yields better resolution of the beamforming algorithm that have problems with frequency below 1000 Hz[1]. Furthermore parameters of the simulations are 3d with a domain depth of 0.04 m, double precision (dp), pressure based Navier-Stokes (pbns), the *Reynolds Number* (Re) is about 7000 [ ], the time step for the CFD simulation was chosen to 5e−06 s and the acoustical time step is 1e−05 s. The radius of rotation is 0.48 m.

Figure 6 shows the sound pressure level of the receiver 48 of the microphone array with a peak at the position of 2300 Hz. In the measurement with the acoustic camera, the peak is at 2336 Hz and the theory (Eq. 11) using *Strouhal Number* of 0.2 gives 2294 Hz.

Figure 7 shows the histogram of the wall Y+ values at the cylinder wall. The maximum Y+ value is stated with one (x-axis). So, most of the cells are below one. This ensures the resolution of the delamination of eddies necessary for Large Eddy Simulations (LES).

Another criterion for a resolved LES is the Courant-Friedrich-Levy Number (CFL). Figure 8 gives a histogram of the CFL Number. Most of the cells are below one.

The one-dimensional acoustic CFL Number is given by

$$CFL = \frac{(u+c)\Delta t}{\Delta x} \tag{12}$$

[1]Valid for standard delay-and-sum beamforming algorithm with the used antenna (size of order 1 m).

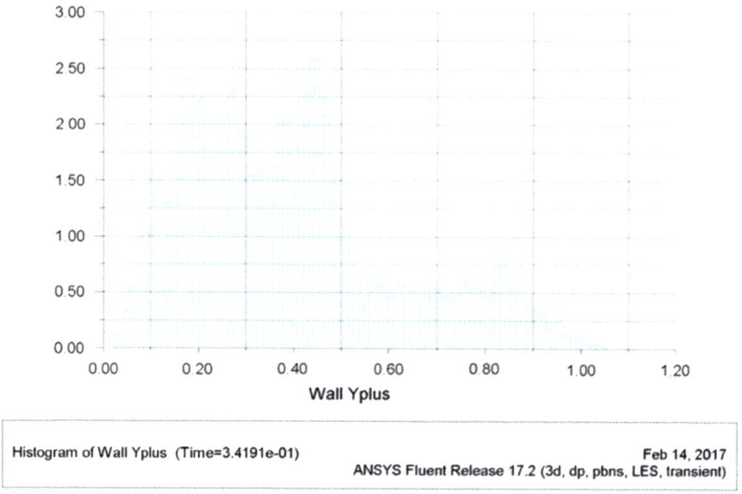

Fig. 7 Histogram of the Y+ value at the cylinder wall

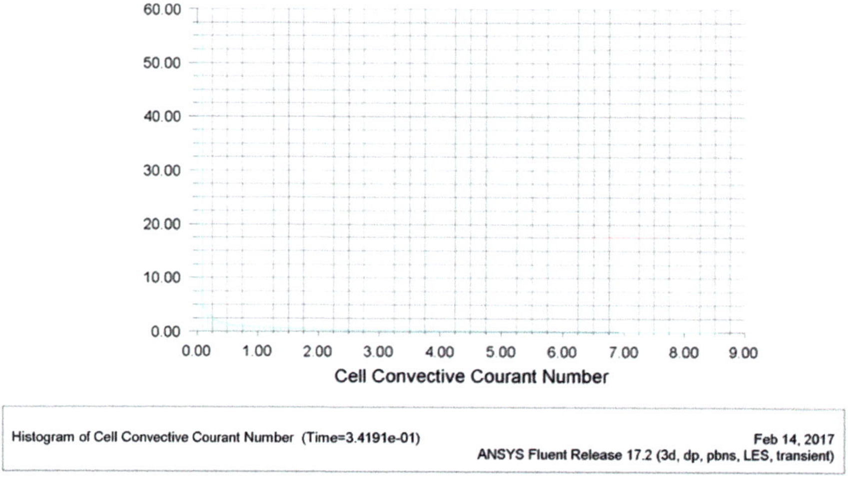

Fig. 8 Histogram of CFL number

where u is the flow velocity, c is the speed of sound, Δt is the time step and Δx represents the mesh size. Equation (12) describes an acoustical wave that travels in a given time step through the mesh. For the condition $0 \leq CFL \geq 1$ it is ensured that the wave is traveled maximum one cell of size Δx for one time step.

Figure 9 shows the plot from the processed microphones from the Fluent simulation from 2200 to 2400 Hz band. These microphone files are processed with the beamforming algorithm with motion compensation for rotation within numerical software *GNU Octave*.

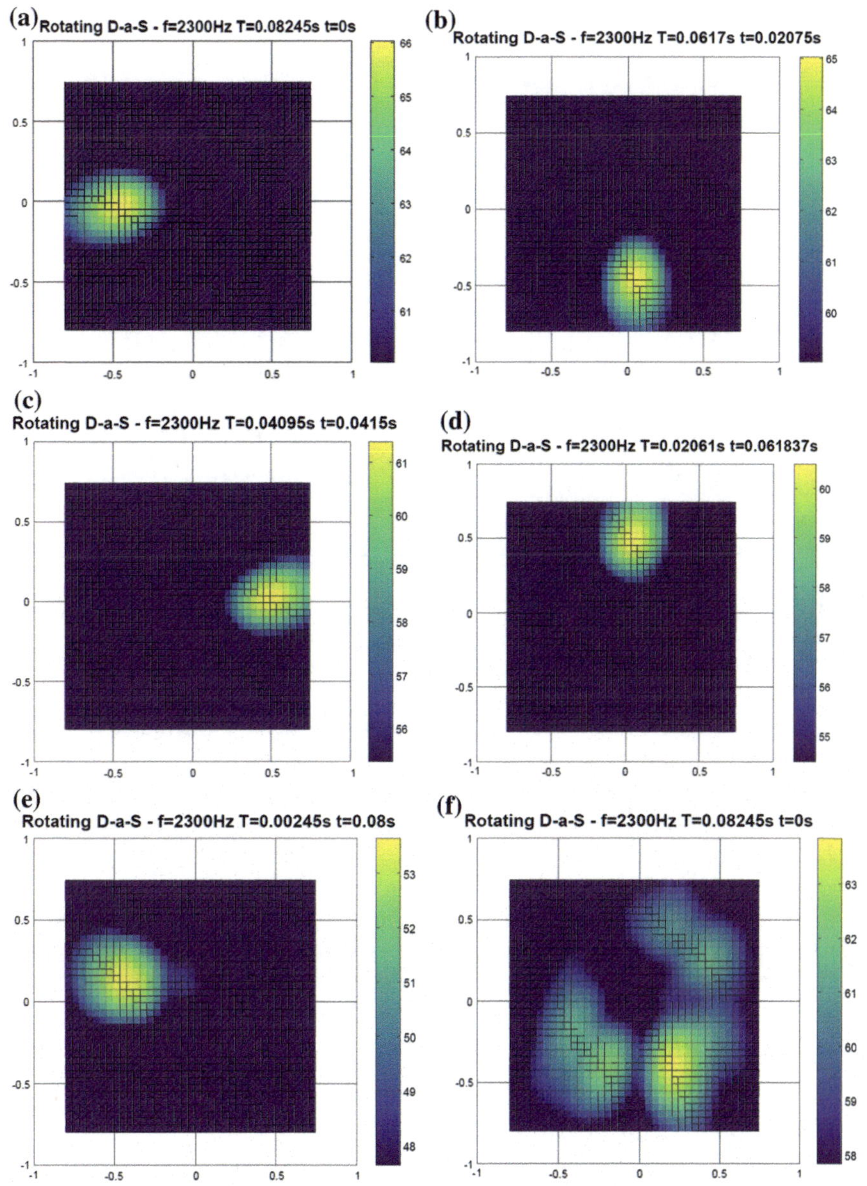

Fig. 9 **a** Beamforming plot at start position, **b** Beamforming plot at time 0.02075 s, **c** Beamforming plot at time 0.0415 s, **d** Beamforming plot at time 0.06225 s, **e** Beamforming plot at time 0.08 s, **f** Beamforming plot

Fig. 10 Rotating beamforming algorithm applied to a rotating cylinder measured with the acoustic camera

The cylinder rotates with 702 rpm mathematical positive at the radius 0.48 m at position 0,0 m. Figure 9a shows the start position after one turn. Almost quarter a turn is given at time 0.02075 s in Fig. 9b. Figure 9c–e shows the position of the cylinder after 0.0415, 0.06225 and 0.08 s. Figure 9f shows the beamforming result with no motion compensation with 0 rpm. The maximum sound source is arranged ring shaped, averaged over one rpm, located at radius 0.48 m.

Figure 10 shows the measurements applied to a rotating cylinder mounted on a bar using the algorithm with the acoustic camera. The optical picture is superimposed with a fixed-image. That allows the clearly identification of the position of the sound sources.

6 Conclusion and Outlook

With the here presented rotating beamforming algorithm it is possible to image stationary sound sources as well as rotating sound sources. This paper is a continuance of the work of Maier [14] and the novelty is the analysis towards rotating sound sources from a rotating CFD simulation using Fluent. In combination with an acoustic camera, it is a helpful tool for optimizing fan geometries to reduce sound emission. With this algorithm, it is possible to locate sound sources at their position on the blades. So it is possible to distinguish between the leading edge and the trailing edge of the blade and study the frequency dependence of the generated noise.

The method of localization of rotating (non-stationary) sound sources is proved with this beamforming method and is applied to a non-trivial rotating fan to show the sound sources at the blades to distinguish the sound source mechanism frequency dependent between trailing edge and leading edge in the next steps.

References

1. Brentner KS, Holland PC (1995) An efficient and robust method for computing quadrupole noise. s.l.: 2nd international aeromechanics specialists, 1995. http://www.ingentaconnect.com/content/ahs/jahs/1997/00000042/00000002/art00007
2. Brentner KS, Holland PC (1996) Numerical algorithms for acoustic integrals—the devil is in the details. Pennsylvania: 2nd AIAA/CEAS aeroacoustics conference (17th AIAA aeroacoustics), 1996
3. Brooks TF, Humphreys WM (2004) A deconvolution approach for the mapping of acoustic sources (DAMAS) determined from phased microphone arrays. s.l.: 10th AIAA/CEAS aeroacoustics conference, 2004
4. Christensen JJ, Hald J (2004) *Technical Review No. 1 2004—Beamforming*. [ed.] Sound & Vibration Measurement. s.l.: Brüel & Kjær, 2004
5. Dougherty RP, Walker BE, Sutliff DL (2010) Locating and quantifying broadband fan sources using in-duct microphones. In: 16th AIAA/CEAS Aeroacoustics conference, 2010
6. Ehrenfried K (2004) Skript zur Vorlesung Strömungsakustik (Script to the teaching lesson Fundamentals of Aeroacoustics). Mensch & Buch Verlag, Berlin, 2004. http://vento.pi.tu-berlin.de/STROEMUNGSAKUSTIK/SCRIPT/nmain.pdf
7. Farassat F (1975) Theory of noise generation from moving bodies with an application to helicopter rotors. Langley Research Center, Harnpton. Vols. NASA TR R-451
8. Ffowcs Williams JE, Hawkings DL (1969) Sound generated by turbulence and surfaces in arbitrary motion, vol 1151, s.l.: The Royal Society, Philosophical Transactions of the Royal Society. Series A, Mathematical and Physical Sciences, pp. 321–342. http://www.jstor.org/discover/10.2307/73790?uid=3737864&uid=2129&uid=2&uid=70&uid=4&sid=47698738473707
9. Fröhlich J (2006) Large Eddy Simulationen Turbulenter Strömungen (engl. LES of turbulent flows). 1st edn. Wiesbaden: Teubner, p 107. ISBN 978-3-8351-0104-3
10. Herold G, Sarradj E (2015) Microphone array method for the characterization of rotating sound sources in axial fans. Noise Control Eng J 63(6):546–551
11. Kirchhoff code—a Versatile CAA Tool—NASA SBIR-Phase I final report (1995) Advanced Rotorcraft Technology, Inc. Mountain View: contract NAS1-20366
12. Lighthill MJ (1952) On sound generated aerodynamically: I. General theory. In: Proc. Royal Society London, pp 564–587
13. Lighthill MJ (1954) On sound generated aerodynamically: II. Turbulence as a source of sound. In. Proceeding Royal Society London, pp 1–32
14. Maier C (2013) Practical and theoretic aero-acoustics. Ph.D. Glasgow: ethos database
15. Minck O et al (2012) Fan noise analysis using a microphone array. s.l.: Fan 2012—international conference on fan noise, technology, and numerical methods, 2012
16. Oswald M (2011) Aeroacoustics simulation with ANSYS CFD, Teaching script, Darmststadt
17. Pannert W, Maier C (2014) Rotating beamforming—motion-compensation in the frequency domain and application of High-resolution beamforming algorithms. J Sound Vib 333(7):1899–1912. http://dx.doi.org/10.1016/j.jsv.2013.11.031
18. Pope S (2000) Turbulent flows. Cambridge University Press, Cambridge. ISBN 978-0-521-59886-6

19. Ramachandran RC, Raman G, Dougherty RP (2014) Wind turbine noise measurement using a compact microphone array with advanced deconvolution algorithms. J Sound Vib 14:3058–3080, Volume 333, 2014
20. Sijtsma P, Oerlemans S, Holthusen H (2001) Location of rotating sources by phased array measurements. National Aerospace Laboratory, Amsterdam

Signal Analysis and Numerical Method Roles in Neural Interfaces Development

I. Kuzmanić, I. Vujović, J. Šoda and M. Rogić Vidaković

Abstract Neural interfaces are developing worldwide owing to the wide spectrum of possible applications, ranging from military to entertainment industry. Future human-machine interaction (HMI) and brain-machine interfaces (BMI) are inconceivable without further development in the field. Possible applications can also be found in traffic, i.e. remote supervision and/or control of vehicles/ships. In order to obtain effective BMI, two tasks should be performed: the calculation of the EM field by numerical methods (FEM (Finite Elements Method), BEM (Boundary Elements Method), and hybrid) and signal analysis, which should tell the machine what is required (command understanding). Examples of brain signal analysis using FFT and various WT transforms are presented. The basics of numerical methods for topic application are also covered. EEG data are obtained by experiments on consenting human subjects. Finally, the implications of neural interface development and introduction to traffic applications are considered.

Keywords Signal processing and analysis · BEM/FEM · BMI · Wavelets · Short-time FT · EMD

1 Introduction

The study of brain neural interfaces can be divided into several stages. The first stage is hardware development, where brain signals are measured, collected and conditioned. The second stage is software and algorithm development, involving the low-level processing of measured data followed by high-level programming. The processing

I. Kuzmanić (✉) · I. Vujović · J. Šoda
Analysis and Advanced Diagnostics Research and Education Laboratory (SPAADREL), Faculty of Maritime Studies, Signal Processing, University of Split, Ruđera Boškovića 37, 21000 Split, Croatia
e-mail: ikuzman@pfst.hr

M. Rogić Vidaković
Department of Neuroscience, Laboratory for Human and Experimental Neurophysiology (LAHEN), School of Medicine, University of Split, Šoltanska 2, 21000 Split, Croatia

A. Öchsner and H. Altenbach (eds.), *Engineering Design Applications II*,
Advanced Structured Materials 113, https://doi.org/10.1007/978-3-030-20801-1_16

of measured data includes various transforming techniques in the time, frequency or time-frequency domain or another spatial domain used to extract appropriate features from collected signals. High-level processing includes the modelling of EM field around the head using numerical methods such as FEM and recently BEM. In this respect, the paper deals with two topics, integrated in their final application, namely:

- signal processing and analysis, and
- numerical methods for EM modelling/calculation.

Both topics are important for the study of neural interfaces and should be addressed accordingly. The obtained signal should be processed, and then analyzed to extract features. EM field in the vicinity of the head should be modelled to obtain reliable input for the signal analysis part of the system.

Brains research is still a mainly uncharted field with many challenges. Many mechanisms are unknown or partially known. Most research is conducted for diagnostic purposes of various kinds of disorders, like stutter (a motor speech disorder which implies understanding motor neural path) [1], epilepsy [2, 3] and other. Monitoring writing disorders [4] is also an advance in the motor path execution understanding. The BMI uses many signal processing and analysis techniques, such as combination of DWT (Discrete Wavelet Transform) and PCA (Principle Component Analysis) [5], neural networks [6], and others.

The prospect of using BMI is considered in various applications, even in games [7]. The following consideration is the consequence of BMI development for traffic applications. The future of maritime traffic could be developed by researching an efficient neural interface, which would be used to operate vessels, ROVs, submarines, etc. In that case, a ROV as in [8] could be operated from land or nearby marine vehicle. Operator could be remotely placed in a virtual reality environment created as an exact replica of the real ROV's environment. The data obtained as in [9], could be obtained using sensor fusion i.e. from various monitoring systems like smart buoys or a smart buoy net, marine vehicles, satellites and such integrated into a single large data feeder. A net of such buoys could be operated and maintained using neural interfaces commanded by ships servicing the buoys network. The research of neural interfaces in traffic applications is reported in i.e. [10], where car could read the driver's emotional state or intoxication level. Hence, it could decide to take over driving based on brains' EM radiation, received by the neural interface. The basis for these applications already exists as P300-based BCI (Brain-Computer-Interface) [11, 12]. In P300 BCI, the interface analyses the user's EEG (Electroencephalogram) and recognizes which stimuli are present (pattern recognition in EEG signal). The recognized stimuli are converted into a corresponding command, which is then executed.

This chapter is organized as follows. The second section deals with basic signal processing and analysis techniques used in the Results section. The third section provides introductory explanations required for BMI. The fourth section presents results. Finally, conclusions are given.

2 Signal Processing and Analysis

In this section, some standard signal processing techniques are reviewed, which are important for this research: FT (Fourier transform), STFT (Short-term FT), WT (Wavelet transform), and ℂWT (complex WT).

FT is a standard tool for stationary signal analysis. The basic definition is expressed with [13]:

$$F(\omega) = \int_{-\infty}^{\infty} f(t)e^{-i\omega t}dt = |F(\omega)| \cdot e^{arctg(F(\omega))} \tag{1}$$

where $f(t)$ is an observed signal in the time-domain (in our case EEG signal), $F(\omega)$ a spectrum of the observed signal, t is time, and ω denotes the frequency. To arrive at the relevant conclusions and deal with non-stationary signals, such as EEG/EMG, STFT was developed, which is FT performed in the limited time interval. This is expressed by [13]:

$$STFT_f^{\omega}(\tau, \omega) = \int_t [f(t) \cdot W(t - \tau)] \cdot e^{-j\omega t}dt \tag{2}$$

where STFT of signal $f(t)$ is computed for each window centered at a $t = \tau$, at $\omega = \omega_0$ frequency, which is called localized spectrum. A $W(t - \tau)$ is referred to as windowing function. An expression $e^{-j\omega t}$ is referred to as the FT kernel. The STFT transform is well-suited for signals in the audio domain (band of frequencies between 0 Hz and 20 kHz).

The WT is defined as [13]:

$$CWT_f(a, b) = \frac{1}{\sqrt{|a|}} \int_{-\infty}^{\infty} f(t)\psi^*\left(\frac{t-b}{a}\right)dt \tag{3}$$

where parameter a represents scale i.e. frequency-shifts, parameter b represents translation i.e. time-shifts, and function ψ is called the mother wavelet.

When using the FT, STFT and WT transforms, the result is expressed by transform coefficients. When wavelet coefficients are complex, a Complex WT (ℂWT) is introduced. It should not be confused with continuous WT (CWT). In case of ℂWT, and discrete case like on computers, wavelets are complex-valued [14]:

$$\psi_C(t) = \psi_r(t) + j\psi_i(t) \tag{4}$$

and wavelet coefficients (5) have magnitude (6) and phase (7):

$$d_C(j, n) = d_r(j, n) + jd_i(j, n) \tag{5}$$

$$|d_C(j,n)| = \sqrt{[d_r(j,n)]^2 + [d_i(j,n)]^2} \tag{6}$$

$$\angle d_C(j,n) = \arctan\left(\frac{d_i(j,n)}{d_r(j,n)}\right) \tag{7}$$

where frequency content is controlled by parameter *j* (scale) at different times, which is controlled by time shift, n. All transforms are nowadays solved in standard software packages like Matlab, Octave or other and there are libraries for C or C++. Some packages even have GUIs (Graphical User Interface) to simplify the usage of specific functions.

3 Numerical Methods for EM Calculation Around the Head

FEM usage has limitations in modelling the forward problem for geometries necessary for the development of effective neural interfaces. The problem occurs due to thin shells describing the human head. BEM is better suited for thin boundaries. Hence, hybrid FEM-BEM methods are better suited for the calculation of fields around the human head.

The electric field outside the head is expressed by the boundary integral equation [15, 16]:

$$\alpha \vec{E}'_{ext} = \vec{E}'_{inc} + \oint_{\partial V} \hat{n} \times \left(\nabla \times \vec{E}_{ext}\right) G dS + \oint_{\partial V} \left[\left(\hat{n} \times \vec{E}_{ext}\right) \times \nabla G + \left(\hat{n} \cdot \vec{E}_{ext}\right)\nabla G\right] dS \tag{8}$$

where:

- ∂V is bounded volume,
- $\hat{n}$ is outer vector to the bounded volume,
- α is the angle subtended at the observation point,
- $\vec{E}_{ext}$ and $\vec{E}_{inc}$ denote the total and the incident electric field, and
- G is the free space Green's function.

In order to derive the final system of matrix equations, the use of the following two equations are required:

$$\int_V \left[\frac{j}{\omega\mu} \nabla \times \delta_i \vec{w}_i \cdot \vec{E}_{\text{int}} - (\sigma + j\omega\varepsilon)\delta_i \vec{w}_i \cdot \vec{E}_{\text{int}}\right] dV$$

$$= \oint_{dV} d\vec{S} \cdot \delta_i \vec{w}_i \times \vec{H}_{\text{int}} \tag{9}$$

where:

- δ_i is the parameter equal to -1 if the local edge direction does not coincide with the chosen global edge direction, and $+1$ in other cases,
- w_i denotes vector based shape function,
- σ [S/m] and ε (constant, no units) express dielectric properties of head tissue,
- $\hat{n} \times \vec{E}_{\text{int}} = \hat{n} \times \vec{E}_{ext}$ and $\hat{n} \times \vec{H}_{\text{int}} = \hat{n} \times \vec{H}_{ext}$ (due to the condition that tangential components must be continuous across the surface).

The second equation required is:

$$\begin{aligned}
\oint_{\partial V} d\vec{S}' \cdot \delta_i \vec{w}_i \times \alpha \vec{E}'_{ext} &= \oint_{\partial V} d\vec{S}' \cdot \delta_i \vec{w}_i \times \vec{E}'_{inc} \\
&- j\omega\mu \oint d\vec{S}' \cdot \delta_i \vec{w}_i \times \oint_{\partial V} \hat{n} \times \vec{H}_{\text{int}} G ds \\
&+ \oint_{\partial V} d\vec{S}' \cdot \delta_i \vec{w}_i \times \oint_{\partial V} \left(\hat{n} \times \vec{E}_{\text{int}}\right) \times \nabla G dS \\
&- \frac{1}{\omega + j\omega\mu} \oint_{\partial V} d\vec{S}' \cdot \delta_i \vec{w}_i \times \oint_{\partial V} \nabla_S \cdot \left(\hat{n} \times \vec{H}_{ext}\right) \nabla G dS
\end{aligned} \tag{10}$$

Hence, the electric field yields magnetic field and vice versa, and the FEM/BEM coupling needs to be identified. If we denote e_{bem} and h_{bem} as unknown coefficients associated with boundary surface, e_{inc} as known coefficients obtained by calculating the incident field, and $[E_{\text{bem}}]$ and $[H_{\text{bem}}]$ matrices obtained from the integral Eq. (10), and matrices $[E_{\text{fem}}]$ and $[H_{\text{fem}}]$ from (9), the system of matrix equations suitable for computer calculations (see [17]) can be obtained:

$$\begin{aligned}
[E_{bem}]\{e_{bem}\} &= \{e_{inc}\} + [H_{bem}]\{h_{bem}\} \\
\left[E_{fem}\right]\left\{e_{fem}\right\} &= \left[H_{fem}\right]\left\{h_{fem}\right\}
\end{aligned} \tag{11}$$

These equations are input to computer program in order to model EM field around human head, and as beginning of understanding of brains output/input data. This will enable what the brains wants that computer performs.

4 Results and Discussion

For experiments with nTMS we used eXimia Navigated Brain Stimulation System 4.0 (Nexstim Ltd., Helsinki, Finland). The system combines non-invasive TMS with MRI-based stereotactic navigation and real-time EMG measurement. Individual MRI images are loaded and aligned with the head, allowing generation of an accurate 3D model of the actual head. Individual MRIs of the head were performed with Philips Magnetic Resonance Achieva 1.5 T A-series, Head Coil 8 channel (Polyclinic Sunce, Split, Croatia). The test subjects have problems with stuttering and obstructive sleep apnea (snoring). This section tries to illustrate the problems in the analysis of TMS-obtained signals without attempting to generalize conclusions and without developing the diagnostics algorithm. Simple research provides experience for further more precise experiments in the exploration of the decision/action path useful for neural interfaces.

Figure 1 illustrates an EEG signal. It is a non-stationary time-domain signal. Periodicity is present in the artificial external excitation signal.

In order to distinguish normal talk from snoring, we first tried to correlate the signal of the same examinee recorded while talking and sleeping. The correlation operation was performed for both a single trail and the entire recorded signal (see Fig. 2). Trail can be easily observed from "Gate in" signal. One trail consists of 4 by 4 impulse sequences colored in pink, followed by a long pause till the next trail. There is a smaller pause between impulse sequences. One sequence with the corresponding pause is called a window. Obviously, nothing useful was accomplished by simple correlation.

Figure 3 illustrates the comparison of FT on a single trial, and can be considered as STFT with physically constrained time period. Time interval was reduced to better cover the signal's MEP (motor evoked potential) marker. Figure 3 illustrates how MEP's spectra looks more regular to the human observer than spectra of the entire trial. However, generalizing with an explicit mathematical expression is difficult.

It is obvious that another analysis should be used to obtain practical results. The next possible method is the use of spectrograms. Figure 4 shows an example of a spectrogram.

It is possible to correlate markers in the right half of Fig. 5 (time-domain) with lower-time part of the corresponding spectrum. It should be noted that this does not mean that this is a relevant detection method, since it has not been performed and confirmed on a large set of subjects. Figure 5 shows the results of ℂWT analysis performed in Matlab's wavelet GUI.

One can recognize patterns in modulus and angle for apnea and normal talk in Fig. 5. This method is potentially promising for further research.

Finally, it is useful to isolate the sources of EEG signals in brain research. One of the ways is to perform an EMD analysis, which is a modern approach in many signal analysis applications. EEG signals analyzed had 21 components at 40 dB resolution, and 31 components at 60 dB resolution. Reconstruction error depends on resolution and, at 40 dB, it is about 10-18 order of magnitude (in comparison, the

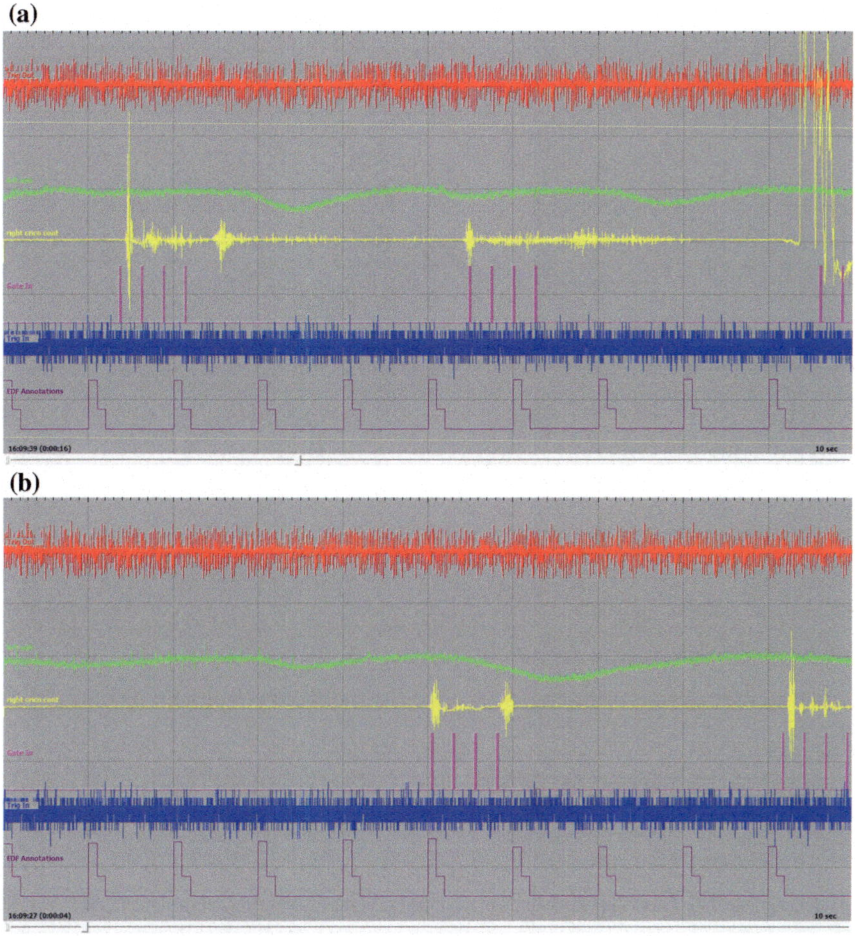

Fig. 1 Example of source signal (in temporal domain), shown in EDF browser (two time intervals of the same signal/patient)

signal's order of magnitude is 10-3). The EMD produces very low reconstruction error (Fig. 6a). However, the results of EMD analysis (see Fig. 6) show that isolating different sources of the brain's signals is not as easy as it might seem.

5 Conclusions

The paper presents results obtained from the experiments on two disorders. Such experiments and results are a step further in the understanding thought processes and brain activities measured using EEG signals. It is necessary to keep in mind that

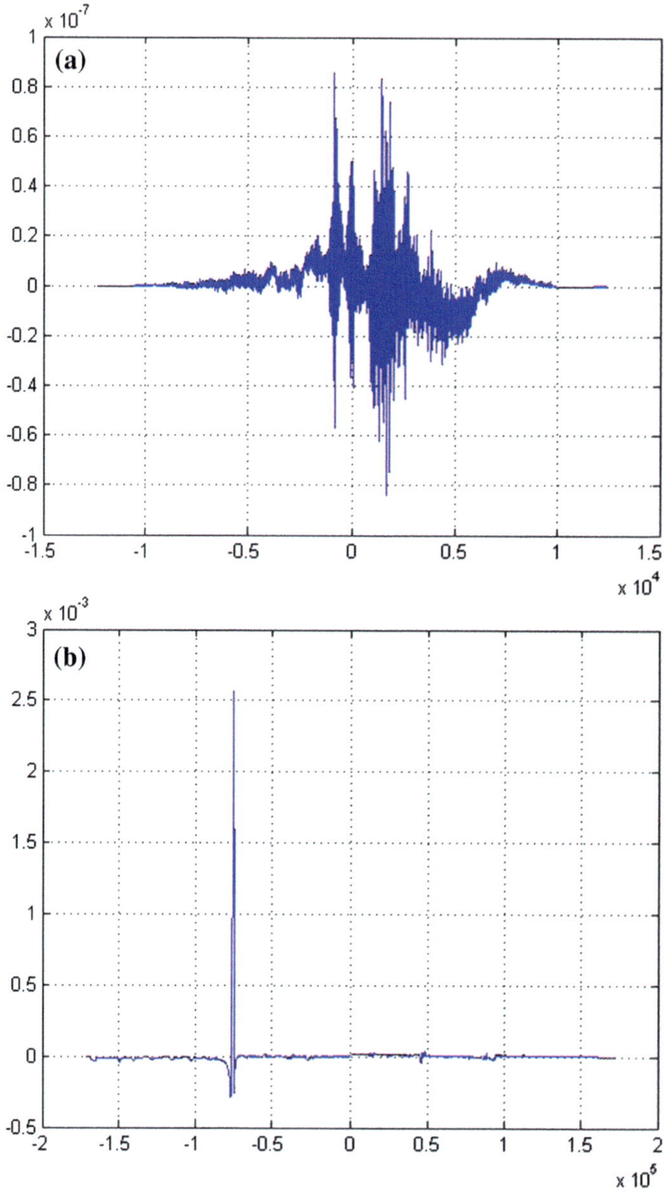

Fig. 2 Correlation of the talk and snore signal for the same examinee: **a** in the first trail, **b** for the entire signal. Correlation magnitude is at y-axis, and serial sample number at the x-axis

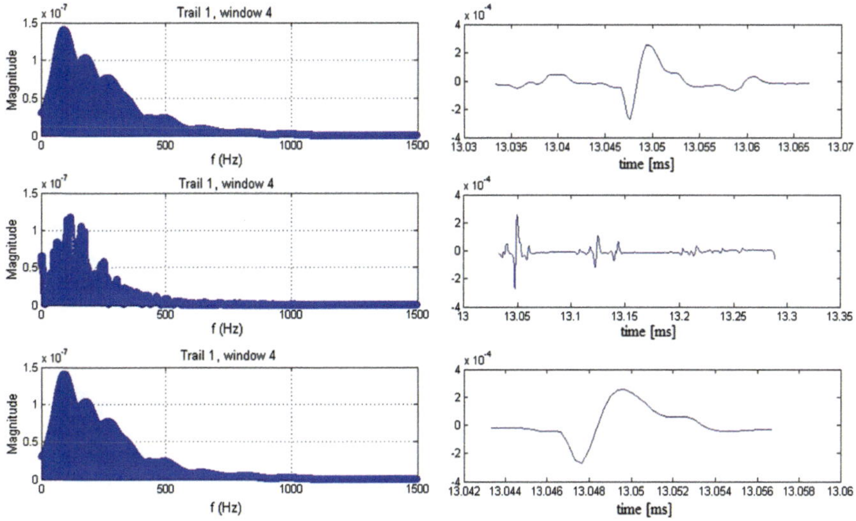

Fig. 3 Investigation of MEP marker in the brain's EMG (time-domain right, frequency domain left)

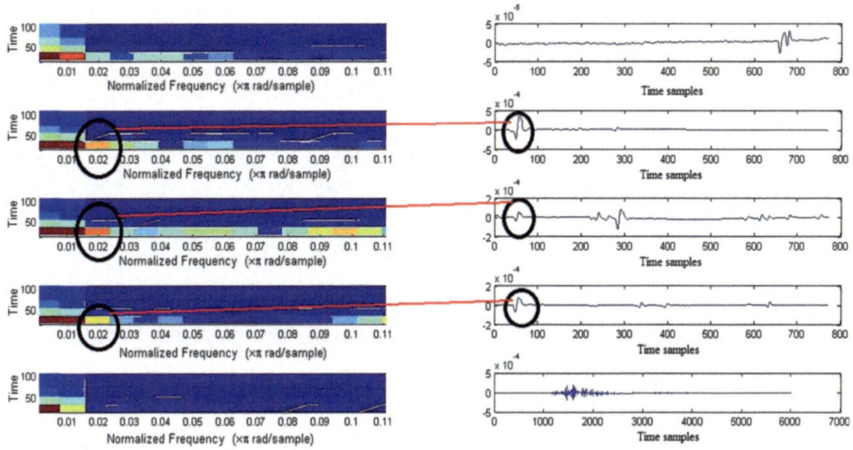

Fig. 4 Spectrogram of 4 windows and the entire trial for one examinee (left) and the corresponding time-domain signal (right)

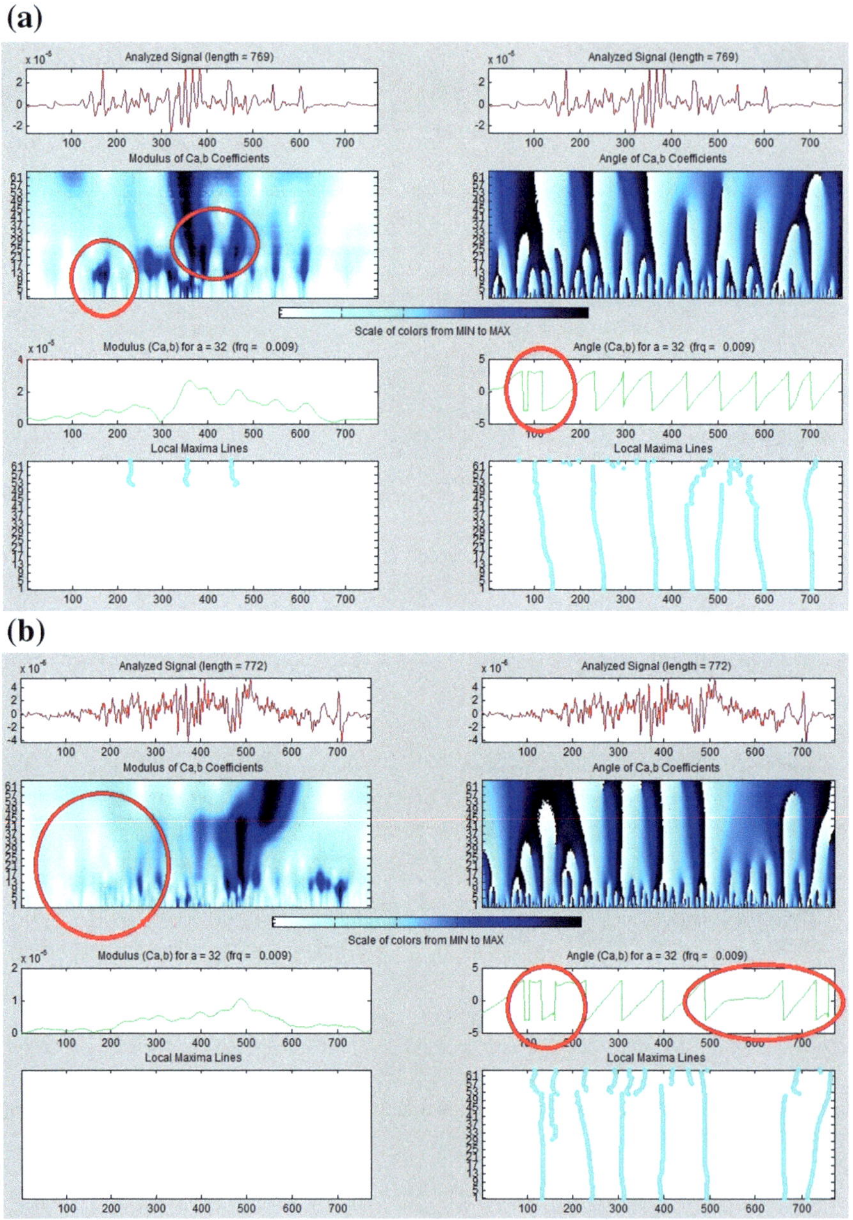

Fig. 5 ℂWT analysis in wavelet GUI with complex Gaussian wavelet of 1st order for the same subject: **a** example of talking (darker blue components in modulus section, red-marked circle), **b** example of snoring detection (white and light blue components in red-marked circle)

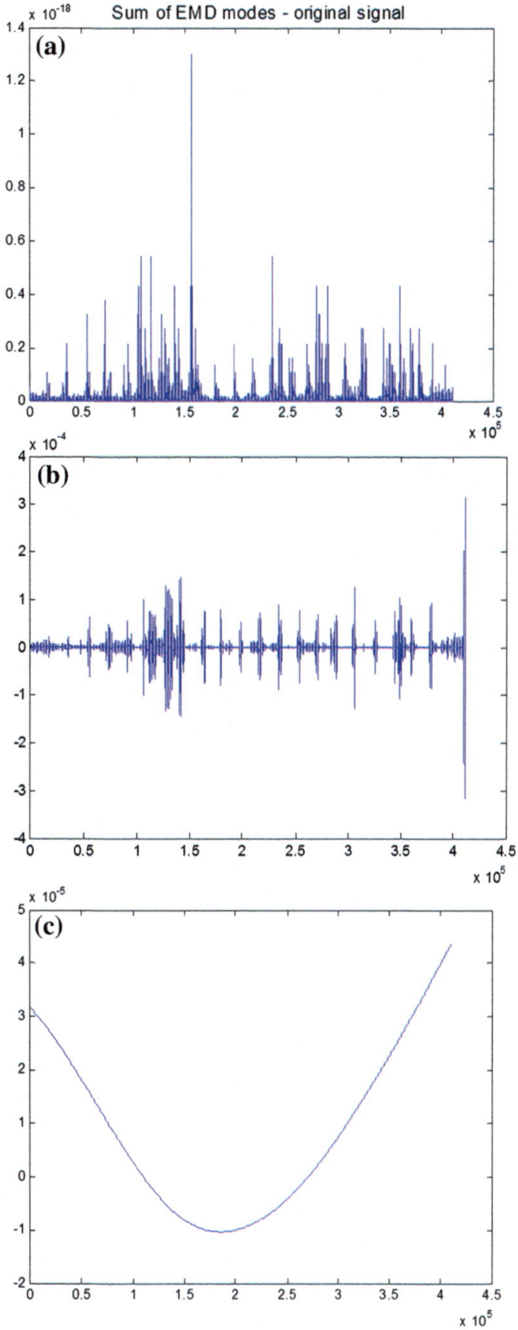

Fig. 6 Example of the EMD analysis: **a** reconstruction error (amplitude is in volts), **b** component 1, **c** component 31, **d** component 30, **e** component 29, **f** component 28, **g** component 27, **h** component 26, **i** component 25, **j** component 19, **k** component 2. Time axis is equivalently shown in sample number

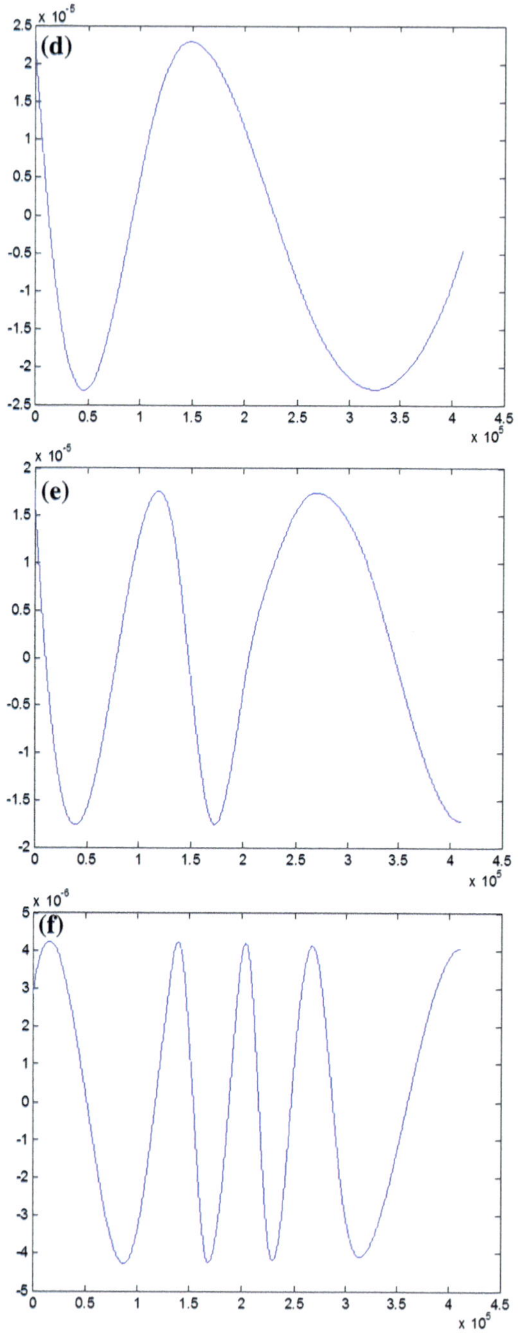

Fig. 6 (continued)

Fig. 6 (continued)

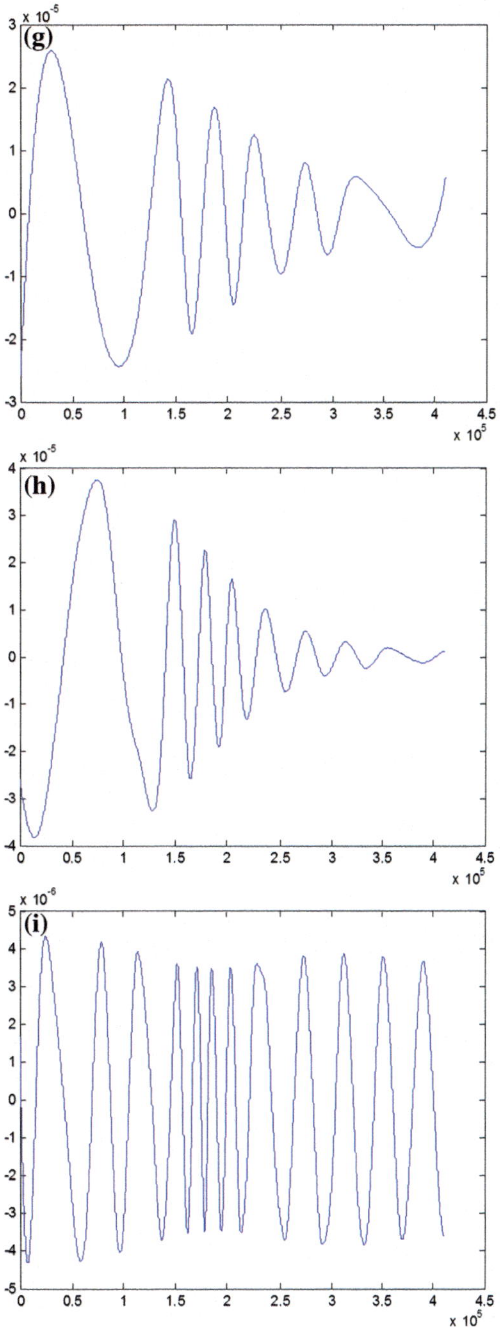

Fig. 6 (continued)

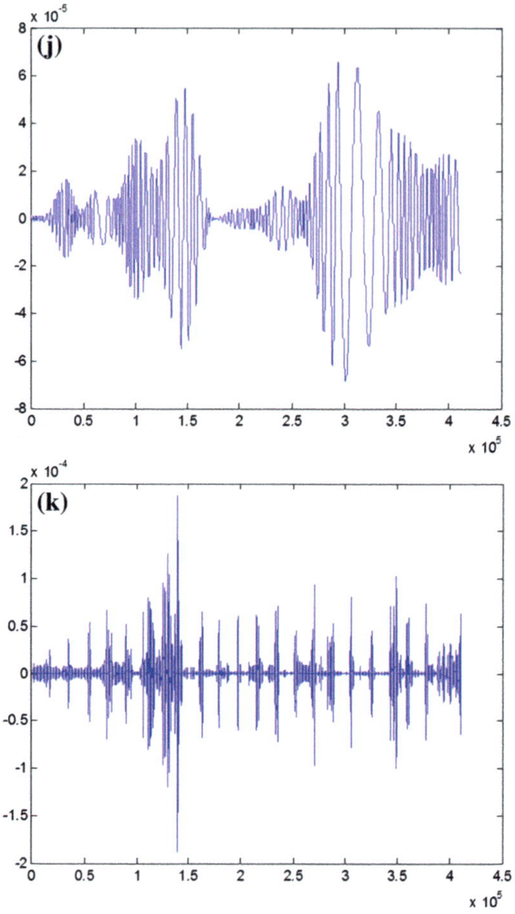

EEG signals are weak (an amplitude of order 40 μV) and very susceptible to noise. The final goal is to obtain a BMI from such a signal.

The potential of using signal analysis on EEG signals is shown by in Fig. 5, since snoring could not be mistaken for talking (giving command). An example of necessary research is the need to recognize what talking in sleep and normal talking actually are. It is obvious that talking in sleep is not intended to be orders for audio input to a computer. Furthermore, this talking would change brain waves and consequently, EM fields, which could be misinterpreted as input to BMI. If BMI is used for ship control, it could be of vital importance to distinguish between talking in sleep, snoring and giving orders to the interface. This could be achieved by the research of EM fields produced by the brain. The next steps should be taken in that direction if we want to get practical BMI for traffic purposes.

References

1. Vidaković MR, Jerlović A, Jurić T, Vujović I, Šoda J, Erceg N, Bubić A, Schönwald MZ, Liumis P, Gabelica D, Đogaš Z (2016) Neurophysiologic markers of primary motor cortex for laryngeal muscles and premotor cortex in caudal opercular part of inferior frontal gyrus investigated in motor speech disorder: a navigated transcranial magnetic stimulation (TMS) study. Cogn Process 17:429–443
2. Kocadagli O, Langari R (2017) Classification of EEG signals for epileptic seizures using hybrid artificial neural networks based wavelet transforms and fuzzy relations. Expert Syst Appl 88:419–434
3. Patidar S, Panigrahi T (2017) Detection of epileptic seizure using Kraskov entropy applied on tunable-Q wavelet transform of EEG signals. Biomed Signal Process Control 34:74–80
4. Vidaković MR, Gabelica D, Vujović I, Šoda J, Batarelo N, Džimbeg A, Schönwald MZ, Rotim K, Đogaš Z (2015) A novel approach for monitoring writing interferences during navigated transcranial magnetic stimulation mappings of writing related cortical areas. J Neurosci Meth 255:139–150
5. Mousaa FA, El-Khoribib RA, Shomanb ME (2016) A novel brain computer interface based on principle component analysis. Procedia Comput Sci 82:49–56
6. Gandhi V, Prasad G, Coyle D, Behera L, McGinnity TM (2015) Evaluating quantum neural network filtered motor imagery brain-computer interface using multiple classification techniques. Neurocomputing 170:161–167
7. Cecílio J, Andrade J, Martins P, Castelo-Branco M, Furtado P (2016) BCI framework based on games to teach people with cognitive and motor limitations. Procedia Comput Sci 83:74–81
8. Vukšić M, Josipović S, Čorić A, Kraljevića A (2017) Underwater ROV as inspection and development platform. Trans Marit Sci. https://doi.org/10.7225/toms.v06.n01.005
9. Mudronja L, Matić P, Katalinić M (2017) Data-based modelling of significant wave height in the adriatic sea. Trans Marit Sci. https://doi.org/10.7225/toms.v06.n01.001
10. Bennakhi A, Safar M, Abdulrasoul J (2017) Homonoia: when your car reads your mind. Procedia Comput Sci 110:135–142
11. Ganin IP, Shishkin SL, Kaplan AY (2013) A P300-based brain-computer interface with stimuli on moving objects: four-session single-trial and triple-trial tests with a game-like task design. PLoS ONE 8. https://doi.org/10.1371/journal.pone.0077755
12. Mak JN, Arbel Y, Minett JW, McCane LM, Yuksel B et al (2011) Optimizing the P300-based brain-computer interface: current status, limitations and future directions. J Neural Eng. https://doi.org/10.1088/1741-2560/8/2/025003
13. Vujović I, Šoda J, Kuzmanić I (2012) Cutting-edge mathematical tools in processing and analysis of signals in marine and navy. Trans Marit Sci. https://doi.org/10.7225/toms.v01.n01.005
14. Selesnick IW, Baraniuk RG, Kingsbury NG (2005) The dual-tree complex wavelet transform. IEEE Signal Process Mag 22:123–151
15. Poljak D, Cavka D, Dodig H, Peratta C, Peratta A (2014) On the use of the boundary element analysis in bioelectromagnetics. Eng Anal Bound Elem 49:2–14
16. Dodig H, Poljak D, Peratta A (2012) Hybrid BEM/FEM edge element computation of the thermal rise in the 3D model of the human eye induced by high frequency EM waves. In: 20th international conference on software, telecommunications and computer networks (SoftCOM), pp 1–5
17. Dodig H, Cvetkovic M, Poljak D, Hirata A, Laakso I (2017) Hybrid FEM/BEM for human head exposed to high frequency electromagnetic radiation. In: 40th international conference on boundary elements and other mesh reduction methods, Brockenhurst, UK, pp 239–250, 12–14 Sept 2017

Optimization of the Head Geometry for a Cable Car Passing over a Support

M. Wenin, A. Windisch, S. Ladurner, M. L. Bertotti and G. Modanese

Abstract In this work we discuss the problem of finding an optimal shape of a cable ropeway support head using optimization techniques. We define a cost function and relevant constraints with the goal to minimize the oscillations of the vehicle when it crosses the support, valid for both driving directions. Our findings reveal potential for practical use by extending Computer Aided Engineering tools by taking this optimization procedure into account.

Keywords Cable ropeway · Optimal support geometry · Numerical optimization · Unwanted vehicle oscillations

M. Wenin (✉)
CPE Computational Physics and Engineering, Weingartnerstrasse 28,
39011 Lana, BZ, Italy
e-mail: markus.wenin@cphysics.com

A. Windisch
Department of Physics, Washington University in St. Louis, St. Louis, MO 63130, USA
e-mail: windisch@physics.wustl.edu

A. Windisch
Department of Physics, University of Graz, Universitätsplatz 5, 8010 Graz, Austria

S. Ladurner
Doppelmayr Italia, Industriezone 14, 39011 Lana, BZ, Italy
e-mail: Siegfried.Ladurner@Doppelmayr.com

M. L. Bertotti · G. Modanese
Faculty of Science and Technology, Free University of Bozen,
Piazza Università 1, 39100 Bolzano, BZ, Italy
e-mail: marialetizia.bertotti@unibz.it

G. Modanese
e-mail: giovanni.modanese@unibz.it

A. Öchsner and H. Altenbach (eds.), *Engineering Design Applications II*,
Advanced Structured Materials 113, https://doi.org/10.1007/978-3-030-20801-1_17

Fig. 1 Left: Areal cable ropeway with two vehicles passing over the same support at different times and different driving directions. Right: a picture sequence to visualize the oscillations when the vehicle passes the support (photo *Doppelmayr*)

1 Introduction

When a cable car vehicle passes over a support, the direction of movement changes and the vehicle begins to oscillate [5, 6]. The unwanted oscillations depend on the deflection angle, the velocity of the vehicle and the length of the hanger (i.e. the reduced length of the pendulum, see Fig. 1). These oscillations can be reduced by the help of various damping systems installed on the vehicles. In this work we employ numerical optimization algorithms [9, 10] to investigate how the shape of the support head, which defines the trajectory of the vehicle's running gear, affects the vehicle oscillations. The geometry of standard support heads are constructed by several circular arcs [3, 4]. However, more sophisticated curves, such as clothoids, could also be of interest. To the best of our knowledge, no research that addresses this important aspect exists in the relevant literature. The present paper contains first results of the support head shape optimization. The paper is organized in three main sections: in Sect. 2 we consider the mathematical description of a class of curves with appropriate geometric properties. Since this is the basis of all following investigations, we provide all equations in detail. Section 3 describes the equation of motion for the vehicle and the procedure by which means we calculate the optimal support head geometry. In Sect. 4, we present a numerical demonstration of our approach. We do not consider optimization of the vehicle velocity in this paper, which can be interesting for advanced studies. Finally, we conclude in Sect. 5.

2 Parametrization of the Vehicle Path

Let l denote the length of the support head (the region of interest for this study) and α_-, α_+ be the tangent angles of the vehicle trajectory (see left part of Fig. 2). We assume that these quantities are known and treat them as external parameters. The starting point of the following mathematical derivations is the idea that if we

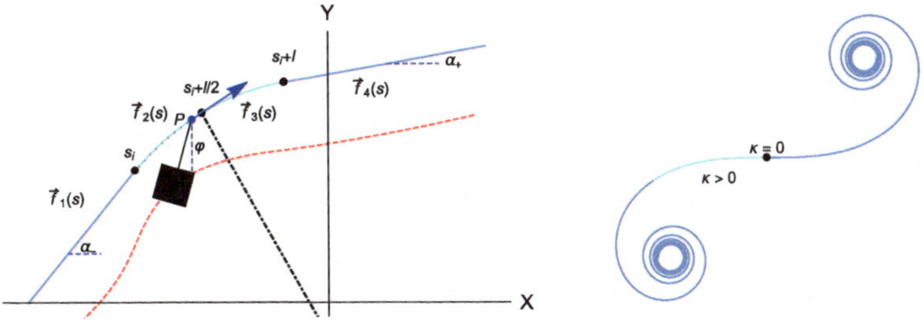

Fig. 2 Left: The system considered in this study: a cable car vehicle, modeled as a damped pendulum, moves over a support. The support is characterized by three fundamental parameters: the tangent angles α_-, α_+ and the length of the support head l. The shape of the support head is computed by solving an optimization problem, where the optimum corresponds to the least amount of vehicle oscillations. Here we use symmetric support geometries only (the black dot–dashed line indicates the symmetry axis), while taking into account both driving directions. Right: A piece of a clothoid (the curvature is proportional to the arc length). A more general ansatz for the curvature gives a class of curves, including the clothoid

use a generalized clothoid (right part of Fig. 2) we can implement some desired properties of the vehicle path in a simple way (curvature of the support head should be greater than and zero at the end-points etc.). In fact the concept of curvature plays an important role in our approach. To describe the geometry of the support head we start by defining an auxiliary function $K(u; c_1, c_2, c_3)$ containing three free parameters $c_{1,2,3} > 0$,

$$K(u; c_1, c_2, c_3) := \arctan(c_1 u) + c_2 u e^{-c_3 u^2}. \tag{1}$$

As one can see later, this ansatz ensures vanishing curvature at the end-points of the support head and covers three important special cases (see Fig. 3). Furthermore, we need the integral (for simplicity we suppress the arguments c_1, c_2, c_3 here and in the following)

$$f(u) := \int_0^u K(u')du' = u \arctan(c_1 u) + \frac{c_2}{2c_3}\left(1 - e^{-c_3 u^2}\right) - \frac{1}{2c_1} \ln(1 + c_1^2 u^2), \tag{2}$$

to define the generalized Fresnel integrals

$$F_s(t) := \frac{1}{\gamma} \int_0^t \sin(f(u))du, \quad F_s(-t) = -F_s(t), \tag{3}$$

$$F_c(t) := \frac{1}{\gamma} \int_0^t \cos(f(u))du, \quad F_c(-t) = -F_c(t). \tag{4}$$

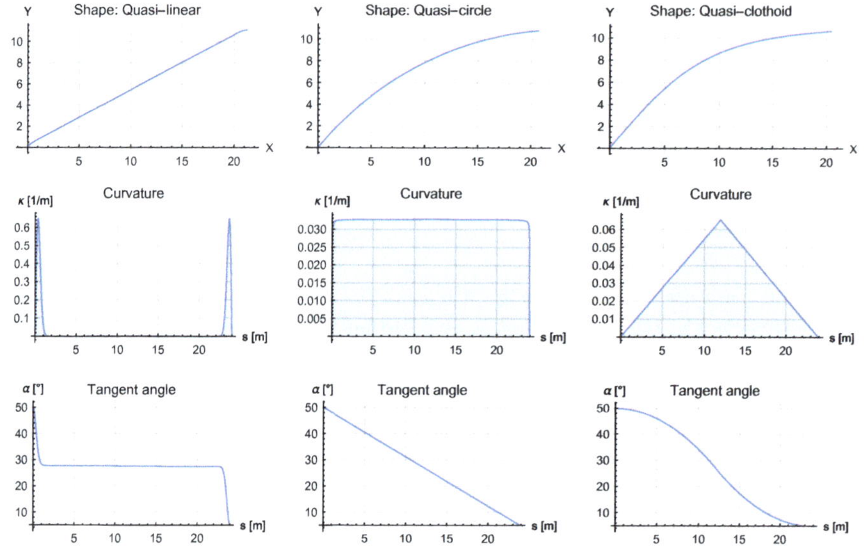

Fig. 3 Limiting cases generated by the ansatz Eq. refansatz for different parameter settings (in this Figure we set $s_i = 0$). For the quasi-linear shape we set $c_1 \to 0$, $c_{2,3} \to \infty$, such that $c_2/c_3 \to \alpha_- - \alpha_+$, leading to $f(u) = c_2/2c_3$ (constant function for $u > 0$). For the quasi-circle case we set $c_1 \to \infty$, $c_2 \to 0$ and $c_3 \neq 0$ arbitrary, $f(u) = u\pi/2$ (linear function for $u > 0$). For the quasi-clothoid case, we set $c_{1,3} \to 0$, $c_2 \to \infty$ and get $f(u) = c_2 u^2/2$ (quadratic function for $u > 0$). The construction ensures that the integral of the curvature over the path $\int_0^l \kappa ds = \alpha_- - \alpha_+$ is constant for all curves, as required by the definition and $\kappa \geq 0$ for all $c_{1,2,3} > 0$

Both functions are odd, which is a property that proves to be useful when simplifying the expressions in the case at hand. The parameter γ is defined implicitly as the zero of the transcendental equation

$$f\left(\gamma \frac{l}{2}\right) = \frac{\alpha_- - \alpha_+}{2}. \tag{5}$$

This equation ensures the continuity of the tangent angle in the middle of the support (we construct the entire support head using two equal length pieces). The curvature $\kappa = d\alpha/ds$, where α is the tangent angle, is given by

$$\kappa = \gamma K, \tag{6}$$

and $R = 1/\kappa$ is the radius of curvature. The generalized Fresnel integrals allow us to construct a generalized clothoid (for clothoids see for example [1])

$$(X, Y) = (F_c(t), F_s(t)), \tag{7}$$

with a horizontal tangent at the origin, which follows from $F_s'(0) = 0$.

2.1 Parametrization of the Path for a Symmetric Support

We obtain the desired parametrization by appropriate rotations and translations of the generalized clothoid Eq. (7). Since we demand that the curvature vanishes at the endpoints of the support head, we divide it into two regions with equal lengths $l/2$ and use Eq. (7) for both pieces. Including two linear paths before and after the support, the complete vehicle path can by written as (see Fig. 2)

$$\mathbf{U}(\alpha_-, \alpha_+; s) = \begin{cases} \mathbf{f}_1(s), & \text{for } 0 \leq s < s_i, \\ \mathbf{f}_2(s), & \text{for } s_i \leq s < s_i + l/2, \\ \mathbf{f}_3(s), & \text{for } s_i + l/2 \leq s < s_i + l, \\ \mathbf{f}_4(s), & \text{for } s_i + l \leq s \leq s_{\max}, \end{cases} \tag{8}$$

with a linear pre-support region

$$\mathbf{f}_1(s) = s \begin{pmatrix} \cos\alpha_- \\ \sin\alpha_- \end{pmatrix}, \tag{9}$$

the first support part

$$\mathbf{f}_2(s) = \begin{pmatrix} \cos\alpha_- F_c\Big(\gamma(s - s_i)\Big) + \sin\alpha_- F_s\Big(\gamma(s - s_i)\Big) + s_i \cos\alpha_- \\ \sin\alpha_- F_c\Big(\gamma(s - s_i)\Big) - \cos\alpha_- F_s\Big(\gamma(s - s_i)\Big) + s_i \sin\alpha_- \end{pmatrix}, \tag{10}$$

the second support part ($\tilde{s} \equiv s_i + l - s$)

$$\mathbf{f}_3(s) = \begin{pmatrix} -\cos\alpha_+ F_c\Big(\gamma\tilde{s}\Big) + \sin\alpha_+ F_s\Big(\gamma\tilde{s}\Big) \\ -\sin\alpha_+ F_c\Big(\gamma\tilde{s}\Big) - \cos\alpha_+ F_s\Big(\gamma\tilde{s}\Big) \end{pmatrix} + \begin{pmatrix} x_A \\ y_A \end{pmatrix}, \tag{11}$$

and a linear post-support region

$$\mathbf{f}_4(s) = (s - s_i - l) \begin{pmatrix} \cos\alpha_+ \\ \sin\alpha_+ \end{pmatrix} + \begin{pmatrix} x_A \\ y_A \end{pmatrix}. \tag{12}$$

The displacement vector is given by

$$\begin{pmatrix} x_A \\ y_A \end{pmatrix} = \begin{pmatrix} (\cos\alpha_- + \cos\alpha_+) F_c\left(\gamma\frac{l}{2}\right) + (\sin\alpha_- - \sin\alpha_+) F_s\left(\gamma\frac{l}{2}\right) + s_i \cos\alpha_- \\ (\sin\alpha_- + \sin\alpha_+) F_c\left(\gamma\frac{l}{2}\right) - (\cos\alpha_- - \cos\alpha_+) F_s\left(\gamma\frac{l}{2}\right) + s_i \sin\alpha_- \end{pmatrix}. \tag{13}$$

For the second derivatives we obtain the expressions

$$\mathbf{f}_2''(s) = \kappa(\gamma(s-s_i)) \begin{pmatrix} \sin\left(\alpha_- - f(\gamma(s-s_i))\right) \\ -\cos\left(\alpha_- - f(\gamma(s-s_i))\right) \end{pmatrix}, \tag{14}$$

$$\mathbf{f}_3''(s) = \kappa(\gamma\tilde{s}) \begin{pmatrix} \sin\left(\alpha_+ + f(\gamma\tilde{s})\right) \\ -\cos\left(\alpha_+ - f(\gamma\tilde{s})\right) \end{pmatrix}. \tag{15}$$

Note that $\mathbf{f}_2''(l/2) = \mathbf{f}_3''(l/2)$ and $\mathbf{f}_2''(s_i) = \mathbf{f}_3''(s_i + l) = (0, 0)^T$ for all parameters $c_{1,2,3}$.

3 Equation of Motion for the Vehicle and Numerical Shape Optimization

3.1 Equations of Motion for the Vehicle

Using a suitable Lagrangian, it is straightforward to derive the equation of motion for the angle $\varphi(t)$ describing the single degree of freedom of the vehicle [7],

$$ml_P\ddot{\varphi}(t) + m\left[\ddot{X}_P\cos(\varphi(t)) + \ddot{Y}_P\sin(\varphi(t))\right] + mg\sin(\varphi(t)) + \eta\dot{\varphi}(t) = 0. \tag{16}$$

Here l_P is the reduced pendulum length, g the gravitational acceleration, m the mass of the pendulum and η accounts for the damping; $\ddot{X}_P$, $\ddot{Y}_P$ are the components of the acceleration of the suspension point P. To solve the optimization problem it is convenient to use the arc length $s \in [0, s_{max}]$ as the independent variable instead of the time. For a constant velocity v one gets the differential equations for $\varphi_{1,2}(s)$ (the $'$ denotes the derivative with respect to s), where $\varphi_1(s)$ accounts for the ascending vehicle and $\varphi_2(s)$ for the descending vehicle respectively,

$$l_P\varphi_1''v^2 + X''v^2\cos(\varphi_1) + Y''v^2\sin(\varphi_1) + g\sin(\varphi_1) + v\frac{\eta}{m}\varphi_1' = 0, \tag{17}$$

$$l_P\varphi_2''v^2 + X_I''v^2\cos(\varphi_2) + Y_I''v^2\sin(\varphi_2) + g\sin(\varphi_2) + v\frac{\eta}{m}\varphi_2' = 0. \tag{18}$$

The initial conditions are

$$\varphi_1(0) = 0,\ \varphi_1'(0) = 0,\ \varphi_2(0) = 0,\ \varphi_2'(0) = 0. \tag{19}$$

The subscripted variables $X_I''(s)$, $Y_I''(s)$ correspond to the accelerations for the opposite direction. These quantities are given by the path parametrization $X_I(s)$, $Y_I(s)$ by a change of the angles $\alpha_- \to -\alpha_+$, $\alpha_+ \to -\alpha_-$ in the original path, $\mathbf{U}_I(\alpha_-, \alpha_+; s) = \mathbf{U}(-\alpha_+, -\alpha_-; s)$.

3.2 Double Kicked Pendulum

It is instructive to consider a double kicked pendulum, i.e. the case with linear support shape, since this special case is available analytically and all quantities of interest can be compute by hand. Without damping, the linearized equation of motion Eq. (16) reads

$$l_P\ddot{\varphi}(t) + g\varphi(t) + v\left[\cos\left(\frac{\alpha_- + \alpha_+}{2}\right) - \cos(\alpha_-)\right]\delta(t) +$$
$$v\left[\cos(\alpha_+) - \cos\left(\frac{\alpha_- + \alpha_+}{2}\right)\right]\delta(t - l/v) = 0, \varphi(0_-) = 0, \dot{\varphi}(0_-) = 0, \quad (20)$$

where $\delta(t)$ denotes the Dirac delta-function. The solution of this initial value problem is given by

$$\varphi(t) = -\frac{2v}{\sqrt{gl_P}}\sin\left(\frac{\alpha_- - \alpha_+}{4}\right)\left[\sin\left(\frac{1}{4}(3\alpha_- + \alpha_+)\right)\sin\left(\sqrt{\frac{g}{l_P}}t\right)\Theta(t) +\right.$$
$$\left.\sin\left(\frac{1}{4}(\alpha_- + 3\alpha_+)\right)\sin\left(\sqrt{\frac{g}{l_P}}(t - l/v)\right)\Theta(t - l/v)\right] \quad (21)$$

where $\Theta(t)$ the step-function. We consider the energy of the pendulum for small amplitudes, where we ignore the translational energy, $E = \frac{1}{2}ml_P^2\dot{\varphi}^2 + \frac{1}{2}mgl_P\varphi^2$. After the first kick, the energy is

$$E^{(1)} = \frac{1}{2}mv^2\left[\cos\left(\frac{\alpha_- + \alpha_+}{2}\right) - \cos(\alpha_-)\right]^2, \quad (22)$$

and does not depend on the pendulum length. After the second kick, the energy is

$$E^{(2)} = 2mv^2\sin^2\left(\frac{\alpha_- - \alpha_+}{4}\right)\left[\sin^2\left(\frac{1}{4}(3\alpha_- + \alpha_+)\right) +\right.$$
$$\sin^2\left(\frac{1}{4}(\alpha_- + 3\alpha_+)\right) + 2\sin\left(\frac{1}{4}(3\alpha_- + \alpha_+)\right) \times$$
$$\left.\sin\left(\frac{1}{4}(\alpha_- + 3\alpha_+)\right)\cos\left(\sqrt{\frac{g}{l_P}}\frac{l}{v}\right)\right]. \quad (23)$$

Now the energy also depends on the pendulum length, because there are two time scales present: the oscillation-period on the one hand and the time between the two kicks on the other hand. Thus, assuming $\sin\left(\frac{1}{4}(3\alpha_- + \alpha_+)\right) \times \sin\left(\frac{1}{4}(\alpha_- + 3\alpha_+)\right) > 0$, the minimum of the absorbed energy is attained for $\sqrt{\frac{g}{l_P}}\frac{l}{v} = \pi$. An evaluation of this equation gives consistent and realistic values: $l_P = 7$ m, $v = 5$ m/s leads to a support length $l = 13.2$ m. Using such parameters, one finds that the pendulum is at $\varphi = 0$ when it receives the second kick.

3.3 Cost Function

The cost function $J(c_1, c_2, c_3)$ is constructed using the concept of the phase–space of the dynamical system. We demand to trap the phase–point of the pendulum near the origin of the phase–space spanned by the variables $\varphi_{1,2}(s)$ and $\varphi'_{1,2}(s)$ for the complete path, i.e. we set

$$J = \int_0^{s_{max}} \frac{ds}{v}\left\{\frac{1}{2}l_P^2 v^2\Big[\varphi_1'(s)^2 + \varphi_2'(s)^2\Big] + g l_P\Big[2 - \cos(\varphi_1(s)) - \cos(\varphi_2(s))\Big]\right\}. \tag{24}$$

We solve this equation together with Eqs. (17) and (18) and seek the minimum

$$J(c_1, c_2, c_3) \to \min \tag{25}$$

by varying the parameters $c_{1,2,3}$. Additional constraints, as e.g. the agreement of least radius of curvature R with the CEN-standards [2] can be taken into account in a standard way.

4 Numerical Results

In this section we present first numerical results of the optimization procedure. We employ the Nelder–Mead algorithm to find the global minimum [8] using the following parameters: reduced pendulum length $l_P = 7$ m, damping parameter $\eta/m = 8.3$ m/s, $l = 25$ m, $\alpha_- = 45°$, $\alpha_+ = 5°$, $s_i = 100$ m, $s_{max} = 325$ m. Figures 4 and 5 show results from the minimization of the cost function Eq. (24) without further constraints, where the parameters are restricted to $10^{-6} \leq c_i \leq 10^4$ for $i = 1, 2, 3$. In Fig. 6 we plotted optimal support shapes obtained by minimization of J subject to the constraint $R > 20$ m. For small velocities $v \approx 5$ m/s, we obtain a quasi-circular solution, whereas for higher velocities the optimal solution becomes more flat in the center region. Similar results were achieved when the optimization was restricted to the support only (without pre- and post-support linear travel) without damping.

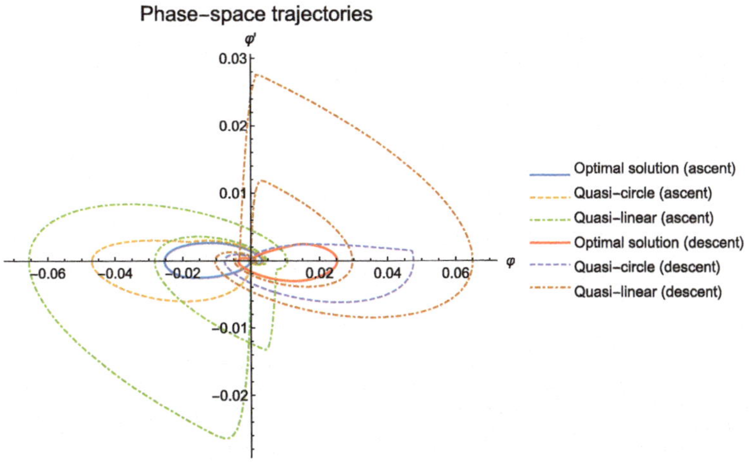

Fig. 4 Phase–space trajectories (φ, φ' stands for both variables) for the vehicle for both driving directions with $v = 5$ m/s. For comparison, we also plotted the trajectories for the quasi-circular and the quasi-linear shape. Since no additional constraints were added, the optimization is quite successful and confines the trajectories close to the origin of the phase–space. The costs J in the three cases satisfy $J_{opt.} : J_{Q-c} : J_{Q-l} = 1 : 1.67 : 3.47$

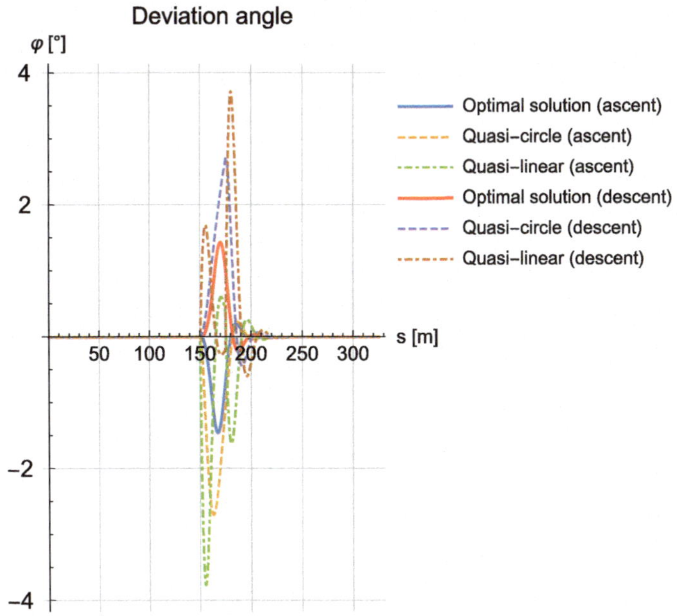

Fig. 5 Deviation angle for both driving directions, corresponding to Fig. 4. The maximum of the oscillation amplitude is reduced significantly

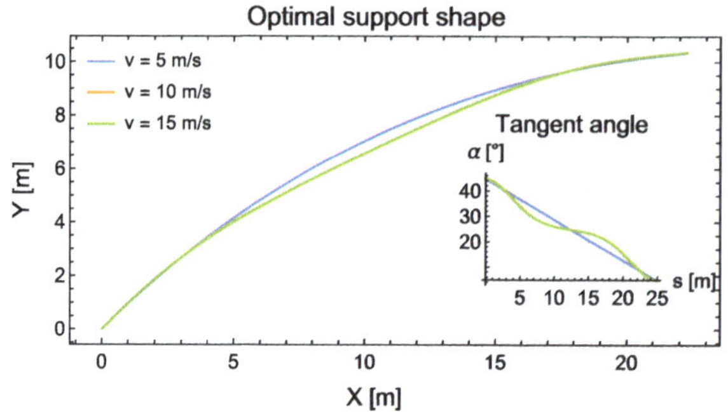

Fig. 6 Optimal support shape for different velocities and both driving directions (X,Y are local coordinates). The curves for $v = 10$ m/s and $v = 15$ m/s coincide. Here we added the constraint according to which the radius of curvature $R > 20$ m for all s. For $v = 5$ m/s we obtain a quasi-circle solution as the optimum. This can be seen by looking at the tangent angle in the inset plot. The radius of the curvature is $R \approx 36$ m for $v = 5$ m/s and $R \approx 160$ m in the center of the support in the other cases.

5 Conclusion

In this paper we presented a novel approach to compute the optimal shape of a cable ropeway support head numerically. The support head geometry is described by a class of curves defined by three parameters which are computed using a Nelder–Mead optimizer, available e.g. in *Mathematica*. The vehicle is modeled as a damped pendulum, whose suspension point moves along the support contour. As a result of the optimization, the corresponding phase–space point is trapped close to the origin of phase–space. Both driving directions are considered. A numerical example reveals promising potential for practical applications: given the tangent angles of the vehicle trajectory (the result of a classical quasi-static cable car calculation), the support head length and the (constant) velocity as input variables, the algorithm gives the optimal shape geometry for the support as output, thereby minimizing the vehicle oscillations considerably. Possible direction for future studies are to not only optimize the shape of the support head, but also allow for a variation of the velocity profile at the same time. The latter has been the focus of an earlier study. Another possible extension would be to take the dynamics of the hauling cable into account.

Acknowledgements S. Ladurner and M. Wenin acknowledge financial support by the Amt für Forschung und Innovation der Provinz Bozen, Italy (this work is a part of the project "Steigerung der Geschwindigkeit und des Fahrkomforts bei der Stützenüberfahrt von Seilbahnanlagen"). A. Windisch acknowledges support by the Austrian Science Fund (FWF), Schroedinger Fellowship J 3800-N27, as well as by the U.S. Department of Energy, Office of Science, Office of Nuclear Physics under Award No. #DE-FG02-05ER41375.

References

1. Bronstein I, Semendjajew K (1991) Taschenbuch der Mathematik, 25th edn. Teubner Verlagsgesellschaft, B.G
2. CEN-Norm: Sicherheitsanforderungen für Seilbahnen für den Personenverkehr. Amtsblatt der EU C51 (2009)
3. Czitary E (1962) Seilschwebebahnen, 2nd edn. Springer, Wien
4. Doppelmayr homepage. https://www.doppelmayr.com/unternehmen/ueber-uns/. Accessed 26 Feb 2018
5. Hoffmann K (2009) Oscillation effects of ropeways caused by cross-wind and other influences. FME Trans 175(37):175–184
6. Kopanakis G (2010) Schwingungen bei Seilbahnen. Internationale Seilbahn-Rundschau 1:22
7. Landau LD, Lifschitz EM (2011) Lehrbuch der Theoretischen Physik I, Klassische Mechanik (14.Auflage). Verlag Harry Deutsch
8. Mathematica homepage. https://www.wolfram.com/mathematica/. Accessed: 16 Mar 2017
9. Thaler H, Wenin M, Brunner J, Reiterer D, Bertotti ML, Modanese G, Oberhuber E (2016) Numerical optimization in ropeway planning. Advanced structured materials. Springer, Heidelberg, pp 113–124
10. Wenin M, Windisch A, Ladurner S, Bertotti ML, Modanese G (2019) Optimal velocity profile for a cable car passing over a support. Eur J Mech 73:366–372

Comparison in Performance of Hybrid and Marvel NoKH Okra/*Abelmoschus esculentus* Fibre Reinforced Polymer Composites Under Tensile Load

Nadendla Srinivasababu

Abstract Fibrous materials obtained from the renewable natural sources of plants, crops are very much essential for the development of green/partially green composite materials. In that line several fibres were extracted from bast, leaves, stem, fruit, seed and stalk etc. by various researchers. Limited work was done on okra fibre and its composites performance under different loading conditions. Now, in this work an attempt was made to explore a new variety of okra fibre i.e. NoKH and was reinforced into epoxy matrix for making cum processing of composites under different conditions of temperature and time. Further, the performance of these composites is compared with hybrid okra fibre reinforced polyester composites conditioned and tested under ASTM standards. Marvel NoKh fibre reinforced epoxy composites processed at T2, t_3 have exhibited highest tensile strength and modulus of 60.51 MPa, 1.05 GPa respectively than other composites. Tested composite specimens were examined under SEM to know the bonding between fibre and matrix.

Keywords Hybrid okra · NoKH okra · Polyester · Aerospace epoxy · Tensile properties · SEM

1 Introduction

Natural fibrous materials have been in universal use from thousands of years, but after the introduction of polymer science lead to study and analyze them in-depth. Structure of the fibre give rise properties, by using properties fibre structure was deduced. Shape and size were most important structural features of fibre. A fibre was a unit of matter characterized by fineness, flexibility and high ratio of length to thickness. Another physical parameter i.e. density will definitely influence weight of the material. Water/moisture uptake behavior has many technical consequences.

N. Srinivasababu (✉)
Fibrous Composites Research Lab, Department of Mechanical Engineering, Vignan's Lara Institute of Technology & Science, Vadlamudi 522213, Andhra Pradesh, India
e-mail: drnsbabu_vlits@vignan.ac.in; cnjlms22@yahoo.co.in

A. Öchsner and H. Altenbach (eds.), *Engineering Design Applications II*,
Advanced Structured Materials 113, https://doi.org/10.1007/978-3-030-20801-1_18

Abelmoschus esculentus fibres were extracted from the stem and analyzed for its morphology, tensile properties [1]. But an exact measurement of the fibre properties is a very important aspect to achieve an agreement between the theoretical approach and practical results. So, some specific problems like elongation correction related to natural fibre tensile testing was studied [2]. Due to hand isolation kink bands were absent in flax fibres and had shown the tensile strength of 1500–1800 MPa [3]. By extrusion-compression molding, injection molding flax fibre reinforced polypropylene granules/composites was tested for tensile and impact properties [4]. *Prosopis juliflora* bark fibre was introduced and belongs to gelatinous or mucilaginous type has 17.11% lignin, possess 580 kg/m^3 density. The fibre had exhibited tensile strength of 558 MPa at a strain rate of 1.77, micro fibril angle of 10.64° [5].

Physical, chemical composition of natural fibres will decide their physical properties such as fibre structure, cellulose content, fibril angle, cross-section, degree of polymerization. Alkali treatment, acetylation resulted in fibre physical structures modification [6]. Bare scouring, pure bleaching, followed by acetylation, alkalization, pure acetylation, sulfuric acid with potassium permanganate treated okra fibres were tested for tensile properties. The results had shown that the mechanical properties were decreased due to chemical treatments. However the treatments lead to variation of void content and fibre dimensions [7].

Using alkaline treated sugar palm fibres at different concentrations and three soaking times, epoxy composites were prepared and tested for tensile properties. The effectiveness of treatment was understood from higher tensile modulus of treated fibre composites than untreated ones [8]. Through batch mixing flax fibre reinforced bionolle polyester composites were prepared. When 25 vol% natural fibre (acetate groups) were reinforced, 30% increase in strength was observed [9].

Kenaf, hemp bast fibres in as received condition, treated with 0.06 M NaOH solution were combined with polyester matrix to form composites by hot pressing. These composites had shown more flexural strength, modulus at low work of fracture [10]. Four different polyester resin formulations were made and used to prepare composites using alkali treated (6% NaOH) kenaf fibres. Out of all the resins, B type formulation had given highest flexural strength, modulus [11]. Coir fibres obtained from disregarded coconut shells were reinforced (80 wt%) into polyester to make, test the composites under flexural load. Obtained composites are rigid up to 50 wt% of reinforcement, but beyond this they performed as flexible agglomerates [12]. Straw fibre reinforced polyester composites had shown flexural stiffness, strength of 7.3 GN/m^2 and 56 MN/m^2 [13]. Ballistic impact performance of flax, hemp, jute fibre reinforced polypropylene composites was described [14].

Mechanical behavior/performance of various composite materials was studied and described that the mercerization, silane treatment was done on sisal fibres to enhance adhesion characteristics along with the moisture resistance [15]. Strips were milled out from bamboo wall core for making laminates [16]. Interfacial bonding ability of coir, glass fibre composites was studied [17]. Agave fourcroydes fibres was alkali treated, role of silane coupling agent in fibre-matrix adhesion was analyzed on high density polyethylene composites [18]. Bio fibres like kenaf, jute, henequen, pineapple leaf and sisal were reinforced into synthetic, biodegradable polymers [19].

Hibiscus sabdariffa fibre reinforced urea formaldehyde composites had shown that the increase in fibre content results increase in tensile, compressive and wear properties when compared with the composites reinforced with particle, short fibre form [20]. With vacuum bagging technique flax, linen and bamboo fabric reinforced epoxy composites were prepared. Here 5 wt% of NaOH solution treatment on tensile properties of the composites was studied [21].

By using agricultural fibres like sugar bars, corn & palm and carbon, barium sulfate, silica, metallic powders & phenol formaldehyde composites were prepared, machined to obtain the cylindrical form of 8 mm diameter. After conducting friction and wear experiments, it was observed that the composites have 30 wt% iron, 25 wt% sugar bar fibre had shown 0.58 friction value [22]. Epoxy based flax composites were prepared with the incorporation of alumina, iron and brass particulates and graphite lubricant [23].

Stiffness of some natural fibre composite formulations was determined from the micromechanical models available in the short fibre composites [24]. Most commonly used biodegradable polymer matrices, and natural fibre reinforcements were described [25]. Ramie fibre reinforced cellulose composites were prepared to test mechanical and thermal properties. The composites had shown a tensile strength, storage modulus of 480 MPa at 25 °C, 20 GPa at 300 °C [26].

An overview on natural fibre composites mechanical behavior, fibre characteristics and bonding effects were discussed [27]. A review was conducted to describe physical structure of plant fibre and its effect on properties of composites [28]. Possible development of advanced fibre composites from macro to nano level was described [29]. AFT, ford motor company-from 1930 has exercised various applications of natural fibres and its composites. In order to ensure the quality of products made more concentration was given on fibre processing [30].

Based on the review of a set of articles, it is understood that the researchers have focused on the following aspects in order to explore natural fibre composites to the world.

- Characterization of fibre (untreated/treated) for physical, chemical, mechanical and thermal properties.
- Introduced modified resins for better composites performance.
- Diversified chemically induced fibres were reinforced to make and test composites performance specifically under mechanical loading.
- Efforts were made to introduce fully green materials.
- Possible applications in automotive filed was described.

So, an opportunity was taken by the author to describe two varieties of okra natural fibres which were grown especially in the field. Care was taken during the plant growth from beginning to to end i.e. till the fibres were extracted by retting and manual peeling method. Then the fibres were introduced into two different resins after assessing the performance of first type composites i.e. hybrid okra-polyester. Conditioning of composites was also changed in order to know the behavior of the composites. Different processing temperature at various heating times depending on the past experience was introduced to know the mechanical performance of okra-epoxy composites. Finally the composites performance was compared and SEM images of later composites were taken to know the bond between reinforcement and matrix.

2 Fibres, Resins, and Fabrication of Composites

In this experimentation two types of okra fibres namely hybrid, marvel NoKH were used for making the composites. Okra is botanically called *Abelmoschus esculentus* and the hybrid seeds of '2405133' variety was supplied by Syngenta India Limited, Pune, India. THIRAM chemical was used to treat the seeds and its characteristics are given in Table 1. Other type of okra i.e. "Marvel NoKH" produced by Nuziveedu seeds was purchased at Guntur. This vegetable plant grows from medium to tall in height with about three branches and five ridges.

Two types of matrices namely Ecmalon 4413 polyester resin of Ecmas Resins Private Ltd., Hyderabad, aerospace resin LY 5052 of Huntsman grade was used to prepare composite specimens with hrbird, marvel NoKH fibre respectively. Before the fibres were reinforced into composite they were dried in an oven at 70 °C up to 4 h for the removal of moisture from them. Polyester resin with catalyst and accelerator were mixed one after another in the proportion of 100 ml: 1 ml: 1 ml (resin: catalyst: accelerator) in a Borosil beaker, twenty turns clock wise and anti-clock wise using glass stirring rod up to ten minutes such that the resin mixture is free from air bubbles. Aerospace resin, LY 5052 and hardner, Aradur 5052 was taken in a proportion of 100:50 by volume in a borosil beaker and thoroughly mixed manually using stirring rod up to fifteen minutes. A mask was wear to the nose for avoiding the inhale of smell come out from the hardner.

Mold with each tensile test specimen size according to ASTM D 5083-02 was coated with a layer of PVC, allowed to settle at room temperature up to 24 h. Then it is further coated with wax before the reinforcement was placed in the mold, such that the composite specimens will easily be detached from it. Now, oven heated and cooled room temperature hybrid okra fibres were placed (0–20.84%) in a pre-prepared, constant thickness i.e. 5 mm rubber mold and the resin was poured into it till the individual fibres were completely wet. The excess resin squeezed out from the mold once a load of 200 N placed over it and this process is most commonly, popularly known as hand lay-up/wet lay-up. However the viscosity build up (500–700 mPa s as

Table 1 Hybrid okra variety 2405133 seeds characteristics

Germination (Min.)	65%
Physical purity (Min.)	99%
Inert matter (Max.)	1%
Moisture (Max.)	8%

per data sheet) in aerospace resin mixture cause the use of mild steel roller of 25 mm diameter, 300 mm long on the mold, which ensure the complete wet of fibres. Again the above said load was placed over the mold which results in spilling of the excess resin from it. Here maximum average weight fraction of 17.77% was used to make all marvel okra aerospace epoxy composites.

Specimens from the mold were taken out after 24 h and are polished with the belt grinding machine for getting straight smooth edges on all sides of it. Dimensions of the specimen were taken with Mitutoyo digital micrometer and are noted for further calculations. Hybrid okra composite tensile specimens were conditioned according to ASTM D 618-05 in an Environmental chamber. Based on literature and with the author's experience marvel okra composites were processed in an oven at temperature of 60 °C (T1), 80 °C (T2) and 100 °C (T3) up to 1 (t_1), 2 (t_2) and 4 (t_3) hours in order to assess the strength, modulus of them.

3 Results and Discussion

Conditioned hybrid okra polyester composites were tensile tested on PC 2000 tensometer at a cross head speed of 5 mm/min and their tensile properties were calculated. With increase in fibre weight fraction from 0 to 10.75%, the tensile strength was increased linearly from 27.47 to 54.9 MPa. A step up increase in tensile strength was observed with further increase in fibre content and is visible form Fig. 1. A maximum tensile strength of 65.6 MPa was achieved at maximum fibre weight fraction of 20.84%. Similar trend was seen from Fig. 2, where the graph was drawn between tensile modulus and fibre weight fraction. Continuous increase in modulus from 0.52 to 0.96 GPa was obtained for hybrid okra polyester composites.

The composites processed at T1, T2 and T3 were tested and tensile strength, modulus was calculated and a graph was represented in Fig. 3, 4 which was drawn between tensile strength, modulus and number of hours the composites were heated respectively. Tensile strength of the composites processed at 60 °C was initially increased and then it is decreased with heating time. Further increase in heating time i.e. 4 h resulted in 14.62% more tensile strength than the initial time (t_1). More or

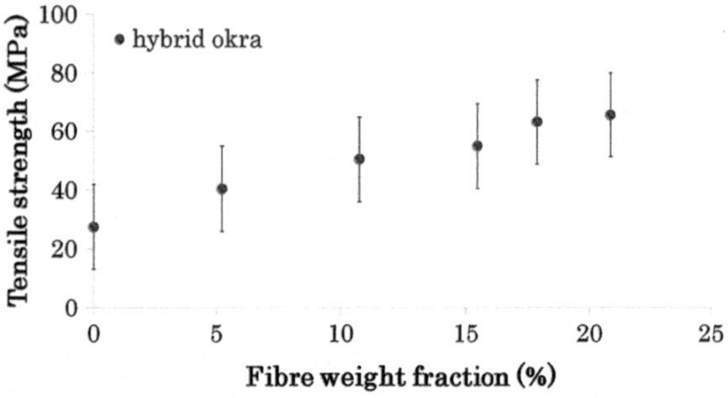

Fig. 1 Effect of fibre weight fraction on tensile strength of hybrid okra fibre reinforced polyester composites

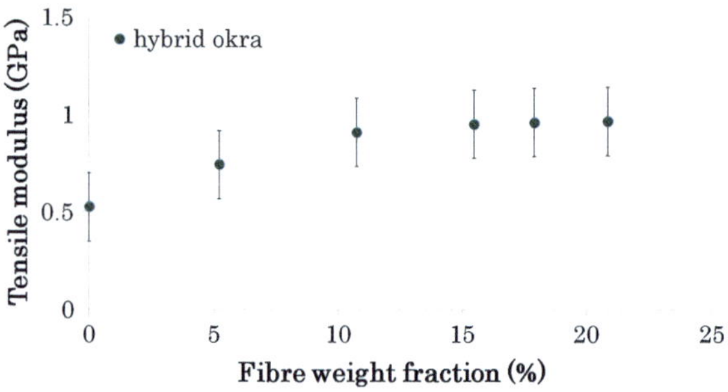

Fig. 2 Effect of fibre weight fraction on tensile modulus of hybrid okra fibre reinforced polyester composites

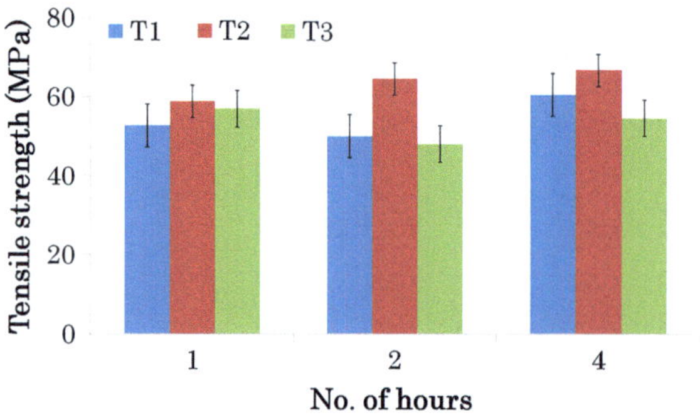

Fig. 3 Effect of temperature, heating time on tensile strength of marvel okra fibre reinforced aerospace epoxy composites at maximum fibre weight fraction

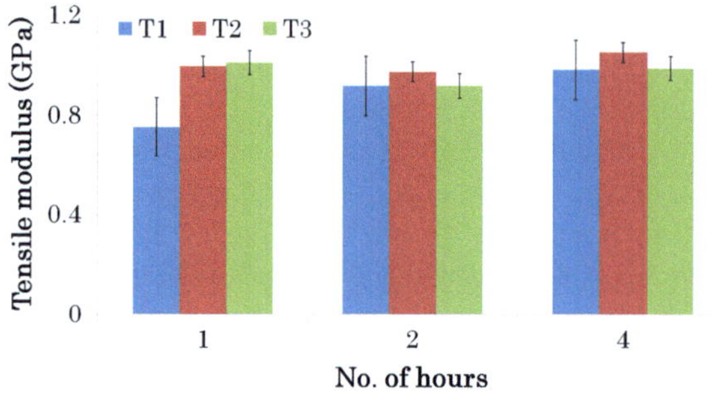

Fig. 4 Effect of temperature, heating time on tensile modulus of marvel okra fibre reinforced aerospace epoxy composites at maximum fibre weight fraction

less linear increase in tensile strength from 58.78 to 66.74 MPa was achieved with increase in heating time. A clear increase in tensile strength was observed among marvel okra epoxy composites processed at T2 than T1 and the increase is in the order of 11.35, 28.86 and 10.3% with respect to increase in heating time. Increase in time of heating the composites not only evaporating the volatile matters present in the resin but also causes the good bond between reinforcement and matrix.

At the processing temperature of T3 the tensile strength of the marvel okra composites lies between T1 and T2 at t_1. But the tensile strength of the composites processed at T3, with heating time t_2, t_3 is less when compared with composites of T1, T2 conditions. These results indicate that there may be fibre burn in composites at this temperature which causes decrease in strength.

With increase in curing temperature from 40 to 100 °C the tensile strength of jute-epoxy composites was increased and then it was decreased with increase in temperature [31]. But in the present work marvel okra epoxy composites had shown this behaviour after 80 °C and peculiar behavior is seen at T3, t_3 conditions. But the tensile strength of hybrid/marvel okra polyester/epoxy composites is much superior to jute-epoxy composites.

Probably a similar trend was seen when tensile modulus was calculated at various process conditions T1, T2 and T3 along with heating time t_1, t_2 and t_3 with minor exceptions. A linear increase in modulus from 0.75 to 0.98 GPa was exhibited by marvel okra composites with increase in heating time $t_1 - t_3$. The highest tensile modulus of 1.05 GPa was achieved at processing temperature of T2, heating time of t_3. The tensile modulus of the composites at T3 are 34.53, 1.51% higher than processed at T1, T2 respectively when they were compared at t_1. However the modulus of marvel okra composites process at T3 is on par with the results of T1 at t_2, t_3 conditions. Similar nature of tensile modulus is experienced in jute-epoxy composites. However the composites had recorded a very good tensile modulus of 2110 MPa [31] and the value obtained for the composites belong to this work is nearly half of its value.

In overall hybrid okra polyester composites tensile performance is little more than marvel okra reinforced composites. But the processing temperature, time of heating of marvel okra epoxy composites result in achieving good tensile strength and modulus. Further the effect of process conditions on bonding behavior of marvel okra composites were understood from Scanning Electron Microscope (SEM) images of tested composites specimens.

SEM images of marvel okra tensile tested composite specimen made at T1, t_1 are shown in Fig. 5a–f. Here number of fibrils stapled together in a fibre and there exist a good bond between fibre and matrix (Fig. 5a, b, d). Majority of the fibres in the composites were failed due to tension only and a clear failure is visible from Fig. 5c–e. But few fibres up to short length were pulled out from the matrix.

When the marvel okra composites fractured specimens at T2, t_2 examined under SEM had given the images shown in Fig. 6a–i. Fibres were broken clearly up to the matrix surface (Fig. 6a), several fibrils, pits in matrix at few locations were observed, Fig. 6b. A broken fibre with pull out is visible from Fig. 6c, h. The fibril diameter was measured at different locations and it varied from 11.46 to 14.69 μm and the pull out length of the fibre is 111.9 μm, Fig. 6d. Matrix was attached with the fibres and is clearly identifiable from Fig. 6e–g, i.

Tensile tested composites of marvel okra were seen under SEM and the images are shown in Fig. 7a–e. Short fibre pull outs were experienced in the composites, Fig. 7a. Diameter of the marvel okra individual fibrils varied from 6.922 to 12.50 μm, Fig. 7b. Fibre pullout and clear matrix failure is recognised from Fig. 7c, d. Various fibrils of definite wall thickness which resumes the squeezed pipe are sighted from Fig. 7e.

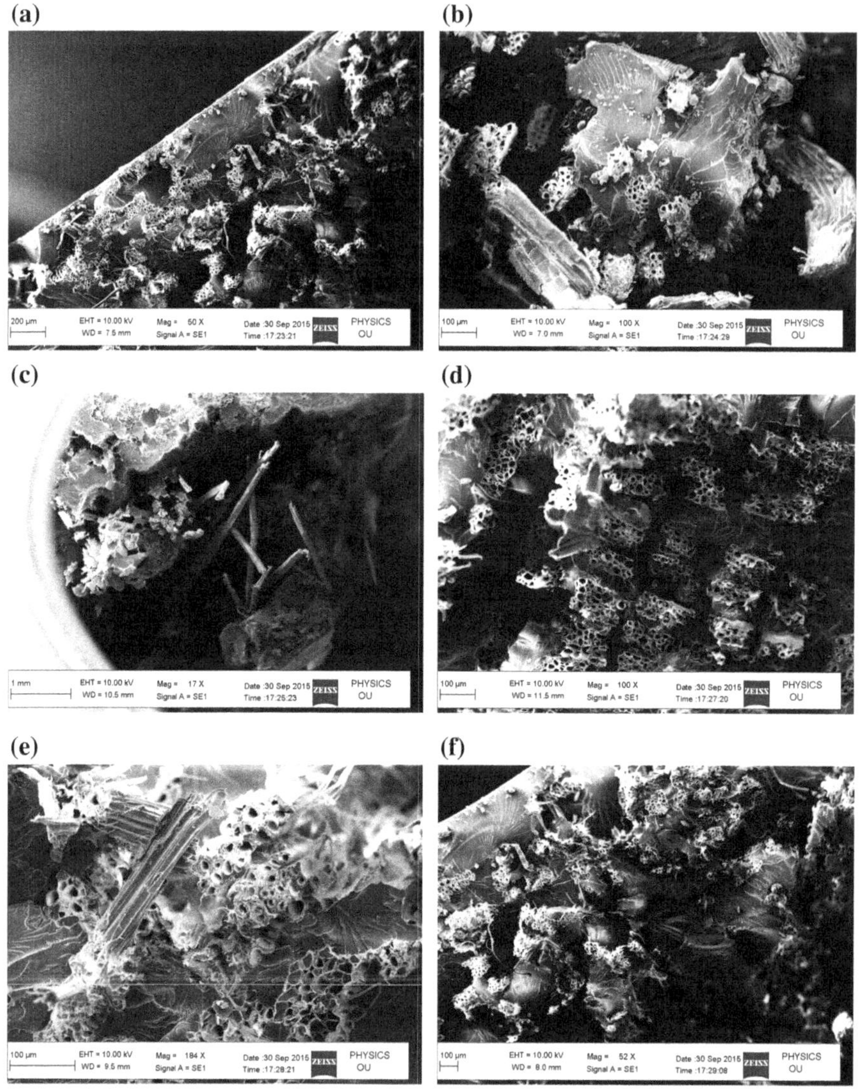

Fig. 5 SEM images of tensile tested marvel okra epoxy composites processed at T1, t_1

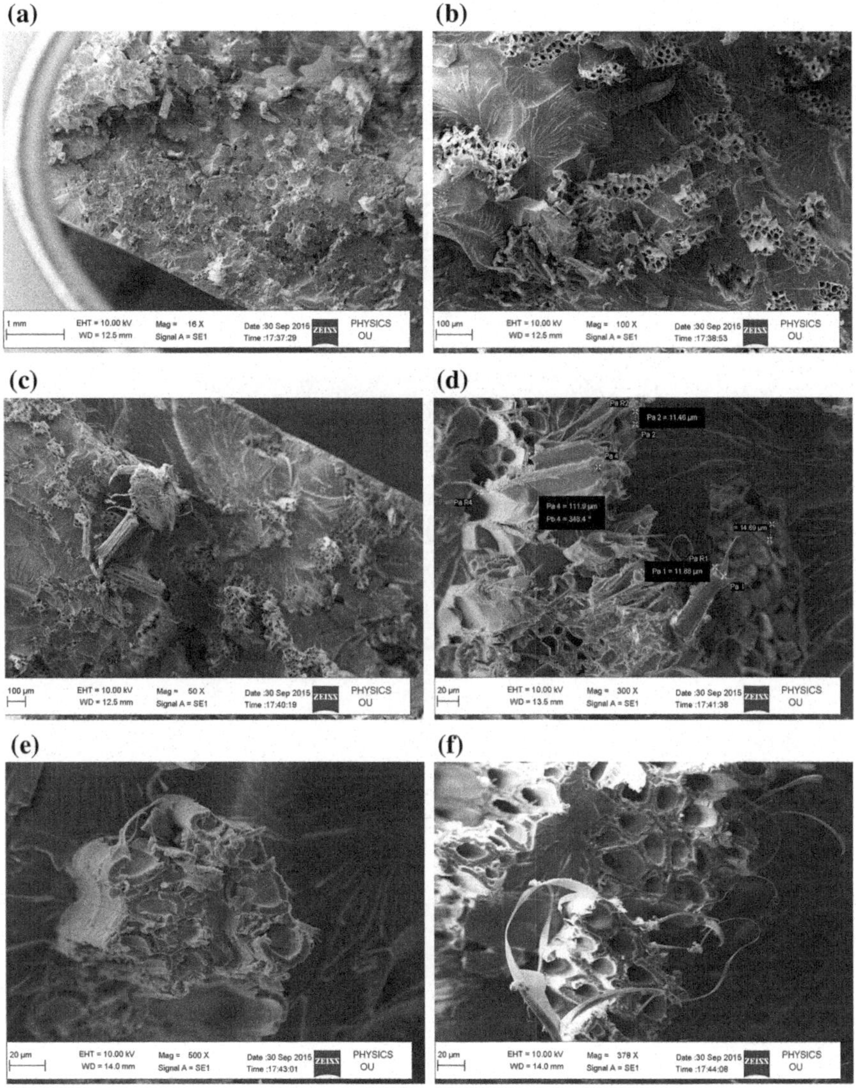

Fig. 6 SEM images of tensile tested marvel okra epoxy composites processed at T2, t_2

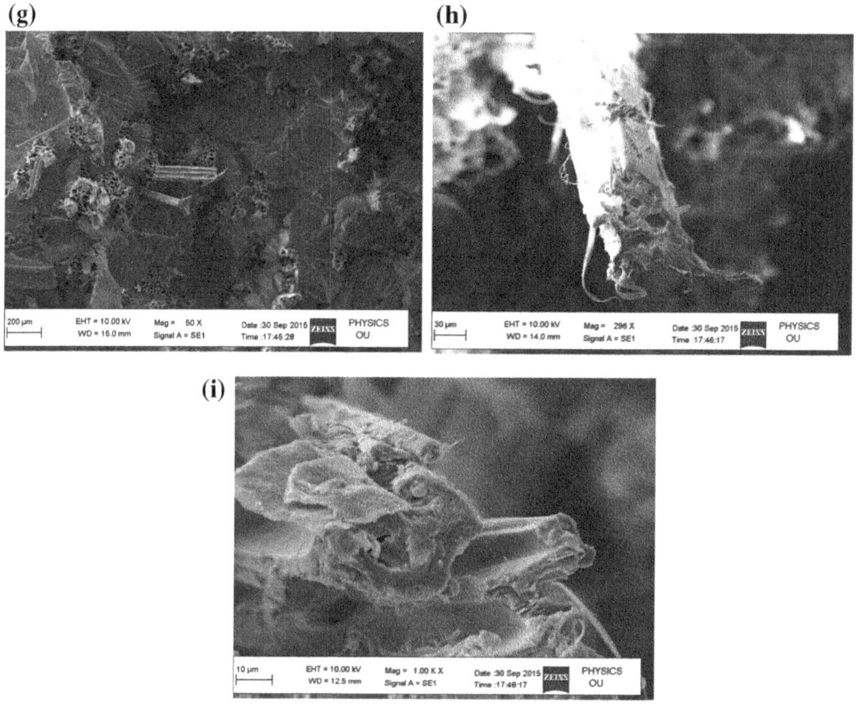

Fig. 6 (continued)

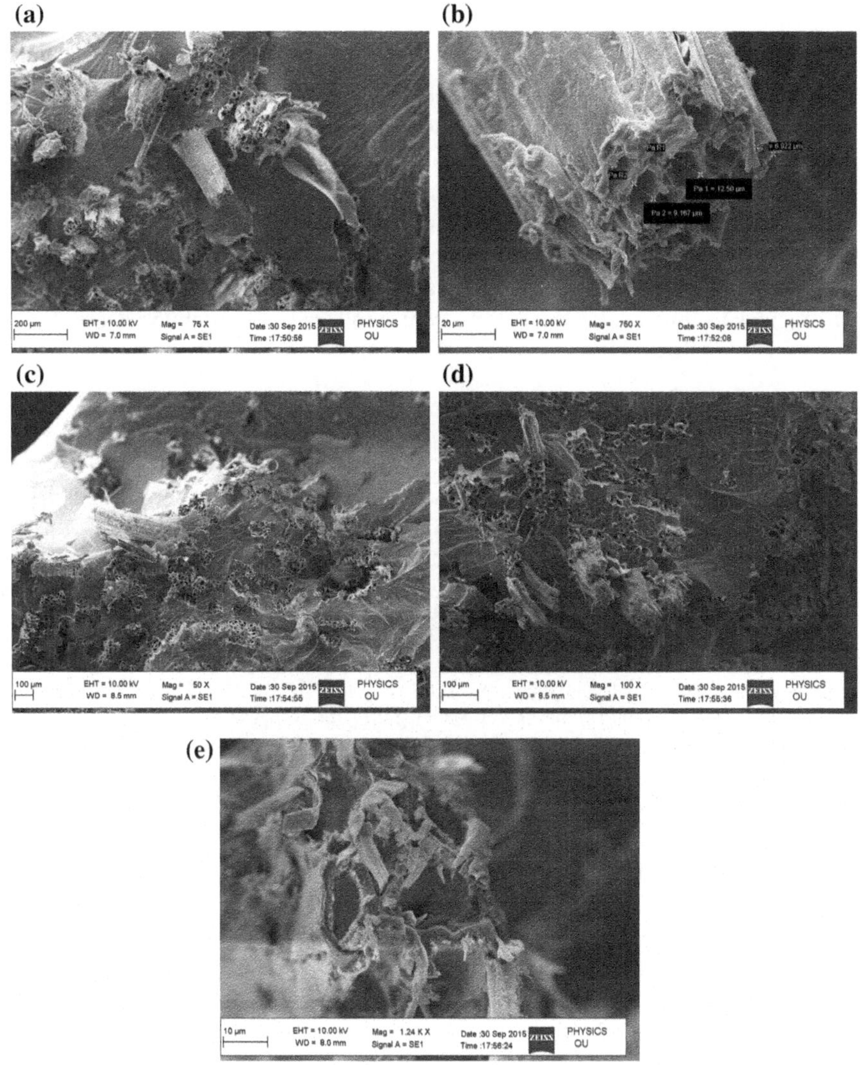

Fig. 7 SEM images of tensile tested marvel okra epoxy composites processed at T3, t_1

4 Conclusions

Using hybrid, marvel okra polyester and aerospace epoxy composites were successfully fabricated through hand lay-up and rolling cum lay-up technique respectively. These two varieties of the fibre composites have exhibited good tensile properties. By applying different processing conditions with different heating times marvel okra epoxy composites performance was experimentally examined and understood that the composites at T2, t_3 exhibited highest tensile strength (66.74 MPa) and modulus (1.05 GPa). After suitable modifications and further extensive studies at full scale okra composites find applications starts from furniture to automobile sectors.

References

1. De Rosa IM, Maria Kenny J, Puglia D et al (2010) Morphological, thermal and mechanical characterization of okra (*Abelmoschus esculentus*) fibres as potential reinforcement in polymer composites. Compos Sci Technol 70:116–122
2. Mieck KP, Reussmann T, Nechwatal A (2003) About the characterization of the mechanical properties of natural fibres. Mater Sci Eng Technol 34:285–289
3. Bos HL, Van Den Oever MJA, Peters OCJJ (2002) Tensile and compressive properties of flax fibres for natural fibre reinforced composites. J Mater Sci 37:1683–1692
4. Nechwatal A, Mieck KP, Reußmann T (2003) Developments in the characterization of natural fibre properties and in the use of natural fibres for composites. Compos Sci Technol 63:1273–1279
5. Saravanakumara SS, Kumaravelb A, Nagarajanc T et al (2013) Characterization of a novel natural cellulosic fiber from *Prosopis juliflora* bark. Carbohydr Polym 92:1928–1933
6. Bledzki AK, Gassan J (1999) Composites reinforced with cellulose based fibres. Prog Polym Sci 24:221–274
7. De Rosa IM, Kenny José M, Maniruzzaman Md et al (2011) Effect of chemical treatments on the mechanical and thermal behaviour of okra (*Abelmoschus esculentus*) fibres. Compos Sci Technol 71:246–254
8. Bachtiar D, Sapuan SM, Hamdan MM (2008) The effect of alkaline treatment on tensile properties of sugar palm fibre reinforced epoxy composites. Mater Des 29:1285–1290
9. Baiardo M, Zini E, Scandola M (2004) Flax fibre–polyester composites. Compos A 35:703–710
10. Aziz Sharifah H, Ansell Martin P (2004) The effect of alkalization and fibre alignment on the mechanical and thermal properties of kenaf and hemp bast fibre composites: part 1—polyester resin matrix. Compos Sci Technol 64:1219–1230
11. Aziz Sharifah H, Ansell Martin P, Clarke Simon J et al (2005) Modified polyester resins for natural fibre composites. Compos Sci Technol 65:525–535
12. Monteiro SN, Terrones LAH, D'Almeida JRM (2008) Mechanical performance of coir fiber/polyester composites. Polym Test 27:591–595
13. White NM, Ansell MP (1983) Straw-reinforced polyester composites. J Mater Sci 18:1549–1556
14. Wambua P, Vangrimde B, Lomov S et al (2007) The response of natural fibre composites to ballistic impact by fragment simulating projectiles. Compos Struct 77:232–240
15. Bisanda ETN, Ansell MP (1991) The effect of silane treatment on the mechanical and physical properties of sisal-epoxy composites. Compos Sci Technol 41:165–178
16. Corradi S, Isidori T, Corradi M et al (2009) Composite boat hulls with bamboo natural fibres. Int J Mater Prod Technol 36:73–89

17. Harisha S, Peter Michael D, Bensely A et al (2009) Mechanical property evaluation of natural fiber coir composite. Mater Charact 60:44–49
18. Herrera-Franco P J, Valadez-Gonza´lez A (2004) Mechanical properties of continuous natural fibre-reinforced polymer composites. Composites Part A 35:339–345
19. Mohanty A K, Misra M, Drzall L T (2002) Sustainable bio-composites from renewable resources: opportunities and challenges in the green materials world. J Polym Environ 10:19–26
20. Singha AS, Thakur Vijay Kumar (2008) Mechanical properties of natural fibre reinforced polymer composite. Bull Mater Sci 31:791–799
21. Yan L, Chouw N, Yuan X (2012) Improving the mechanical properties of natural fibre fabric reinforced epoxy composites by alkali treatment. J Reinf Plast Compos 31:425–437
22. Bakry M, Mousa MO, Ali WY (2013) Friction and wear of friction composites reinforced by natural fibres. Mater Sci Eng Technol 44:21–28
23. Taha I (2012) Investigation of flax fibre reinforced epoxy friction composites. Mater Sci Eng Technol 43:1059–1066
24. Facca AG, Kortschot Mark T, Yan N (2006) Predicting the elastic modulus of natural fibre reinforced thermoplastics. Compos A 37:1660–1671
25. Bogoeva-Gaceva G, Avella M, Malinconico M et al (2007) Natural fiber eco-composites. Polym Comp 28:98–107
26. Nishino T, Matsuda I, Hirao K (2004) All-cellulose composite. Macromolecules 37:7683–7687
27. Nabi Saheb D, Jog JP (1999) Natural fiber polymer composites: a review. Adv Polym Tech 18:351–363
28. Ruys D, Crosky A, Evans WJ (2002) Natural bast fibre structure. Int J Mater Prod Tech 17:2–10
29. Thomas S, Pothan LA, Cherian BM (2009) Advances in natural fibre reinforced polymer composites: macro to nanoscales. Int J Mater Prod Tec 36:317–333
30. Mougin G, Magnani M, Eikelenberg N (2009) Natural-fibres composites for the automotive industry: challenges, solutions and applications. Int J Mater Prod Tec 36:176–188
31. Singh JIP, Singh S, Dhawan V (2018) Effect of curing temperature on mechanical properties of natural fiber reinforced polymer composites. J Nat Fibers 15:687–696

Analysis of Stress-Strain State of Multi-wave Shell on Parabolic Trapezoidal Plan

V. N. Ivanov, Timur Soibnazarovich Imomnazarov, Ismael Taha Farhan Farhan and Daou Tiekolo

Abstract The paper presents the calculation of stress-strain state of shell with carved middle surface (Krivoshapko and Ivanov in Encyclopedia of analytical surfaces. Springer International Publishing Switzerland, 752p, 2015 [1], Ivanov and Krivoshapko in Analytical methods of analyses of shells of non canonic form. RUDN, Moscow, 542p, 2010 [2], Ivanov and Rynkovskaya in MATEC web of conferences 95, 5p, 2017 [3]) with horizontal parabolic curve as directrix and multi-wave sine curve with axis in the plane of parabola as generatrix. We obtain the multi-wave parabolic sine shell on a curved trapezoidal plan (Fig. 1). There is a comparative calculation of two shells with different amplitudes of the sine generatrix—1 m and 0.5 directrix is a quadratic parabola—$y = y_0 - a \cdot x^2$, $y_0 = 11.5$; $a = -0.05$ m^{-1}; $-15 \leq x \leq 15$ (m).

Keywords Modelling multi-wave shells · Parabolic trapezoidal plan · Stress-strain state · Shell with carved middle surface

1 Introduction

Generatrix is a sine curve—$z = b\sin\left(p\pi\frac{t}{c}\right)$, c = 18 m—overall length of a sine curve; number of half waves $p = 6$; sine wave amplitudes $b = 1$ m (Ill. 2a) and $b = 0.5$ m (Fig. 2b). The half-wave length is 3 m. Shell thickness h = 10 cm. On the outer contour of the shell, a fixed hinged support is adopted [6–15]. The calculation

V. N. Ivanov · T. S. Imomnazarov (✉) · I. T. F. Farhan · D. Tiekolo
Department of Civil Engineering, Engineering Academy, Peoples' Friendship University of Russia (RUDN University), 6 Miklukho-Maklaya Street, Moscow 117198, Russian Federation
e-mail: timur-imomnazarov@mail.ru

V. N. Ivanov
e-mail: i.v.ivn@mail.ru

I. T. F. Farhan
e-mail: ismael.civilengineer@gmail.com

D. Tiekolo
e-mail: daout88@gmail.com

A. Öchsner and H. Altenbach (eds.), *Engineering Design Applications II*, Advanced Structured Materials 113, https://doi.org/10.1007/978-3-030-20801-1_19

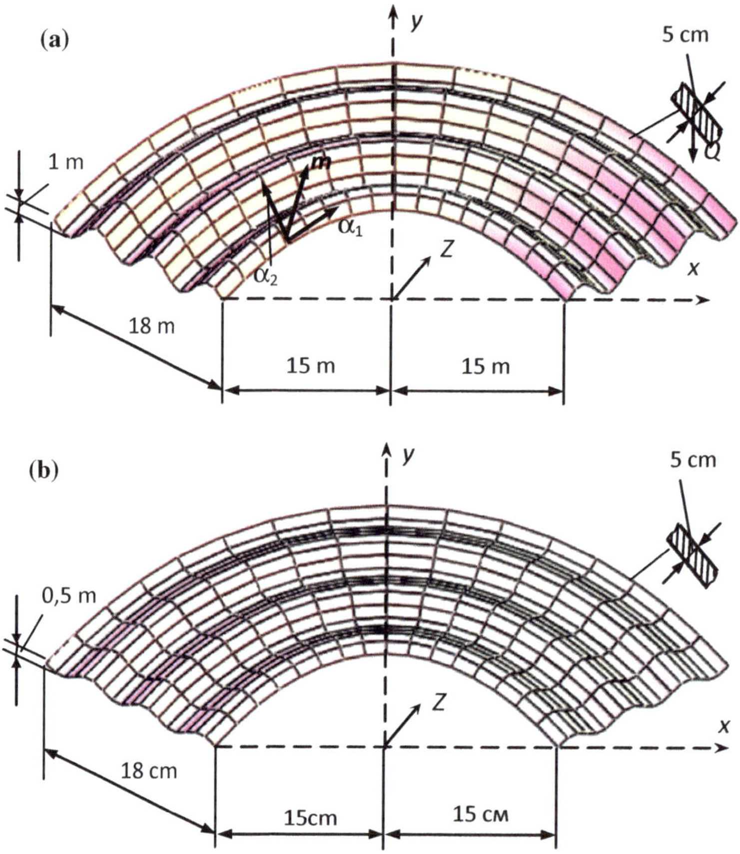

Fig. 1 Multi-wave sine shells on parabolic trapezoidal plan

is based on the gravity load (self-weight of the structure) q = 1 kPa. Shell modulus of elasticity $E = 3.5 \times 10^4$ mPa, the Poisson ratio $\nu = 0.15$.

2 The Modeling of Multi-wave Shell on Parabolic Trapezoidal Plan

The software complex "VRM Shell", developed on the description is used for calculation. The software complex implements the algorithm of the variational difference method for calculating thin-walled structures [4,5]. The complex includes a library of curves that allows automatically designing the required shell and calculating the necessary geometric characteristics of the middle surface of the shells such as the coefficients of the basic quadratic forms and the radii of curvature of the surface.

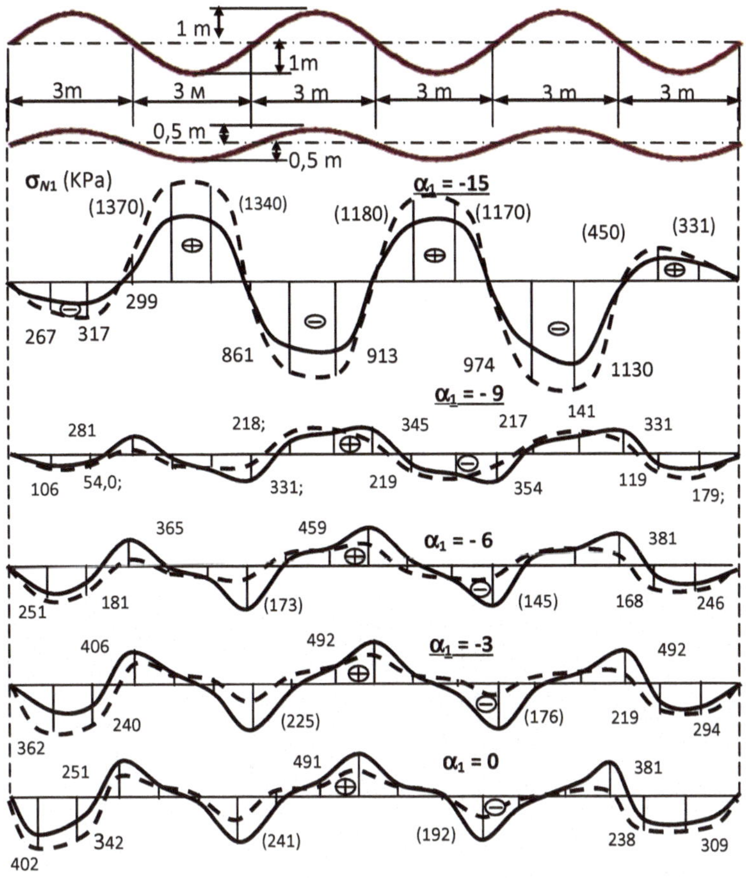

Fig. 2 **a** Normal tangential shear stresses Sine curve amplitude: solid line (nm)—1 m; dashed line (nm)—0.5 m. **b** Normal bending stresses Sine curve amplitude: solid line (nm)—1 m; dashed line (nm)—0.5 m

The software complex allows you to calculate plates and shells, including shells of complex geometry and shapes, the geometry of the middle surfaces of which is described in curvature lines. In particular, it is possible to calculate shells in the form of Joachimsthal channel surfaces and Monge carved surfaces.

A difference grid with a constant or variable interval is applied to the area of the shell.

To consider the symmetry of the shell, half the shell is introduced into the calculation [16–23]. We take a uniform difference grid 80×180, while on the half-wave of the sine wave we have 30 steps of the difference grid.

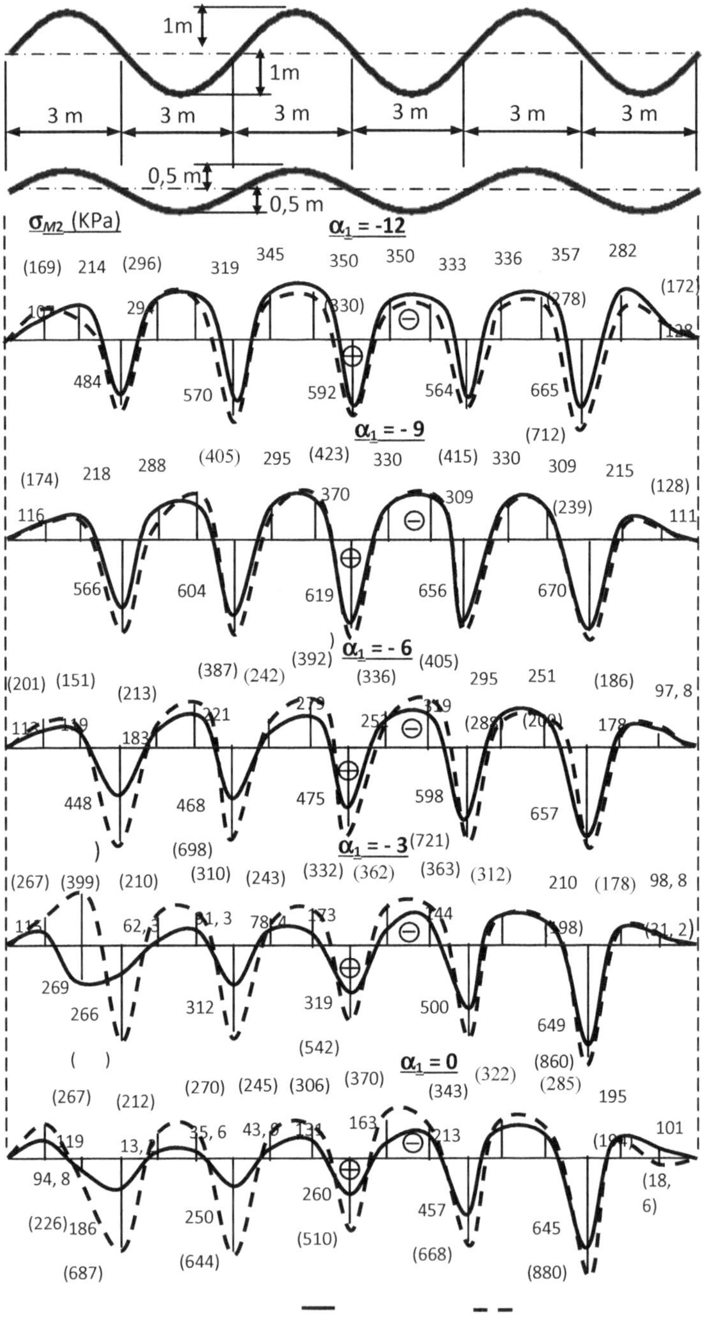

Fig. 2 (continued)

3 Discussion and Conclusions

Based on the results of the calculation, displacements in the grid nodes, stress, normal and tangential forces N_1, N_2, shear forces S, flexural and bending moment M_1, M_2 and twisting moment H are displayed for printing. Normal stresses from normal forces and bending moments and tangential stresses from shearing forces and twisting moment can be also displayed for printing.

As a result of calculation, there are graphs of normal stresses from tangential forces—σ_{N1}(Fig. 2a) and moment forces—σ_{N1} (Fig. 2b). For clarity, the stresses from the same type of effort for shells with amplitudes of a sine curve 1 m (solid line) and 0.5 m (dashed line) are combined in one drawing. It follows from the calculation, that the stresses σ_{N2} and σ_{M1} are an order of magnitude or less smaller than the stresses σ_{N1} and σ_{M2}, respectively, and their diagrams are not given here.

Solid lines show the stresses in the shell with a sine curve amplitude of 1 m; for a shell with a sine curve amplitude of 0.5 m lines are dashed, the stress values are given within the brackets.

From the graphs shown in Fig. 2a, b, it can be seen that the greatest normal stresses σ_{N1} from tangential normal forces N_1 arise near the reference section $\alpha_1 = \pm 15$ m. The stresses are alternating along the sine wave. In the support zone, the upper half-waves are compressed and the lower ones are stretched. In the middle zone, the inversion of normal stress always occurs at the inflection of the sine wave. The maximum tensile stresses are on average 10% higher than the maximum compressive stresses. In the middle zone of the shell, the stresses on the average are 30% of the stress in the reference zone. However, for a shell with a smaller amplitude of the sine curve the stresses in the reference zone are approximately 40% higher. In the middle zone of the shell, the greatest stresses in the shell with a smaller Amplitude of the sine curve are basically smaller than the corresponding shell stresses with a bigger amplitude of the sine curve.

Stresses from bending forces arise in the transverse direction σ_{M2} from bending moment's M_2. In the reference sections, the bending moments and, accordingly, the stresses equal to zero, which corresponds to the specified boundary support conditions. With a slight distance from the reference zones, bending moments appear and, accordingly, bending,

As in the case of normal tangential forces, the disposition of moment forces along the sine wave is alternating. However, unlike the tangential forces, the maximum negative moments appear on the upper and lower crests of the shell sine curve, and the maximum moments opposite in sign are on the sine curve inflections. However, the maximum bending stresses at the inflection points are 1.5–2 times higher than the stresses at the peaks of the sine curve. The maximum normal stresses for a shell with a smaller amplitude in most of the shell are only 10–20% higher than the corresponding stresses in the shell with a higher amplitude shell sine curve. The greater difference is in the middle zone of the shell.

Acknowledgements The publication was prepared with the support of the «RUDN University Program 5-100».

References

1. Krivoshapko SN, Ivanov VN (2015) Encyclopedia of analytical surfaces. Springer International Publishing Switzerland, 752p
2. Ivanov VN, Krivoshapko SN (2010) Analytical methods of analyses of shells of non canonic form. RUDN, Moscow, 542p
3. Ivanov V, Rynkovskaya M (2017) Analysis of thin-walled wavy shell of Monge type surface with parabola and sinusoid curves by variation-difference method. In: MATEC web of conferences 95, 12007, Art. № 12007, 5p
4. Ivanov VN, Rizvan M (2002) Geometry of Monge surfaces and construction of the shell. Structural mechanics of s and buildings. Construction: collection of science works, vol 11, ACB, pp 27–36
5. Ivanov VN, Nasr Y (2000) Analyses of the shells of complex geometry by variation differential difference method. In: Collection of science works, vol 11, ACB, pp 25–34
6. Mesnil R, Santerre Y, Douthe C, Baverel O, Leger B (2015) Generating high node congruence in freeform structures with Monge's surfaces. In: Proceedings of the international association for shell and spatial structures (IASS)
7. Ivanov VN, Mathieu G (2014) Some aspects of the geometry of surfaces with a system of flat coordinate lines. Int J Soft Comput Eng (IJSCE), pp 77–82
8. Lagu R (1986) A variational finite-difference method for analyzing channel waveguides with arbitrary index profiles. IEEE J Quant Electron 22(6):968–976
9. Smit GD (1986) Numerical solution of partial differential equations by finite difference methods. In: Oxford applied mathematics and computing science series, 2nd edn. UK, 350p
10. Mitchell AR, Griffiths DF (1980) The finite difference method in partial differential equations. Wiley, 284p. 11. Curant R (1943) Variational methods for the solution of problems of equilibrium and vibrations. Bull Am Math Soc 49:1–23
11. Maksimyuk VA, Storozhuk EA, Chernyshenko IS (2012) Variational finite-difference methods in linear and nonlinear problems of the deformation of metallic and composite shells (review). Int Appl Mech 48(6):613–687
12. Singh JP, Dey SS (1990) Variational finite difference method for free vibration of sector plates. J Sound Vib 136(1):91–104
13. Curant R (1943) Variational methods for the solution of problems of equilibrium and vibrations. Bull Am Math Soc 49:1–23
14. Reddy JN (2002) Energy principles and variational methods in applied mechanics. Willy. New Jersey, 542p
15. Andujar Moreno IR, Roset J, Kilar V (2014) Variational mechanics and stochastic methods applied to structural design, 173p
16. Aginam CH, Chidolue CA, Ezeagu CA (2012) Application of direct variational method in the analysis of isotropic thin rectangular plates. ARPN J Eng Appl Sci 7(9):1128–1138
17. Lanczos C, (1970) The variational principles of mechanics. University of Toronto Press, Canada, 418p
18. Argyris JH (1960) Energy theorems and structural analysis. Butterworths Scientific Publications, London, 85p
19. Lapidus L, Pinder GF (1982) Numerical solution of partial differential equations in science and engineering. Wiley-Interscience, New York, 677p
20. Wunderlich W, Pilkey W (2002) Mechanics of structures. Variational and computational methods. CRC Press, pp 852–877
21. Ivanov VN (2001) Variation principals and methods of analyses of the problems of theory of elasticity. RUDN, Moscow, 176p
22. Ivanov VN (2008) The base of the finite elements method and the variation-difference method. RUDN, Moscow, 170p
23. Ivanov VN, Kushnarenko IV (2013) The variational-difference method for the analyses of the shell with complex geometry. In: International association for shell and spatial structures proceedings of the IASS 2013 symposium "beyond the limits of man", 6p

General Boundary Element Method for the Dual-Phase Lag Equations Describing the Heating of Two-Layered Thin Metal Films

Ewa Majchrzak

Abstract Thermal interaction between two-layered thin metal film and the ultra-short laser pulse is considered. The problem is described by the system of the dual-phase lag equations supplemented by the appropriate boundary and initial conditions. To solve these equations the general boundary element method (GBEM) is proposed. At first, the GBEM for one layer is presented. Next, using the ideal contact condition at the interface between the layers (the form of this condition in the case of DPLE differs significantly from the classical condition occurring in the Fourier-type models), the final system of algebraic equations is formulated. At the stage of numerical computations the different intensities of the laser beam and the different exposure times are taken into account. The results are compared with the repeatedly verified explicit scheme of the finite difference method. The impact of the time step and length of internal cells on the results of computations is also discussed.

Keywords Dual-phase lag equation · Thin metal films · General boundary element method · Ultra-short laser pulse

1 Introduction

Heat transfer through films subjected to an ultrashort laser pulse is of vital importance in microtechnology applications and it is a reason that the problems connected with fast heating of solids have become a very active research area [1].

The traditional Fourier's law of heat conduction becomes questionable in the microscale regime [1–3], because the characteristic lengths of the various heat carriers are comparable to the characteristic length of the system considered and/or the characteristic times of the various heat carriers are comparable to the characteristic energy excitation time. Considering the problem discussed, one can use the different models of heat conduction. In this place the parabolic [4] and hyperbolic [5] two-step models can be mentioned. The two-step models describe the nonequilibrium

E. Majchrzak (✉)
Silesian University of Technology, Konarskiego 18a, 44-100 Gliwice, Poland
e-mail: ewa.majchrzak@polsl.pl

A. Öchsner and H. Altenbach (eds.), *Engineering Design Applications II*,
Advanced Structured Materials 113, https://doi.org/10.1007/978-3-030-20801-1_20

phenomena between the electrons and lattices. Currently, these models are used for numerical modeling of thermal processes occurring in the laser treated materials e.g. [6–8]. In the case of high laser intensity the problems concerning the melting, resolidification and evaporation are also considered e.g. [9–11].

The second group of models is based on the modified Fourier law [1–3]. The Cattaneo-Vernotte equation can be derived under the assumption that heat flux vector (cause) can precede temperature gradient (effect), while in the dual-phase lag model (DPL) additionally it is assumed that the temperature gradient (cause) can precede heat flux vector (effect). In the Cattaneo-Vernotte equation the phase lag called the relaxation time appears, in the dual-phase lag equation two parameters: relaxation time and thermalization time are taken into account. These models (especially DPL) are widely applied to the simulation of the laser-material interactions, e.g. [12–16].

Numerical solutions of equations describing the heat transfer in the microscale are mainly based on the finite difference method. In this paper the general boundary element method for the dual-phase lag equation is proposed. At first, the algorithm for thin metal film subjected to the laser pulse is presented (1D problem). Next, two layers with ideal contact between them are considered. The results are compared with the finite difference method solution. The problems related to the choice of time step and length of internal cells are also discussed. In the final part the results of computations for different intensities of the laser beam and the different exposure times are presented and the conclusions are formulated.

2 Dual-Phase Lag Equation

The 1D dual-phase lag equation has the following form [1–4]

$$c\frac{\partial T(x,t)}{\partial t} + c\,\tau_q\frac{\partial^2 T(x,t)}{\partial t^2} = \lambda\frac{\partial^2 T(x,t)}{\partial x^2} + \lambda\tau_T\frac{\partial^3 T(x,t)}{\partial t\,\partial x^2} + Q(x,t) + \tau_q\frac{\partial Q(x,t)}{\partial t} \tag{1}$$

where c is a volumetric specific heat of material, λ is a thermal conductivity, τ_q and τ_T are the phase lags (relaxation and thermalization times), Q is the capacity of internal heat sources, T, x, t denote the temperature, geometrical co-ordinate and time.

The laser irradiation is described by a source term introduced in Eq. (1)

$$Q(x,t) = \sqrt{\frac{\beta}{\pi}}\,\frac{1-R}{t_p\,\delta}\,I_0\exp\left[-\frac{x}{\delta} - \beta\frac{(t-2t_p)^2}{t_p^2}\right] \tag{2}$$

where I_0 is the laser intensity, t_p is the characteristic time of laser pulse, δ is the optical penetration depth, R is the reflectivity of the irradiated surface and $\beta = 4\ln 2$.

Equation (1) is supplemented by the modified Neumann boundary conditions for $x = 0$ and $x = L$

$$q_b(x,t) + \tau_q \frac{\partial q_b(x,t)}{\partial t} = -\lambda \left[\frac{\partial T(x,t)}{\partial x} + \tau_T \frac{\partial}{\partial t} \left(\frac{\partial T(x,t)}{\partial x} \right) \right] \tag{3}$$

where $q_b(x, t)$ is the boundary heat flux. The heat losses from front and back surface could be neglected and then $q_b(x, t) = 0$.

The initial conditions are also given

$$t = 0: \quad T(x,0) = T_p(x), \quad \left. \frac{\partial T(x,t)}{\partial t} \right|_{t=0} = w(x) \tag{4}$$

where $T_p\ (x)$ is the initial temperature, $w(x)$ is the initial heating rate.

3 General Boundary Element Method for One Layer

At the first stage, for the purpose of numerical stability, the DPL Eq. (1) is discretized in the time domain in terms of a fully implicit form with the backward finite-difference method for both the first-order and second-order time derivatives [17, 18]

$$c\frac{T^f(x) - T^{f-1}(x)}{\Delta t} + c\,\tau_q \frac{T^f(x) - 2T^{f-1}(x) + T^{f-2}(x)}{(\Delta t)^2}$$
$$= \lambda \frac{\partial^2 T^f(x)}{\partial x^2} + \frac{\lambda \tau_T}{\Delta t} \left[\frac{\partial^2 T^f(x)}{\partial x^2} - \frac{\partial^2 T^{f-1}(x)}{\partial x^2} \right] + Q^f(x) + \tau_q R^f(x) \tag{5}$$

where $T^f(x) = T(x, f\Delta t)$, Δt is the time step, $f = 2, 3, \ldots, F$ and $T^{f-1}(x)$, $T^{f-2}(x)$ are known temperature distributions at $(f - 1)$th and $(f - 2)$th time steps, respectively, while $R^f(x) = \partial Q(x,t)/\partial t$ for $t = t^f$. It should be noted that taking into account the initial conditions (4) one has: $T^0(x) = T_{0,} T^1(x) = T_0 + w(x)\Delta t$.

The boundary conditions (3) are also transformed ($q_b(x,t) = 0$)

$$-\lambda \left[\frac{\partial T^f(x)}{\partial x} + \frac{\tau_T}{\Delta t} \left(\frac{\partial T^f(x)}{\partial x} - \frac{\partial T^{f-1}(x)}{\partial x} \right) \right] = 0 \tag{6}$$

In this way one obtains the equations that present a boundary-value problem at each time step.

Equation (5) can be written in the form

$$\frac{\partial^2 T^f(x)}{\partial x^2} - BT^f(x) + C\frac{\partial^2 T^{f-1}(x)}{\partial x^2} + DT^{f-1}(x)$$

$$+ET^{f-2}(x)+F\left[Q^{f}(x)+\tau_{q}R^{f}(x)\right]=0 \tag{7}$$

where

$$\begin{aligned} B &= \frac{c(\Delta t+\tau_q)}{\lambda\Delta t(\Delta t+\tau_T)}, \quad C = -\frac{\tau_T}{\Delta t+\tau_T}, \quad D = \frac{c(\Delta t+2\tau_q)}{\lambda\Delta t(\Delta t+\tau_T)} \\ E &= -\frac{c\tau_q}{\lambda\Delta t(\Delta t+\tau_T)}, \quad F = \frac{\Delta t}{\lambda(\Delta t+\tau_T)} \end{aligned} \tag{8}$$

Equation (7) is supplemented by boundary conditions for $x=0$ and $x=L$

$$-\lambda\frac{\partial T^{f}(x)}{\partial x}=w_{b}^{f}(x) \tag{9}$$

where [cf. Eq. (6)]

$$w_{b}^{f}(x)=-\frac{\lambda\tau_{T}}{\Delta t+\tau_{T}}\frac{\partial T^{f-1}(x)}{\partial x} \tag{10}$$

Using the homotopy analysis method [17–22], one obtains [cf. Eq. (7)]

$$\frac{\partial^{2}U^{[1]}(x)}{\partial x^{2}}-BU^{[1]}(x)+R[U(x)]=0 \tag{11}$$

where

$$\begin{aligned} R[U(x)] = {} & \frac{\partial^{2}U(x)}{\partial x^{2}}-B\,U(x)+C\frac{\partial^{2}T^{f-1}(x)}{\partial x^{2}}+DT^{f-1}(x) \\ & +ET^{f-2}(x)+F\left[Q^{f}(x)+\tau_{q}R^{f}(x)\right] \end{aligned} \tag{12}$$

and

$$U^{[1]}(x)=\left.\frac{\partial\,\Phi(x;\,p)}{\partial p}\right|_{p=0} \tag{13}$$

while $\Phi(x;p)$ is the function (homotopy) associated with the family of partial differential equations [17, 18, 21, 22], $p \in [0, 1]$ is the embedding parameter and $U(x)$ is an initial approximation of temperature distribution $T^f(x)$ [e.g. $U(x) = T^{f-1}(x)$].

Equation (11) is supplemented by boundary conditions [21, 22]

$$-\lambda\frac{\partial U^{[1]}(x)}{\partial x}=w_{b}^{f}(x)+\lambda\frac{\partial U(x)}{\partial x} \tag{14}$$

The problem (11), (14) can be solved using standard boundary element method [23, 24] for 1D steady state task. Thus, the following system of equations should be solved

$$\begin{bmatrix} G_{11} & G_{12} \\ G_{21} & G_{22} \end{bmatrix}\begin{bmatrix} W^{[1]}(0) \\ W^{[1]}(L) \end{bmatrix} = \begin{bmatrix} H_{11} & H_{12} \\ H_{21} & H_{22} \end{bmatrix}\begin{bmatrix} U^{[1]}(0) \\ U^{[1]}(L) \end{bmatrix} + \begin{bmatrix} Z(0) \\ Z(L) \end{bmatrix} \tag{15}$$

where

$$\begin{aligned} G_{11} &= -\tfrac{1}{2\lambda\sqrt{B}}, & G_{22} &= -G_{11} \\ G_{21} &= -\tfrac{1}{2\lambda\sqrt{B}}\exp\left(-L\sqrt{B}\right), & G_{12} &= -G_{21} \end{aligned} \tag{16}$$

$$\begin{aligned} H_{11} &= H_{22} = -0.5 \\ H_{21} &= H_{12} = \frac{1}{2}\exp\left(-L\sqrt{B}\right) \end{aligned} \tag{17}$$

and

$$Z(\xi) = \int_0^L R[U(x)]T^*(\xi, x)\mathrm{d}x \tag{18}$$

while $W^{[1]}(x) = -\lambda\,\partial U^{[1]}(x)/\partial x$.

Next, the values of function $U^{[1]}(x)$ at the internal points x_i, $i = 1, 2, \ldots, n$ are calculated

$$\begin{aligned} U_i^{[1]} = &\frac{1}{2}\exp\left[-(L - x_i)\sqrt{B}\right]U^{[1]}(L) + \frac{1}{2}\exp\left[-x_i\sqrt{B}\right]U^{[1]}(0) \\ &- \frac{1}{2\lambda\sqrt{B}}\exp\left[-(L - x_i)\sqrt{B}\right]W^{[1]}(L) \\ &+ \frac{1}{2\lambda\sqrt{B}}\exp\left[-x_i\sqrt{B}\right]W^{[1]}(0) + Z(x_i) \end{aligned} \tag{19}$$

Finally, the temperature distribution in the all nodes is determined using the formula [17, 18]

$$T^f(x_i) = U(x_i) + U^{[1]}(x_i) \tag{20}$$

4 Dual-Phase Lag Model for Two-Layered Thin Metal Films

In the case of two layers, the system of equations of type (1) for each sub-domain should be taken into account and on the contact surface the appropriate boundary condition should be assumed. In the case of ideal contact this boundary condition takes a form

$$x = L_1 \colon \begin{cases} T_1(x,t) = T_2(x,t) \\ q_1(x,t) = q_2(x,t) \end{cases} \tag{21}$$

It should be noted that using the relationship (3) between the heat flux and temperature gradient, the second part of boundary condition (21) should be properly formulated, namely [12, 25]

$$\begin{aligned} & -\lambda_1 \frac{\partial T_1(x,t)}{\partial x} - \lambda_1(\tau_{T1} + \tau_{q2}) \frac{\partial^2 T_1(x,t)}{\partial t\, \partial x} - \lambda_1 \tau_{T1} \tau_{q2} \frac{\partial^3 T_1(x,t)}{\partial t^2\, \partial x} \\ & = -\lambda_2 \frac{\partial T_2(x,t)}{\partial x} - \lambda_2(\tau_{T2} + \tau_{q1}) \frac{\partial^2 T_2(x,t)}{\partial t\, \partial x} - \lambda_2 \tau_{T2} \tau_{q1} \frac{\partial^3 T_2(x,t)}{\partial t^2\, \partial x} \end{aligned} \tag{22}$$

As previously, for $x = 0$ and $x = L$ the no-flux boundary conditions are accepted. The initial conditions are also assumed.

5 General Boundary Element Method for Two-Layered Thin Metal Films

At first, for each layer the system of Eq. (15) should be created, this means

$$\begin{bmatrix} G_{11}^1 & G_{12}^1 \\ G_{21}^1 & G_{22}^1 \end{bmatrix} \begin{bmatrix} W_1^{[1]}(0) \\ W_1^{[1]}(L_1) \end{bmatrix} = \begin{bmatrix} H_{11}^1 & H_{12}^1 \\ H_{21}^1 & H_{22}^1 \end{bmatrix} \begin{bmatrix} U_1^{[1]}(0) \\ U_1^{[1]}(L_1) \end{bmatrix} + \begin{bmatrix} Z_1(0) \\ Z_1(L_1) \end{bmatrix} \tag{23}$$

and

$$\begin{bmatrix} G_{11}^2 & G_{12}^2 \\ G_{21}^2 & G_{22}^2 \end{bmatrix} \begin{bmatrix} W_2^{[1]}(L_1) \\ W_2^{[1]}(L) \end{bmatrix} = \begin{bmatrix} H_{11}^2 & H_{12}^2 \\ H_{21}^2 & H_{22}^2 \end{bmatrix} \begin{bmatrix} U_2^{[1]}(L_1) \\ U_2^{[1]}(L) \end{bmatrix} + \begin{bmatrix} Z_2(L_1) \\ Z_2(L) \end{bmatrix} \tag{24}$$

Let

$$g_1 = -\lambda_1 \frac{\partial T_1}{\partial x} - \lambda_1(\tau_{T1} + \tau_{q2}) \frac{\partial}{\partial t}\left(\frac{\partial T_1}{\partial x}\right) - \lambda_1 \tau_{T1} \tau_{q2} \frac{\partial^2}{\partial t^2}\left(\frac{\partial T_1}{\partial x}\right) \tag{25}$$

$$g_2 = -\lambda_2 \frac{\partial T_2}{\partial x} - \lambda_2(\tau_{T2} + \tau_{q1}) \frac{\partial}{\partial t}\left(\frac{\partial T_2}{\partial x}\right) - \lambda_2 \tau_{T2} \tau_{q1} \frac{\partial^2}{\partial t^2}\left(\frac{\partial T_2}{\partial x}\right) \tag{26}$$

Then, the boundary condition (21) can be written as [cf. Eq. (22)]

$$x = L_1 \colon \begin{cases} T_1(x,t) = T_2(x,t) \\ g_1(x,t) = g_2(x,t) \end{cases} \tag{27}$$

The following approximation of dependence (25) with respect to time is proposed

$$g_1^f = -\lambda_1 \frac{\partial T_1^f(x)}{\partial x} - \lambda_1(\tau_{T1} + \tau_{q2}) \frac{1}{\Delta t}\left[\frac{\partial T_1^f(x)}{\partial x} - \frac{\partial T_1^{f-1}(x)}{\partial x}\right] - \lambda_1 \tau_{T1}\tau_{q2} \frac{1}{(\Delta t)^2}\left[\frac{\partial T_1^f(x)}{\partial x} - 2\frac{\partial T_1^{f-1}(x)}{\partial x} + \frac{\partial T_1^{f-2}(x)}{\partial x}\right] \quad (28)$$

or

$$-\lambda_1 \frac{\partial T_1^f(x)}{\partial x} = \frac{1}{A_1} g_1^f + \frac{A_2}{A_1}\lambda_1 \frac{\partial T_1^{f-1}(x)}{\partial x} + \frac{A_3}{A_1}\lambda_1 \frac{\partial T_1^{f-2}(x)}{\partial x} \quad (29)$$

where

$$A_1 = \frac{(\Delta t)^2 + \Delta t(\tau_{T1} + \tau_{q2}) + \tau_{T1}\tau_{q2}}{(\Delta t)^2}, \quad A_2 = -\frac{\Delta t(\tau_{T1} + \tau_{q2}) + 2\tau_{T1}\tau_{q2}}{(\Delta t)^2}, \quad A_3 = \frac{\tau_{T1}\tau_{q2}}{(\Delta t)^2} \quad (30)$$

In a similar way one obtains [cf. formula (26)]

$$-\lambda_2 \frac{\partial T_2^f(x)}{\partial x} = \frac{1}{A_4} g_2^f + \frac{A_5}{A_4}\lambda_2 \frac{\partial T_2^{f-1}(x)}{\partial x} + \frac{A_6}{A_4}\lambda_2 \frac{\partial T_2^{f-2}(x)}{\partial x} \quad (29)$$

where

$$A_4 = \frac{(\Delta t)^2 + \Delta t(\tau_{T2} + \tau_{q1}) + \tau_{T2}\tau_{q1}}{(\Delta t)^2}, \quad A_5 = -\frac{\Delta t(\tau_{T2} + \tau_{q1}) + 2\tau_{T2}\tau_{q1}}{(\Delta t)^2}, \quad A_6 = \frac{\tau_{T2}\tau_{q1}}{(\Delta t)^2} \quad (32)$$

Taking into account the formulation of boundary conditions in the general boundary element method [17, 21, 22], one has

$$x = L_1: \quad \begin{cases} U_1^{[1]}(x) = T_1^f(x) - U_1(x) \\ W_1^{[1]}(x) = \frac{1}{A_1} g_1^f(x) - \frac{A_2}{A_1} V_1^{f-1}(x) - \frac{A_3}{A_1} V_1^{f-2}(x) - W_1(x) \end{cases} \quad (33)$$

and

$$x = L_1: \quad \begin{cases} U_2^{[1]}(x) = T_2^f(x) - U_2(x) \\ W_2^{[1]}(x) = \frac{1}{A_4} g_2^f(x) - \frac{A_5}{A_4} V_2^{f-1}(x) - \frac{A_6}{A_4} V_2^{f-2}(x) - W_2(x) \end{cases} \quad (34)$$

where $W_e^{[1]}(x) = -\lambda_e\, \partial U_e^{[1]}(x)/\partial x$, $W_e(x) = -\lambda_e\, \partial U_e(x)/\partial x$, $V_e^s(x) = -\lambda_e\, \partial T_e^s(x)/\partial x$, $e = 1, 2$ and $s = f - 1$ or $s = f - 2$.

Basing on the condition of ideal contact (27), one obtains

$$x = L_1: \quad \begin{cases} U_1^{[1]}(x) + U_1(x) = U_2^{[1]}(x) + U_2(x) \\ A_1 W_1^{[1]}(x) + A_2 V_1^{f-1}(x) + A_3 V_1^{f-2}(x) + A_1 W_1(x) \\ = A_4 W_2^{[1]}(x) + A_5 V_2^{f-1}(x) + A_6 V_2^{f-2}(x) + A_4 W_2(x) \end{cases} \tag{35}$$

or

$$x = L_1: \quad \begin{cases} U_2^{[1]}(x) = U_1^{[1]}(x) + U_1(x) - U_2(x) \\ W_2^{[1]}(x) = \frac{A_1}{A_4} W_1^{[1]}(x) + \frac{A_2}{A_4} V_1^{f-1}(x) + \frac{A_3}{A_4} V_1^{f-2}(x) \\ \quad + \frac{A_1}{A_4} W_1(x) - \frac{A_5}{A_4} V_2^{f-1}(x) - \frac{A_6}{A_4} V_2^{f-2}(x) - W_2(x) \end{cases} \tag{36}$$

The systems of Eqs. (23), (24) can be written in the form

$$\begin{bmatrix} -H_{11}^1 & -H_{12}^1 & G_{12}^1 \\ -H_{21}^1 & -H_{22}^1 & G_{22}^1 \end{bmatrix} \begin{bmatrix} U_1^{[1]}(0) \\ U_1^{[1]}(L_1) \\ W_2^{[1]}(L_1) \end{bmatrix} = \begin{bmatrix} -G_{11}^1 W_1^{[1]}(0) \\ -G_{21}^1 W_1^{[1]}(0) \end{bmatrix} + \begin{bmatrix} Z_1(0) \\ Z_1(L_1) \end{bmatrix} \tag{37}$$

and

$$\begin{bmatrix} -H_{11}^2 & G_{11}^2 & -H_{12}^2 \\ -H_{21}^2 & G_{21}^2 & -H_{22}^2 \end{bmatrix} \begin{bmatrix} U_2^{[1]}(L_1) \\ W_2^{[1]}(L_1) \\ U_2^{[1]}(L) \end{bmatrix} = \begin{bmatrix} -G_{12}^2 W_2^{[1]}(L) \\ -G_{22}^2 W_2^{[1]}(L) \end{bmatrix} + \begin{bmatrix} Z_2(L_1) \\ Z_2(L) \end{bmatrix} \tag{38}$$

Introduction of the dependences (36) to the system of Eq. (38) gives

$$\begin{gathered} \begin{bmatrix} -H_{11}^2 & \frac{A_1}{A_4} G_{11}^2 & -H_{12}^2 \\ -H_{21}^2 & \frac{A_1}{A_4} G_{21}^2 & -H_{22}^2 \end{bmatrix} \begin{bmatrix} U_1^{[1]}(L_1) \\ W_1^{[1]}(L_1) \\ U_2^{[1]}(L) \end{bmatrix} \\ = \begin{bmatrix} H_{11}^2 [U_1(L_1) - U_2(L_1)] - G_{11}^2 R(L_1) - G_{12}^2 W_2^{[1]}(L) \\ H_{21}^2 [U_1(L_1) - U_2(L_1)] - G_{21}^2 R(L_1) - G_{22}^2 W_2^{[1]}(L) \end{bmatrix} + \begin{bmatrix} Z_2(L_1) \\ Z_2(L) \end{bmatrix} \end{gathered} \tag{39}$$

where

$$\begin{aligned} R(L_1) = & \frac{A_2}{A_4} V_1^{f-1}(L_1) + \frac{A_3}{A_4} V_1^{f-2}(L_1) + \frac{A_1}{A_4} W_1(L_1) \\ & - \frac{A_5}{A_4} V_2^{f-1}(L_1) - \frac{A_6}{A_4} V_2^{f-2}(L_1) - W_2(L_1) \end{aligned} \tag{40}$$

After coupling the systems of Eqs. (37) and (39) one obtains

$$\begin{bmatrix} -H_{11}^1 & -H_{12}^1 & G_{12}^1 & 0 \\ -H_{21}^1 & -H_{22}^1 & G_{22}^1 & 0 \\ 0 & -H_{11}^2 & \frac{A_1}{A_4}G_{11}^2 & -H_{12}^2 \\ 0 & -H_{21}^2 & \frac{A_1}{A_4}G_{21}^2 & -H_{22}^2 \end{bmatrix} \begin{bmatrix} U_1^{[1]}(0) \\ U_1^{[1]}(L_1) \\ W_1^{[1]}(L_1) \\ U_2^{[1]}(L) \end{bmatrix}$$
$$= \begin{bmatrix} -G_{11}^1 W_1^{[1]}(0) \\ -G_{21}^1 W_1^{[1]}(0) \\ H_{11}^2[U_1(L_1) - U_2(L_1)] - G_{11}^2 R(L_1) - G_{12}^2 W_2^{[1]}(L) \\ H_{21}^2[U_1(L_1) - U_2(L_1)] - G_{21}^2 R(L_1) - G_{22}^2 W_2^{[1]}(L) \end{bmatrix} + \begin{bmatrix} Z_1(0) \\ Z_1(L_1) \\ Z_2(L_1) \\ Z_2(L) \end{bmatrix} \quad (41)$$

Thus, from the system of Eq. (41), the boundary values $U_1^{[1]}(0)$, $U_1^{[1]}(L_1)$, $W_1^{[1]}(L_1)$, $U_2^{[1]}(L)$ are determined. Next, the values $U_2^{[1]}(L_1)$, $W_2^{[1]}(L_1)$ are calculated using the formulas (36). The values of functions $U_1^{[1]}(x)$, $U_2^{[1]}(x)$ at the internal points are calculated using formulas similar to the (19) for each layer separately. Finally, the temperature distribution is determined based on the formulas similar to the (20) also for each layer separately.

6 Results of Computations

The double-layered thin film (gold and chromium) subjected to a laser pulse is considered. The thicknesses of layers are equal to 50 nm. The initial temperature equals 300 K and initial heating rate $w(x) = 0$. Thermophysical parameters of materials are equal to $\lambda_1 = 315$ W/(mK), $c_1 = 2.4897 \times 10^6$ J/(m^3 K), $\tau_{q1} = 8.5$ ps, $\tau_{T1} = 90$ ps (gold) and $\lambda_2 = 93$ W/(mK), $c_2 = 3.21484 \times 10^6$ J/(m^3 K), $\tau_{q2} = 0.136$ s, $\tau_{T2} =$ 7.86 ps (chromium) [25].

The computations are performed for the laser intensity $I_0 = 40$ J/m^2 and the characteristic time of the laser pulse $t_p = 0.1$ ps [cf. formula (2)]. It is assumed that the optical penetration depth is equal to $\delta = 15.3$ nm and the reflectivity of the irradiated surface equals $R = 0.93$.

In Fig. 1 the temperature history at the irradiated surface for different number of internal cells n and different time steps Δt is shown. Four variants of computations have been done: $n = 200$, $\Delta t = 0.002$ ps; $n = 2000$, $\Delta t = 0.002$ ps; $n = 200$, $\Delta t = 0.01$ ps; $n = 2000$, $\Delta t = 0.01$ ps. In the last two cases the curves overlap. The differences between the remaining solutions are also very small which confirms that in the GBEM the number of internal cells and time step can be selected in a relatively wide range even when the source function is dependent on time and spatial coordinate. Thus, the next calculations were carried out for $n = 200$ and $\Delta t = 0.002$ ps.

Figures 2 and 3 illustrate the temperature distribution in the domain considered for the different moments of time. The curves presented in Fig. 2 correspond to the stage when the source function related to the laser operation increases, while the curves in Fig. 3 correspond to the stage when the source function decreases. It should be

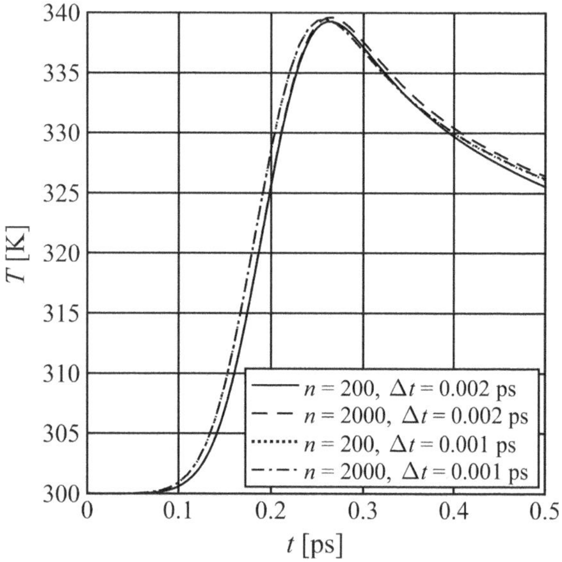

Fig. 1 Temperature history at the irradiated surface for different number of internal cells n and different time steps Δt ($I_0 = 40$ J/m^2, $t_p = 0.1$ ps)

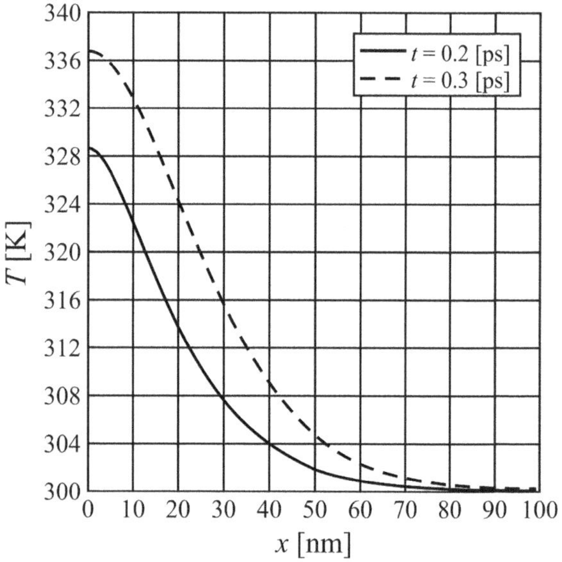

Fig. 2 Temperature profiles after 0.2 ps and 0.3 ps ($I_0 = 40$ J/m^2, $t_p = 0.1$ ps)

noted that in [25] the same problem has been solved using the explicit scheme of the finite difference method and the results are practically the same.

The computations have been also carried out for different laser intensities and different characteristic times of laser pulse. In Figs. 4, 6 and 8 the histories of temperature at the irradiated surface for laser intensities $I_0 = 200$ J/m^2, $I_0 = 400$ J/m^2, $I_0 = 600$ J/m^2, $I_0 = 800$ J/m^2 and characteristic times of laser pulse $t_p = 0.05$ ps, $t_p =$

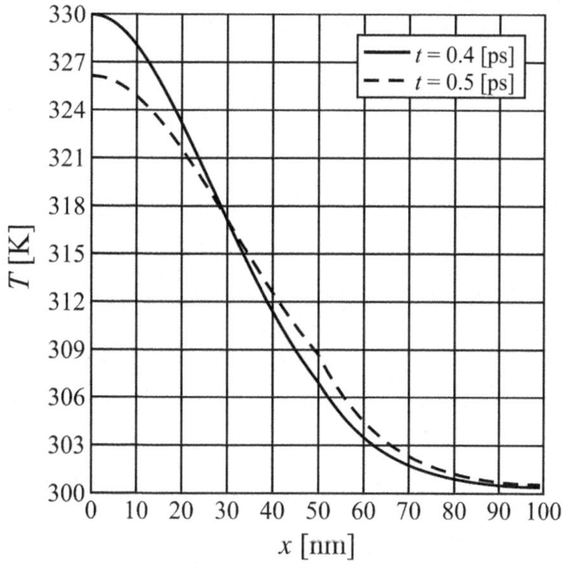

Fig. 3 Temperature profiles after 0.4 ps and 0.5 ps ($I_0 = 40$ J/m^2, $t_p = 0.1$ ps)

0.1 ps, $t_p = 0.2$ ps are shown. Figures 5, 7 and 9 present the temperature profiles for the same values of I_0 and t_p after a specified time that corresponds to the maximum temperature on the heated surface. Thus, for each laser intensity and $t_p = 0.05$ ps the maximum temperatures appear for a time equal to 0.13 ps (Fig. 4), for $t_p = 0.1$ ps this time is about 0.25 ps (Fig. 6), while for $t_p = 0.2$ ps this time is equal to 0.5 ps. As can be seen, for the same laser power, the shorter laser pulse generates the higher temperature on the heated surface (Figs. 4, 6 and 8). On the other hand, for a longer laser pulse, the heat penetrates deeper into the interior of the domain (Figs. 5 and 9).

7 Conclusions

The general boundary element method for the dual-phase lag equations describing the heating of two-layered thin metal films has been presented. The testing computations related to the selection of the time step and the number of internal cells were carried out. The results have also been compared with those obtained using the explicit scheme of the finite difference method and presented in [25]. The very good agreement has been observed.

In the calculations the different laser intensities and different characteristic times of laser pulse have been taken into account. Basing on the results of computations, appropriate conclusions were formulated.

In the paper the gold-chromium thin metal film has been considered. The other materials can also be analyzed under the assumption that the values of thermal parameters occurring in the mathematical model are known [26].

Fig. 4 Temperature histories at the irradiated surface for different laser intensities ($t_p = 0.05$ ps)

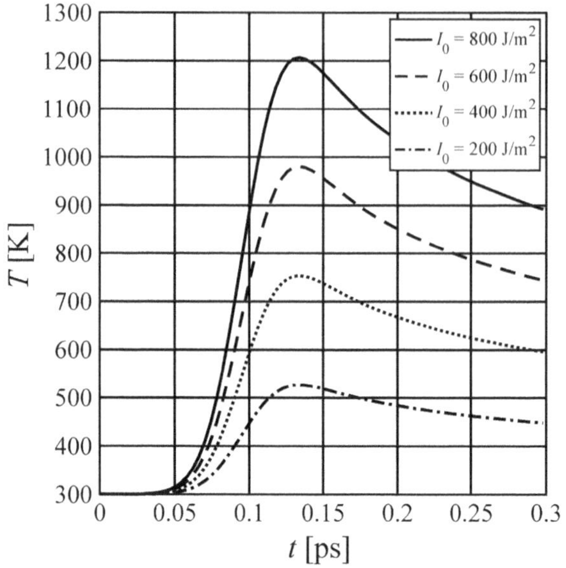

Fig. 5 Temperature profiles for different laser intensities after 0.13 ps ($t_p = 0.05$ ps)

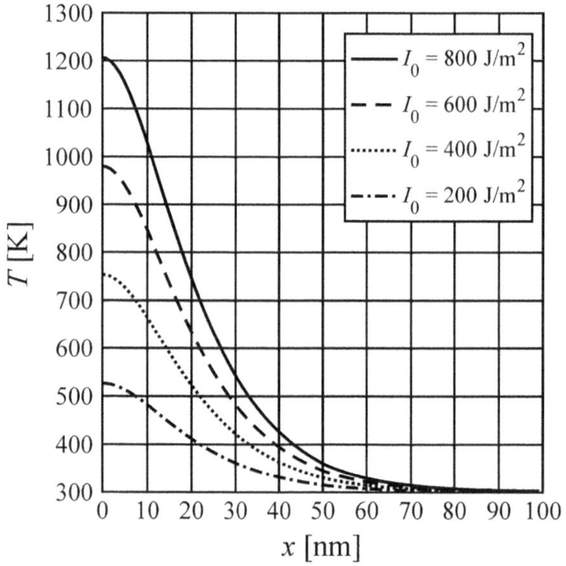

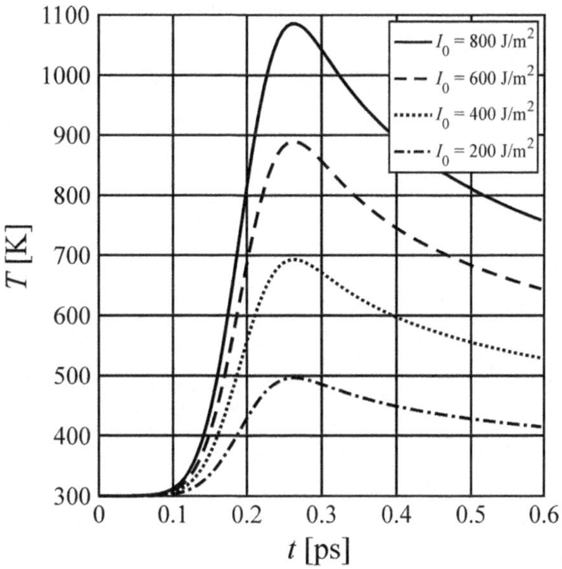

Fig. 6 Temperature histories at the irradiated surface for different laser intensities ($t_p = 0.1$ ps)

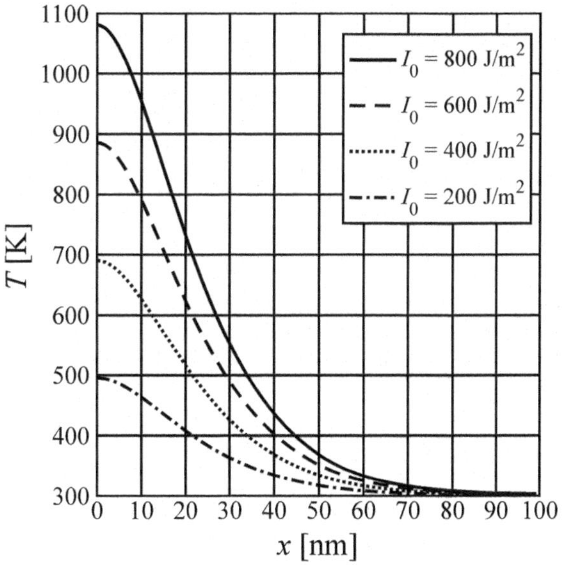

Fig. 7 Temperature profiles for different laser intensities after 0.25 ps ($t_p = 0.1$ ps)

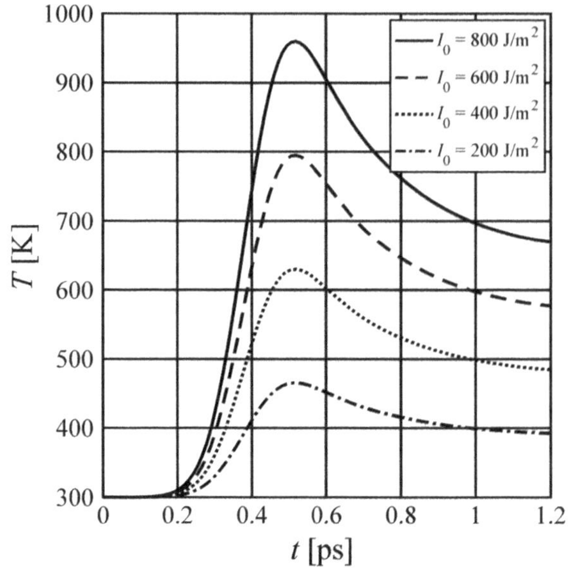

Fig. 8 Temperature histories at the irradiated surface for different laser intensities ($t_p = 0.2$ ps)

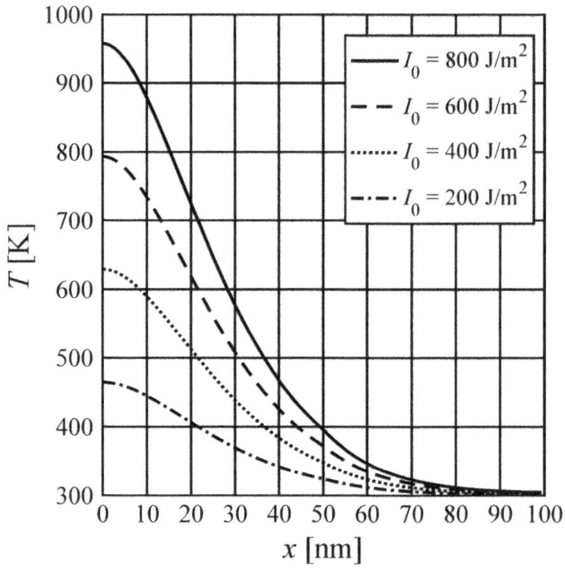

Fig. 9 Temperature profiles for different laser intensities after 0.5 ps ($t_p = 0.2$ ps)

The presented approach can be easily extended to the case of multilayered thin metal films [12, 15, 27, 28].

Acknowledgements The paper and research are financed within the project 2015/19/B/ST8/01101 sponsored by National Science Centre (Poland).

References

1. Zhang Y, Tzou DY, Chen JK (2009) Micro- and nanoscale heat transfer in femtosecond laser processing of metals. In: Barret PH, Palmerm M (eds) High-power and femtosecond lasers: properties, materials and applications. Nova Science Publishers, Inc., Hauppauge, pp 159–206 (Chapter 5). arXiv:1511.03566 [physics.comp-ph]
2. Zhang ZM (2007) Nano/microscale heat transfer. McGraw-Hill
3. Tzou DY (2015) Macro to microscale heat transfer: the lagging behavior. Wiley
4. Anisimov SI, Kapeliovich BL, Perelman TL (1974) Electron emission from metal surfaces exposed to ultrashort laser pulses. JETP 66:375–377
5. Qiu TQ, Tien LC (1993) Heat transfer mechanisms during short-pulse laser heating of metals. J Heat Transfer 115(4):835–837
6. Chen JK, Tzou DY, Beraun JE (2006) A semiclassical two-temperature model for ultrafast laser heating. Int J Heat Mass Transf 49(1):307–316
7. Niu T, Dai W (2009) A hyperbolic two-step model based finite difference scheme for studying thermal deformation in a double-layered thin film exposed to ultrashort-pulsed lasers. Int J Therm Sci 48(1):34–49
8. Dziatkiewicz J, Kus W, Majchrzak E, Burczynski T, Turchan L (2014) Bioinspired identification of parameters in microscale heat transfer. Int J Multiscale Comput Eng 12(1):79–89
9. Huang J, Baheti K, Chen JK, Zhang Y (2011) An axisymmetric model for solid-liquid-vapor phase change in thin metal films induced by an ultrashort laser pulse. Front Heat Mass Transf 2(1):1–10
10. Majchrzak E, Dziatkiewicz J, Turchan L (2017) Analysis of thermal processes occurring in the microdomain subjected to the ultrashort laser pulse using the axisymmetric two-temperature model. Int J Multiscale Comput Eng 15(5):395–411
11. Baheti K, Huang J, Chen JK, Zhang Y (2011) An axisymmetric interfacial tracking model for melting and resolidification in a thin metal film irradiated by ultrashort pulse lasers. Int J Therm Sci 50(1):25–35
12. Ho JR, Kuo CP, Jiaung WS (2003) Study of heat transfer in multilayered structure within the framework of dual-phase-lag heat conduction model using lattice Boltzmann method. Int J Heat Mass Transf 46(1):55–69
13. Mochnacki B, Paruch M (2013) Estimation of relaxation and thermalization times in microscale heat transfer model. J Theoret Appl Mech 51(4):837–845
14. Ramadan K, Tyfour WR, Al-Nimr MA (2009) On the analysis of short-pulse laser heating of metals using the dual phase lag heat conduction model. J Heat Transf 131(11):111301
15. Majchrzak E, Mochnacki B, Suchy JS (2009) Numerical simulation of thermal processes proceeding in a multi-layered film subjected to ultrafast laser heating. J Theoret Appl Mech 47(2):383–396
16. Mochnacki B, Ciesielski M (2016) Dual phase lag model of melting process in domain of metal film subjected to an external heat flux. Arch Foundry Eng 16(4):85–90
17. Liao S (1997) General boundary element method for non-linear heat transfer problems governed by hyperbolic heat conduction equation. Comput Mech 20:397–406
18. Liao SJ (1998) On the general boundary element method. Eng Anal Boundary Elem 2:39–51

19. Hetmaniok E, Nowak I, Slota D, Witula R (2012) Application of the homotopy perturbation method for the solution of inverse heat conduction problem. Int Commun Heat Mass Transfer 39:30–35
20. Slota D (2010) The application of the homotopy perturbation method to one-phase inverse Stefan problem. Int Commun Heat Mass Transfer 37:587–592
21. Majchrzak E (2010) Numerical solution of dual phase lag model of bioheat transfer using the general boundary element method. CMES Comput Model Eng Sci 69(1):43–60
22. Majchrzak E, Turchan L (2015) The general boundary element method for 3D dual-phase lag model of bioheat transfer. Eng Anal Boundary Elem 50:76–82
23. Brebbia CA, Telles JCF, Wrobel LC (1984) Boundary element techniques. Springer, Berlin
24. Majchrzak E, Mochnacki B (2017) Numerical model of thin metal film heating using the boundary element method. Comput Methods Mater Sci 17(1):12–17
25. Majchrzak E, Kałuża G (2017) Analysis of thermal processes occurring in the heated multi-layered metal films using the dual-phase lag model. Arch Mech 69(4–5):275–287
26. Iida T, Guthrie RIL (2015) The thermophysical properties of metallic liquids, volume 1: fundamentals. Oxford
27. Al-Nimr MA, Naji M, Abdallah RI (2004) Thermal behavior of a multi-layered thin slab carrying periodic signals under the effect of the dual-phase-lag heat conduction model. Int J Thermophys 25(3):949–966
28. Karakas A, Tunc M, Camdali U (2010) Thermal analysis of thin multi-layer metal films during femtosecond laser heating. Heat Mass Transf 46:1287–1293

Numerical and Experimental Analysis of a Personalized Prosthesis for a Patient with Unilateral Hip Osteoarthritis

Juan Alfonso Beltrán-Fernández, Omar Rolando Ruiz-Muñoz, Luis Héctor Hernández-Gómez, Alejandro González-Rebattú y González, Itzel Bantle-Chávez, Carolina Alvarado-Moreno, Pablo Moreno-Garibaldi and Nefi Pava Chipol

Abstract In this work, the design and numerical evaluation of a personalized hip prosthesis for a 31-year-old male patient, who suffered of unilateral coxarthrosis, was performed. Based on the information obtained of the patient's computerized axial tomography and the employment of specialized design platforms such as Solidworks®, a numerical analysis was performed for the prosthetic model in conjunction with the anthropometric data of the patient following the finite element mathematical method. Furthermore, the natural posture of the patient was taken into consideration every time the patient balanced the total weight on one leg. Furthermore, the structural mechanical evaluation

J. A. Beltrán-Fernández (✉) · O. R. Ruiz-Muñoz · L. H. Hernández-Gómez · I. Bantle-Chávez · C. Alvarado-Moreno · P. Moreno-Garibaldi · N. P. Chipol
Instituto Politécnico Nacional - Escuela Superior de Ingeniería Mecánica y Eléctrica - Sección de Estudios de Posgrado e Investigación Edificio 5, 2do Piso, Unidad Profesional Adolfo López Mateos "Zacatenco" Col. Lindavista, C.P. 07738 Ciudad de México, Mexico
e-mail: jbeltranf@hotmail.com

O. R. Ruiz-Muñoz
e-mail: omarrolandocv@hotmail.com

L. H. Hernández-Gómez
e-mail: luishector56@hotmail.com

I. Bantle-Chávez
e-mail: itzi.bantle@gmail.com

C. Alvarado-Moreno
e-mail: aly-oroshiwa@hotmail.com

P. Moreno-Garibaldi
e-mail: pmg170588@gmail.com

N. P. Chipol
e-mail: sowbran@gmail.com

A. González-Rebattú y González
Hospital Regional 1 De Octubre – ISSSTE - Av. Instituto Politécnico Nacional #1669, Lindavista, 07300 Gustavo A. Madero, CDMX, Mexico
e-mail: alexrebattu@hotmail.com

A. Öchsner and H. Altenbach (eds.), *Engineering Design Applications II*,
Advanced Structured Materials 113, https://doi.org/10.1007/978-3-030-20801-1_21

of the prostheses was set under evaluation, considering the osseous integration of the femoral bone and the acetabulum before and after surgery.

Keywords Personalized hip prosthesis · Unilateral coxarthrosis · Computerized axial tomography · Numerical analysis · Femur · Acetabulum

1 Introduction

Although the main cause of the hip osteoarthritis is unknown, the best way to define this particular disease is taking into account the whole clinical picture that includes a group of heterogenic pathologies with common clinical manifestations.

Arthrosis or hip coxarthrosis is defined as a disease characterized by alteration between the formation and degradation cycle of the joint cartilage and the subchondral bone, causing morphological lesion areas that provoke pain and incapacity, it is a relatively common disease found in patients between 50 and 55 years old and affects 20–30% of the population. Taking into account that the hip plays a major role in the balance and support of the body, the coxarthrosis can be considered as one of the most disabling diseases in its category.

Due to its insidious start, the patient isn't initially aware of the beginning of the pain, the intensity also, is variable and is accompanied with joint rigidity each time a movement begins. The disease is the direct consequence of a constant usage of the joint when walking. The pain, however, has a tendency to vanish when the patient rests. This pathology also occurs after excessive use of the joint, even when walking or standing up.

Other particular symptom of hip osteoarthritis is the limitation of mobility with functional impotence and progressive lameness, in addition to limitations in everyday activities, such as difficulty sitting and getting up from a chair, crossing the legs to wear shoes, impossibility to straddle or when trying to pick something up off the ground. The disease has also two variants; it can be unilateral or bilateral. When it is unilateral it can be easily confused with secondary arthritis due to the appearance of previous articular defects. According to Rozadilla et al. (2002), this disorder can be classified as unilateral and bilateral. Patients who suffer of unilateral coxarthrosis present previous articular lesions in the affected area, forcing the osseous system to compensate and balance the total patient's weight on one side, having as consequence, a tendency to develop similar lesions in the opposite hip joint over the years". [1]

Studies performed by Murray et al. (1979) and Olsson et al. (1985), suggested that total hip arthroplasty (THA) as a surgical procedure, usually results in a marked pain decrease and an improvement in the functional capacity of the patients [2, 3]. Under other conditions, Illyés et al. (2006), refers that THA can reverse the adverse influence of other joints before the symmetric normalization of the hip's movement is achieved [4].

Due to the evolution of the design for whole joint prostheses, there is a constant increment of the need to enhance the role of the biomechanical nature of the body's joints. The mechanical problems associated to a total replacement of the joint include aspects that concern load surfaces, mechanical failures, detachments and dislocations of the implant over the joint's surface. Having as goal the obtention of this information, multiple experiments in vitro have been conducted using joints and tissues that belong to human corpses, simulating natural movements and loads to which the joints are usually submitted. These experiments were also complemented with studies of human locomotion that were performed in vivo, Nordin & Frankel (2000) [5].

Generally, gait analyses have been used for the quantitative measurement and evaluation of human locomotion. In the case, gait measurement protocols and extensive physical analyses are included in order to determine the characterization of the anthropometry of each subject, Bronzino (2000) [6].

Seireg and Arvikar (1975), determined the minimization of muscle forces as well as the reactions of the corresponding joints through a quasi-static walking cycle treated as a sequence of static postures in successive intervals. The final result was the description of a model of lower extremities with a system of 7 segments connected in 31 main muscles, as shown in Fig. 1 [7].

Bergmann et al. (2001) measured the contact forces of the hip in instrumented implants and performed synchronous analyses of walking patterns to measure the reactions of the forces on the ground. For this experiment, 4 patients with hip diseases were studied during their daily activities. Specialists reported that the presence of dysfunctions in a muscle increases the contact force of the joint. Additionally, they indicated that some biological optimizations with respect to the torsion of the femur or the implant could be implemented and concluded that the design of an implant with high performance in the resistance of the force moment is a fundamental requirement [8]. Figure 2 shows the coordinate system used to measure the contact forces of the hip.

2 Patient and Methods

In this study, the clinical case of a 31-year-old patient who suffers of an advanced case of degeneration in the right pelvic area, particularly the zone of the acetabulum and the femoral head was analyzed. The male patient weights 48 kg and is 1.64 m high. For the development of this project, the tomography and personal data was used under the patient consent. According to Chaffin (1969), "The digital technology has provided the necessary computational capacity to create a wide range of biomechanical models" [9]. This is the reason why the design of the prostheses was carried out employing the ScanIP software, which allowed designing an accurate model of the patient's clinical characteristics as shown in Fig. 3.

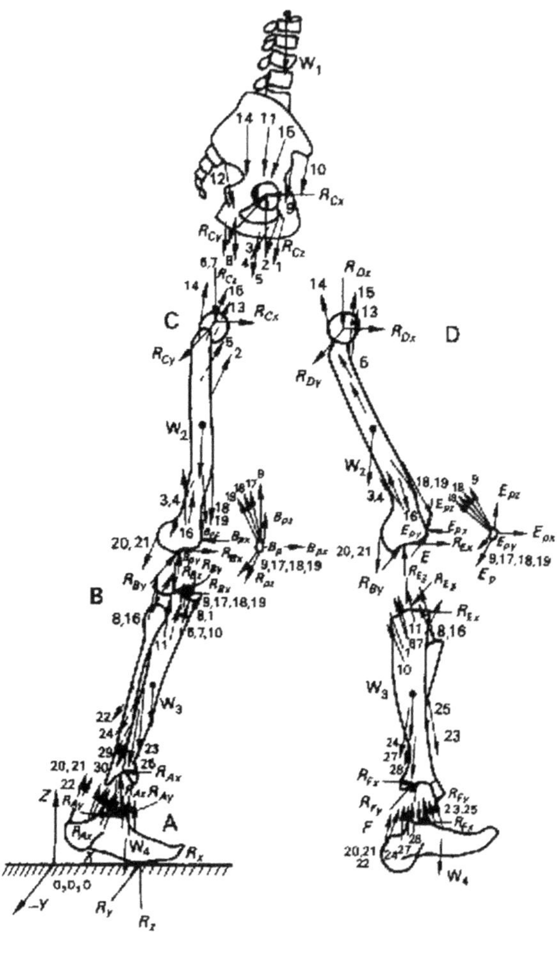

Fig. 1 Free-body diagram of several segments for a typical posture during a quasi-static walking cycle by Seireg and Arvikar [7]

Based on the tridimensional model, specialists analyzed the right side of the pelvic system where the damage was located. According to the Wolff's Law, the natural mechanical loads which are applied over the osseous surface have a great impact on the development of the bone tissue, Cowin (2000) [10].

After exporting the tridimensional model's data in form of a. stl archive to the Rhinoceros platform, 4.0 version, the number of elements which compose the model's surface was reduced in order to enhance the manipulation and design of the final archive with the usage of other types of software. Subsequently, the archive was converted into a.iges file and modified through the software Solidworks® with the extension *.iges for its further manipulation and analysis (Figs. 5 and 6).

Once exported, the "Personalized hip prosthesis system" was designed. The first part of the model consisted of the addition of a device that needed to be fit inside the acetabulum's body, known as Burch-Shnider's® reconstruction ring (Zimmer). This

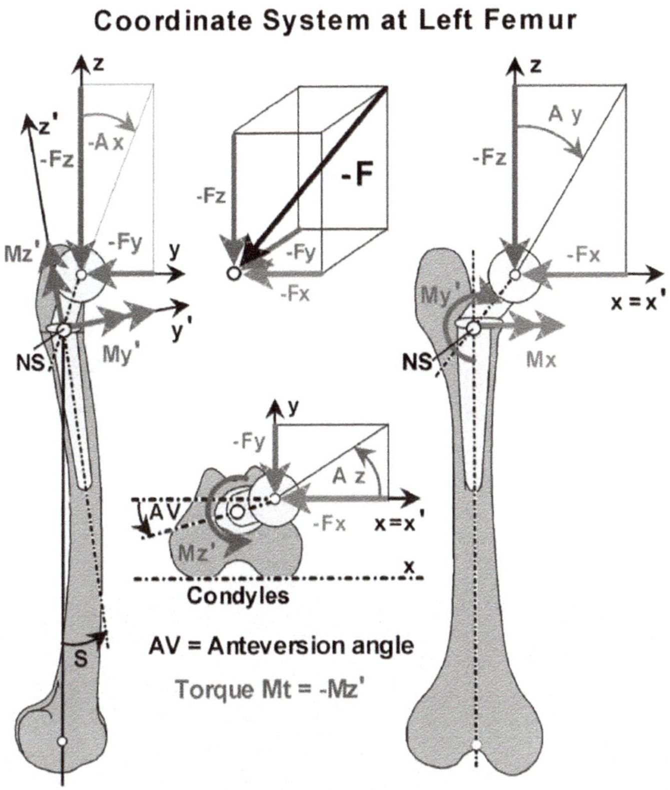

Fig. 2 Coordinate system used by Bergmann and collaborators to measure contact forces in the hip

Fig. 3 3D dimensional representation of the patient's right and left hip after rendering

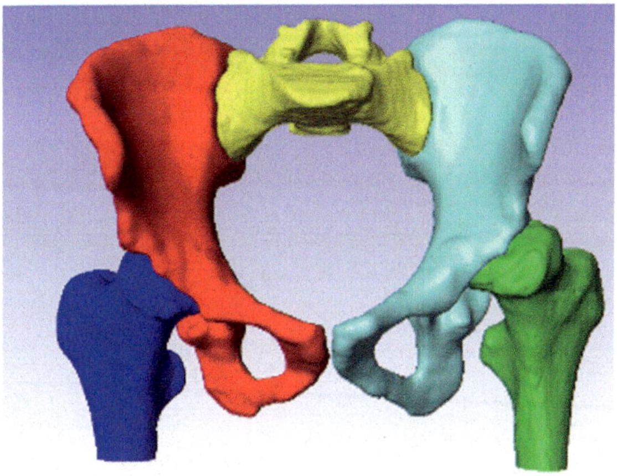

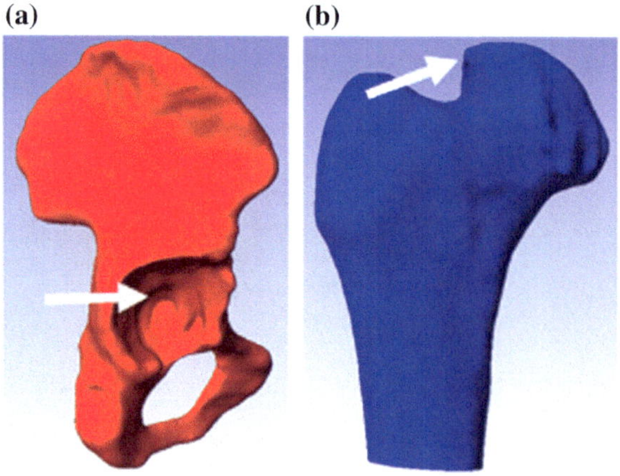

Fig. 4 Damage caused by the disease on **a** acetabulum's fossa and **b** femoral head

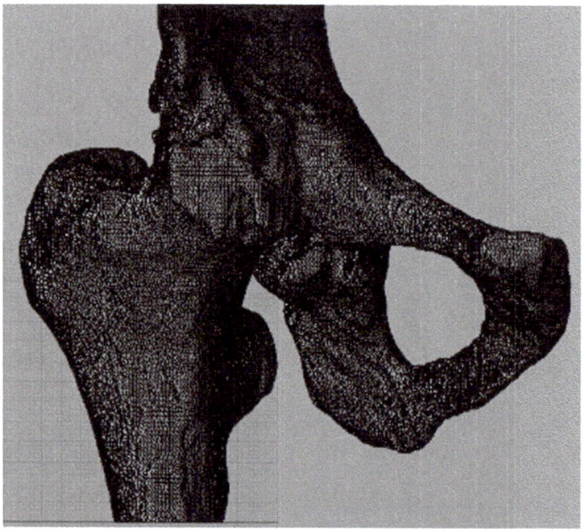

Fig. 5 Excess of elements on the hip joint's surface

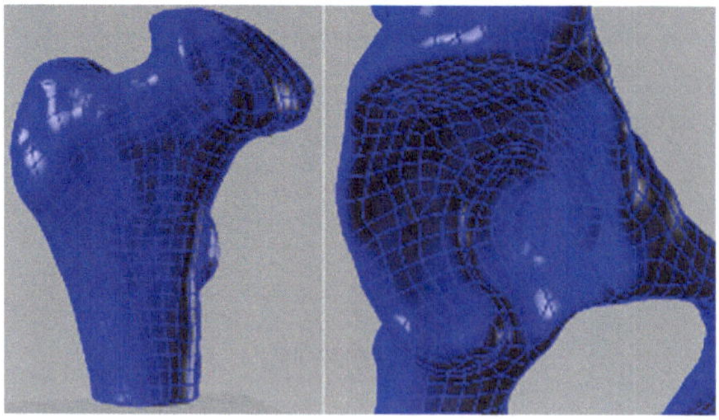

Fig. 6 Parameterization of the completed model

Fig. 7 Burch-Shnider's® reconstruction ring (Zimmer), Ruíz Muñoz (2016) [11]

procedure was performed according to the indications and observations given by the surgeon specialist, Dr. Alejandro Rebatú, of the "Hospital Regional 1° de Octubre". The design of the ring can be appreciated in Fig. 7.

However, on works carried out by Paprosky et al. (1997) and Perka et al. (2001) it was described that the Burch-Schnider's® ring, which for many years has been considered the principal implant for the reconstruction of large bone defects, fails after 6–8 years of use in 30% of the cases [12, 13].

Having as objective the replacement of the rotation's center, the acetabulum's spare part was adjusted to the right hip's bone surface. Areas that presented central and oval cavitary defects of different sizes (type IIb Paprosky), as shown in Fig. 8, lessons were reconstructed through the implementation of a Trabecular Metalic Technology System, known as® TMT (Fig. 9). This system presents advantages as: a high coefficient of friction that ensures a primary stability, good osteointegration and a low tendency to be reabsorbed by the body.

Unfortunately, some disadvantages are the fact that a long-term positive result is not guaranteed and also the behavior of the metal-metal interface in cases of advanced infection has not been fully studied, to mention a few, Hasart et al. (2011) [14].

To continue with the clinical case, it was designed a device that fusion both of the previously exposed systems. For the acetabular cup, the first element that needed to be considerate was the angle in which the device needed to be positioned. For that reason, it is crucial to manage an angle of abduction of 45° as shown in Fig. 10, according to the indications of the study performed by Udofia et al. (2007) [15].

After defining the angle of inclination, it was proceeded to define the design of the acetabular component's support. The dimensions are shown in Fig. 11, with $\phi =$ 51.51 mm outer, $\phi = 43.51$ mm inner and a thickness of 4 mm.

Subsequently, it was pretended simulated on the surface of the acetabulum in order to place the support of the cup. Unfortunately, the roughness of the affected area represented a difficulty for the placement of the ring. An optimal solution it was opted to made extrusions at a distance of 10 mm from the affected area, as shown in Fig. 12.

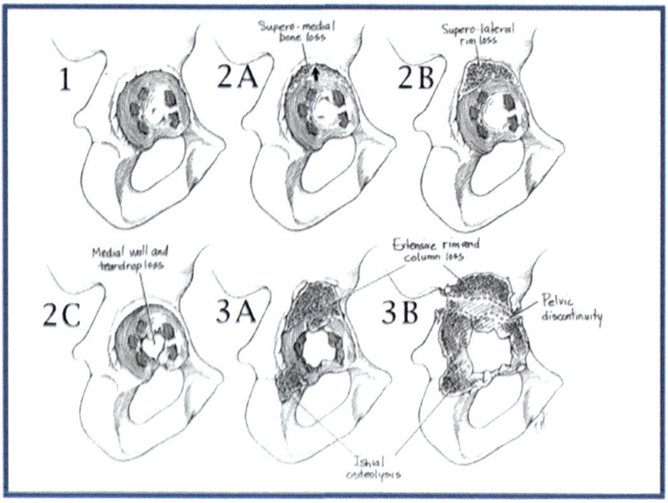

Fig. 8 Acetabular bone defect according to Paprosky, Sancho Navarro (2006)

Fig. 9 Trabecular Metalic Technology System, TMT (Zimmer), Hasart et al. (2011)

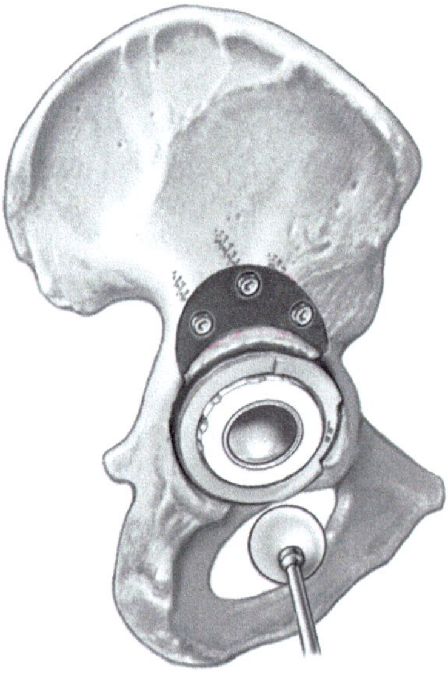

Finally, the body of the cup was generated with a hemispherical surface with an outer $\phi = 25.75$ mm and an inner $\phi = 21.75$ mm, but not before generating the necessary remaining place to put it, in the acetabulum zone, as shown in Fig. 13.

Once the cup was placed, it was proceeded to design the acetabular reconstruction system, which begins from the base of the cup, this dimensions are specified in Figs. 14, 15 and 16. The device was elaborated with fins located in the lower, lateral and superior part of the acetabulum. These were set in the lateral and superior part

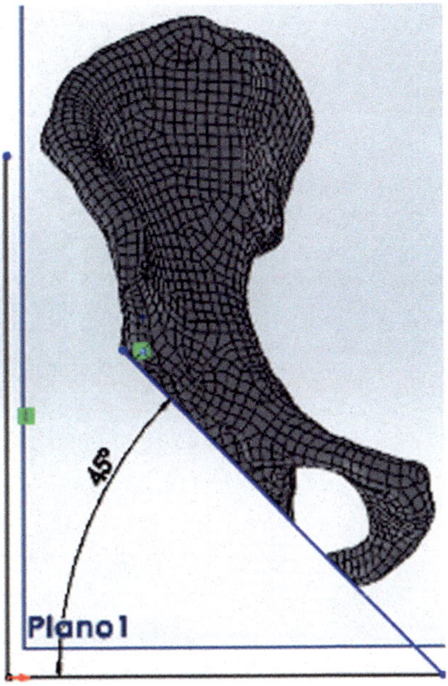

Fig. 10 Inclination angle (45°) used to place the acetabular cup

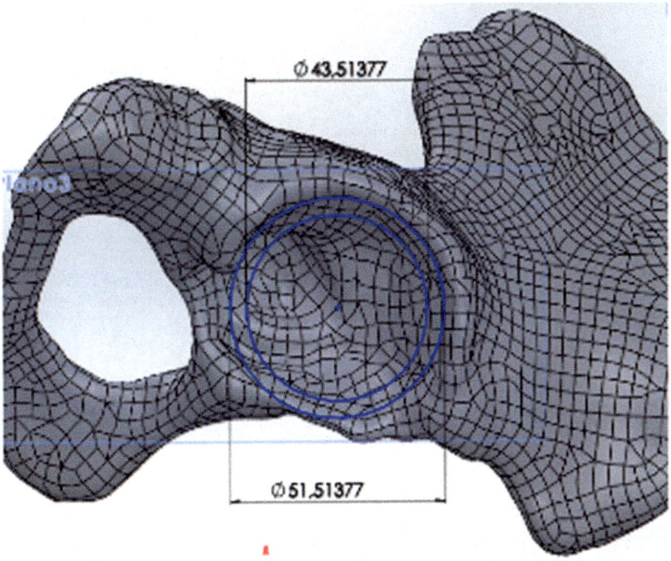

Fig. 11 Dimensions of the acetabular cup support

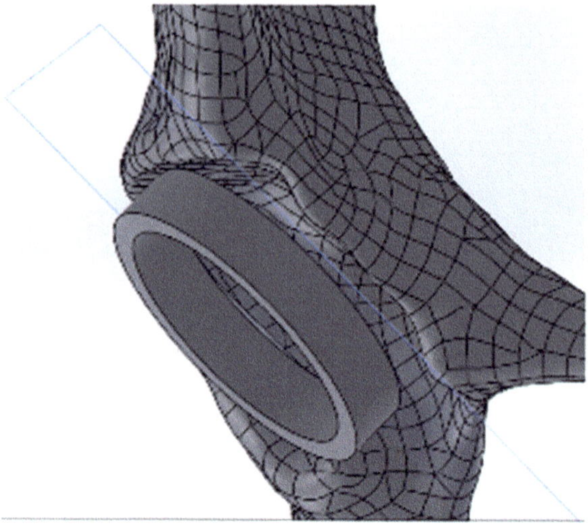

Fig. 12 Support design, of the acetabular cup

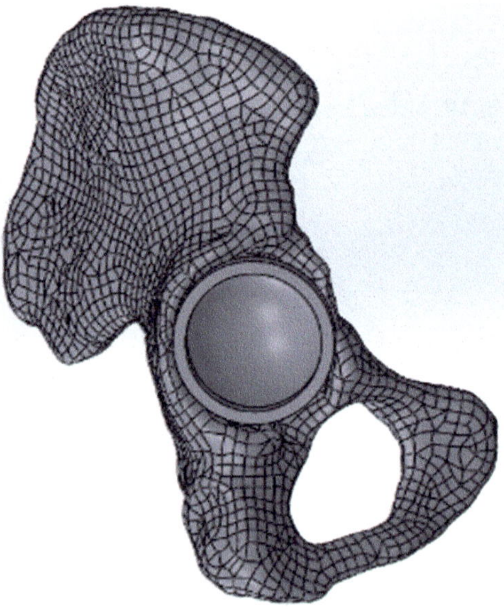

Fig. 13 Acetabular cup placed inside the body of the acetabulum zone

of the acetabular cup with the usage of screws and thin metric thread M3—0.35 mm × 16 mm and M3—0.35 mm × 18 mm, respectively.

Hollow hexagonal countersunk heads crossed the bone and in the lower part of the ischium area, a spongy surgical screw of 6.5 mm × 25 mm was collocated according to the indicated Standard 060—Healing Material Osteosynthesis (1991) [8].

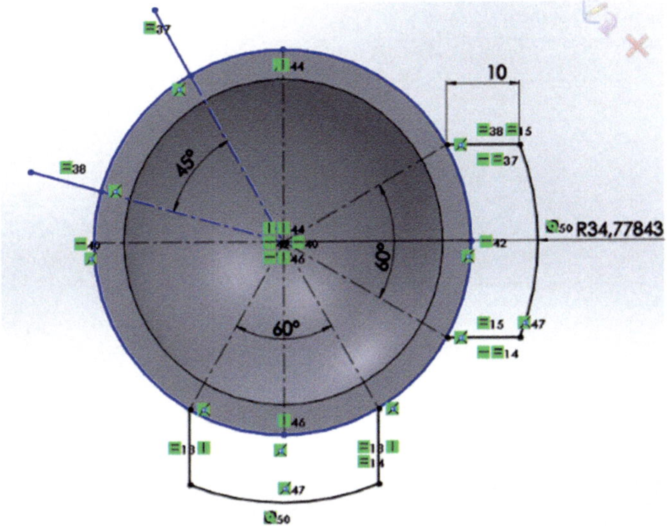

Fig. 14 Angles used for the generation of the anchoring device

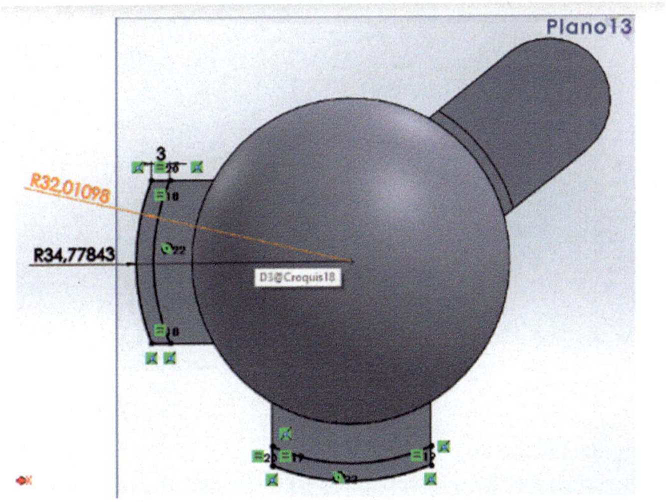

Fig. 15 Dimensions for the generation of the lateral and upper supports

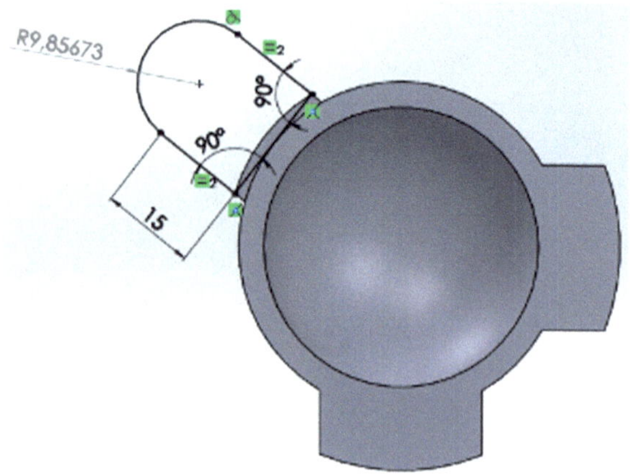

Fig. 16 Dimensions for the support placed in the area of the ischium

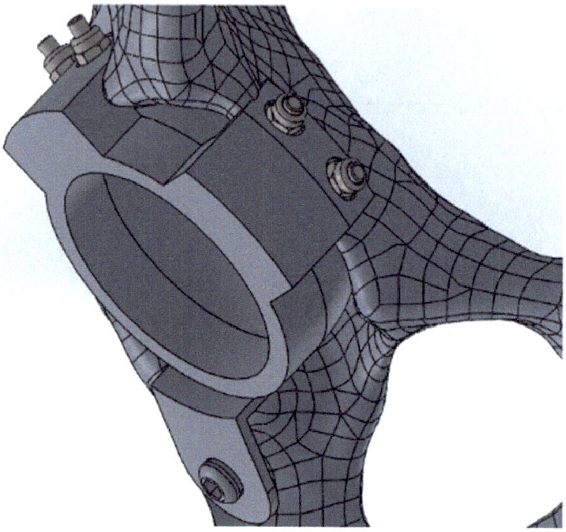

Fig. 17 Device fixed with screws

The final model can be appreciated in Fig. 17.

For the generation of the non-cemented stem model, the design of a Taperloc® Fe-moral stem (BIOMET) is taken as the reference model illustrated on Fig. 18. As a starting point of the model, the sections of the femur were modified to project the contour of the stem on the internal walls of the femur. This, in order to have sufficient anatomical contact, as seen in Fig. 19a. Subsequently, a series of geometric lines were determined to make the implant more stable. The first element that needed to be defined was the cervical-diaphyseal angle (CCD), for which an angle of 135° is considered, taking into account the range of tolerance proposed by Schidlo and cols. (1999) [16]. Furthermore, the stem was designed with a length of 129.19 mm (Fig. 19b).

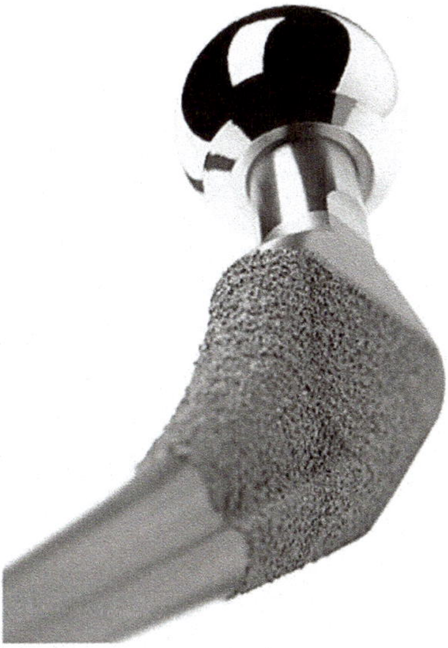

Fig. 18 Taperloc® Fe-moral stem (BIOMET)

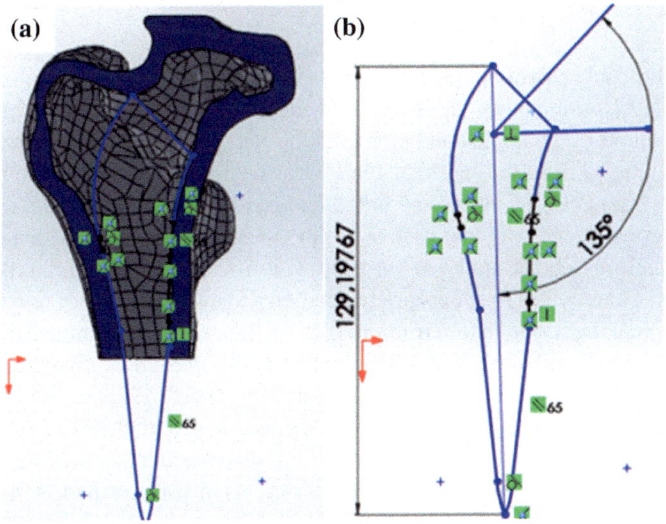

Fig. 19 **a** Diagram of the stem following the contour of the diaphysis channel **b** diagram with the angle of CCD of 135° and length of 129.19 mm

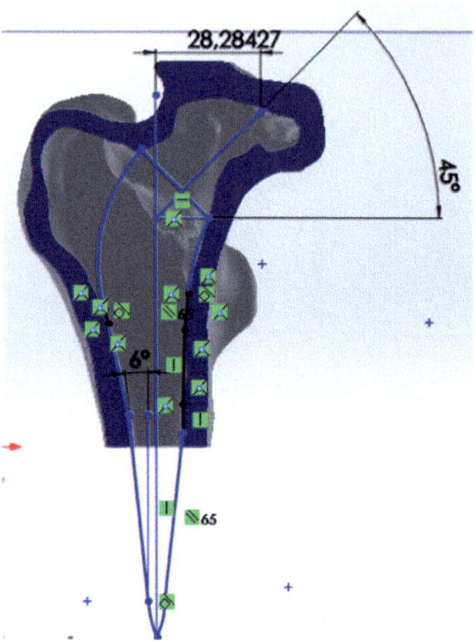

Fig. 20 Diagram shows the offset length of 28.28 mm and conicity angle of 6°

Thereafter, other parameters were defined such as the taper's angle of 6°, with the purpose of establishing the relation of the friction coefficient in the interface with the tangent of the angle that belongs to the stem's conicity. This procedure was performed considering the fact that the lowest friction coefficient in the interface reaches the value of 0.15 during the first days after the operation. This phenomenon is a direct consequence of the presence of blood and marrow remains that culminate in inflammation. Schultz et al. [17] in conjunction with Ojeda (2009), reported that the angle of conicity has the appropriate geometry to avoid the risk of rejection since it complies that the coefficient of friction (0.15) is greater than the tangent of the angle of conicity (tag $6° = 0.105$). Regarding the offset, this was 28.28 mm to compensate for the pathological deterioration, as shown in Fig. 20 [18, 19].

Finally, the so-called Morse cone was designed, with a diameter of 10 mm and a length of 29.82 mm. At its base, a radius of 6 mm was added, as shown in Fig. 21.

The design of the alumina cup began after the stem was finished, with a base of 43.51 mm. The hemisphere has a diameter of 21.75 mm, the modular head of 28 mm and the extrusion of the figure has an internal diameter of 10 mm, and a depth of 20 mm, as shown in Fig. 22.

Finally, all the components were assembled (Fig. 23) and the material's assignment was performed. Biomaterials such as titanium and its alloys, offer better resistance to corrosion than Stainless Steel or Chromium-Cobalt alloys, they also don't show evidence of stings or intragranular corrosion in biological conditions. The most commonly alloy used is Ti-6Al-4V, which was used as princiapl component of the proposed hip prosthesis system [10].

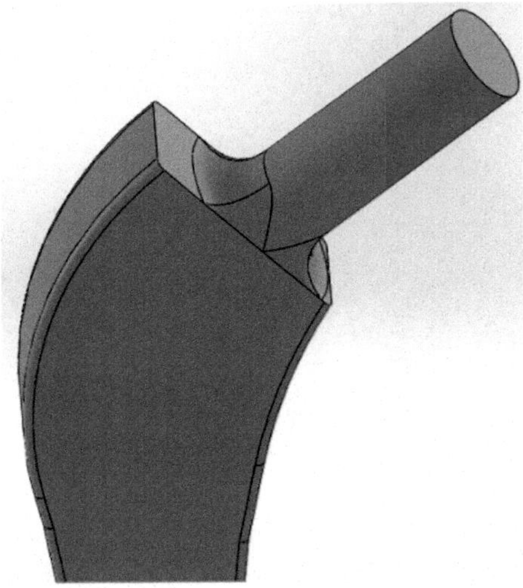

Fig. 21 A circle of radius 0.5 mm was collocated on the surface of the stem. The second circle of radius 6 mm was positioned in the area of the Morse cone

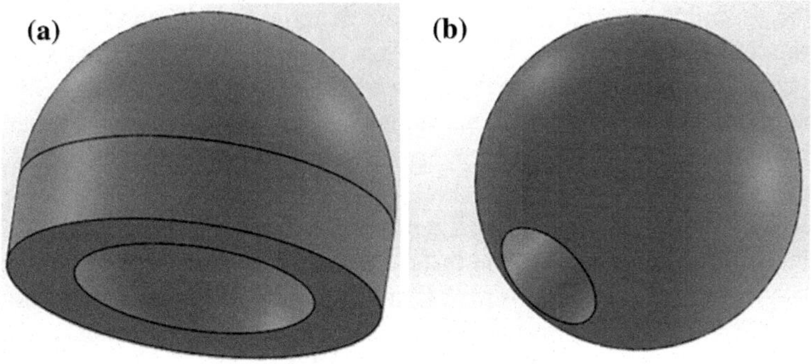

Fig. 22 **a** Alumina cup **b** modular head

In view of these, the alumina cup was chosen due to the poor results that the use of high-density polyethylene entails; since, as is commonly described, it is the weakest link in total hip arthroplasty because of its short durability. The prolonged wear releases nonbiocompatible particles that produce inflammatory reactions and osteolysis, having as a final consequence the loosening of the implants [20]. Alumina, on the other hand, has greater rigidity and hardness than metals and a better wear resistance [10].

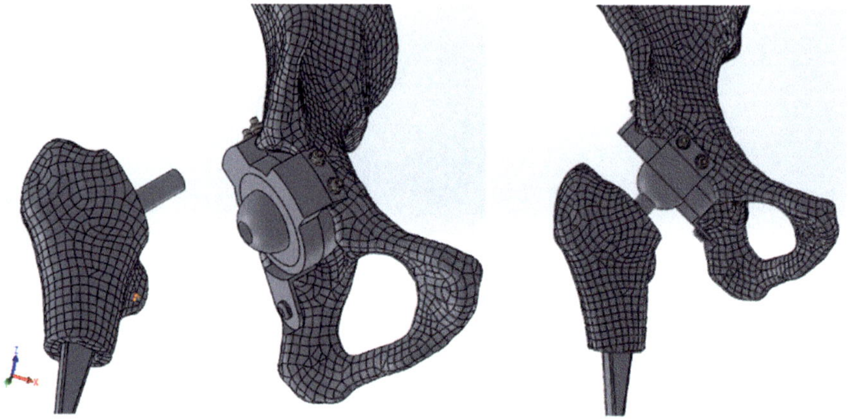

Fig. 23 Assembly of the "Personalized hip prosthesis"

3 Results

Once the assembly was completed, the functioning of the proposed system was tested through a MEF analysis. With the usage of the Solidworks® force simulation function and based on previous works done by Beltrán et al. (2013, 2014), based on the anthropometric data of the patient, the model was submitted to simulated force vectors with a magnitude that approaches the natural forces and tensions executed by the body [21].

As it was mentioned, the biomaterial of the reconstruction system of the acetabulum, molar head, and femoral stem is the titanium, aluminum, and vanadium alloy Ti-6Al-4V; which mechanical properties are according to the recommendations of ASTM F136: Modulus of elasticity = 121 GPa, Poisson's coefficient = 0.34 and density = 4.3×10^{-6} kg/mm^3(4.3 g/cm^3) [18].

For the pressure impact cup, the selected biomaterial was alumina, which properties are summarized in Table 1.

Table 1 Bioceramics physical features of alumina, specified in the study performed by Calambas [22]

Properties of Al_2O_3	
Content (% p/p)	>99.8
Densidad (g/cm^3)	>3.93
Average grain size (μm)	3–6
Hardness (Vickers)	2300
Mechanical resistance to bending (MPa)	550
Mechanical resistance to compression (MPa)	4000
Elastic modulus (GPa)	380

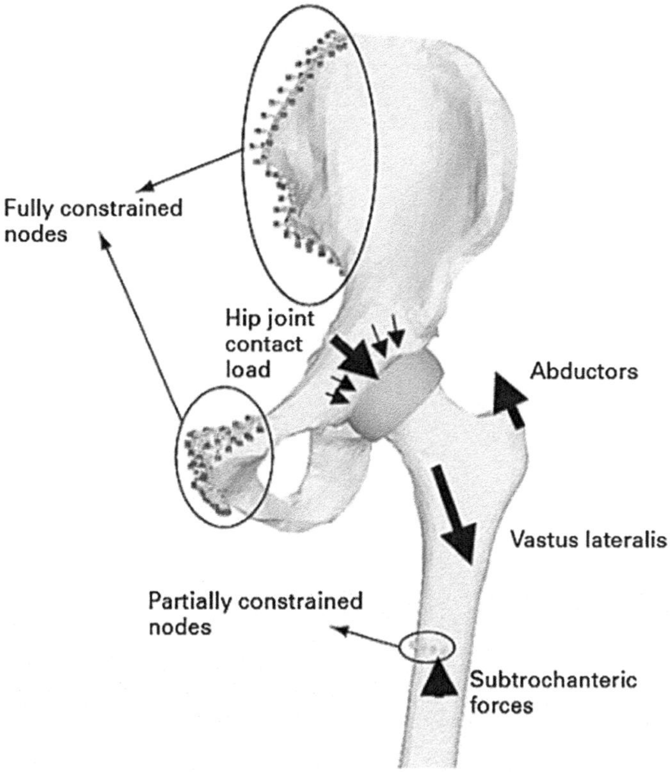

Fig. 24 Loading conditions and boundary constraints [15]

While for the pelvis the modulus of elasticity and the Poisson's coefficient were of 17 GPa and 0.3, respectively [15], for the femur, the modulus of elasticity was 13.6 GPa and the same Poisson coefficient for the pelvis [23].

After select the materials for the study, it was defining the load application conditions and movement restrictions, based on the movement conditions of the hip joint.

An example that illustrates how restrictions and loads are applied in a hip prosthesis model is shown in Fig. 24, which was proposed by Udofia et al. (2007) [15], in which it was applied a finite element analysis, with a contact force of 3200 N in the hip joint, simulating the sacral support. The border conditions were established with nodes that were located in the sacroiliac joint and pubic symphysis to limit them completely and prevent movement in any other direction. In the lower part around the femur cross section, the nodes partially limited the area, allowing one or two degrees of freedom.

The results obtained are specified below based on the boundary shown conditions, in order to obtain the behavior for four different cases of gait.

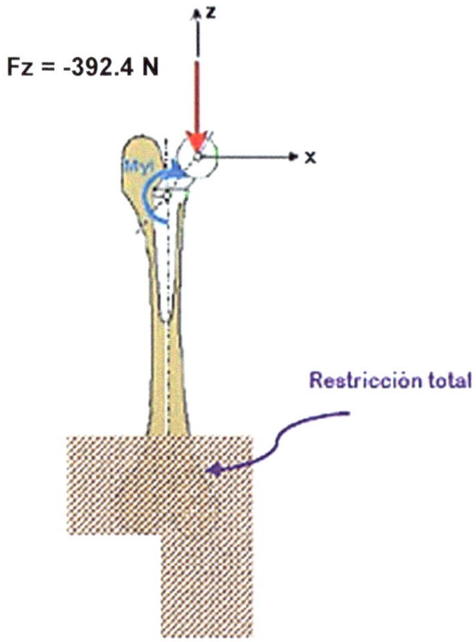

Fig. 25 Border condition: position on one foot, without walking, used by Ruíz Muñoz (2016) [11]

3.1 Border Condition, Case 1, Standing on One Foot Without Walking

For this case, the load direction is only vertical and coincides with the femoral axis Fz = −392.4 N remembering that the load is a 5/6% of the weight of the body and the resulting force direction of the weight is descending, its sign turns negative.

Figure 25 illustrates the boundary condition for a position on a foot without walking, it should be noted that for this case the applied load is 5/6% of the body's weight because the weight of the support member is being rested of the total body's weight.

The type of mesh that was applied to all the models of the femur-stem-head-cup system during this investigation was a solid tetrahedral type, as shown in Fig. 26, with a size per element of 2 mm and a tolerance of 0.1 mm. Furthermore, it was not necessary to apply mesh control techniques and for this first case, a model with 116,461 nodes and 74,808 elements was obtained. Final results of the simulations can be seen in Figs. 27, 28 and 29.

Considering that in the case of the acetabular reconstruction system and the pelvis, the same type of solid tetrahedral meshing was used (Fig. 26), the size per element was changed to 3 mm with a tolerance of 0.1 mm, mesh techniques, also weren't applied. All the models followed these specifications, as a result, a model with 187,956 nodes and 126,537 elements was obtained, shown in Figs. 30, 31 and 32.

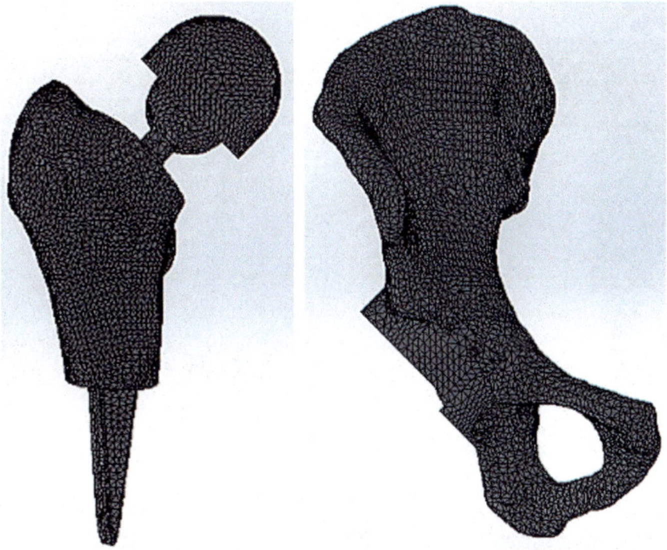

Fig. 26 3D model for finite elements of the hip joint

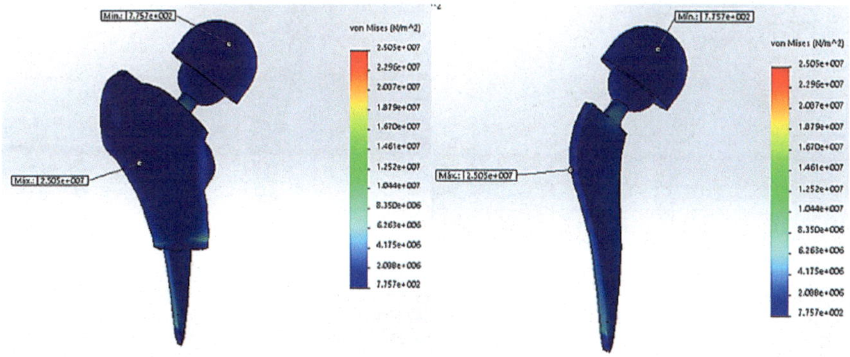

Fig. 27 Maximum stress concentration around 25.05 MPa, on the implant's body

3.2 *Border Condition, Case 2, Walking with a Stick*

According to the results published by Davy et al. (1988), on which the natural hip joint movements and forces were measured using telemetry in daily activities during the first month after the operation, it was determined that the action of the abductor muscles is minimal due to the use of the walking stick. For that reason, for this case the use of the walking stick was highly recommended in order to project the primary stability [24].

Therefore, the load on the hip is reduced on the side on which the stick is used. It is located on the left hand in order to reduce the pressure on the right femoral head,

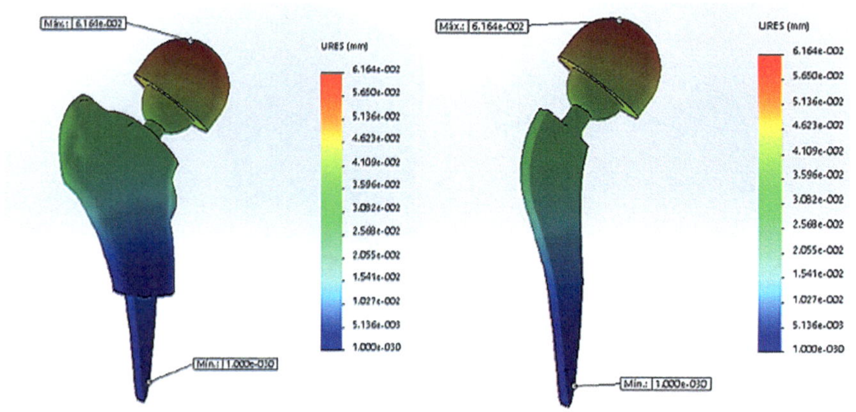

Fig. 28 Maximum micro-displacement of 61.64 μm in the cotyle impact, under pressure

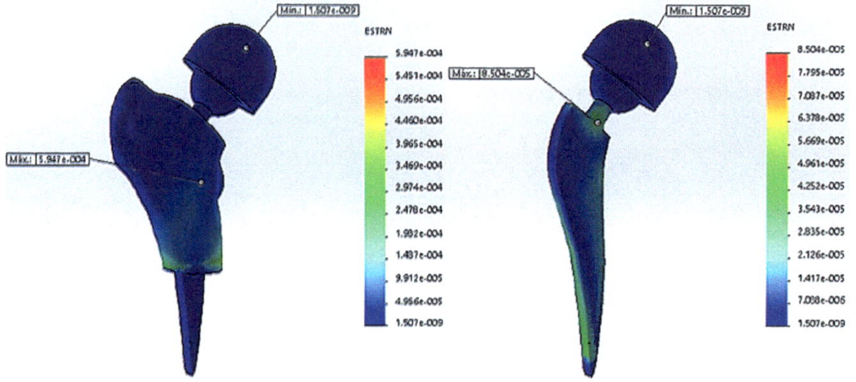

Fig. 29 Maximum unitary deformity of 5.947×10^{-4} located in the femur's diaphysis. A maximum unit deformation of 8.504×10^{-5} is observed on the Morse cone

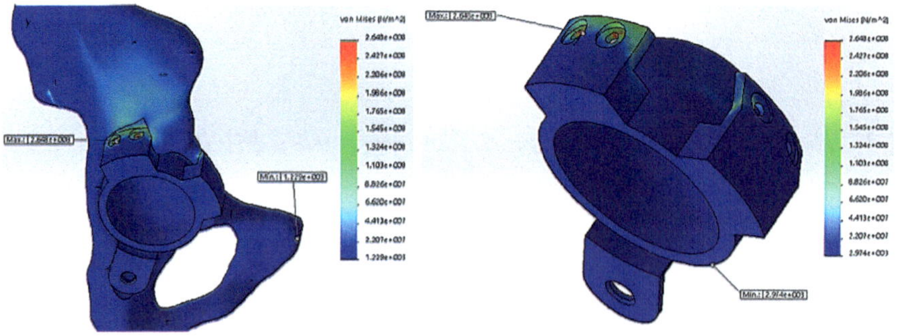

Fig. 30 Maximum stress concentration of 264.8 MPa located on the holes for the screws

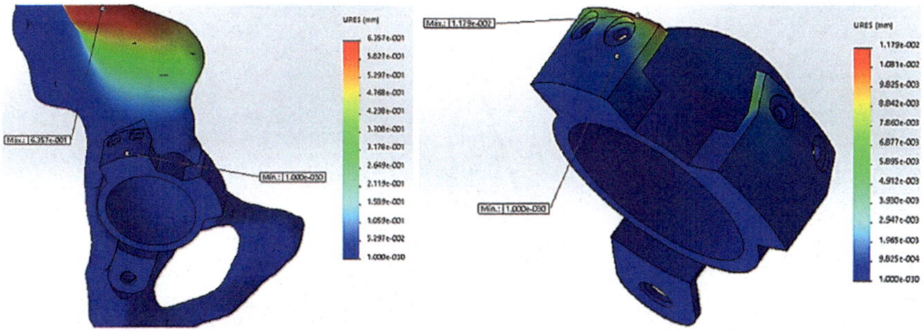

Fig. 31 Maximum micro-displacement of 635.7 μm in the right pelvis; in contrast, on the implant, there is a maximum micro-displacement of 11.79 μm

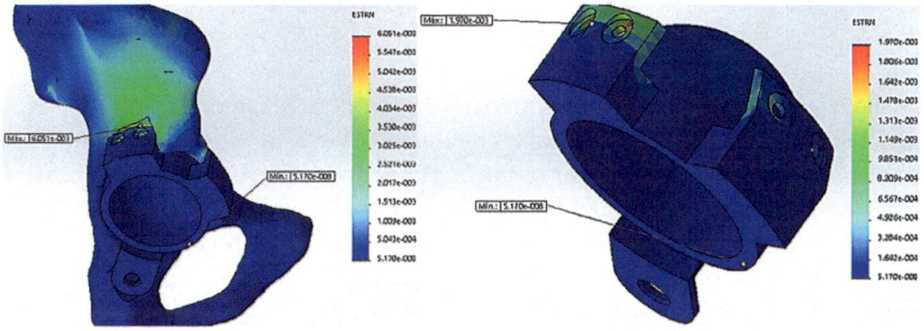

Fig. 32 Maximum unitary deformation of 6.051×10^{-3} located in the upper part of the acetabulum; while in the implant there is a maximum unitary deformation of 1.97×10^{-3} located in the area of the holes for the screws

Table 2 Pressure on the cane, Blount (1956)

Fuerza designada	Pressure on cane [kg]	Pressure on femoral head [kg]
R	0	174
1	9	99
2	14	51
3	17	29

eliminating the hobble along, as it is illustrated in Fig. 33. In turn, the support offers the traction required for the abductor's muscles and as a support for the body weight, in addition to working through a lever arm, which moderates the support to reduce the deformation on the hip.

The reaction under the stick was determined according to relative forces shown in Table 2, established by Pauwels, who estimated that, during walking phase without a support, an average person performs a static force of 174 kg on the hip. In turn, he found that the force can be reduced to 99 kg by pushing down on a support, in the opposite hand with a load of up to 9 kg, Blount (1956) [25].

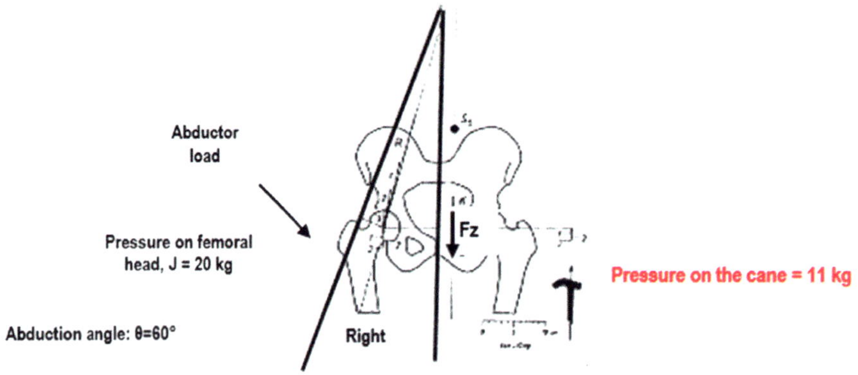

Fig. 33 Border condition: walking with a stick, used by Blount (1956) [25]

The first parameter that was determined is the force on the femoral head. This value was obtained from the data established by Inman (1947), in which it was estimated that the forces W and M, and their respective lever arms were the static compression forces on the femoral head and its magnitudes were approximately 2.4–2.6 times the body weight during unilateral support [26]. From these values, Blount (1956), estimated that an average person whose body mass is 69.84 kg when leaning with a force of 174 kg, walking with a stick held with the opposite hand reduced the force imposed on the femoral head from 174 to 29 kg, as shown in Table 2.

For this purpose, if the force imposed on the femoral head is 2.5 bigger than the body weight, then $2.5 * 48\ \text{kg} = 120\ \text{kg}$, the value will be a force assigned to the femoral head. Therefore, if this value is associated with Table 2, it is obtained that, for a load on the 11 kg heavy stick, the force on the femoral head is 20 kg (196.2 N) considering the patient's body mass of 48 kg. For the pelvis the load is the same as 20 kg (196.2 N), remembering that it transmits the weight of the body to the femoral head.

The suggested load on the hip joint of 20 kg (196.2 N) is reasonable taking into account what was reported by Hasart et al. (2011) [14], who explained that, if the reconstruction of the acetabular and femoral defects is stable, it can proceed to the total load. If the bone defects are very extensive or in case of pelvic and trochanteric insufficiency, the ideal recommendation was a partial load of 196,133 N for a 6-week period. As soon as the values were obtained, the numerical validation was conducted. For the femur with the placed implant, a model with 116,424 nodes and 74,791 elements was obtained. The results of the simulations are shown in Figs. 34, 35 and 36.

In the case of the acetabular reconstruction system and the pelvis, a model with 187,864 nodes and 126,465 elements was obtained. The results of the simulations are shown in Figs. 37, 38 and 39.

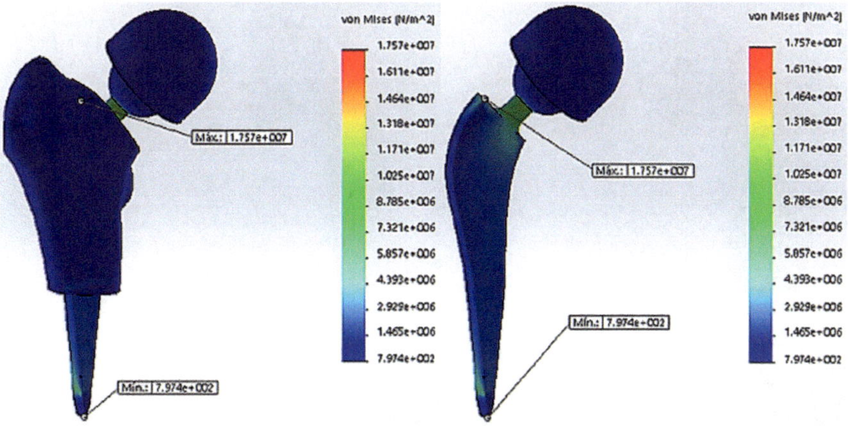

Fig. 34 Maximum stress concentration of 17.57 MPa located at the base of the Morse cone

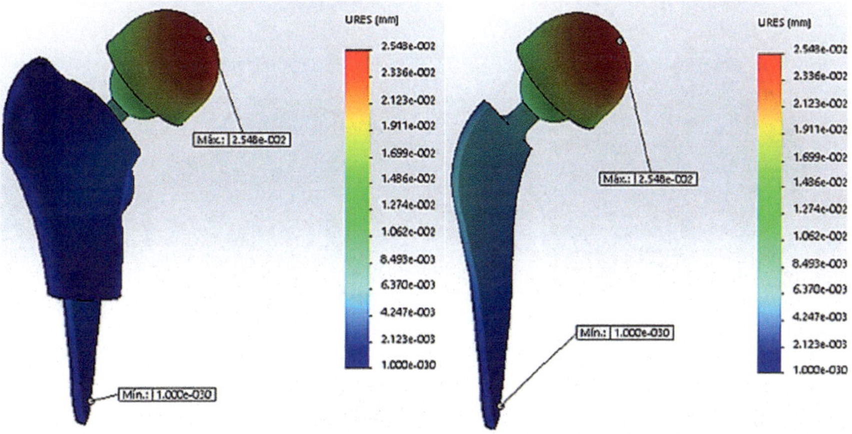

Fig. 35 Maximum micro-displacement of 25.48 μm in the cotyle impact, under pressure

3.3 Border Condition, Case 3, Walking Without a Stick

For this case, the load on the femoral head was determined from the direction of the resulting abductor muscle group of the hip that acts on the greater trochanter, forming an angle of 71°, illustrated on Fig. 40. The load obtained was a descending force of 135.18 kg (1326.11 N) that acted to the left and on the horizontal region with an angle of 75°, as a result in the equations described below.

The reaction force of the floor over the whole body is 48 kg; by adding the weight of the entire limb through its center of gravity, the X, Y components of the reaction on the hip joint and the muscular force on the greater trochanter complete the limb diagram. For this case, general forces system and equilibrium equations are:

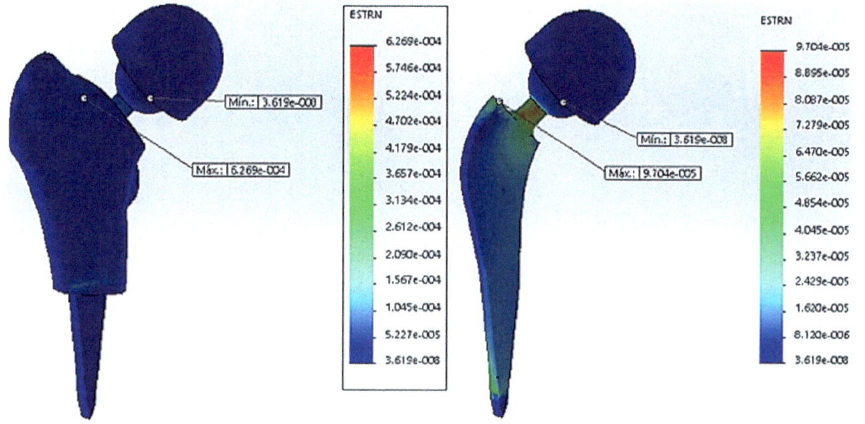

Fig. 36 Maximum unitary deformation of 6.269×10^{-4} located in the proximal area of the femur; while in the implant there is a maximum unit deformation of 9.704×10^{-5} at the base of the Morse cone

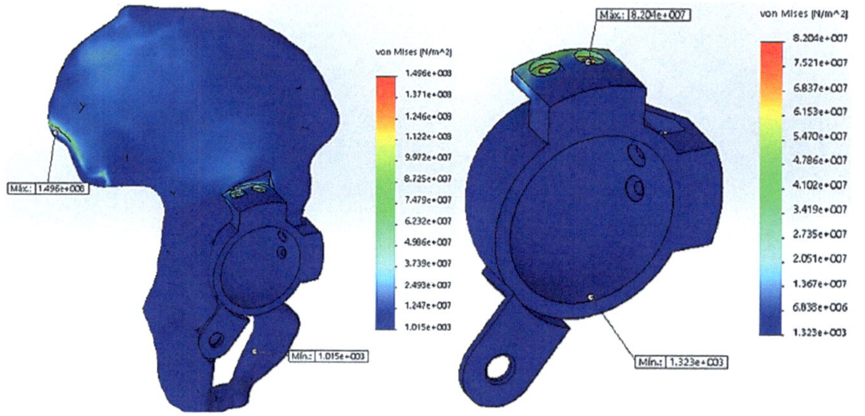

Fig. 37 Maximum stress concentration of 149.6 MPa located in the iliac crest; whereas in the implant the maximum forces located in the holes for the screws are 82.04 MPa

$$\sum Fx = 0$$
$$\sum Fy = 0$$
$$\sum M = 0$$

By substituting the equations and taking the center moment (O) at the point where the muscle force is applied to the greater trochanter, the following equation is obtained:

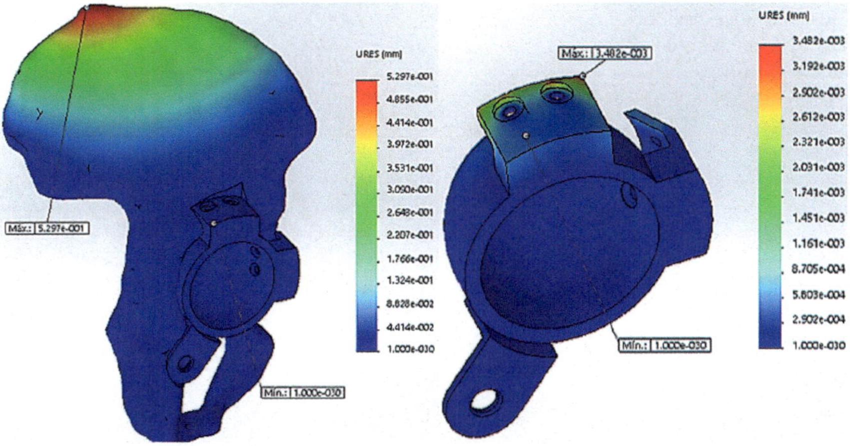

Fig. 38 Maximum micro-displacement of 52.97 μm in the right pelvis. In contrast, the implant has a maximum micro-displacement of 3482 μm

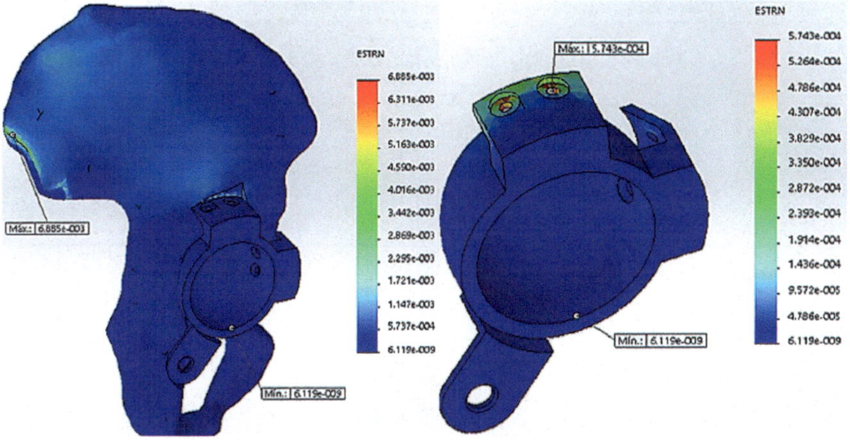

Fig. 39 Maximum unitary deformation of 6.885×10^{-3} located in the iliac crest, while in the implant there is a maximum unitary deformation of 5.743×10^{-4} located in the holes for the screws

$$\sum Fx = 0$$

$$M \cos 71^{\circ} + J_x = 0$$

As the weight of the leg is descending, the value is negative.

$$\sum Fy = 0$$

$$8\,\text{kg} + Msen71^{\circ} + (-14\text{kg}) + J_y = 0$$

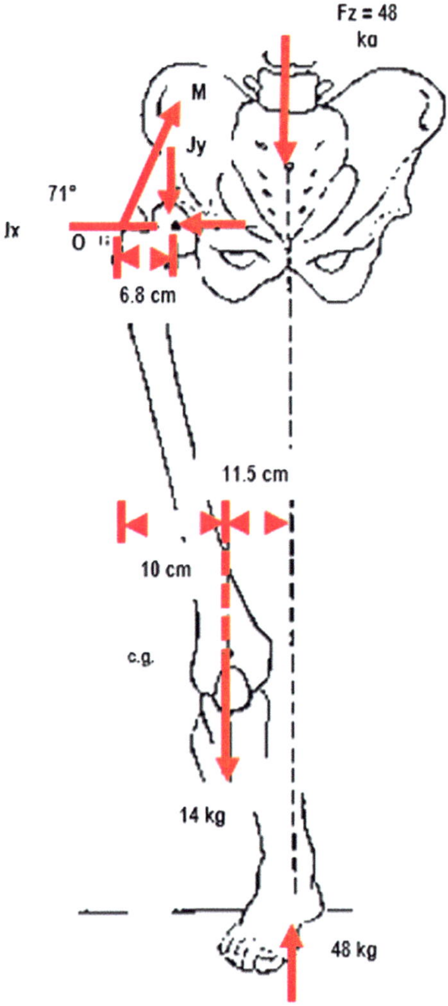

Fig. 40 Border condition, walking without a stick, Ruíz Muñoz (2016)

A negative moment in the opposite direction to the clocks hand was produced by the reaction of the floor, resulting in:

$$\sum M = 0$$

$$6.8J_y + (14\,\text{kg} * 10\,\text{cm}) + (-48\,\text{kg} * 21.5\,\text{cm}) = 0$$

$$J_y = 131\,\text{kg}$$

Consequently, Jy acts down, and the first equilibrium condition is negative:

$$\sum Fy = 0$$

$$48\,\text{kg} + Msen71^{\circ} + (-14\,\text{kg}) + (-131\,\text{kg}) = 0$$
$$M = 102.58\,\text{kg}$$

Therefore, it gives:

$$\sum Fx = 0$$
$$M\cos 71^{\circ} + J_x = 0$$
$$J_x = 33.39\,\text{kg}\ (towards\ the\ left)$$

Through the Pythagoras theorem it is obtained:

$$J = \sqrt{J_x^2 + J_y^2}$$
$$J = \sqrt{(33.39)^2 + (131)^2}$$
$$J = 135.18\,\text{kg}\ (descending\ and\ towards\ the\ left)$$

Finally,

$$\tan\theta = \frac{J_y}{J_x}$$
$$\tan\theta = \frac{131}{33.39}$$
$$\theta = 75$$

The femoral head supports a force of 135.18 kg and acts at an angle of 75° with the horizontal.

For the numerical analysis, a model with 116,424 nodes and 74,791 elements was obtained, as in the previous case. The results of the simulations are shown in Figs. 41, 42 and 43.

In the case of the acetabular reconstruction system and the pelvis, a model with 187,864 nodes and 126,465 elements of the previous case was obtained. The results of the simulations are shown in Figs. 44, 45 and 46.

3.4 Border Condition, Case 4, Carrying a Backpack

As in the previous case, the load on the femoral head was determined from the direction of the abductor muscle group acting on the greater trochanter, resulting on having 71° grades, with an additional load of 6 kg, illustrated in Fig. 47. The load obtained on the femoral head was 193.97 kg (1902.84 N), acting on the opposite direction (descent) on the left direction. It also acts on the horizontal direction with an 74° angle, shown in the calculations exposed below.

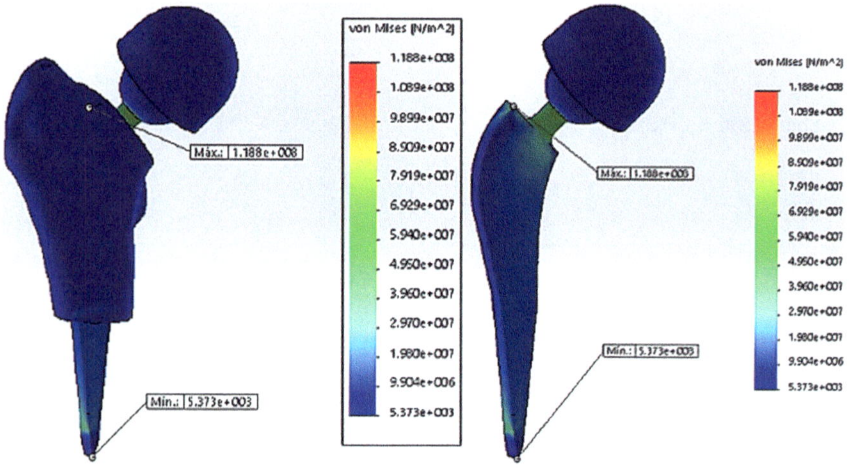

Fig. 41 Maximum stress concentration of 118.8 MPa located at the base of the Morse cone

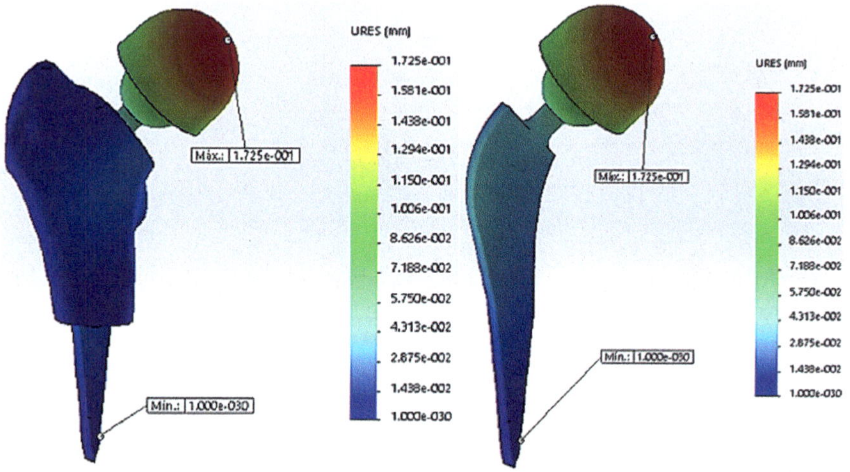

Fig. 42 Maximum micro-displacement of 172.5 μm in the cotyle impact

For this case, the patient carries a 6 kg backpack in his left hand at the time, are resting on his right foot. The center of mass of the backpack is 42 cm from its center of mass.

By taking the moments around the center of mass, its obtained:

$$6 * 42 = 54X$$

As resultant force R, is 54 kg, then:

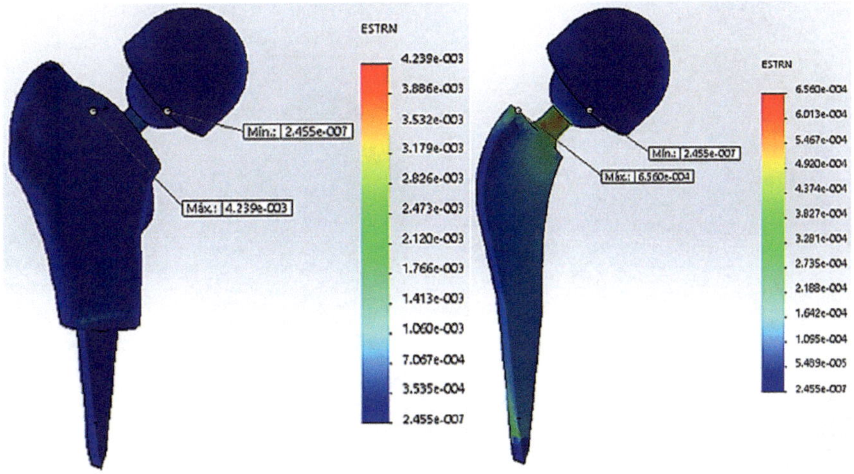

Fig. 43 Maximum unitary deformity of 4.239×10^{-3} located in the proximal area of the femur. In contrast, there is a maximum unit deformation of $6.56 \times$@@@ 10^{-4} at the base of the Morse cone

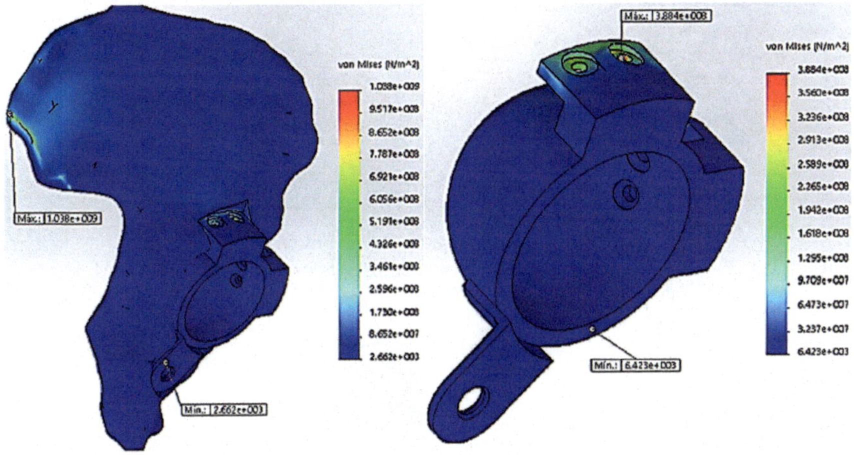

Fig. 44 Maximum stress concentration of 1038 GPa located in the iliac crest. Whereas in the case of the implant the maximum stresses are 388.4 MPa and are located in the holes for the screws

$$X = \frac{252}{54}$$
$$X = 4.67\,\text{cm}$$

Now the reaction force through the foot should be at 4.67 cm on the left direction of the body's central line. Substituting the results in the equilibrium equations, it is obtained.

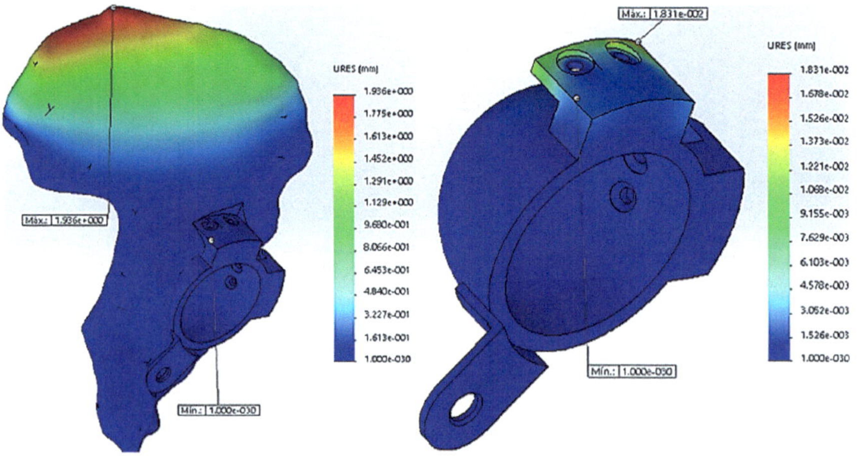

Fig. 45 Maximum micro-displacement of 1936 μm in the pelvis. In contrast, the implant has a maximum micro-displacement of 18.31 μm

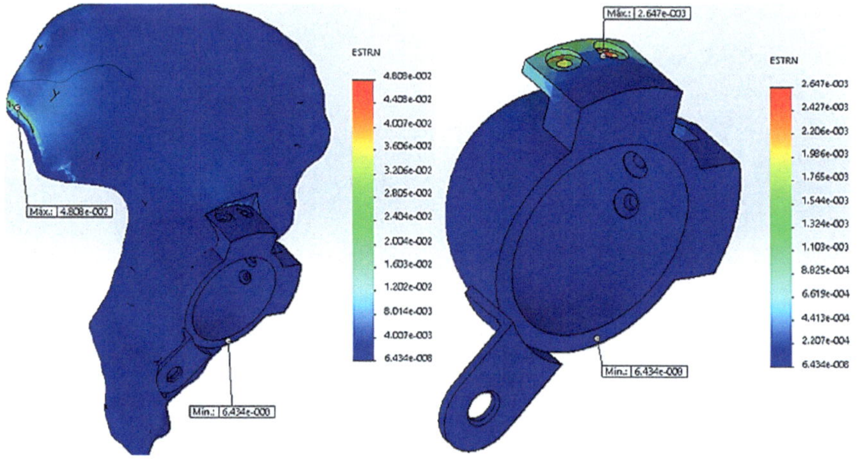

Fig. 46 Maximum unitary deformity of 4.808×10^{-2} located in the iliac crest. On the implant, there is a maximum unitary deformation of 2.647×10^{-3} located in the area of the holes for the screws

$$\sum Fx = 0$$

$$M \cos 71^\circ + J_x = 0$$

The weight of the leg is directed downwards, resulting on a negative sign.

$$\sum Fy = 0$$

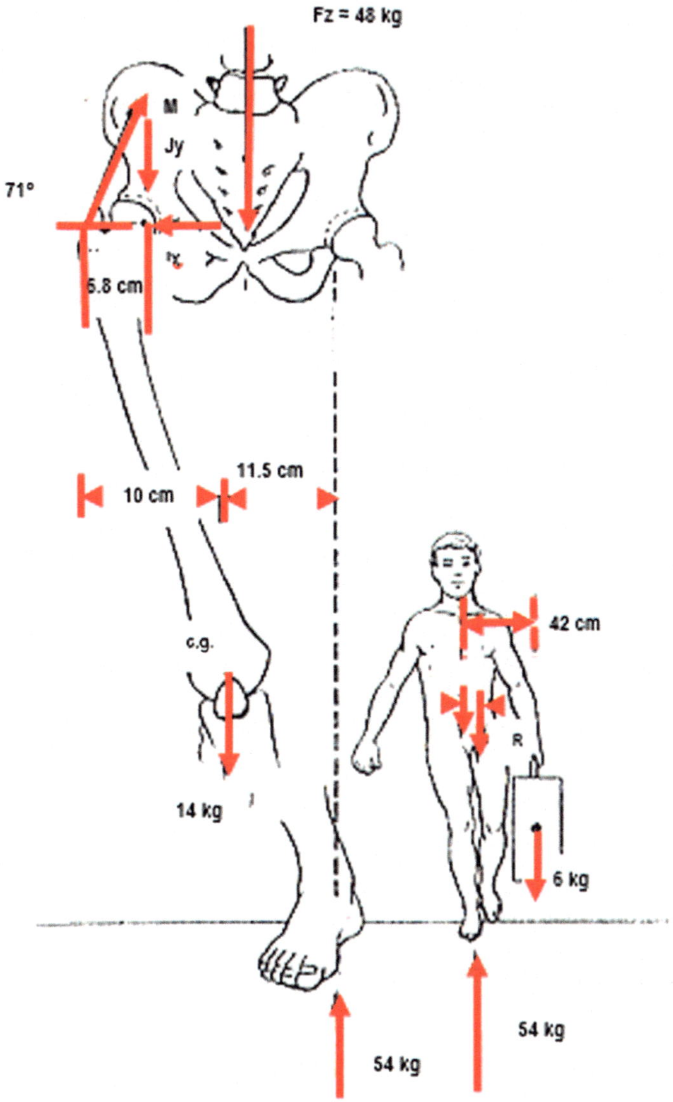

Fig. 47 Border condition: carrying a backpack, Ruiz Muñoz (2016)

$$54\,\text{kg} + Msen71^{\circ} + (-14\,\text{kg}) + J_y = 0$$

A negative sign is given to the moment produced by the reaction of the floor in the opposite direction to the clock hands.

$$\begin{aligned}
&\sum M = 0\\
&6.8J_y + (14\,\text{kg} * 10\,\text{cm}) + (-54\,\text{kg} * 26.17\,\text{cm}) = 0\\
&J_y = 187.23\ (in\ the\ direction\ of\ the\ hands\ clock)
\end{aligned}$$

As J_y acts down, the equation corresponding to the first equilibrium condition is negative.

$$\begin{aligned}
&\sum Fy = 0\\
&54\,\text{kg} + Msen71^{\circ} + (-14\,\text{kg}) + (-187.23\,\text{kg}) = 0\\
&M = 155.71\,\text{kg}
\end{aligned}$$

Therefore, it gives:

$$\begin{aligned}
&\sum Fx = 0\\
&M\cos 71^{\circ} + J_x = 0\\
&J_x = 50.69\ \text{kg}\ (towards\ the\ left)
\end{aligned}$$

Through the Pythagoras theorem it is obtained:

$$\begin{aligned}
&J = \sqrt{J_x^2 + J_y^2}\\
&J = \sqrt{(50.69)^2 + (187.23)^2}\\
&J = 193.97\ \text{kg}\ (descending\ and\ toward\ the\ left)
\end{aligned}$$

Finally,

$$\begin{aligned}
\tan\theta &= \frac{J_y}{J_x}\\
\tan\theta &= \frac{187.23}{50.69}\\
\theta &= 74^{\circ}
\end{aligned}$$

Resulting that the femoral head supports a force of 193.97 kg and acts at an angle of 74° with the horizontal.

For the numerical analysis of the femur with the placed implant, a model with 116,424 nodes and 74,791 elements was obtained, as in cases 2 and 3. The results of the simulations are shown in Figs. 48, 49 and 50. In the case of the acetabular

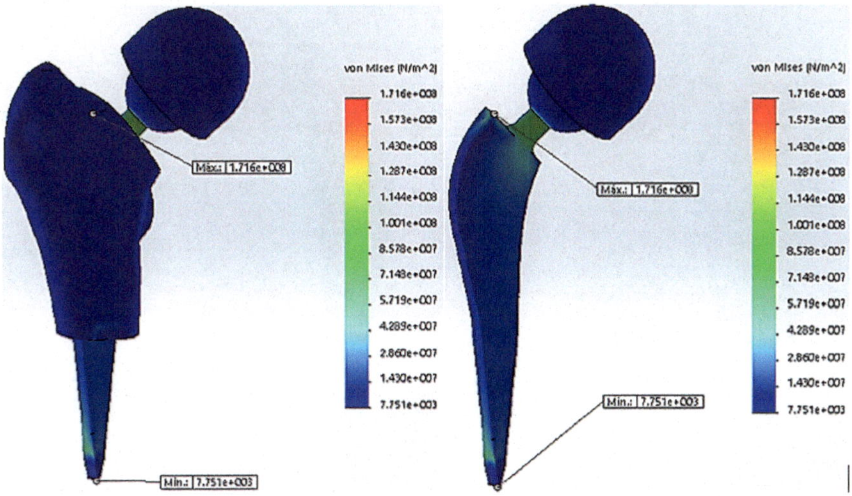

Fig. 48 Maximum stress concentration of 171.6 MPa located at the base of the Morse cone

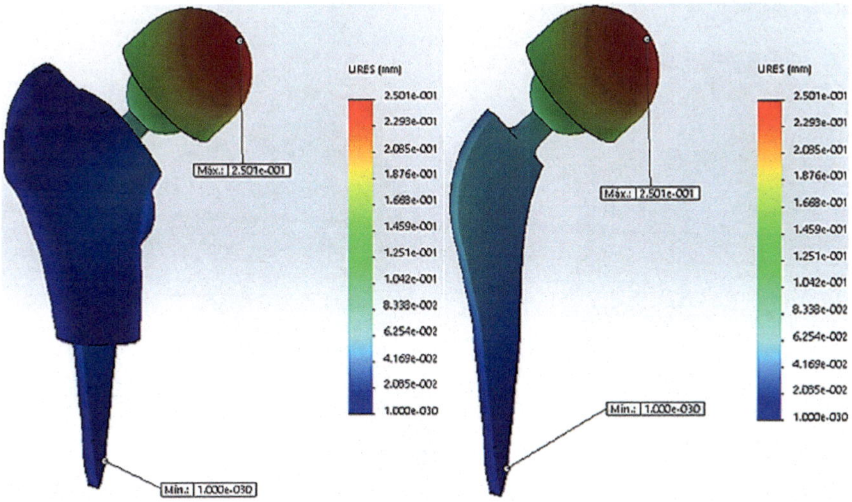

Fig. 49 Maximum micro-displacement of 250.1 μm in the cotyle impact under pressure

reconstruction system and the pelvis, a model with 187,864 nodes and 126,465 elements was obtained, as in cases 2 and 3. The results of the simulations are shown in Figs. 51, 52 and 53.

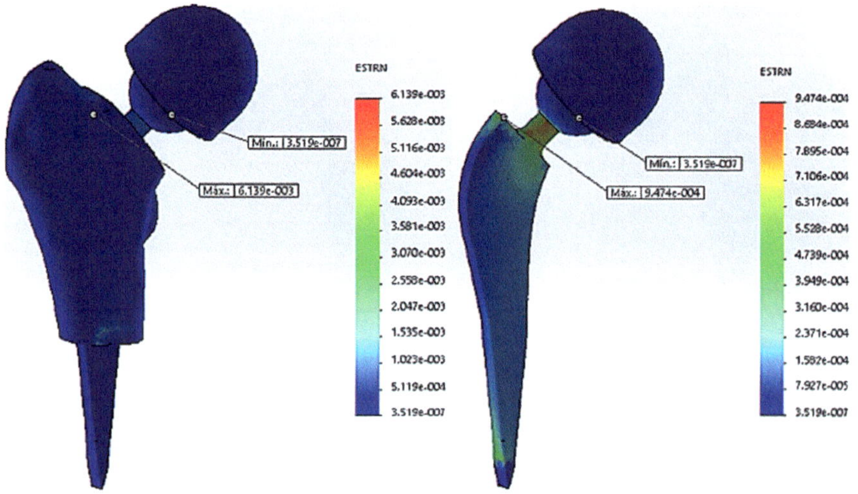

Fig. 50 Maximum unitary deformation of 6.139×10^{-3} located in the proximal area of the femur. In contrast in the implant, there is a maximum unit deformation of 9474×10^{-4} at the base of the Morse cone

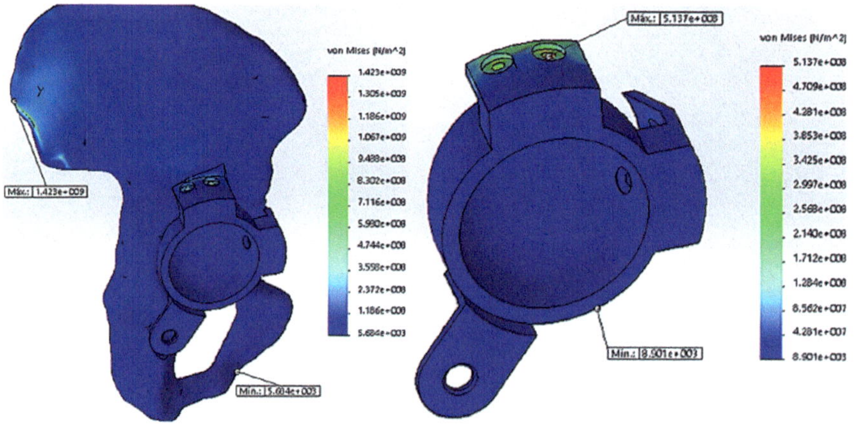

Fig. 51 Maximum stress concentration of 1423 MPa located in the iliac crest. In contrast, in the case of the implant, the maximum forces are 513.7 MPa located in the holes for the screws

4 Discussion

Based on the numerical performed analysis, the distribution of the loads in the areas near the implants of the femur and the pelvis manifested a significant reduction of the charges provoked by the natural loads of the body in each one of the cases of study.

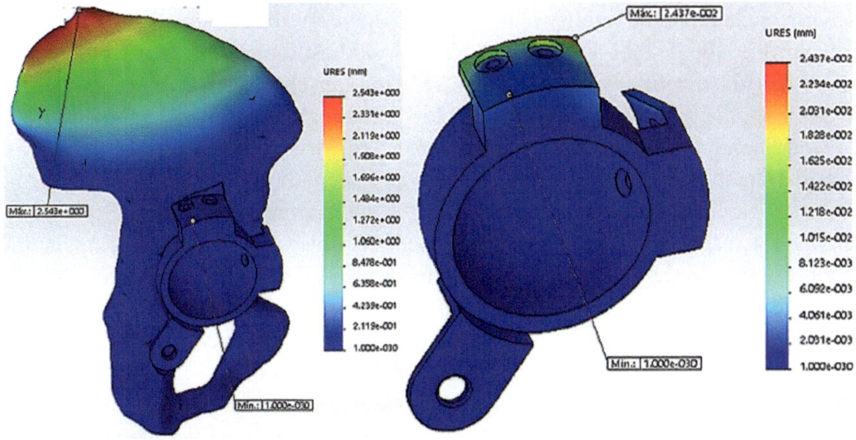

Fig. 52 Maximum micro-displacement of 2543 μm in the pelvis, whereas the implant has a maximum displacement of 24.37 μm

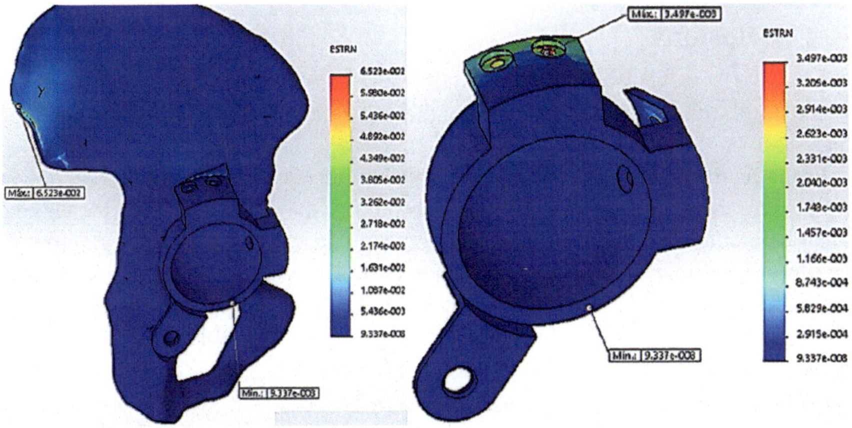

Fig. 53 Maximum unitary deformity of 6.523×10^{-2} located in the iliac crest. In contrast in the implant, there is a maximum unitary deformation of 3.497×10^{-3} located in the area of the holes for the screws

However, Weinans et al. [27] realized a study in which they estimated the effects of bone restructuring in different materials for cemented and uncemented prostheses, with a high and low modulus of elasticity using Cobalt-Chromium-Molybdenum alloy (CoCrMo), Titanium alloy (Ti-6Al-4V) and an isoelastic material. In this study it was reported that for the isoelastic material could become a favorable option only when the shape of the stem was designed to ensure that the efforts at the proximal interface were minimal. In contrast with CoCrMo and Ti-6Al-4V, it is considered that the osseointegration of the implant-bone interface tends to be much greater than in the implant interface cement and cement-bone. Taking everything into account, it was concluded that the rigidity of the CoCrMo stem for the cemented case is a better choice while the Ti-6Al-4V titanium alloy is the best option for the non-cemented case where the overall rigidity of the CoCrMo implant causes an adverse remodeling. In addition, the titanium alloy generates less bone reabsorption.

Additionally, the micro displacements obtained during the numerical simulations, considering the projection of the primary and secondary stabilization during the patient's march with and without the help of a stick, were analyzed. The specialists obtained a result of 3.185 μm displacement for the stem and 3.482 μm for the reconstruction system when the patient made use of a stick in order to walk. In comparison, the values which were obtained for both components when the patient walked without a stick were of 21.56 and 18.31 μm, respectively.

In works elaborated by Udofia et al. (2007) [5], a required limit of 50 μm in the micro displacements to guarantee osseointegration was reported. Therefore, the aforementioned values when walking with or without a stick are within that required limit, which guarantees the osseointegration of the implant-bone interface.

5 Conclusions

According to the results obtained during the realization of this work, it is concluded that.

From the damage located in the right joint of the patient's hip, caused by one of the most disabling osteoarthritides as is indicated on the clinal resume of the patient, it can be mentioned that due to the slow course and without previous symptom of the illness, multiple areas of the morphological lesion with pain are originated.

At the same time, it is mentioned that these injuries are caused by mechanical loads acting on the right hip of the patient; conducting the articulation to adopt a new form, size, and structure, as indicated by the Law of Wolff. This was also reflected in the models of the femur and pelvis generated by the ScanIP® 3.2 program, as shown in Fig. 4, where the pathological deterioration in the femoral head and pelvis can be observed, as well as the damage located in the acetabular fossa. Therefore, from the scale of Paprosky defects, such defects can be classified, as:

Type IIB, for the case of the acetabulum; for oval cavities and large defects.
Type I, for the case of the femur; for defects located in the proximal epiphyseal area of the femur, specifically in the area of the femoral head.

An important aspect to be considered was the geometry of the implant for being a critical factor that determines the stress distribution in the bone, mainly in the interface of the implant with the surroundings.

As shown in the reconstruction system that is placed in the body of the acetabulum and absorbs the most critical efforts in the holes located in the upper part, the femoral stem located in the area of the cone Morse enhances the effort reduction. The angle of conicity suggested was 6°in order to prevent the risk of rejection due to its low coefficient of friction in the interface, being of 0.15 in the first days after the operation.

Additionally, the conicity angle has the appropriate geometry to avoid the risk of rejection. It complies, that the coefficient of friction (0.15) is greater than the tangent of the taper angle (tg $6^\circ = 0.105$).

In view of this, for the selection of the biomaterials involved in the hip prosthesis, it was considered that for the surgical procedures and associated instrumentation, some must be changed with the passage of time while others must be designed to improve the osseo integration.

The rigidity of the hip prosthesis' components is an important factor that guarantees the stability of the implant. A low modulus of elasticity induces to have less reabsorption in the bone but a greater effort in the interface that originates a higher failure rate due to mechanical loosening.

Having an implant with a high modulus of elasticity, as it was for this investigation, the titanium alloy Ti-6Al-4V, caused that the efforts in the interface were minimal, ensuring that in the long term osseointegration of the implant-bone interface and avoiding adverse remodeling were maximal.

Regarding the projection of primary and secondary stability, it can be concluded that for this research the use of a stick was suggested as an efficient mechanical device, to achieve the primary stability in a period of 1–6 weeks and to reduce the load on the affected part. After those 6 weeks it is suggested to start walking without the use of the stick, giving way to the second projection for two main reasons.

Mobility in the hip can cause loss of stability of prosthesis or the stability that will require secondary integration to complete primary initial stability.

As seen in the micro displacement analyzes, these are below the required limit value of 50 μm, which guarantees osseointegration of the implant-bone interface.

For this reason, it can be concluded that the "Personalized Hip Prosthesis System" is optimal.

References

1. Rozadilla A, Mateo L, Romera M (2002) Artrosis de cadera. La medicina hoy 62(1433):31–37
2. Murray MP, Gore DR, Brewer BJ, Gardner GM, Sepic SB (1979) A comparison of the functional performance of patients with Charnley and Muller total hip replacement. Acta Orthop Scand 50:563–569
3. Olsson E, Goldie I, Wykman A (1985) Total hip replacement. Scand J Rehabil Med 18:107–116
4. Illyés A, Bejek Z, Szlávik I, Paróczai R, Kiss RM (2006) Three-dimensional gait analysis after unilateral cemented total hip arthroplasty. Phys Educ Sport 4(1):27–34
5. Nordin M, Frankel VH (2004) Biomecánica Básica del Sistema Musculoesquelético. McGraw Hill–Interamericana
6. Bronzino J (2000) Biomedical Engineering Handbook. CRC Press
7. Serieg A, Arvikar RJ (1975) The prediction of muscular load sharing and joint forces in the lower extremities during walking. J Biomech 8(2):89–102
8. Bergmann G, Deuretzbacher G, Heller M, Graichen F, Rohlmann A, Strauss J, Duda GN (2001) Hip contact forces and gait patterns from routine activities. J Biomech. 34(7):859–871
9. Chaffin B (1969) A computerized biomechanical model development of and use in studying gross body actions. J Biomech 2:429–441
10. Cowin S (2001) Bone mechanics handbook. CRC Press

11. Ruíz Muñoz OR (2016) Diseño de prótesis personalizada de cadera para paciente con coxartrosis unilateral, Tesis de maestria (en Español), Instituto Politécnico Nacional. México
12. Paprosky WG (1997) Cementless sockets: optimums and outcomes. Orthopedics 20:777–779
13. Perka C, Ludwig R (2001) Reconstruction of segmental defects during revision procedures of the acetabulum with the Burch-Schneider anti-protusio cage. J Arthroplasty 16:568–574
14. Hasart O, Perka C, Lehnigk R, Thotz S (2011) Reconstrucción de grandes defectos acetabulares con aumentos metálicos tecnología de metal trabecular. Tec Quir Otop Traumatol 20(4):228–238
15. Udofia I, Liu F, Jin Z, Roberts P, Grigoris P (2007) The initial stability and contact mechanics of a press-fit resurfacing arthroplasty of the hip. J Bone Joint Surg (Br) 89-B(2):549–556
16. Schidlo C, Becker C, Jansson V, Reflor J (1999) Change in the CCD angle and the femoral anteversion angle by hip prosthesis implantation. Z Orthop Ihre Grenzgeb 137(3):259–264
17. Schultz T, Blaha R, Gruen JD, Norman TA (2006) Cortical bone viscoelasticity and fixation strength of press-fit femoral stems: a finite element model. J Biomech Eng 128:7–12
18. Ojeda Díaz CJ (2009) Estudio de la influencia de estabilidad primaria en el diseño de vástagos de prótesis femorales personalizadas: aplicación a paciente específico, Tesis doctoral (en Español), Universidad Politécnica de Madrid
19. Norma 060. Material de Curación (Osteosintesis) (1991) Tornillos para esponjosa
20. Sánchez VM Ilizaliturri, Pariente G Mangino, Galindo J Camacho (2007) Tratamiento quirúrgico de la osteoartritis en la cadera: actualidades en artroplastía total de cadera. Reumatología clínica 3(3):57–62
21. Beltrán-Fernández JA, González Rebatú y González M, Hernández-Gómez LH, González Rebatú y González A, Urriolagoitia-Calderón G (2013) Biomechanical prosthesis design of an orbicular cranial cavity, vol 40, pp 87–94. Springer
22. Calambas Pulgarín HL (2014) Materiales compuestos de alumina-zirconia para aplicaciones biomédicas, Tesis doctoral, (en Español), Universidad Nacional de la Plata. Argentina
23. Beltrán-Fernández JA, Romo-Escalante E, López-Saucedo F, Moreno-Garibaldi P, Hernández-Gómez LH, Urriolagoitia-Calderón G, Camacho-Tapía, N (2014) Biomechanical assembled prosthesis of a temporo mandibular joint disorder using biocompatible materials, vol 54, pp 135–146. Springer
24. Davy T, Kotzar MG, Heiple KG, Goldberg WM, Berilla J, Burstein AH (1998) Telemetric force measurements across the hip after total hip arthroplasty. J Bone Surg 70(1):45–50
25. Blount WP (1956) Don't throw away the cane. J Bone Joint Surg 38-A(3):695–708
26. Inman VT (1947) Functional aspects of the abductor muscles of the hip. J Bone Joint Surg 29A:607–619
27. Weinans H, Huiskes R, Grootenboer HJ (1992) Effects of material properties of femoral hip components on bone remodeling. J Orthop Res 10(6):845–853

Biomechanical Fundamentals for Designing Machines that Allows Wear Testing

Iván González-Uribe, Liliana Gutiérrez-Lonche, Diego Ivan Islas-Jiménez, Guillermo Manuel Urriolagoitia-Calderón and Guillermo Urriolagoitia-Sosa

Abstract There are few machines developed to produce wear on a prosthesis so they are basically to evaluate the behavior of the knee and hip. Most of this equipment are either prototypes or incomplete devices, this is because this machines do not reproduce the movements of the body. The investigation presented in this article, focuses the fundaments of biomechanics to design and manufacture a machine that allows wear at the ankle joint. So that, the use of tomography or magnetic resonances with combination with computational programs allows the generation of an ankle joint model. This was realized in order to obtain all geometries of a personalized model for an endoprosthesis design. Used as a base to develop a personalized machine able to simulate movements of this joint, which consist of 3° of freedom; flection-extension, inversion-eversion and abduction-adduction. Likewise, it permit the simulation of the body load during the motion process.

Keywords Biomechanical · Wear testing · Design · Ankle join · Model

I. González-Uribe (✉) · L. Gutiérrez-Lonche · D. I. Islas-Jiménez ·
G. M. Urriolagoitia-Calderón · G. Urriolagoitia-Sosa
Instituto Politécnico Nacional, Escuela Superior de Ingeniería Mecánica Eléctrica, Sección de Estudios de Posgrado e Investigación, Unidad Profesional "Adolfo López Mateos", Edificio 5, 3er piso, Colonia Lindavista, Delegación Gustavo A. Madero, CP 07738 Mexico City, MX, Mexico
e-mail: Ing.mec.igu@hotmail.com

L. Gutiérrez-Lonche
e-mail: Lilian_lonche@yahoo.com.mx

D. I. Islas-Jiménez
e-mail: islas-jimenez_17@hotmail.com

G. M. Urriolagoitia-Calderón
e-mail: urrio332@hotmail.com

G. Urriolagoitia-Sosa
e-mail: guiurri@hotmail.com

A. Öchsner and H. Altenbach (eds.), *Engineering Design Applications II*,
Advanced Structured Materials 113, https://doi.org/10.1007/978-3-030-20801-1_22

1 Introduction

There are few machines developed to reproduce wear on a prosthesis and they are basically used to evaluate the joint behavior of the knee and hip [1]. Some machines to reproduce knee wear are; the pin on disk machine, the AMTI Boston, simulator of knee "Kansas" and the axial load system. For the hip prosthesis there is only one equipment known as triboadhesion device (Fig. 1) [2]. Up to now most of this equipment are either prototypes or incomplete devices, this is because this machines do not fully reproduce the movements of the body. The research presented in this article, focuses the fundaments of biomechanics applied to the design and manufacture of a machine that produces wear at the ankle joint.

The use of tomography or magnetic resonances complemented with computational programs allows the generation of an ankle joint model to simulate the performance of the ankle joint [3]. Through the software ScanIP® and using files of magnetic resonance, it can be obtained a shape based on points of the ankle joint. Using as well the software Delcam Power Shape® to obtain the surfaces and volumes involved in the tridimensional model of the ankle joint.

This procedure was realized in order to obtain all geometries of a personalized model for the endoprosthesis design. This leads to produce a personalized machine able to simulate movements of the ankle joint, which considers 3° of freedom;

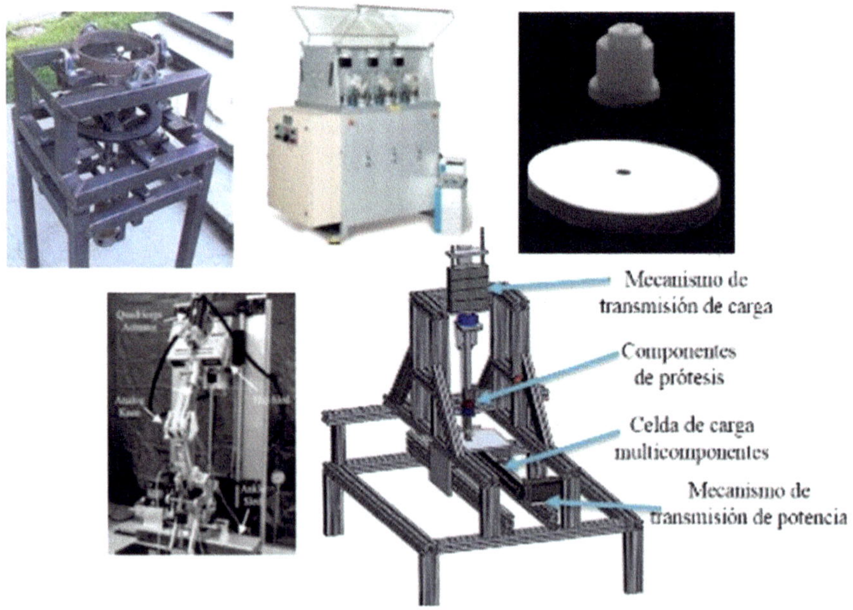

Fig. 1 Machines to reproduce wear on a prosthesis

flection-extension, inversion-eversion and abduction-adduction. Likewise, it allows to simulate the whole body loading during the motion process.

2 Methods

The first step to take in order to design a machine for wear tests for an ankle prosthesis is to obtain the geometries and surfaces existing of the joint. This process is made through the use of an ankle tomography, which is obtained from a patient 30 years old with a completely healthy joint his tomography is shown in Fig. 2.

Once the tomography is obtained, with the help of computer softwares, the computational model of this joint is developed. The first step, is to introduce the tomography files in the SCAN IP® software, to obtain a point model, later this point model is used with the Delcam Power Shape® program, to generate the hull of the articulation model and to finally it is generated the solid and three-dimensional joint model (Fig. 3).

The previous model has been obtained, by using the Solidworks® software, considering volumes, characteristic dimensions, as well as all the surfaces that form the ankle joint the design of the machine can be made to perform wear tests on the ankle prosthesis.

As soon as obtained the previous parameters, the design of the mechanical elements is performed by the fault analysis using the deformation-life method. For the application of this method, it's necessary to have the limits to the fatigue strength of the elements, the factors that affect the limits to the fatigue strength, the media and

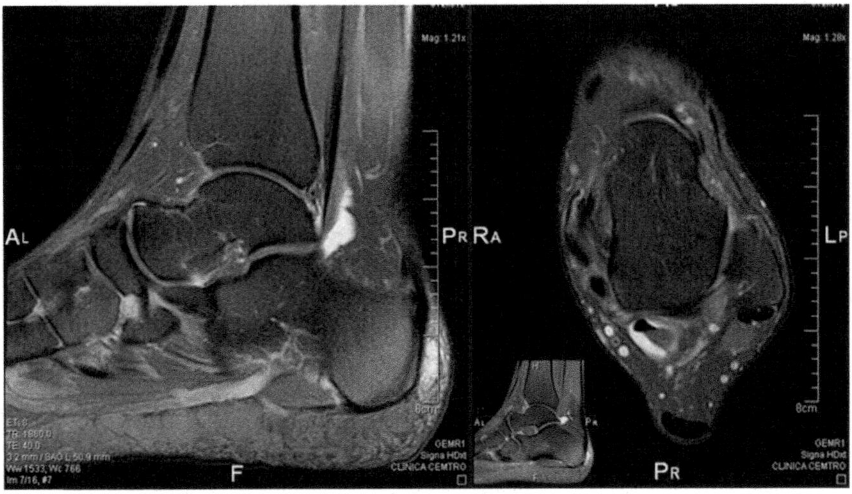

Fig. 2 Tomography of the ankle joint

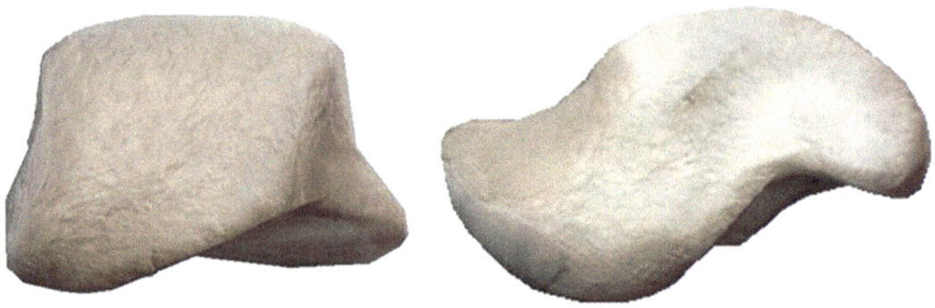

Fig. 3 Three-dimensional model of an ankle joint

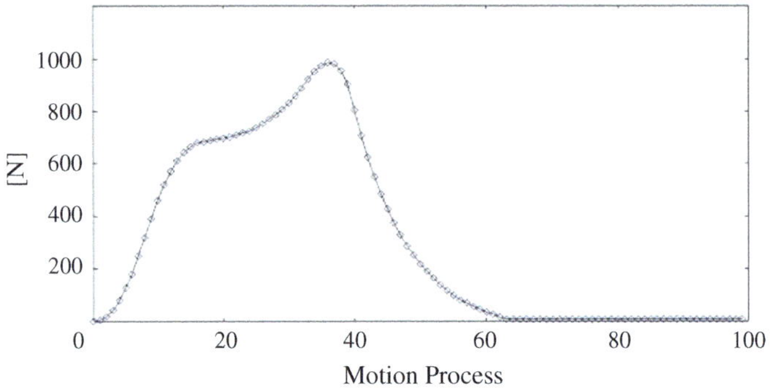

Fig. 4 Load analysis during the running process

alternating stress which can be combined with the failure theory to obtain the safety factors needed to each element.

Last but not least, the speed and load determination (Fig. 4) which are produced the motion process showed above. Because of this we can place the mechanisms in the appropriate positions and allow the movements generated by said articulation to be carried out.

3 Results

When performing the biomechanical analysis of the ankle joint, are determined the positions of all the mechanisms to perform the wear tests, once knowing these locations, the design of the elements is made by fatigue failure criterion analysis of mechanical elements. It should be noted that to design a machine is necessary the interaction of different areas such as control, electronics, manufacturing, pneumatics, among others, which will allow a better functionality of the machines, in addition we must know the parameters of speeds and loads applied in the motion process. Once joint all these areas, you get a machine that performs the movements of the joint which are Flexion–Extension, Investment–Eversion and Abduction–Adduction and which is shown in Fig. 5.

Fig. 5 Machines that allows wear testing

4 Conclusions

The use of the fundamentals of biomechanics has not been of great importance in recent years for the design of the machinery, when perform the interaction of this area with other engineering not only allows to make equipment that can simulate the movements of the body, also this application allows the production of personalized prostheses with greater functionality that improves the life of the individuals they use.

Due to the design made, wear tests can be performed on the ankle prosthesis with greater security than possible to test biocompatible materials that will achieve a longer life for the inserts that are counted. Just as the machine can reproduce all the movements of the ankle, it should be expected in the future that it is not very far, generate a prosthesis that allows the reproduction of these movements and not only the Flexion–Extension.

Acknowledgements The authors thank the Instituto Politècnico Nacional and the Consejo Nacional de Ciencia y Tecnología(CONACYT) for the support provided in the preparation of this work.

References

1. López EA, Lesso AR, Esquivel GJ (2012) Desarrollo de Bancos de Pruebas para Desgaste Usando Metodologías KANO- TRIZ. Memorias del XVIII Congreso Internacional Anual de la SOMIN., del 19 al 21 de Septiembre 2012. Salamanca Guanajuato, México
2. Rodríguez LJ, Abundez PA, Abundez PC, Chinchilla GS (2006) Evaluación del desgaste de prótesis cerámicas de cadera recubiertas con DLC por triboadhesión Parte I - Diseño de la máquina de deposición y desgaste. Revista Mexicana de Ingeniería Biomédica, pp 23–30
3. González UI (2017) Diseño y construcción de una máquina para análisis de desgaste de prótesis para la articulación de tobillo. Instituto Politécnico Nacional, Ciudad de Mexico. pp 10–18

Mechanical Design and Numerical Analysis of a Femorotibial Implant in Patients with Medial Knee Osteoarthritis

José María Aburto-Barrera, Mildred Egure-Hidalgo, Christian Díaz-León, Juan Alejandro Vázquez-Feijoo and Guillermo Urriolagoitia-Sosa

Abstract Medial knee osteoarthritis (OA) is a progressive and degenerative disease characterized by a gradual and progressive destruction of the cartilage that covers the articular surface. OA has increased dramatically in recent years, mainly among young people basically due to sports injuries and overweight. It is one of the leading causes of the musculoskeletal pain and disability in adulthood. For those cases in which the damage at the medial knee zone with a slight to moderate articular wear, it has been proposed a femorotibial implant design which should stop the progression of the degenerative process and alleviate the pain without altering the anatomy and biomechanics of the knee joint. This study is made through computational programs, using the *Finite Element Method* (*FEM*), for modeling a three-dimensional assembly which is applied to analyze an implant with a mechanical function and the behavior of the device is studied. Once the femorotibial implant is designed and modeled, a structural numerical lineal-elastic analysis through FEM was made, with the objective to determine the total displacements, as well as, the stress field on the materials. A suitable device was obtained which produced very good results for the safety factor obtained from the stress analysis of the pieces. Finally, it is expected the patient is to be able to carry on his daily activities without pain and improving the quality of life.

J. M. Aburto-Barrera (✉) · M. Egure-Hidalgo · C. Díaz-León · G. Urriolagoitia-Sosa
Unidad Profesional "Adolfo López Mateos", Escuela Superior de Ingeniería Mecánica Eléctrica, Sección de Estudios de Posgrado e Investigación, Instituto Politécnico Nacional, Edificio 5, 3er Piso, Colonia Lindavista, Delegación Gustavo A. Madero, 07738 Ciudad de México, CP, Mexico
e-mail: josemaria620@hotmail.com

M. Egure-Hidalgo
e-mail: eguremil@gmail.com

C. Díaz-León
e-mail: christiandiazleon1804@gmail.com

G. Urriolagoitia-Sosa
e-mail: guiurri@hotmail.com

J. A. Vázquez-Feijoo
Instituto Tecnológico de Tuxtla Gutiérrez, División de Posgrado E Investigación, Tuxtla Gutierrez, Chiapas, Mexico
e-mail: javazquezfeijoo@yahoo.com.mx

A. Öchsner and H. Altenbach (eds.), *Engineering Design Applications II*,
Advanced Structured Materials 113, https://doi.org/10.1007/978-3-030-20801-1_23

Keywords Knee osteoarthritis · Implant · Mechanical design · Finite element method

1 Introduction

The incidence of medial knee osteoarthritis is overly increasing. This increase has been notorious even in young people among 35 and 45 years old, so the replacement of the knee joint is being questioned more and more in the biomedical area, especially in patients who want to maintain high levels of activity.

Regardless the type of endoprostheses (total or unicompartmental), it has been shown that patients under age of 65 have an increased risk to undergoing revision by arthroplasty, compared with patients older than 75 years [1, 2].

The development of the research covers the mechanical design and the numerical analysis of a femorotibial implant with the function to mitigate the pain caused by medial knee osteoarthritis. In the mechanical implant design, each of the components is shaped coupling to the femur and tibia bones, the main mechanism of operation is determined, the dimensions of the implant are specified and the biocompatible materials to be used are established. Furthermore, a numerical analysis of the femorotibial implant was made, using the *Finite Element Method* (*FEM*) by a computational tool, establishing the boundary conditions and applying external agents to analyze the field of stresses and displacements of each one of the components of the femorotibial implant.

The biggest breakthrough biomedical advance to combat medial knee osteoarthritis has been leading by the US company *Moximed*, whom have developed a mechanical device called *Kinespring*, able to mitigate the pain caused by the medial knee OA [3]. Which, despite the great medical development and acceptance, presents two mechanical design failures that require a deeper engineering study.

The first problem was the presence of high stress concentration in the main mechanism (a helical spring) causing the rupture of the spring, and second, the presence of metallosis in the rotation axes joint of the device that causes inflammation and damage to the epithelial tissue [4, 5].

Therefore, this research is focused on the design of a femorotibial knee implant to mitigate medial osteoarthritis, analyzing the problems presented by *Kinespring* unloading system and proposing a solution that provide the functional requirements of joint unloading, without putting at risk each of the mechanical components under *Von Mises* stresses without exceeding the elastic limit of the materials [6].

2 Method

The implant is inserted through a subcutaneous tissue, this allows implantation of the device without resection of bone, muscle or ligaments, keeping the structure of

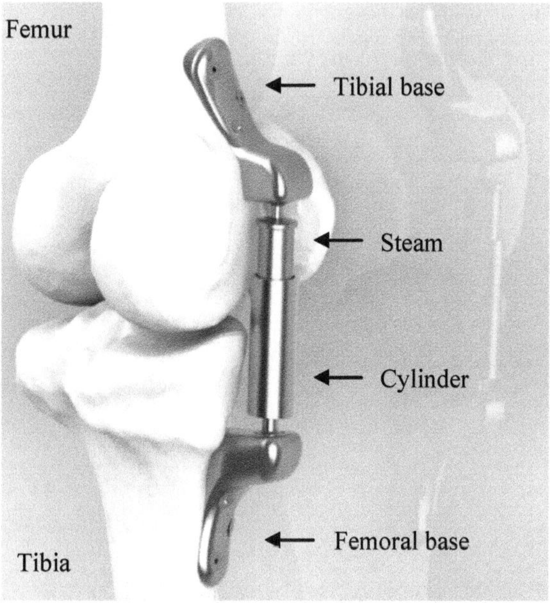

Fig. 1 Components of the medial knee implant

the knee intact. The main components of the femorotibial implant are shown (Fig. 1). The knee solid model was constructed from a computer tomography (CT) by a 3D image processing and model generation software, once the solid knee model was finished, the femoral and tibial bases were designed keeping the bones shape.

The femoral and tibial bases are attached to the periosteal cortices of the medial femur and tibia, fixed by cortical screws, both are the support of the main mechanism, where in the cylinder is placed a helical spring that produces an axial force of 127 N to the steam, enough to relieve pain in the medial knee zone according to mechanical in vitro study performed by *Kinespring* team researchers [7, 8].

The spring is inserted in the cylinder and pre-compressed with the steam (Fig. 2).

Some factors were taken into consideration according to the spring design, such as the type of wire section, the material, the external diameter and the load ratio to unloading the respective force of 127 N. It was decided to use a helical spring with a circular wire section instead a rectangular section, in order to reduce the stresses concentration owing to the geometry of the spring, where in a circular wire section the stresses concentrations are reduced notoriously in repetitive loads. The selected spring has a load ratio of 28.29 N/mm, an external diameter of 12.19 mm, a wire diameter of 2.06 mm and the selected material was Stainless Steel 316L.

It is important to evaluate the shear stress in order to ensure the working conditions of the spring to avoid the possibility of failure. Applying the dimensions and the force acting in the spring, it is possible to obtain the maximum shear stress by the Eq. 1.

The maximum shear stress was obtained by the following equation:

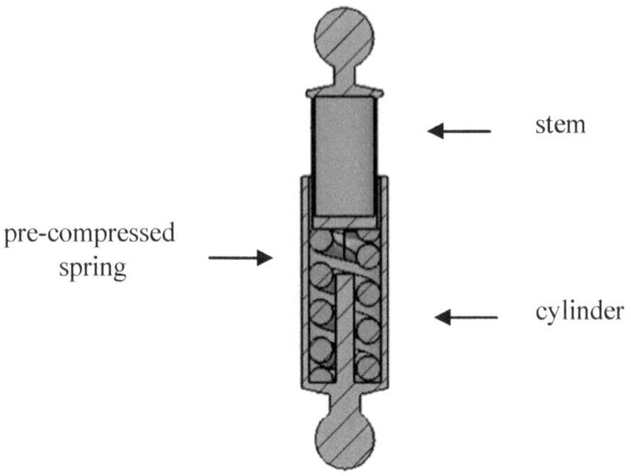

Fig. 2 3D model of section view of the spring system

$$\tau_{max} = \frac{8FD}{\pi d^3} + \frac{4F}{\pi d^2} \tag{1}$$

Obtaining:

$$\tau_{max} = 414.58\,\text{MPa} \tag{2}$$

In standard specifications for ***ASTM A313*** stainless steel it is possible to work in a severe service without failure with the maximum shear stress obtained, which means that the spring will work for more than a million cycles according to test performed with 316 L wire material [9, 10].

Below is a table with the mechanical properties for each component analyzed by FEM, these properties are essential to perform the structural analyzes to simulate the most real behavior, and if is necessary make the redesign and optimization before starting manufacturing production (Table 1).

The boundary conditions and the external agents applied in each component are shown (Table 2). In the figures are represented with *purple lines* the restriction conditions and the *red* ones represent the pressure applied on each component. As we proposed, the pressure is the relation between force and contact area. The boundary conditions indicate where rotation and displacement are restricted. In this case, the steam and cylinder is able to rotate in any direction, this allows the free movement of the knee and the implant.

3 Results

Subsequently, the numerical analyzes of each element were carried out, obtaining the *Von Mises* stresses, the total displacements and the safety factors (Table 3). The

Table 1 Mechanical properties of the main components of the femorotibial implant

Component	Material	Young modulus (MPa)	Poisson ratio	Density (g/mm^3)	Mesh Figure
Tibial base	ASTM B265 (Ti6Al4 V Gr 5)	113,800	.342	.00443	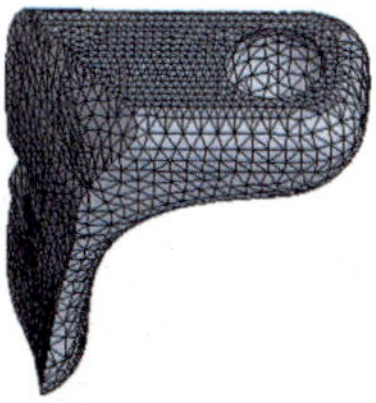
Femoral base	ASTM B265 (Ti6Al4 V Gr 5)	113,800	.342	.00443	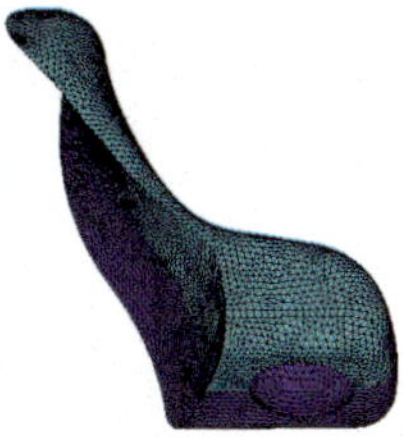
Steam	ASTM B265 (Ti6Al4 V Gr 5)	113,800	.342	.00443	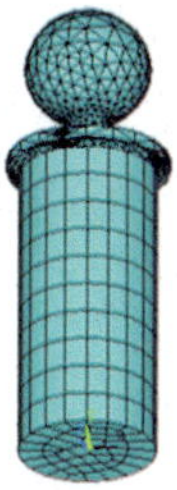
Cylinder	ASTM B265 (Ti6Al4 V Gr 5)	113,800	.342	.00443	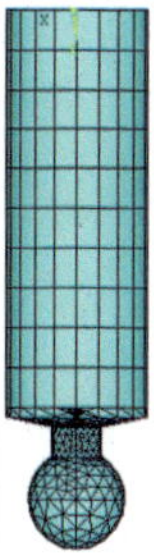

Table 2 Boundary conditions and pressure applied on each component

Components	Boundary conditions	Pressure applied (N/mm^2)	Figure
Tibial base	UX, UY, UZ = 0 Rot X, Rot Y, Rot Z = 0	.364	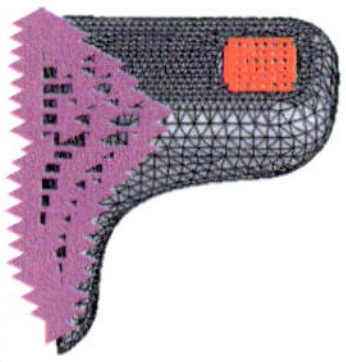
Femoral base	UX, UY, UZ = 0 Rot X, Rot Y, Rot Z = 0	.364	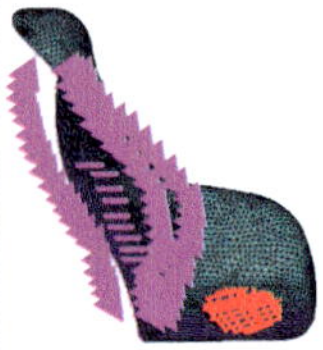
Steam	UX, UY, UZ = 0	1.34	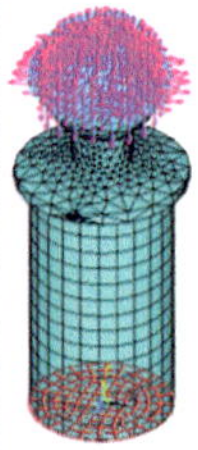
Cylinder	UX, UY, UZ = 0	1.34	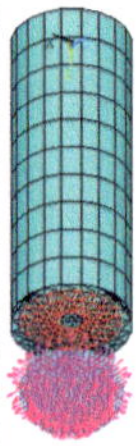

results of the elastic linear structural type analysis will allow us to predict the behavior and integrity of the materials.

In the previous Table 3 are shown the figures marked in circles where the maximum *Von Mises* stress is located.

4 Discussions

In the results, the *Von Mises* stresses shown in the components of the femorotibial implant are almost insignificant due to the maximum stresses are working under the

Table 3 *Von Mises* stresses, total displacements and safety factor of the implant

Components	Maximum stresses (MPa)	Total displacements (mm)	Safety factor	Figure
Tibial base	.995	.0051	762	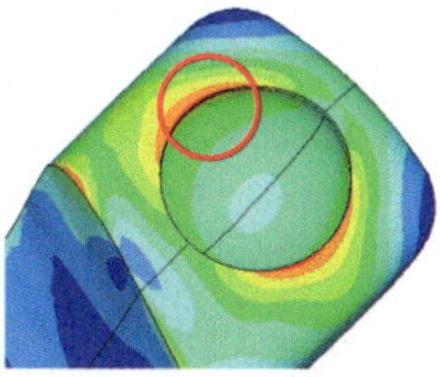
Femoral base	1.821	.0093	421	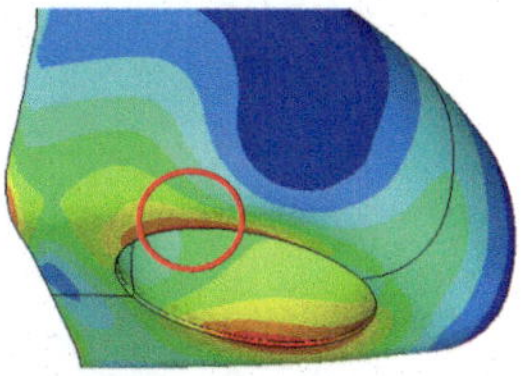
Steam	40.909	0.0021	18.5	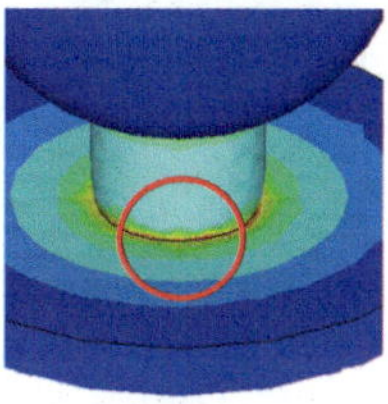
Cylinder	30.1042	0.0012	25.2	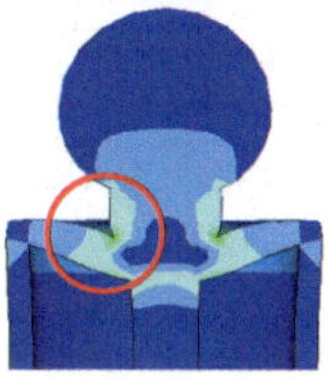

elastic linear zone which means consequently, the safety factors are higher. The force applied by the spring does not affect the integrity of the implant and assures it wont fail under normal working conditions. In the same way, the total displacements of the components are almost zero, which indicates that we are structurally in the elastic linear zone and emphasize the implant is well designed avoiding the mechanical failure under normal conditions.

According to the Table 3, the figures marked in the femoral and tibial bases, the maximum stresses are due to the free rotational movement of the cylinder, the steam over these components and the change of geometry in the corners. In the cylinder and steam also the abrupt change of geometry and thicknesses are the reason of the maximum stresses.

The safety factors in the bases are too high, so its possible to assume these components are overdesigned, but the results are statically analyzed and this allows an error range due to the implant is working dynamically regarding to the human gait.

Owing to the change of geometry on the helical spring, the external diameter of the implant turns out to be larger than expected, so the way to make the implant thinner must be looked for reducing the stretching of the epithelial tissue when the implant is fixed under it. However, the spring has a very high lifecycle but it is necessary experimental tests to estimate with more precise the lifecycle.

5 Conclusions

It is obtained a mechanical design of a femorotibial implant able to unload a medial knee osteoarthritis by a mechanical spring system with a proven force able to mitigate the pain of osteoarthritis in patients with slight to moderate radiological osteoarthritis.

The implant is only structurally analyzed, considered as linear, isotropic, continuous and homogeneous, also, it is necessary to analyze aspects such as wear, impact loads and dynamics of the human gait. In the same way, the implant just only mitigates the pain when it is pre-compressed with the knee in extension, so it wont work when going up or down stairs owing to decompression of the spring in knee flexion.

It is important to emphasize that the implant is a method to prolong the quality of life of the patient because it is a degenerative disease and for now there is no reversible cure. Likewise, the implant can be removed if you need a knee endoprosthesis later without affecting the integrity of the femoral and tibial bones.

Finally, in this development it is not mention other component which will allows to reduce the rate of wear between metal-metal components but it also very important, the cover material *Ultra High Molecular Weight Polyethylene* (UHMWPE) will reduce the wear between steam–femoral base and therefore, avoid the metallosis presented in the epithelial tissue.

Acknowledgements The authors gratefully acknowledge the financial support from the Mexican government by the Consejo Nacional de Ciencia y Tecnología and the Instituto Politécnico Nacional.

References

1. Otto R, W-Dahl A, Sundberg M, Stefansdottir A, Knutson K, Lidgren L, Nilsson C, Ranstam J, Czuba T (2014) The Swedish Knee Arthroplasty Register Report, Part 1. ISBN: 978-91-980722-7-3
2. Meehan JP, Danielsen B, HK Sunny, Jamali AA, White RH (2014) Younger age is associated with a higher risk of early periprosthetic joint infection and aseptic mechanical failure after total knee arthroplasty. J Bone Joint Surg https://doi.org/10.2106/jbjs.m.00545
3. Clifford AG, Gabriel SM, O'Connell MK, Lowe D, Miller LE, Block JE (2013) The Kinespring Knee Implant System: an implantable joint-unloading prosthesis for treatment of medial knee osteoarthritis. Med Dev Evid Res https://doi.org/10.2147/mder.s44385

4. Schüttler KF, Roessler M, Winkelmann SF, Efe T, Heyse TJ (2015) Failure of a knee joint load absorber: pain, metallosis and soft tissue damage. Musculoeskeletal J Hosp Spec Surg https://doi.org/10.1007/s11420-015-9436-x
5. Citak M, Kendoff D, O'Loughlin PF, Klatte TO, Gebauer M, Gehrke T, Haasper C (2013) Failed joint unloading implant system in the treatment of medial knee osteoarthritis. Arch Orthopaedic Trauma Surg https://doi.org/10.1007/s00402-013-1830-6
6. Wolff M, Haasper C, Zahar A, Gauck C, Gherke T, Citak M (2017) Severe metallosis and elevate chromium in serum following implantation of the joint unloading implant system. Arch Orthopaedic Trauma Surg https://doi.org/10.1007/s00402-017-2813-9
7. Gabriel SM, Clifford AG, Maloney WJ, O'Connell MK and Tornetta P (2013) Unloading the osteoarthritic knee with a novel implant system. J Appl Biomech https://doi.org/10.1123/jab.29.6.647
8. Bode G, Kloos F, Feucht MJ, Fleischer B, Südkamp N, Niemeyer P, Becker C (2017) Comparison of the efficiency of an extra articular absorber system and high tibial osteotomy for unloading the medial knee compartment: an in vitro study. Knee Surg Sports Traumatol Arthrosc https://doi.org/10.1007/s00167-016-4358-9
9. ASTM A313/ A313 M-18 (2018) Standard specification for stainless steel spring wire. ASTM International, West Conshohocken, PA, http://www.astm.org. Accessed 24 July 2018
10. Mott RL (2006) Diseño de Elementos de Máquinas. Ed. Pearson. ISBN: 9789702608127

Design and Comparative Numerical Analysis of Designs of Intramedular Telescopic Systems for the Rehabilitation of Patients with Osteogenesis Imperfecta (OI) Type III

J. L. Rueda, C. R. Torres, V. Ramírez, L. Martínez and G. Urriolagoitia

Abstract The Fassier-Duval® telescopic intramedullary (IM) system is the most used in the treatment of Osteogenesis Imperfecta (OI) to provide support and stiffness in long bones. Researchers have documented several complications that reduce the percentage of success in rehabilitations between 50 and 60%, this is because of the loss of fixation on the threaded ends with the epiphyses, the rotation and the mechanical failure of the male implant become apparent when exceeding 1/3 of length outside of the female implant causing the need of multiple post-operations on the patient. Therefore, this work proposes innovative designs with different mechanical and geometrical changes, to fight these complications, increasing the percentage of patients with a successful rehabilitation. The ASTM-1264 standard mention that the IM systems must be analyze in 4-points bending and torsion, so numerical analysis of this test were carried out to compare the stresses and strains behavior to conclude which design has the best mechanical characteristics against the actual telescopic system and thus avoid more complications. Likewise, the authors propose a change in the threads on both ends of the system. These modifications are analyzed to improve the grip between the bone and the IM system preventing injuries to the bone.

Keywords FEM · OI · Telescopic device

J. L. Rueda · C. R. Torres (✉) · V. Ramírez · G. Urriolagoitia
Unidad Profesional Adolfo López Mateos, Escuela Superior de Ingeniería Mecánica y Eléctrica, Sección de Estudios de Posgrado e Investigación, Instituto Politécnico Nacional, Edif. 5, 2do piso. Col. Lindavista, C.P. 07320, Ciudad de Mexico, Mexico
e-mail: ctorress@ipn.mx

J. L. Rueda
e-mail: gemini_92_19@hotmail.com

V. Ramírez
e-mail: vramirezv0700@gmail.com

G. Urriolagoitia
e-mail: guiurri@hotmail.com

L. Martínez
Universidad Politécnica de Madrid, INSIA, Campus Sur, Carretera de Valencia, km 7, C.P. 28031 Madrid, Spain
e-mail: luis.martinez@upm.es

A. Öchsner and H. Altenbach (eds.), *Engineering Design Applications II*, Advanced Structured Materials 113, https://doi.org/10.1007/978-3-030-20801-1_24

1 Introduction

OI also known as crystal bone disease is caused by a disorder in the production of collagen, which results in bones with low bone density and reduced mechanical strength [1]. Its treatment requires the intervention of a multidisciplinary team, due to the clinical variables that occur in the disease, among which physical rehabilitation, dental treatment, pharmacological therapy, orthopedics, and surgical intervention are the ones that stand out [2].

Throughout history, fractures in long bones have been treated by osteosynthesis. Which consists of fixing the bone segments with the help of various fastening elements, depending on the severity of the fracture, such as metal plates, intramedullary devices (nails, screws and wires), needles and pins, among others. These methods are not very efficient due to the principle used for their fixation, which consists of the penetration of the bone in different areas for the accommodation of the device used [3].

Subsequently, due to the excellent results obtained when treating fractures using intramedullary devices, it was thought to apply the same method in patients with OI. However, quickly identified the problem of these devices to treat this disease, which was that the device did not grow along with the bone, which came to cause a displacement of the nail into the bone forcing several surgeries for the rearrangement of this [4]; example of the above occurs in 1959, when Solfield and Millar published the results of the surgical technique used at the Chicago-Shriners Hospital since 1948, whose title is Fragmentation, realignment and intramedullary and rod fixation of deformities of the long bones in children, where several patients were children affected with LE, and a high percentage of the infants showed new fractures when treated with intramedullary nails [5].

Knowing the problem treated the fractures with elastic intramedullary nails, Elastic Stable Intramedullary Nail, (ESIN) for its acronym in English. However, a new problem arose in these nails to be over passed by the bone due to the growth of the infant, so to ensure proper functioning it was decided to fix the device in the distal and proximal epiphysis of the bone [6]. In the 1960s, telescopic nails were developed, which, when fixed in the epiphyses, dramatically reduced the number of surgeries needed in a child. However, there were still complications such as the fact that the child's growth exceeded the telescopic range established for the nail [2].

Initially the Bailey-Dubow ™ device was the telescopic nail used for the treatment of OI [7], documenting good success rates, this implant consists of the insertion of its two components by each of the epiphyses they are performed 2 perforations in the bone. Subsequently, the Fassier-Duval ™ device was developed, which incorporates 2 threads at both ends of the implant allowing it to be introduced through a single hole, thus reducing the number of operations to perform for insertion and removal [8]. At present, the most commonly used implants for the treatment of long bone rehabilitation in patients with OI are the two telescopic nails, although elastic or fixation intramedullary nails may also be used depending on the type of OI [2].

In the present work, different premises for the design of a telescopic nail specialized in patients with type III OI will be reviewed, to reduce the existing problems and achieve a better percentage of success in the rehabilitation of these patients. Designs of preliminary telescopic nails were made to which they were made a numerical analysis using the ASTM 1264 standard to generate comparisons between them and prevent greater consequences in the treatment where these nails are used.

2 Methodology

Different cross sections were proposed with which the problems can be attacked. However, the modification of the cross section could generate a decrease in the mechanical properties of the nail since the generation of stress concentrators is involved within the telescopic channel of the intramedullary device. The geometries proposed for the design of a new telescopic nail are that of a hexagonal polygon, a circular with tab and a semicircular as shown in Fig. 1.

First, the numerical analysis was carried out using four-point flexion conditions under the conditions established by ASTM 1264 for intramedullary devices. The material used for the analysis was a medical grade 316L Stainless Steel, from which most of the intramedullary devices are made.

To know the applied load, within the literature you can find different works done with these patients in which, based on the weight of the patient and the curvature of the bone, the loads are determined. In the study patients with whom the work was performed, there were 2 patients with OI type III of 3 and 6 years old, both weighing 11 kg. One of them with implanted telescopic nails, so the curvature of the bone is 0 °. Knowing this, it is determined that the load that must be applied within the analysis is 100 N, this simulating the most extreme case of the patient, in case it is on both legs. In Fig. 2 boundary conditions and meshing can be observed.

Results of stresses shows in Fig. 3, displacements and strain of the proposed redesigns were obtained and compared with analyzes made in the commercial telescopic device, with the purpose of comparing the results and defining if the geometric change of the core of the male implant decreases and in what degree the mechanical properties. The results are compiled in Table 1.

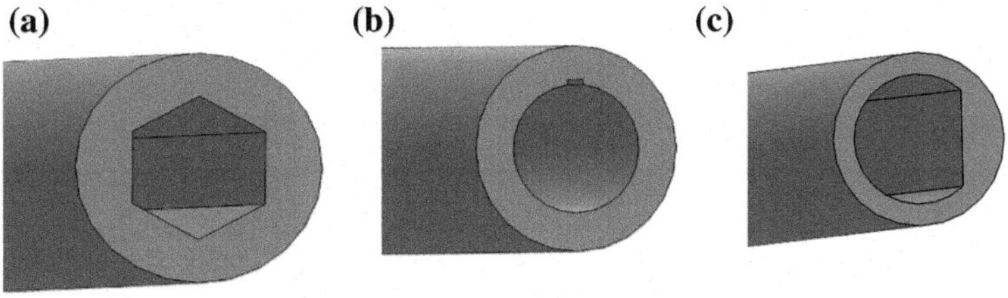

Fig. 1 Case studies. **a** Hexagonal. **b** Circular with tab. **c** Semicircular

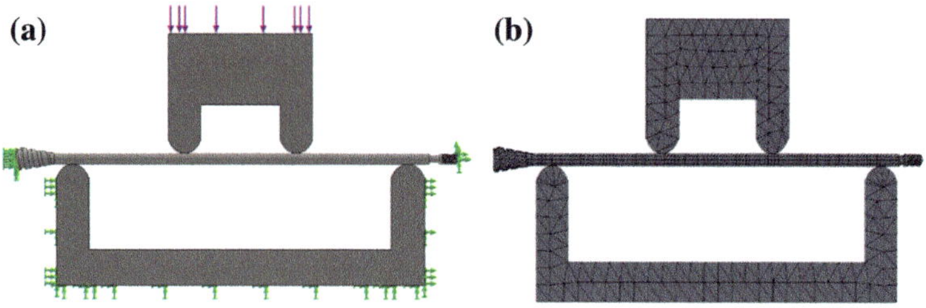

Fig. 2 **a** Boundary conditions. **b** Mesh of system

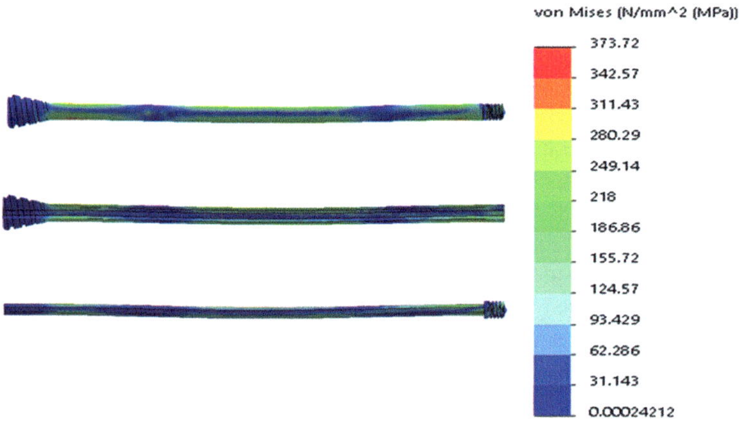

Fig. 3 Von Misses Stresses

Table 1 Result of numerical analysis in 4 points bending

Cases	Stress [σ] N/mm^2	Displacements [δ] mm	Strain [ε]
Fassier-Duval®	433.32	0.63742	0.003382
Hexagonal	373.72	0.63196	0.00103
Circular with tab	448.88	0.55171	0.0008546
Semicircular	338.844	0.5663	0.001218

Likewise, torsion tests established within the standard were carried out, to the 3 proposed designs to determine its resistance to it, and to prevent rotations that cause a bad rehabilitation in the long bones of the patients with OI.

The conditions under which the study will be carried out will consist of applying a clamping point at the end of the external implant and a torque of 0.6 N/m at the end of the internal implant. This magnitude is used because, when performing the movement of the ankle, the only element that causes resistance is the weight of the foot, which is considered equivalent to 1.5% of the total weight of a person, in this case the patient weighs 11 kg, so that the foot would equal 0.165 kg.

Considering that the foot acts as a lever with respect to the tibia, the weight of the foot is considered as a point mass in the center of the foot, which is located 2.5 cm from the axis of the tibia, generates a torque of 0.6 N/m over the male implant. Boundary conditions and the mesh of the IM system are shown in Fig. 4. The results of the numerical analysis were compiled in Table 2.

Once the numerical analysis in the telescopic device was carried out, numerical analysis was carried out considering the bone-device assembly. For this, we obtained three-dimensional models of bone sections, to analyze different thread models and determine the type of thread that would generate less stress in the bone and therefore a better fixation in the epiphyses.

In Figs. 5 and 6 can observe the virtual modeling of the telescopic nails considering the threads currently used for the treatment of the OI was carried out. Additionally, redesigns of the nails were made modifying the profile of the thread, with the purpose of comparing the stresses generated both in the screw thread, as well as in the patient's bone.

The three-dimensional model of the distal epiphysis and the male implant of the telescopic nail were imported into the SolidWorks® software, the assembly is fixed as exemplified in Fig. 8, and a cavity in the bone with the geometry of the implant.

For the analysis, the mechanical properties of the bone with OI are introduced into the computer program, which within the literature indicates that it has a Young's modulus with values close to 12 GPa, a Poisson's coefficient of 0.24 ± 0.08 and an elastic limit of 49 MPa, while the intramedullary device was assigned the mechanical properties of 316L Stainless Steel, which has a Young's modulus of 200 GPa, a Poisson's coefficient of 0.268 and an elastic limit of 170 MPa.

In the literature, it is indicated that patients with type III OI, during their childhood, the most notable growth they suffer is between 10 and 12 cm of their total height, which represents an average growth of between 3 and 4.5 cm. However, as explained

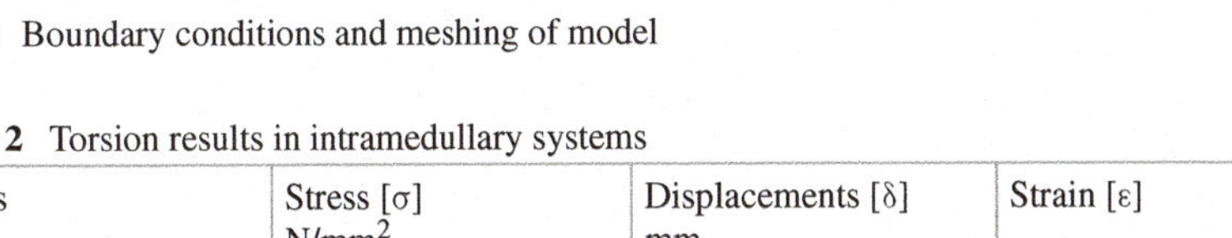

Fig. 4 Boundary conditions and meshing of model

Table 2 Torsion results in intramedullary systems

Cases	Stress [σ] N/mm^2	Displacements [δ] mm	Strain [ε]
Hexagonal	973.47	0.2321	0.003111
Circular with tab	1289.941	0.1731	0.004584
Semicircular	575.76	0.19966	0.001648

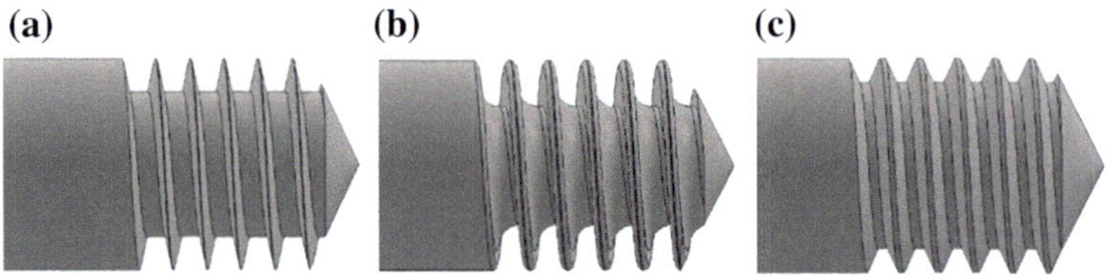

Fig. 5 Thread models for distal epiphysis. **a** Original thread. **b** Cortical thread. **c** ACME thread

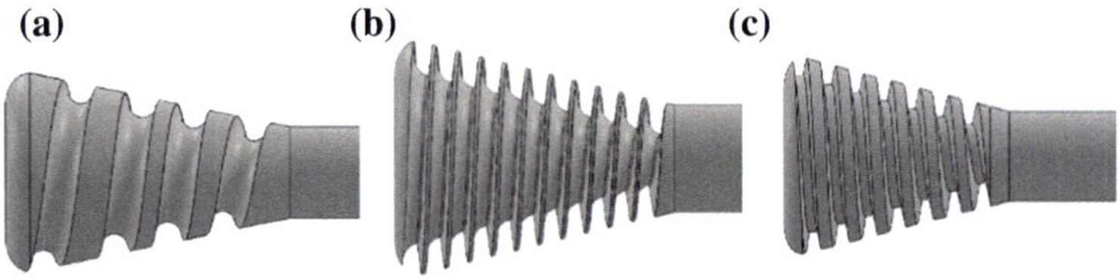

Fig. 6 Thread models for proximal epiphysis. **a** Original thread. **b** Cortical thread. **c** ACME thread

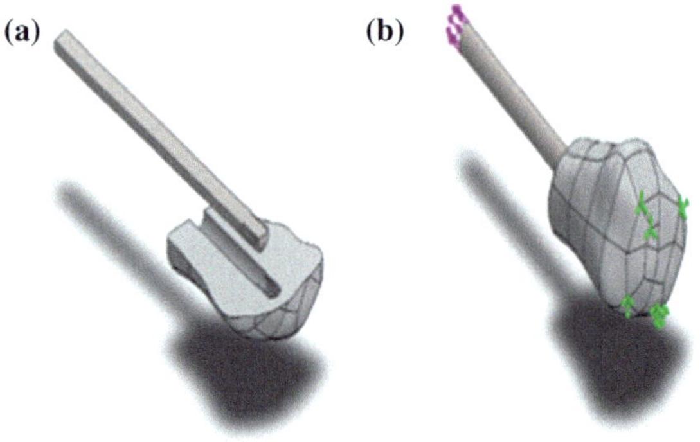

Fig. 7 Assembly model and boundary conditions

above, these devices, being telescopic are lengthening at the same pace as the bone, so that the reaction generated in the thread is minimal. For this analysis, the border conditions to be used consisted of placing a fixation on the end of the bone, while a force of 50 N was exerted, which will be the same in all the analyzes to make the comparisons, in the end opposite of the nail.

Once the elements were mesh and boundary conditions applied (Fig. 7), the analysis of the models was carried out, which were carried out with a waiting time of less than 1 min, thanks to the computer equipment used.

Finally, an analysis was carried out considering a total assembly of the bone and the device, subjecting it to flexion, considering the bone as a cantilever beam, simulating the movements of a small child with the implants placed.

Table 3 Results in distal epiphysis

Threads	Stress [σ] N/mm^2	Displacements [δ] mm	Strain [ε]
Fassier-Duval®	23.425	0.00203	0.000394
Cortical	42.579	0.002157	0.00293
ACME	18.8	0.00204	0.000495

Table 4 Results in proximal epiphysis

Threads	Stress [σ] N/mm^2	Displacements [δ] mm	Strain [ε]
Fassier-Duval®	16.139	0.00217	0.000348
ACME	30.565	0.002094	0.001355
Cortical	55.464	0.00228	0.001411

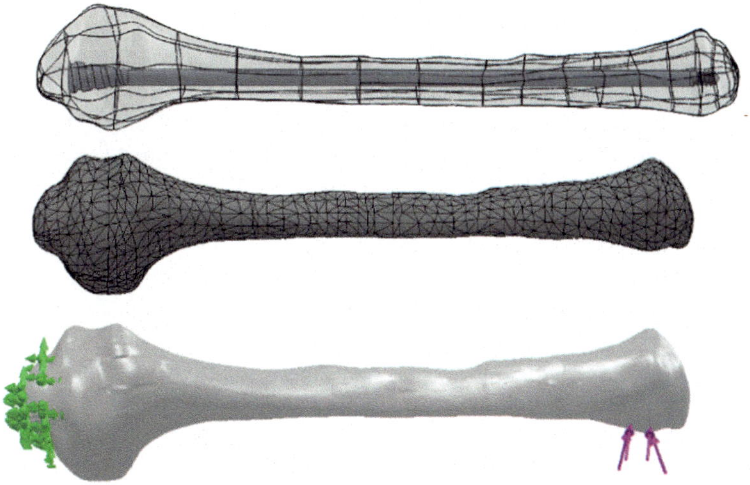

Fig. 8 Assembly model, meshing and boundary conditions

In Tables 3 and 4 are the results of the numerical analysis in both epiphysis, in order to compare the value of the stresses, displacements and strains, to conclude which one improve the fixation with the bone.

For the simulation of the movement, the boundary conditions were defined as seem in Fig. 8, which consist of fixing the tibia at its proximal epiphysis and applying a force on the distal epiphysis. The magnitude of the force is calculated based on the lever arm that is exerted on the tibia to move the infant's body, which is 350 N for the specific case of the 6-year-old patient with whom the support was provided. obtaining the virtual models.

3 Conclusion

With the data obtained it can easily be seen that the proposal with the greatest stiffness to the torsion is the semicircular nail. While the nail that undergoes a greater strain is the circular with tab, reason it is possible to be concluded that, although the circular with tab offers a greater bend stiffness, is easily deformable to the torsion.

Based on the results obtained from the proposals in both mechanical tests, it should be concluded that, within the proposed geometries, the semicircular section is the one that presents a lower stress concentration, as well as a lower strain, under this hypothesis would be the redesign with which one can attack the existing weaknesses within the commercial nail that is currently used to treat OI.

By the results of the analyzes performed on the threads, the ACME presents a smaller amount of stresses around the distal epiphysis, with which it can be argued that this type of threading without suffering stress concentrations does not damages the integrity of the bone, therefore, the attachment of the implant with the bone is greater than with the thread that currently dunes the device.

In the case of the proximal epiphysis, the type of thread with the lowest stress concentration is the one currently found in the commercial intramedullary device, although the stresses generated between the ACME thread and the commercial nail increase twice as much. section due to the conical shape of the threaded end provides greater support between the nail and the epiphysis, as far as the tibial analysis is concerned.

Finally, it is observed that the telescopic nail, provides a better resistance to displacement, as well as is responsible for receiving all the stresses that are generated during the application of the load and thus reduce stress on the bone (Fig. 9).

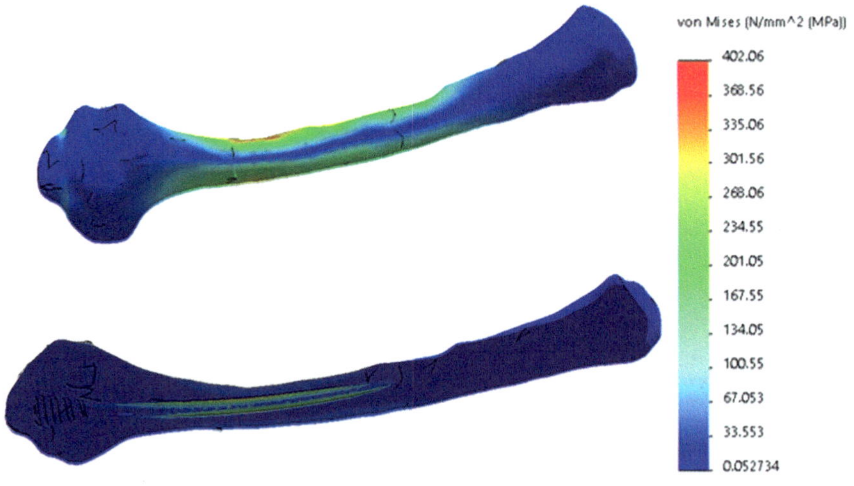

Fig. 9 Comparison between the bone with OI and bone with nail fixed

Acknowledgements The authors thanks for the support to National Polytechnic Institute (IPN) and National Council for Science and Technology (CONACYT) to make this work possible.

References

1. Shapiro JR, Byers PH, Glorieux FH, Sponseller PD (2014) Osteogenesis imperfecta: a translational approach to brittle bone disease. Academic Press, United States of America, pp 5–14
2. Rommens PM, Hessmann MH (eds) (2015) Intramedullary nailing. A comprehensive guide, Springer, pp 565–572
3. Rohrbach M, Giunta C (2012) Recessive osteogenesis imperfecta: clinical, radiological, and molecular findings. Am J Med Genet Part C Semin Med Genet 160C
4. Shapiro JR, Byers PH, Glorieux FH, Sponseller PD (2014) Osteogenesis imperfecta: a translational approach to brittle bone disease, Academic Press, United States of America, pp 421–426
5. Sofield HA, Millar A (1959) Fragmentation, realignment, and intramedullary rod fixation of deformities of the long bones in children: a ten year appraisal. J Bone Joint Surg 41
6. Porat S, Heller E, Seidman DS, Meyer S (1991) Functional results of operation in osteogenesis imperfect: elongating and non-elongating rods. J Pediatr Orthop 11
7. Ingman AM (2000) Percutaneous intramedullary fixation of trochanteric fractures of the femur. Clinical Trial New Hip Nail Injury 31:483–507
8. Sillence FVDD (2004) Osteogenesis imperfecta; clinical diagnosis, nomenclature and severity assessment. Am J Med Genet 164A:1470–1481

Finite Element Analysis of 3D Models of Upper and Lower Limbs of Mexican Patients with Osteogenesis Imperfecta (OI) Type III

V. Ramírez, C. R. Torres, J. L. Rueda, L. Martínez, B. Romero and G. M. Urriolagoitia

Abstract Osteogenesis Imperfecta (OI) is a congenital and incurable disease known as the "Crystal Bones" disease. In most cases, people with OI suffer bone fractures due to minimal causes, although in drastic cases, a bone deformation is presented, consequently causing some physical disability. In open literature, there is no information concerning models for patients belonging to the Mexican phenotype. Hence, this paper is devoted to report advances in the development of finite element models for long bones of OI patients; the model was carried out from a computerized axial tomography. As well, stress analysis is shown by taking into account specific properties of the affected bone tissue. Our results obtained will be used for manufacturing orthopaedic implants and the construction of virtual dummies for crash test.

Keywords Osteogenesis imperfecta · Biomechanics · Biomodelling

V. Ramírez · C. R. Torres (✉) · J. L. Rueda · B. Romero · G. M. Urriolagoitia
Instituto Politécnico Nacional, Escuela Superior de Ingeniería Mecánica y Eléctrica, Sección de Estudios de Posgrado e Investigación, Unidad Profesional Adolfo López Mateos, Edif. 5, 2do piso. Col. Lindavista, C.P. 07320 Ciudad de México, Mexico
e-mail: ctorress@ipn.mx

V. Ramírez
e-mail: vramirezv0700@gmail.com

J. L. Rueda
e-mail: gemini_92_19@hotmail.com

B. Romero
e-mail: romerobeatriz97@hotmail.com

G. M. Urriolagoitia
e-mail: urrio332@hotmail.com

L. Martínez
Universidad Politécnica de Madrid, INSIA, Campus Sur, Carretera de Valencia, km 7, C.P. 28031 Madrid, Spain
e-mail: luis.martinez@upm.es

A. Öchsner and H. Altenbach (eds.), *Engineering Design Applications II*, Advanced Structured Materials 113, https://doi.org/10.1007/978-3-030-20801-1_25

1 Introduction

The condition of Osteogenesis Imperfecta (OI) covers a whole of diseases produced by a heterogeneous connective tissue disorder. Usually, it is of a hereditary character and Type I collagen production is the main cause. These disorders might include: abnormalities in the bone primary structure due to the amount of collagen. As well as modifications of folding and intracellular transport or its incorporation into the bone matrix in different degrees. Together, generate an increased susceptibility to fracture [1].

Treatments for OI patients lead to entirely multidisciplinary methods due to the complexity and the variability of the disease: collagen gene mutations, the anthropometric characteristics and the social situation of each individual [2]. Currently, the anthropometry of people affected by OI have been published, in the studies it is shown a compilation between the types I, III y IV of OI. The papers report the weight, age, dimension of height, head circumference, sitting height, the length of the arms, etc. [3–7].

In the medical field, the anatomical reproduction of the human body, especially for patients with a certain pathology, has been very useful, the development of these computational models is called biomodelling. In these technique, X-rays studies are used in order to obtain trim or section images of some biological structure, the process is carried out by a computerized axial tomography (CAT) [8]. A customized cranioplasty implant design method by employing a CAT of the skull and the MIMICS® computer program was presented in 2003 [9]. In 2008, was developed in C programming language a tomographic image processing software, that generates a point cloud of the bone structure, allowing 3D reconstruction with the ProENGINEER WILDFIRE 3.0® computer program [10]. In 2012, the 3D model of an equine tooth using the AMIRA® computer program was performed [11]. Three-dimensional models representing a section of mandibular bone with implant and crown were designed in 2013, in this case, a tomography and the Inversalius® computer program were used to generate the 3D modeling [12]. In the same year, a femur 3D reconstruction was performed with the MIMICS® computer program [13]. In 2014, a methodology for the generation of a model of the lumbar segment L4-L5 of a pig and a human, was presented using the Scan IP® computer program [14]. In 2017, custom CAD models of the cortical and trabecular surfaces of a femur were developed employing the MIMICS® computer program [15].

The present work is focused on the anthropometric characterization of 3 and 6-year-old two infants with OI type III. Also, was performed the long bones 3D model of the both infants employing different computer packages with the aim to compare anthropometric measurements and 3D model. Consequently, the aim of this research is devoted to implement an effective procedure allowing to estimate the dimensions of intramedullary implants for people affected with this disease. Finally, finite element method is used to assesses the left femur of the 3-year old infant with OI type III, in order to determine the most critical areas of the bone.

2 Methodology

2.1 Anthropometric Data

The anthropometric measurements of infants with OI type III were performed with the equipment briefly described as follows: initially, a weighing scale was used to calculate the weight of children, a height-rod for measuring their height. To measure their lengths, body heights and circumferences were used an anthropometric tape, as well as a goniometer to perform depth measurements.

The measurements were classified into five groups: (a) Basic, (b) Heights, (c) Lengths, (d) Circumferences, and (e) Measurements with respect to the coronal axis [16]. The data plotted in Table 1 corresponds to the weight and height of the infants:

Table 2 shows the data corresponding to the different anthropometric heights according to the anatomical reference place for the measurement. Each case is different, in the first column, is presented the reference anthropometric measure. The result of the second column correspond to the measurement made to the infant without implants, that measurements were made of both right and left sides because the anatomy of the affected body with OI is considerably asymmetric. Finally, in the third column is shown the measurements corresponding to the infant with implants.

The corresponding lengths to the extremities of the infant's body are plotted in Table 3. The distribution of this table is the same as mentioned before in the description of Table 2.

In Table 4 are shown the measurements corresponding to the various circumferences of the infant's bodies. The columns have the same distribution as the Table 2.

Table 1 Basic anthropometric measurements of the infant with OI type III

Anthropometric measures	Infant without implants	Infant with implants
Weight (kg)	11	11
Height (cm)	70	82

Table 2 Measures anthropometric heights of the infant with OI type III

Anthropometric measures	Infant without implants [cm]		Infant with implants [cm]	
	Right	Left	Right	Left
Shoulder height	57	59	56	63
Armpits height	54	56	53	57
Waist height	40	42	33	41
Trochanteric height	32	34	32	37
Tibial height	14	16	14	16
Ankle height	4	4	4	4
Top sitting height	40		49	
Lower sitting height	24		18.5	

Table 3 Measures anthropometric lengths of the infant with OI type III

Anthropometric measures	Infant without implants [cm]		Infant with implants [cm]	
	Right	Left	Right	Left
Length of the foot	5	5	5	5
Length of the arm	12	11	16	15
Length of the forearm	22	23.5	17	19
Length of the vertex	40		48	
Length of the head	17		17.5	
Length of the chin	18		19	

Table 4 Measures anthropometric circumferences of the infant with OI type III

Anthropometric measures	Infant without implants [cm]		Infant with implants [cm]	
	Right	Left	Right	Left
Armpit-shoulder circumference	21	22	20	20
Biceps circumference	13	12.5	15.5	14
Elbow circumference	14	14	14.5	14.5
Forearm circumference	12	12	11	11
Wrist circumference	10	10	9	9
Thigh circumference	21	21	20	21.5
Ankle circumference	12	12	16.5	17.5
Knee circumference	17.5	17	14.5	14
Neck circumference	27.5		23.5	
Chest circumference	59		55	
Waist circumference	56		51	

Table 5 Anthropometric measurements to the coronal axis of the infant with OI type III

Anthropometric measures	Infant without implants [cm]	Infant with implants [cm]
Width of the chest	18	22
Width of the waist	17	16
Width of the hip	21	16
Width of the shoulder	22	23
Width of the head	14	13

In Table 5 are shown the anthropometric measurements that have been done with the coronal axis of the infant's bodies. As in the previous tables, the distribution of the columns is the same.

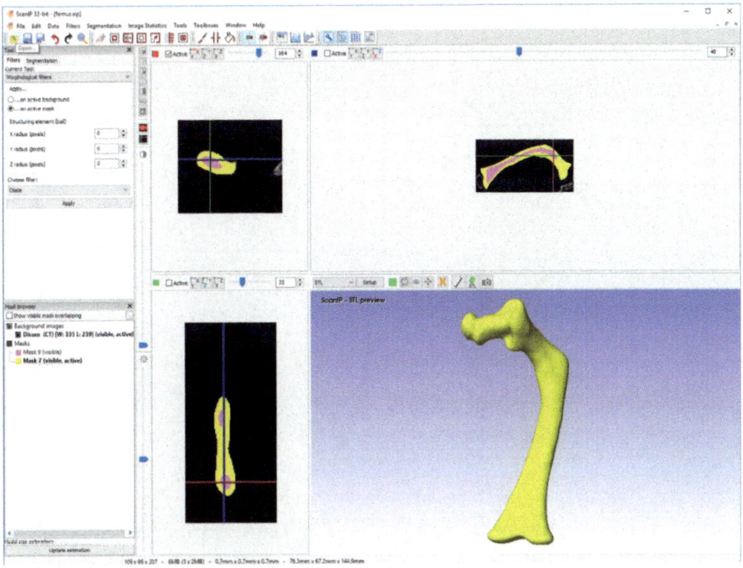

Fig. 1 Model of the trabecular tissue (pink) and cortical (yellow) of the femur of an infant with OI type III obtained from the DICOM file

2.2 *Tomographic Study*

A computerized axial tomography (CAT) was taken for both infant on a Philips Brilliance CT 64 scanner system. The virtual models corresponding to the bones of the infants affected with OI were generated from the DICOM files, which consisted of 1569 tomographic sections each.

2.3 *Procedure to Obtain the Bone Virtual Model*

The ScanIp® software was used to manipulate the 1569 tomographic sections acquired before from the infant CAT. Firstly, the work area must be delimited by marking off only the bone section required to construct, then, is determined the range of the grayscale that represents the bone tissue, in this particular case, two masks were created, one corresponding to the trabecular tissue and the other to the cortical tissue, each one was edited applying many filters (morphological, Gaussian and binary) in order to soft the surface and avoid gaps, finally, the file with extension *.stl. was generated. Figure 1 represents the working environment of ScanIp® software, also two subsections of the infant's femur without implants with OI type III were exposed; two of them correspond to the editable slice and the other two belong to the 3D virtual model of the trabecular and cortical tissues.

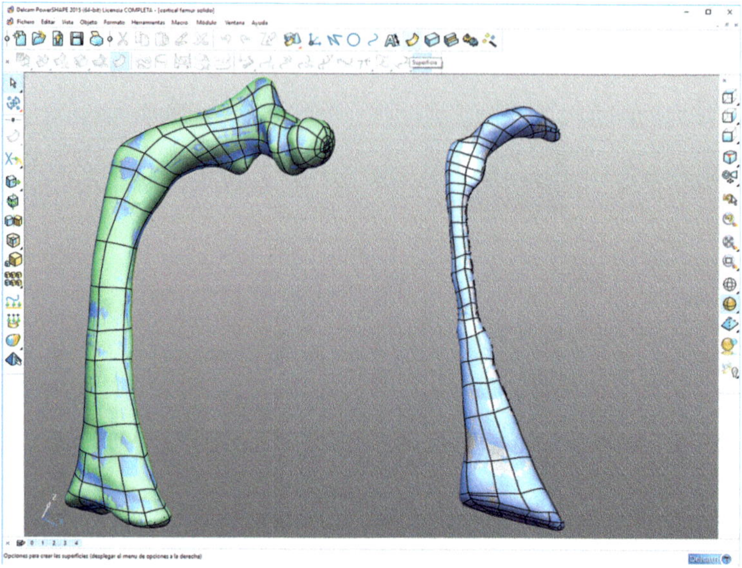

Fig. 2 Models of the meshing of tissues cortical and trabecular of the femur of an infant with OI type III without implants in the Power Shape® software

2.4 Solid Volume Creation of the Bone

Once the tissue model of the femur was obtained, the following step is to clean deeper and soft was carried out with the help of PowerShape®-e Student Edition, this step allowed to correct cavities generated during the creation of the bone system by the modeling in 3D.

Besides, with this computer program, it is possible to generate a cap for delimiting the volume of a desired work space, as well as eliminates areas, points and lines that can affect or create more than one volume within the piece. Figure 2 depicts the working environment of the PowerShape® software, displaying a section where bone tissue models are independently edited.

Finally, the model files which can be identified by programs CAD and analysis using the finite element method, were converted to extension communally named parasolid. The same procedure was used for each reconstruction of both infants' long bones affected by OI Type III with and without implants.

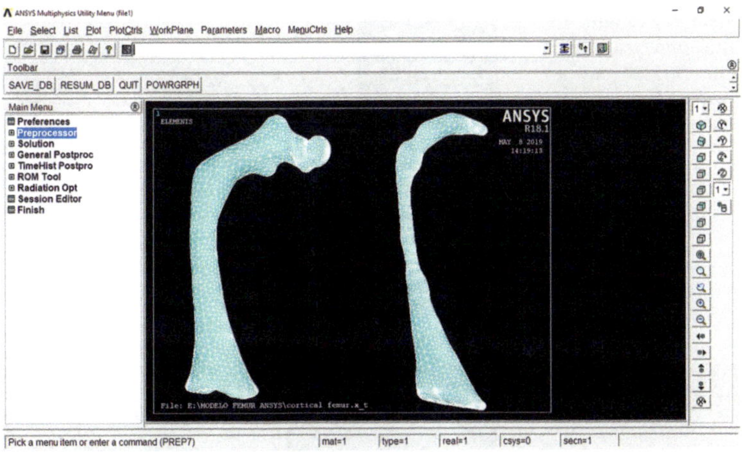

Fig. 3 Files with extension *.parasolid of tissues, cortical and trabecular of the femur of an infant with OI type III

2.5 3D Right Femur Model of the Patient with OI and Numerical Evaluation

Files with extension parasolid were imported in the computer program employed to perform the numerical analysis. Subsequently, the models were configured in order to be recognized as solid volumes, configured to be recognized as solid volumes, this process can be clearly seen in Fig. 3.

Structural analysis of the femur model was performed taking into account a continuous and isotropic behavior are considered the biomechanical properties bones with OI shown in Table 6.

Finite element selected for the complete 3D model of the femur was SOLID186, 3-D Structural Solid 20 Nodes, outstanding to this element can have any spatial orientation. Subsequent, to discretized model, were restricted all the degrees of freedom in the distal epiphysis of femur of this model (Fig. 4).

The model of distributed load of Pauwells for the analysis of the femur, used It is the study on the frontal plane of the system of stabilization of the hip in the monopodal support [21].

Table 6 Properties of the cortical tissue of a bone with OI [17–20]

Tissue	*Young* Module (GPa)	*Poisson* Coefficient
Cortical	12	0.28
Trabecular	3	0.3

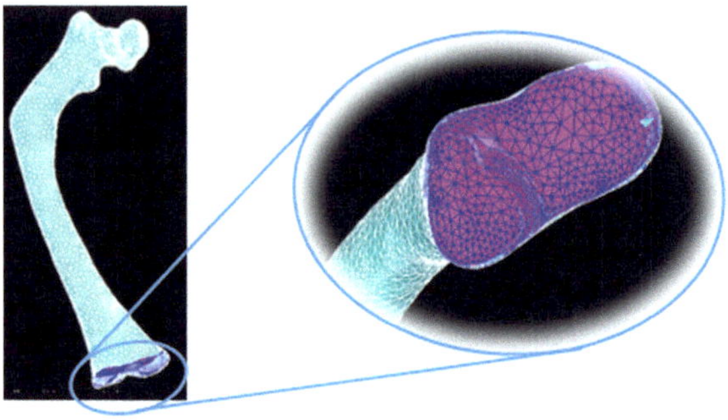

Fig. 4 Boundary condition on the distal epiphysis of the femur, restricting movement and rotation in all the axes X, Y, Z

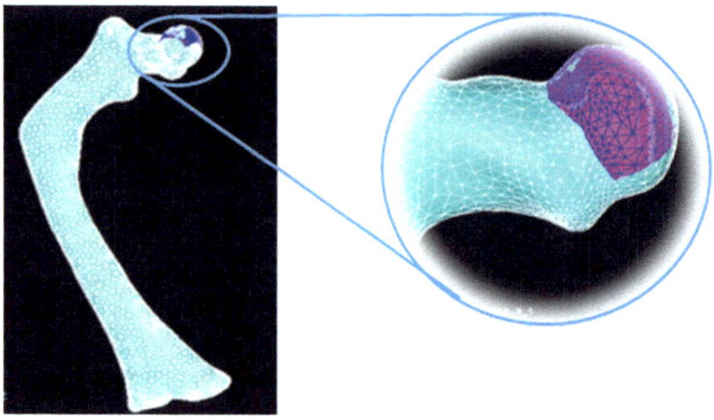

Fig. 5 Force applied on the proximal epiphysis of the femur with value of 431.64 N

Given the weight of the patient with OI 11 kg, the force exerted by abductor muscles will be 4 times the weight of the patient, i.e., 44 kg equivalent to 431.64 N. the force applied on the proximal epiphysis of the femur where is considered the femoral head [22] (Fig. 5).

3 Results

Was performed a comparison between of the 3D reconstruction of the long bones of infants with OI without and with implants. In Table 7 are shown the biomodels corresponding to of the humerus, femurs and shinbones of infants that were developed with the procedure described above.

First numerical analysis by the finite element method of the femur of the infant with OI type III without implants.

The theory of the maximum energy of distortion, also known as the theory of fault of von Mises only presents positive values since it does not indicate compression or tension areas but areas where there is a greater concentration of power. The Red areas shown in Fig. 6 exhibit the most critical points of the bone.

Table 7 Models of bony limbs of children with OI type III

Bone tip	Infant without implants	Infant with implants
Shinbone right	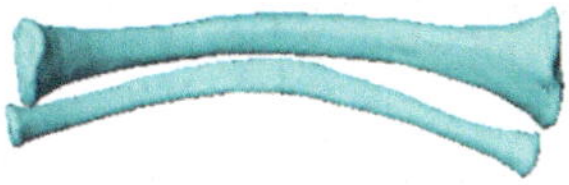	
Femur right		
Humerus right		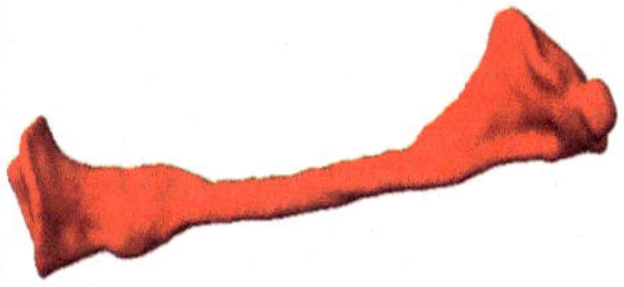

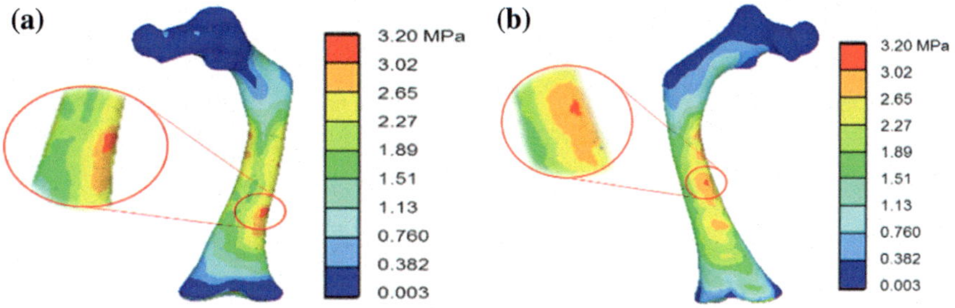

Fig. 6 The analysis of finite element results. **a** Rear view of the von Mises stress, **b** Frontal view of the von Mises stress

4 Discussions

In the present work was developed a methodology different from the open literature reported, this methodology allows to analyses the bone tissue of infants affected with OI type III. As mentioned above, several software has been used to process DICOM files.

In this study is used the Scan Ip® software that has a working environment that allows 3D reconstruction of the bone tissue completely or by sections, as well as the export to CAD programs. The 3D reconstruction of the bone presents great complexity for its interpretation, since, the bone tissue affected with OI is more porous and the Hounsfield Units in the tomographic images are dispersed due to their low density.

The result of the analysis shows that the critical points are located in the central area of the femoral diaphysis. A study developed by Friz [23] also performed a numerical analysis of a femur and, as in this paper, showed that the central area of the diaphysis is the most critical and prone to injury.

5 Conclusions

In the present work anthropometric measurements were performed at patients with OI, at the same time, they were conducted, tomographic studies that facilitated bone models generation and processing for numerical analysis. 3D models reconstructed from the TAC's of the skeletal system of infants with OI type III are displayed in order to observe the bony characteristics infants affected with OI and the influence of the implants on the bones. Also, infants with OI type III are more often placed intramedullary nails, due to the clinical nature of the illness that tend to fracture long bones, continuously humerus, radius, ulna, shinbone and femur.

Reflecting the structural behaviour of both tissues present in long bones, in the latter obtaining an effort of Von Mises of 3.40 MPa.

Acknowledgements The authors thanks for the support to National Polytechnic Institute (IPN) and National Council for Science and Technology (CONACYT) to make this work possible.

References

1. Dragan P, Anticevic D, Barisic I, Hudetz D, Ivkovic A (2014) Osteogenesis imperfecta—multi-systemic and life-long disease that affects whole family. Collegium Antropologicum 38:767–772
2. Fano V, Rodríguez CM, Del Pino M, Buceta S, Obregón MG, Primomo C, Gracía H, Miscione H, Lejarraga H (2010) Osteogenesis imperfecta. Clinical, functional and multidisciplinary evaluation of 65 patients. Anales de Pediatría 72(5):324–30
3. Aglan MS, Zaki ME, Hosny L, El-Houssini R, Oteify G, Temtamy SA (2012) Anthropometric measurements in egyptian patients with osteogenesis imperfect. Am J Med Genet 158A(11):2714-2718
4. Fano V, Rodríguez CM, Del Pino M, Buceta S, Obregón MG, Primomo C (2010) Osteogenesis imperfecta. Clinical, functional and multidisciplinary evaluation of 65 patients. Anales de Pediatría 72(5):324–330
5. Lund AM, Müller J, Skovby F (1999) Anthropometry of patients with osteogenesis imperfect. Arch Dis Childhood 80(6):524–528
6. LoMauro A, Pochintesta S, Romei M, D'Angelo MG, Pedotti A, Turconi AC, Aliverti A (2012) Rib cage deformities alter respiratory muscle action and chest wall function in patients with severe osteogenesis imperfecta. Plos One 7(4):359–365
7. Chang PC, Lin SY, Hsu KH (2007) The craniofacial characteristics of osteogenesis imperfecta patients. Eur J Orthod 29(3):232–237
8. Oliveira M, Sooraj N, Dias AG, Lopes MA, Azevedo L, Zenha H, Costa H, Santos JD (2008) 3-D biomodelling technology for maxilofacial reconstruction. Mater Sci Eng, C 28:1347–1351

9. Hieu LC, Bohez E, Vander J, Phien HN, Vatcharaporn E, Binh PH, Oris P (2003) Design for medical rapid prototyping of cranioplasty implants. Rapid Prototyping J 9(3):175–186
10. Isaza JF, Correa S (2008) Methodology for 3D reconstruction of craniofacial structures and their use in finite element method. Ingeniería y Ciencia 4(7):129–149
11. Cordes V, Gardemin M, Lüpke M, Seifert H, Borchers L, Staszyk C (2012) Finite element analysis in 3-D models of equine cheek teeth. Vet J 193:391–396
12. Santiago JF, Piza E, Ramos F, Perri de Carvalho PS (2013) Stress analysis in bone tissue around single implants with different diameters and veneering materials: a 3-D finite element study. Mat Sci Eng C 33:4700–4714
13. Zhang Y, Zhong W, Zhu H, Chen Y, Xu L, Zhu J (2013) Establishing the 3-D finite element solid model of femurs in partial by volume rendering. Int J Surg 11:930–934
14. Fuerte A, Rodríguez RG, Susarrey O, Merchan EA, Sandoval JM, Pérez E (2014) Application of 3D modeling methodology using CT Scans for numerical analysis. Revista Facultad de Ingeniería Universidad de Antioquía 72:116–126
15. Bosh JA, Ramos YM, Estrada R, Morales Y, Beltran RF (2017) Cad models custom cortical surfaces and a femoral trabecular human from tomographic images. Revista Iberoamericana de Ingeniería Mecánica 21(1):45–51
16. Chamorro GR (2005) Manual de Antropometría, 1ra. Edición, Wanceulen, España, pp 8–25
17. Imbert L, Aurégan JC, Pernelle K, Hoc T (2014) Mechanical and mineral properties of osteogenesis imperfecta human bones at the tissue level. Bone 65:18–24
18. Philippe BJ, Baron C, Pithioux M, Launay F, Chabrand P, Lasaygues P (2014) In vitro ultrasonic and mechanic characterization of the modulus of elasticity of children cortical bone. Ultrasonics 54:1270–1276
19. Albert C, Jameson J, Toth JM, Smith P, Harris G (2013) Bone properties by nanoindentation in mild and severe osteogenesis imperfecta. Clin Biomech 28:110–116
20. Imbert L, Aurégan JC, Pernelle K, Hoc T (2015) Microstructure and compressive mechanical properties of cortical bone in children with osteogenesis imperfecta treated with bisphosphonates compared with healthy children. J Mech Behav Biomed Mat 46:261–270
21. Kvederas G, Porvaneckas N (2002) Biomechanical analysis of the hip before and after total arthroplasty for idiopathic and dysplastic arthrosis. Revista Española de cirugía ortopédica y traumatología. 46(1):12–19
22. Jimenez VE, Urriolagoitia SG, Urriolagoitia CG (2004) Diseño de exoesqueleto de apoyo a la motricidad para la articulación de cadera. Tesis IPN-ESIME-SEPI, pp 63–64
23. Fritz JM, Guan Y, Wang M, Smith PA, Harris GF (2009) A fracture risk assessment model of the femur in children with osteogenesis imperfecta (OI) during gait. Med Eng Phys 31(9):1043–1048

Manufacturing of a Human's Hand Prosthesis with Electronic Movable Phalanges Based on a CT Image: An Amputation Case

Juan Alfonso Beltrán-Fernández, Itzel Alejandrina Aguirre Hernández, Itzel Bantle-Chávez, Carolina Alvarado-Moreno, Luis Héctor Hernández-Gómez, Pablo Moreno-Garibaldi, Nefi Pava Chipol, Juan Carlos Hermida Ochoa, Adolfo Lopez Lievano and Guillermo Manuel Urriolagoitia-Calderón

Abstract In the present work, the creation of a tridimensional electronic prosthesis which intends to replace the left hand's phalanges of a patient who suffered severe burns and needed to undergo a partial amputation surgery was performed. Parting of a CT tomography composed of 539 images, the contour delimitation, and separation

J. A. Beltrán-Fernández (✉) · I. A. A. Hernández · I. Bantle-Chávez · C. Alvarado-Moreno · L. H. Hernández-Gómez · P. Moreno-Garibaldi · N. P. Chipol · G. M. Urriolagoitia-Calderón
Instituto Politécnico Nacional - Escuela Superior de Ingeniería Mecánica y Eléctrica - Sección de Estudios de Posgrado e Investigación Edificio 5, 2do Piso, Unidad Profesional Adolfo López Mateos "Zacatenco" Col. Lindavista, C.P. 07738, Ciudad de México, Mexico
e-mail: jbeltranf@hotmail.com

I. A. A. Hernández
e-mail: itzel25.aguirre@gmail.com

I. Bantle-Chávez
e-mail: itzi.bantle@gmail.com

C. Alvarado-Moreno
e-mail: aly-oroshiwa@hotmail.com

L. H. Hernández-Gómez
e-mail: luishector56@hotmail.com

P. Moreno-Garibaldi
e-mail: pmg170588@gmail.com

N. P. Chipol
e-mail: sowbran@gmail.com

G. M. Urriolagoitia-Calderón
e-mail: urrio332@hotmail.com

J. C. H. Ochoa · A. L. Lievano
Centro de Investigación y Laboratorio Biomecánico, Carmen #18, Chimalistac San Ángel, 01070 Ciudad de México (CDMX), Mexico
e-mail: jhermida@gmail.com

A. L. Lievano
e-mail: k_lievano@hotmail.com

A. Öchsner and H. Altenbach (eds.), *Engineering Design Applications II*,
Advanced Structured Materials 113, https://doi.org/10.1007/978-3-030-20801-1_26

of each phalanx was carried out through the usage of computerized platforms, such as 3D Slicer and Scan IP. Both software allow the user to analyze every single image in three different planes (axial, sagittal and coronal) and segment the areas of interest, in this case, the index, middle, ring and little fingers' phalanges. Subsequently, with the support of the platform SolidWorks©, each phalanx was detailed for its connection and posterior 3D printing, implementing an electronic mechanism capable of reproducing flexion-extension movements. The computerized final model was also submitted to mechanical force tests and force simulations based on the analysis of finite elements to ascertain the selected material's and model's resistance.

Keywords Biomechanic · Hand phalanges · Prosthesis · Manufacturing · Amputation · Computed tomography · Magnetic resonance imaging · 3D printing

1 Introduction

Processes known as image reconstruction and medical imaging are based on the creation of visual representations of the human body's anatomical components. Multiple clinical studies performed with imaging techniques such as computed tomography (CT) and magnetic resonance imaging (MRI) have led to a more accurate recreation of clinical pictures, having as main goal the creation of a tool with which specialists are able to carry out noninvasive methods, this means, avoiding causing damage to tissues and organs.

The imaging field is broad, some current applications of these techniques are the extraction of tumors caused by mutation of neoplastic cells [1], segmentation of tumors and pathologic brain tissues and also early detection of abnormalities [2, 3].

A wide range of specialized type of software allows specialists to acquire signals detected and emitted by imaging devices and convert that information through complex algorithms into images, thus, the employment of those tools is more common in the medical field.

The "3D Slicer Software" is an open source platform which uses multiple filters in order to rebuild different areas of interest such as the osseous system, the nervous system or muscular tissue.

Another tool used for this purpose is the software "Scan IP", which allows the user to "paint" slide by slide several anatomical areas of a digitalized tomography, isolating the components that have to be treated or analyzed and creating through the rendering technique a 3D model with a high accuracy with respect to the anatomical components. Both types of software were used for the segmentation of the patient's phalanges of the index, middle, ring and little finger.

Using the 3D Slicer platform, the segmentation of the hand's phalanges was performed. This program was the first choice since it offers the advantage of a higher data acquisition and a more accurate creation of the structure's symmetry. The procedure started with the definition and isolation of the anatomical areas, based on a digital archive obtained from the conjunction of 539 images created by a computed tomography scan. One of the main challenges was to achieve a clear model creation of the separated phalanges due to the high sensibility of the program. The final tomographic image depends directly on the patient's hand position, making, as a consequence, the final model susceptible to the spatial hand's posture. However, there were several limitations to zoom in on the part of the image that needed to be delimited in spite of the advantages an MRI scanner offers.

Some of them are a higher soft tissue contrast than CT, a better differentiation between soft tissues (fat, water, muscle tissue, organs) and also a clear imaging of osseous structures. Furthermore, the time used to separate and define each phalanx made the whole project more complex and less optimal, although the accuracy with the patient's hand was higher.

In regard of this difficulty, it was opt for the usage of the Scan IP platform, which allows to manipulate the whole conjunction of the images that conform the tomography in three different planes (axial, sagittal and coronal). Manipulating slide by slide the personalized and accurate separation of the phalanges was possible.

In addition, based on the MRI tomography, a movement study of the phalange's articulation was conducted in order to analyze the grades of movement the interphalangeal joints present when the patient grabs a cylindrical object.

The present work is conformed of the phalange's segmentation with the employment of differently specialized software, such as 3D Slicer and Scan IP. Each analysis corresponds to Chapter II and Chapter III.

Subsequently, in Chapter IV the usage of a third software, SolidWorks-CAD program, will be introduced, through which each phalanx is manipulated for its posterior union and 3D printing. Finally, in Chapter V making use of each component, the natural extension-flexion movement of the phalanges can be reproduced. For this purpose, an open loop control system, which includes the implementation of servomotors was built.

2 Statement of the Problem

In 1973, the British Journal of Plastic Surgery reported a total number of 63 cases of implantations, known as aesthetic fingers, of patients who suffered total, hemi-hand and four-finger amputations. Approximately, 29 of these procedures were successful, giving an overall success rate of 45%.

In Mexico, people with a disability represent 5.1% of the total population. From this percentage, 33% often have problems to move or use their upper limbs, arms or hands. Among cases of disability, 10% acquired it from birth and 23% belong to the male gender [4].

Each year 270 thousand people increase the statistics of disability, of which 67 thousand are disabled due to congenital malformations [1].

Different factors affect the needs of the individuals for the minimal restoration of the limb. Some of them feel perfectly comfortable and don't express the need for prostheses in order to fulfill their working duties or for their everyday life's activities.

However, other cases present a punctual difficulty with the loss of a limb, e.g., a multi-instrumentalist who operates using the ten fingers to manipulate the necessary tools. If only the distal phalanx of the small finger would be amputated, it could represent a problem for the specialist to do his or her job properly. This means several factors such as the morphology, occupation, and health of the patient have to be considered (TAC).

2.1 Specifications of the Case of Study

- The patient presents 4 amputated fingers with no functionality in his left hand.
- There are complications when trying to use the hand and grab objects, (the patient is unable to perform certain common activities such as grabbing a spoon or use a knife).
- Because of the difficulties the patient has to confront on a daily basis, it is necessary to design an anatomical accurate and useful model using DICOM files and 3D specialized software.

Based on the design and creation of an accurate model, it was considered as a starting stage a tomographic study. It is intended that the patient:

- will be able to hold larger objects by means of grasping (currently, he is limited of holding objects with a diameter of 1 cm or more).
- will acquire the ability to perform planar and cylindrical grip.
- will acquire the ability to perform "hook action" when holding objects.
- will have an esthetic prosthesis by the use of different quick prototypes (through the employment of ABS, 3-D printing systems and 3D scanners in order to carry out the final adjustments).

2.2 Materials and Methods

The following methods and materials were used for the development of this project.

1. Use of a Phillips Brilliance Tomographer (Magnetic Resonance)
2. Segmentation data from a CT using 3D Slicer software
3. Definition of CT small bones using Scan IP Software 3.8 Version (533 images)
4. 3D printing models to define the affected zones.

A schematic figure in order to comprehend the last explanation is shown in Fig. 1.

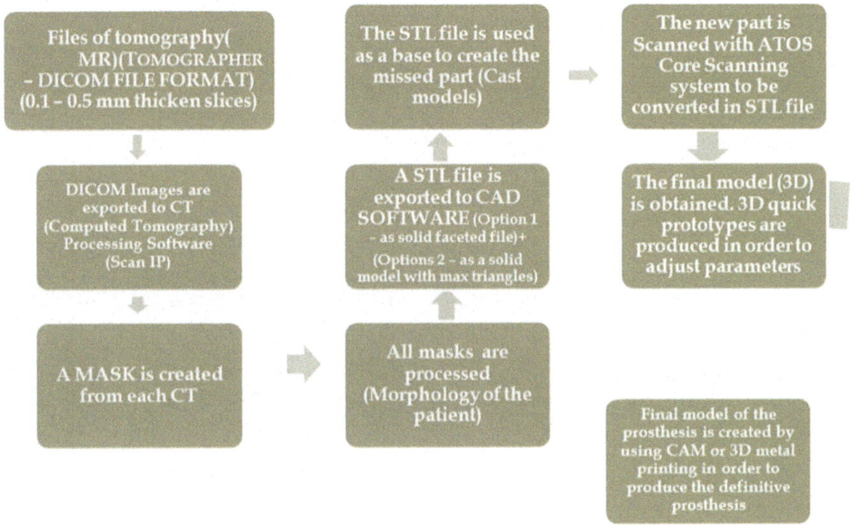

Fig. 1 Steps required for the development of the prosthesis

3 Reconstruction of a Human Hand Based on the 3D Slicer Platform

Starting with the transformation of the tomographic archive into a DICOM file, the osseous structure of the patient was analyzed with the 3D Slicer software. The platform allows specialists to analyze the TC tomography in three planes: sagittal, coronal and axial. After isolating the area of interest and defining the anatomical structure slide by slide, a fourth screen with the rendering model can be appreciated. In the upper part of the screen, the software shows the demographic information of the patient such as, name, gender, the type of clinical study which was performed, etc.

Firstly, the archive's data is loaded through the order "Load Dicom data", as shown in Fig. 1 or by clicking the symbol "DCM" in the right superior beam of the 3D Slicer software screen. After the data is loaded, all the information can be manipulated in form of multiple images (Fig. 2).

The "Editor" tool (Fig. 3) offers a wide range of colors and pallets, which are used to differentiate the anatomical elements shown in the tomography. The range color "Generic Anatomy Colors" was used for this purpose.

For the posterior volume reconstruction, the different sections of the patient's hand need to be marked in each plane with a specific label, paying special attention to the designation of all the elements. This means the same label needs to be assigned to one specific element in all three planes for its posterior volume rendering. The input called ROI or "Region of Interest" has to be the same for the selected elements.

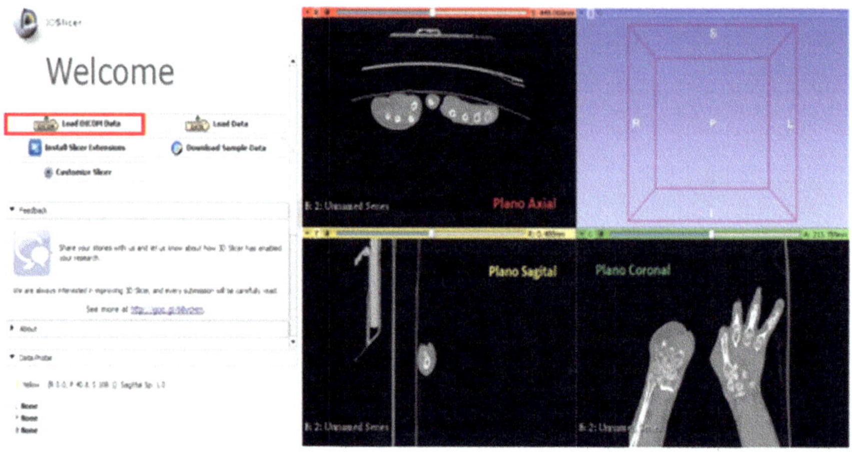

Fig. 2 Loaded data which shows both hands of the patient in three different planes: sagittal (dyed in yellow), coronal (marked in green color) and axial (highlighted in red color)

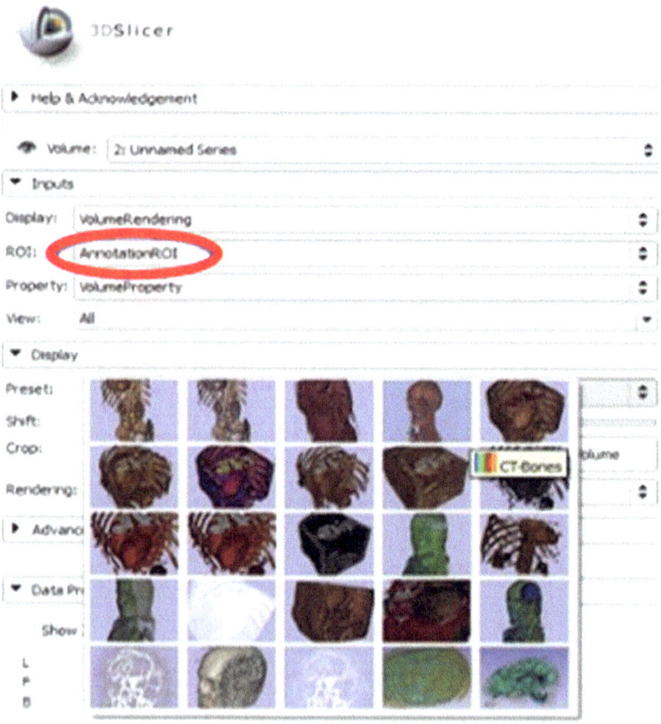

Fig. 3 "Display" and "Advanced" tools used for the selection of the area of interest

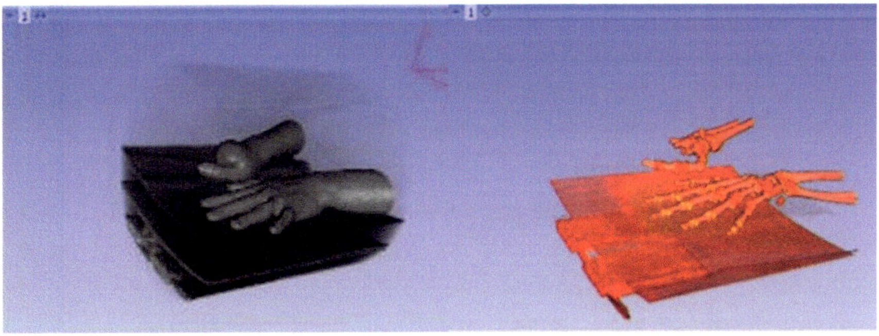

Fig. 4 Tridimensional representation of both hands (dyed in grey) and obtention of the final model after applying the CT-Bones filter (shown in red color)

Table 1 "Threshold Effect ranges"

Threshold range	
Minimum	Maximum
191.99	2110.00

Another useful tool is "Preset", which can be found in the submenu "Display" and allows the user to visualize the region that has to be analyzed, unfolding special anatomical structures and substances like water, osseous tissue, and fat. For instance, the filter CT Bone needs to be applied to obtain the 3D model of both the hand's and arm's bones of the patient, as shown in Fig. 3.

After the segmentation process, a new image needs to be obtained based on the cleaning the ROI filter achieved. For this purpose, the tool "Crop Volume" is applied. It is important to mention that the following parameters need to be checked: the inputs "Input Volume" and "Input ROI" need to coincide with both anterior inputs which are "Unnamed Series" and "Annotation ROI". Then the "ROI" visibility needs to be deactivated and bring forth the interpolation process so that new data points can be built within the set of discrete points. This is achieved with the command "Isotropic output voxel and linear" and the tool "Crop", as shown in Fig. 4.

The new image will be displayed in the axial plane. Based on the tool "Edit Selected Label Map", the different images of the osseous tissue which are intended to be rendered can be selected. Then the tool "Threshold Effect" with the label "bone" is applied (Fig. 5). The following table exhibits the characteristics of the tool "Threshold Effect", which the main function is to give fill ranges to the bone structures. The values are given within the "Threshold Range" that better worked are the followings (Table 1).

This range gives the opportunity to create a more solid representation of the hand's bones (Figs. 6 and 7).

Taking into account that part of the platform of the TC scan appears in the image, this causes the problem of not acquiring a clean image of the hand's bones. This

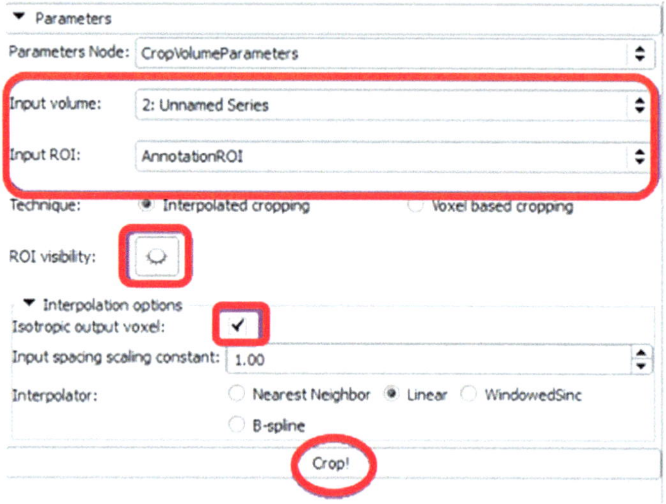

Fig. 5 Menu crop volume

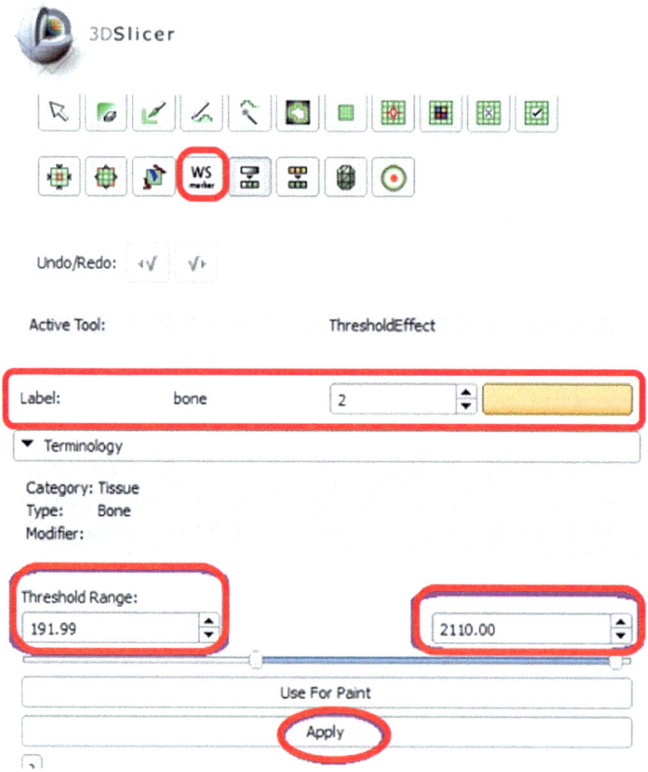

Fig. 6 Threshold effect tool

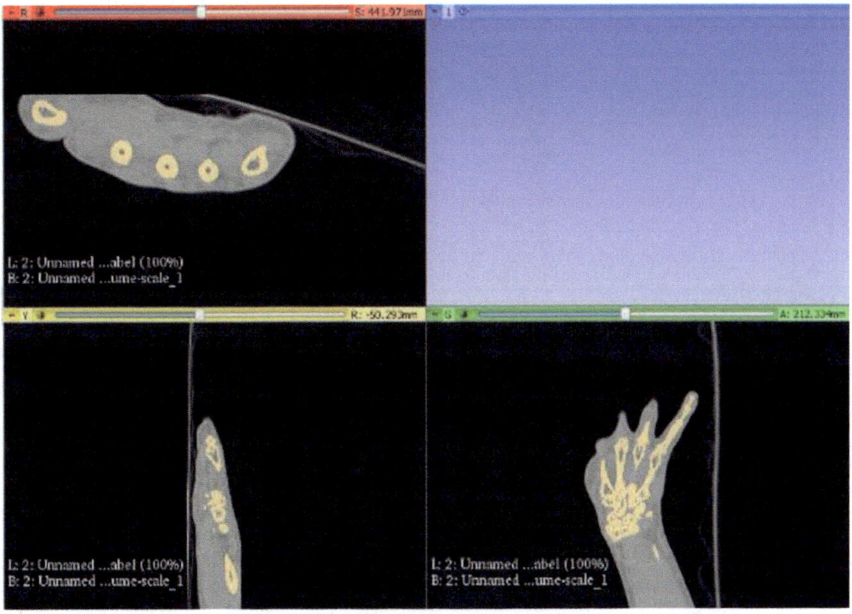

Fig. 7 Filled bone structure shown in three planes: axial, sagittal and coronal planes

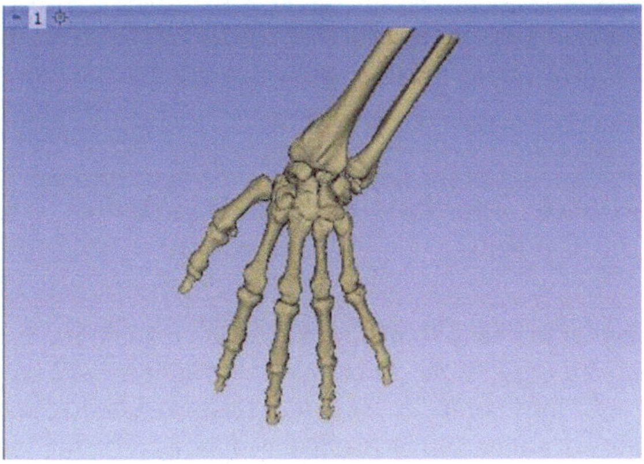

Fig. 8 Final 3D model created with the tool "Make Model Effect"

means this section needs to be eliminated. In order to select these structures, the tool "Change Island Effect" is used, which allows the user to define the bones (solid structures) in function of the given ranges given by the "*Threshold Effect*" tool. If the range is minor, the separation between the elements will have a higher and better definition. Unfortunately, the density of the structures can be compromised, leaving holes in the interior of the osseous structure instead of reproducing them as solid structures. The obtained tridimensional model is shown in Fig. 8.

Fig. 9 Printed model of the patient's hand

The 3D Slicer platform also allows the user to save the created models as polydata (.vtk), XML polydata (.ytp),.stl and wavefront object (.obt) archives.

From the mentioned types, the archive's extension used for the 3D printing process was the.stl archive. The printed model is shown in Fig. 9.

4 Segmentation of the Phalange's Distal, Medial and Proximal Areas

Through the usage of the Scan IP platform and the creation of a new.iges archive, the file to be analyzed acquires the property of compatibility with a different type of CAD workstations. This in order to aid in the creation and design of the individual phalange's prostheses, which will be manipulated for its posterior manipulation and mechanical analysis. The Scan IP platform allows to define in a more accurate way each individual phalanx, this through the manipulation of each slide which composes the tomographic archive. In comparison with the first workstation (3D Slicer), the contour of each individual component can be better traced and, as a consequence, each element can be better differentiated.

4.1 Implementation of the Scan IP Platform

The first step consists of importing the DICOM archives, which contain a larger number of slides to the platform so that a wider image spectrum can be analyzed. The tomographic archive will be uploaded and shown in three different planes (sagittal, axial and coronal) for its manipulation (Fig. 10). The workstation's tools "floodfill" and "Paint" were applied for the contour's definition of each element by painting slide by slide the whole phalange's extension.

The command Floodfill establishes the ranges for the image's grey scale values, being 255 the upper value and 47 the lower value of the +/− Delta parameters. This allows the automatic election of the existent elements dyed in white in the tomography. Subsequently, the function "mask operation" allows the user to literally paint the phalange's surface and the command "merge with mask" juxtapose all the painted slides in order to create a unique model. The painting process can be appreciated in Fig. 11.

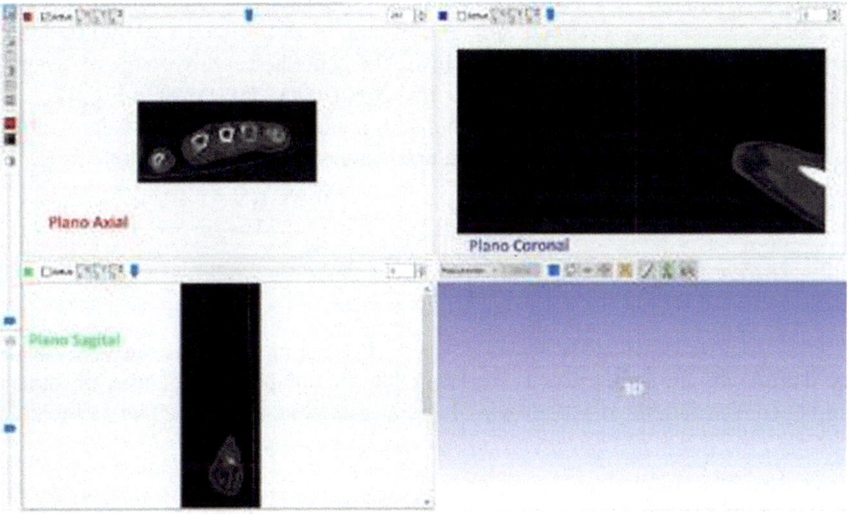

Fig. 10 View of the imported tomography in three planes: sagittal, axial and coronal

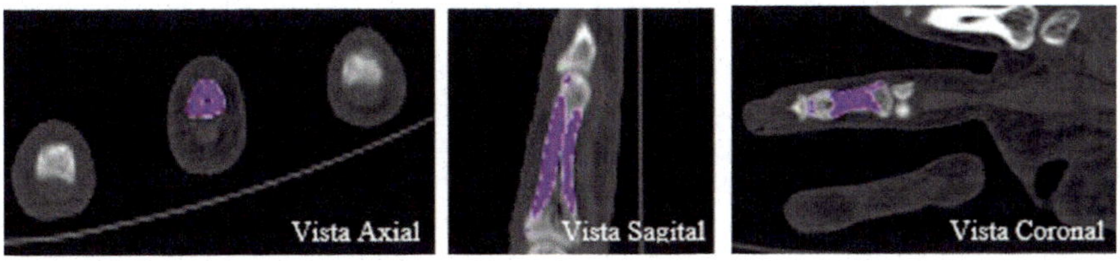

Fig. 11 View of the painted and defined phalange's elements in three planes: sagittal, axial and coronal

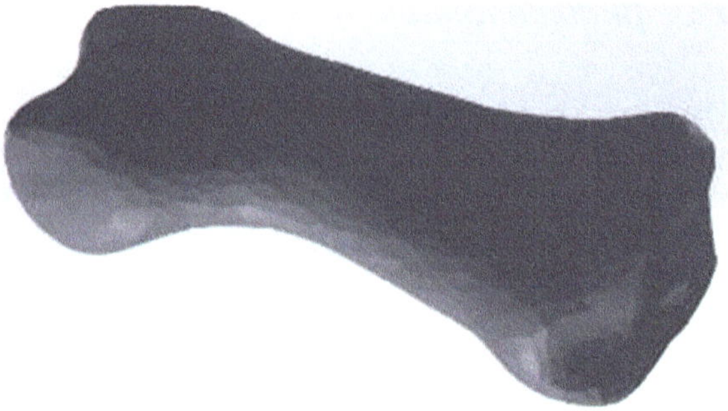

Fig. 12 Obtained phalanx after the rendering process of the Scan IP platform is applied and the archive is opened in solid works

Table 2 Used parameters +/− DELTA

Lower value	96
Upper value	202

After defining the four fingers, the archives are saved as a.iges file with the following specifications to smooth the surface of the phalanges: pre-smoothing (value of 20), post-smoothing (value of 10), "Decimate" with a value of 50,000 triangles per part to enhance the quality of the archive. The generated archive in iges extension can be opened in all CAD platforms, such as SolidWorks, as shown in Fig. 12.

The next step is the creation of each distal, medial and proximal phalanx of the index, middle, ring and little finger. The used values are specified in Table 2.

4.2 Index Finger

For the index finger, the 3D building was performed in different archives. This is the reason why the same mask is shown in the three planes. The number of images employed for each, the distal, proximal and medial phalanx is specified in Table 3.

4.2.1 Middle Finger

The same archive was used for the middle finger's design. The +/− Delta ranges employed for this purpose are specified in Table 4, being the upper vale: 36 and the lower value: 16.

4.2.2 Index Finger

The same process was repeated for the reconstruction of the other finger's phalanges (Table 5).

Table 3 Reconstruction of the distal, proximal and medial phalanx of the index finger

View	Sagittal plane	Coronal plane	STL
Distal phalanx of the index finger			
Number of images	75–82	73–85	–
Distal phalanx	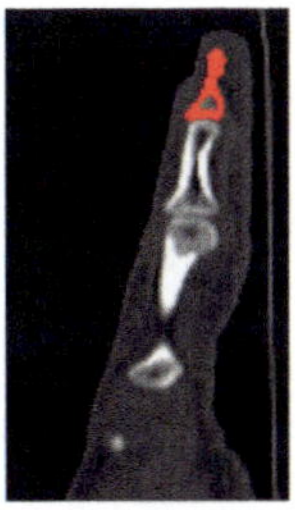	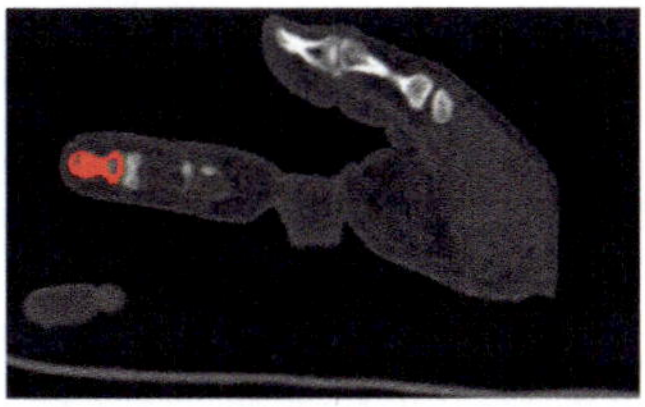	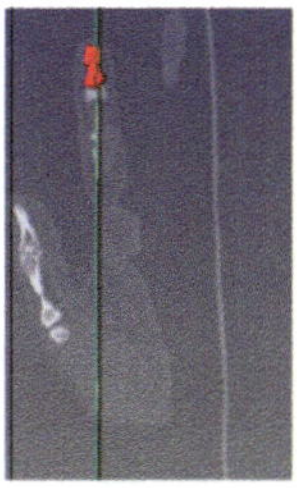
Medial phalanx of the index finger			
Number of images	62–85	43–61	–
Medial phalanx	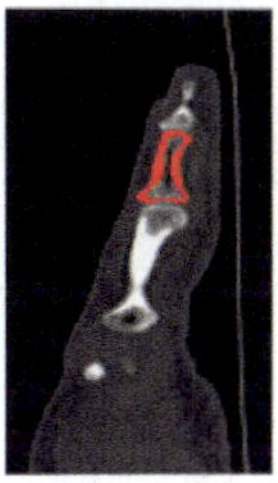	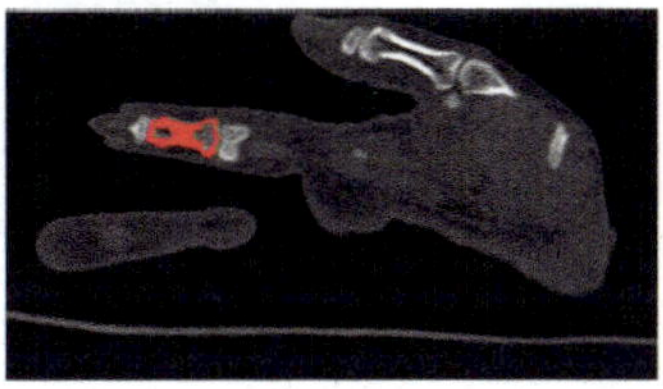	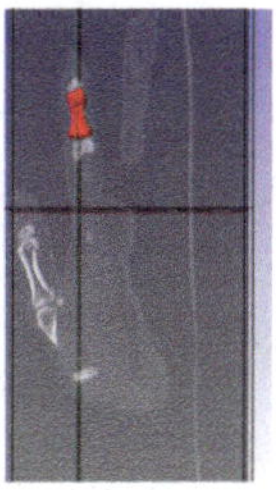
Proximal phalanx of the index finger			
Number of images	70–101	32–64	–
Proximal phalanx	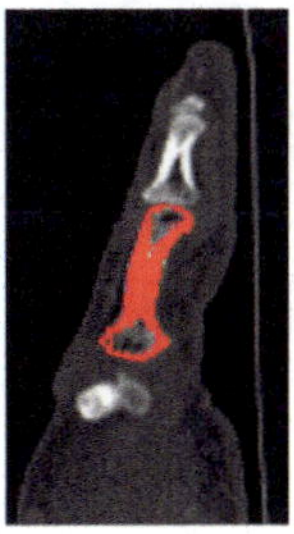	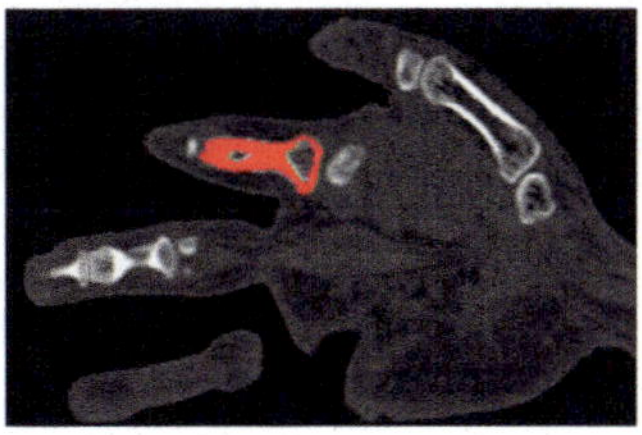	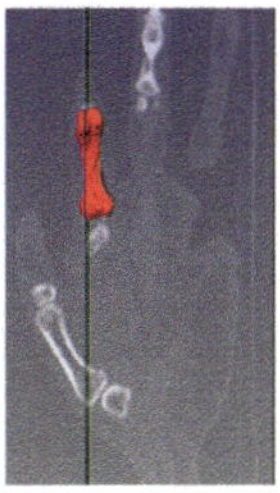

Table 4 Reconstruction of the distal, proximal and medial phalanx of the middle finger

View	Sagittal plane	Coronal plane	STL
Number of images [distal]	133–153	58–71	–
Number of images [medial]	130–154	48–70	–
Number of images [proximal]	125–152	35–65	–
Image	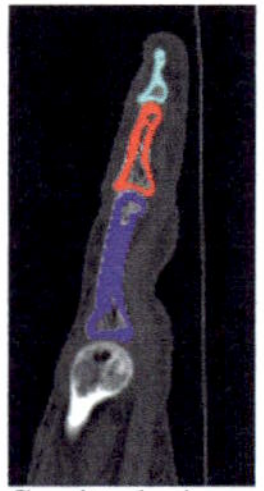 Sagittal view of the middle finger's phalanx	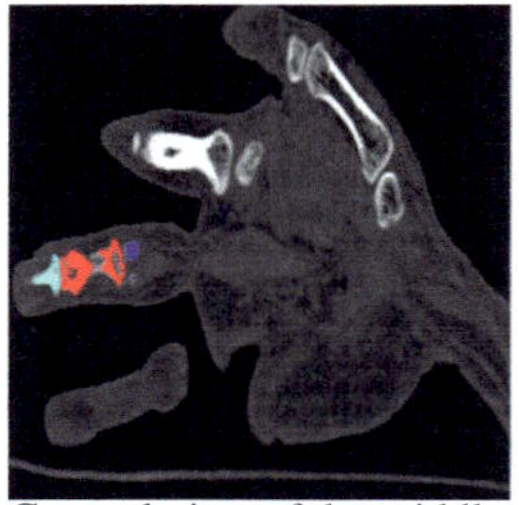 Coronal view of the middle finger's phalanx	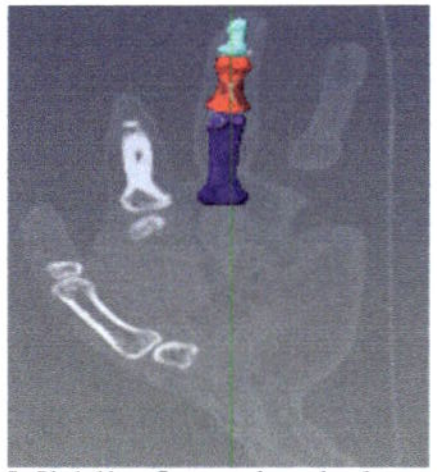 Middle finger's phalanges after applying the 3D rendering process

Table 5 Reconstruction of the distal, proximal and medial phalanx of the index finger

View	Sagittal plane	Coronal plane	STL
Number of images [distal]	192–213	40–52	–
Number of images [medial]	181–208	34–52	–
Number of images [proximal]	160–199	25–51	–
Image	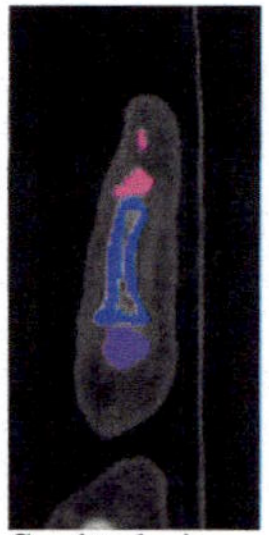 Sagittal view of the index finger's phalanx	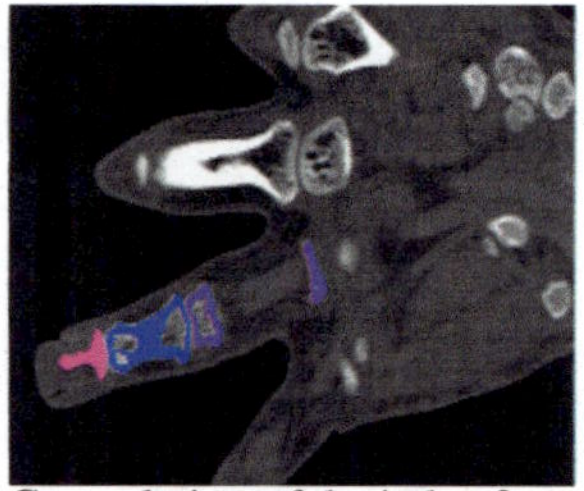 Coronal view of the index finger's phalanx	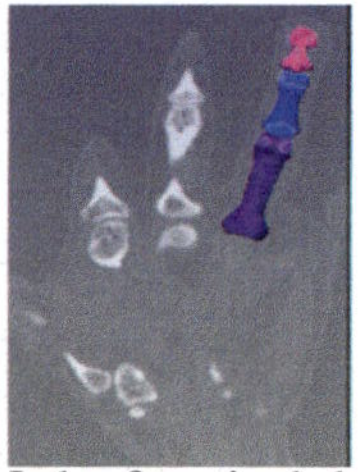 Index finger's phalanges after applying the 3D rendering process

4.2.3 Little Finger

See Table 6.

5 CAD Modeling for the Phalange's Model Modification and Its Assembly

After the reconstruction process for each phalanx with the usage of the Scan IP Software and its conversion into an object format, known as.iges file, each individual piece was manipulated with a CAD Software in order to design an articulation that allowed the user to simulate the natural flexion-extension movement of the hand. In this project, the CAD Software used is the SolidWorks platform. For the manipulation of the pieces, the generation of multiple planes was required so that the needed perforations and extrusions over the pieces could be done. Hereafter, the design process of each piece will be described.

5.1 Index Finger

Starting with the distal phalanx, the exported Scan IP archive could be opened in the Scan IP Software as an object (Fig. 13).

Figure 13b shows the multiple modifications which were performed at the basis of the bone in order to assemble the medial phalanx. For this process two planes were required.

In regards of the piece's edition, for the middle phalanx five planes were required in order to modify the model's surface. Because of being a middle phalanx, the head and base was modeled and perforated, as shown in Fig. 14.

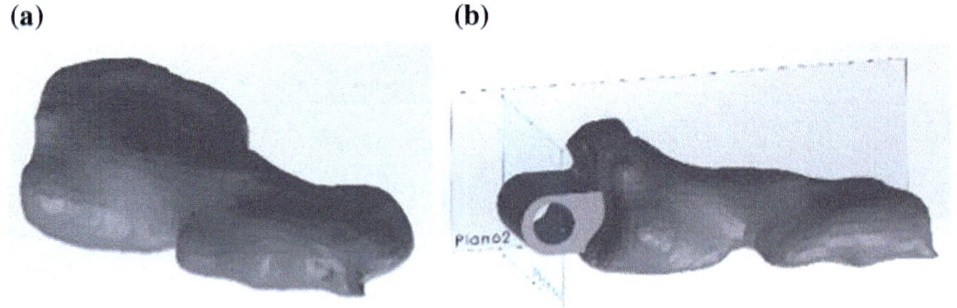

Fig. 13 **a** Distal phalanx of the Index Finger, **b** edited index finger in solid works

Table 6 Reconstruction of the distal, proximal and medial phalanx of the little finger

View	Sagittal plane	Plano coronal	STL
Number of images [Distal]	244–263	109–118	–
Number of images [Medial]	227–255	109–123	–
Number of images [Proximal]	190–238	107–126	–
Image	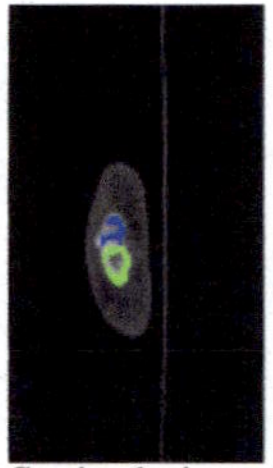 Sagittal view of the little finger's phalanx	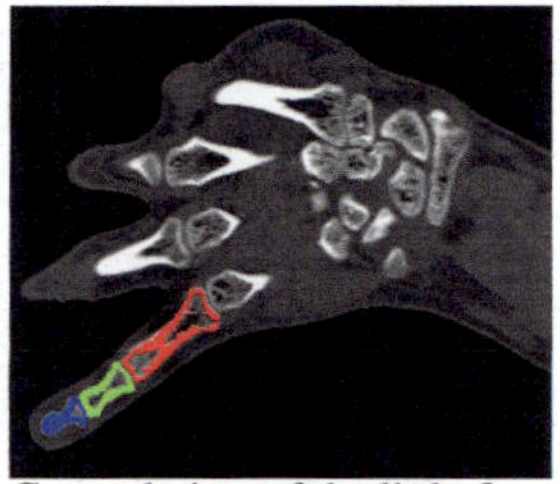 Coronal view of the little finger's phalanx	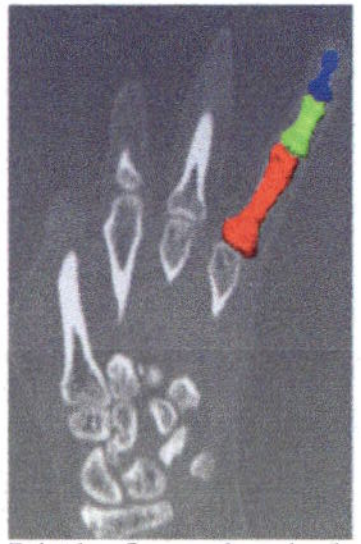 Little finger's phalanges after applying the 3D rendering process

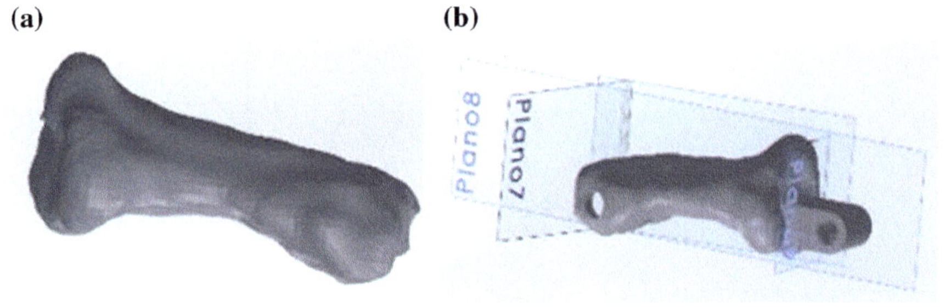

Fig. 14 **a** Medial phalanx of the index finger. **b** Medial phalanx, edited index finger in solid works

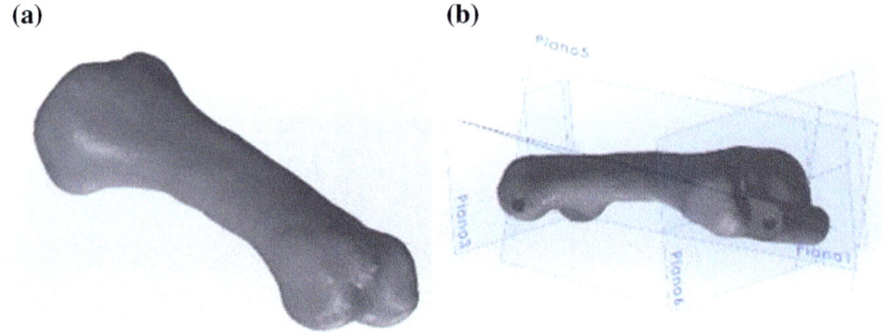

Fig. 15 **a** Proximal phalanx of the index finger. **b** Proximal phalanx, edited index finger in solid works

Finally, Fig. 15 shows the proximal phalanx which has to perform the flexion-extension movement and needs to be assembled to the medial phalanx. The creation of the pieces needed five planes to edit the base and head of the phalanx (Fig. 15).

5.2 *Middle Finger*

In order to perform the assembly of the middle finger, the distal, middle and proximal phalanges needed to be modified. Figure 16a shows the distal phalanx that was obtained with the usage of the Scan IP platform. Particularly, this piece is the final effector and needs only one degree of movement for the flexion-extension. Only one plane, which is parallel to the base of the phalanx, was needed to be created (Fig. 16).

Just like the case of the middle finger, the medial phalanx presents two points of connection at its base and head. Parallel plane to this particular zone were required, in conjunction with a transversal plane to create the necessary perforations for the assembly of all the pieces (Fig. 17).

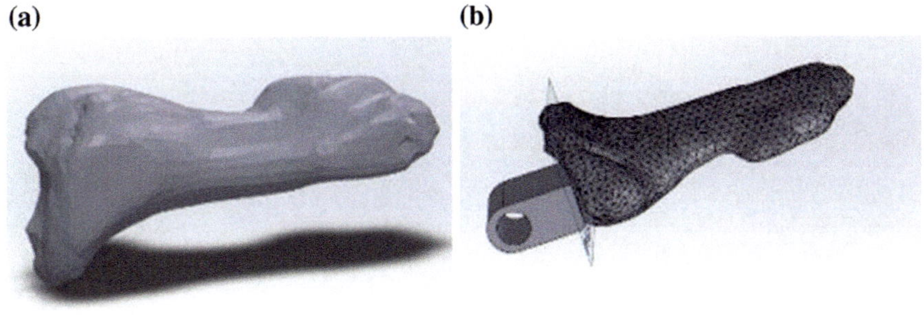

Fig. 16 **a** Distal phalanx of the middle finger. **b** Distal phalanx, edited middle finger in solid works

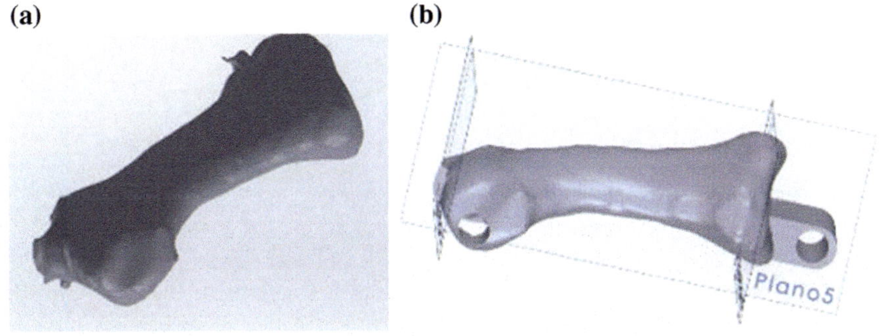

Fig. 17 **a** Middle phalanx of the middle finger. **b** Middle phalanx edited middle finger in solid works

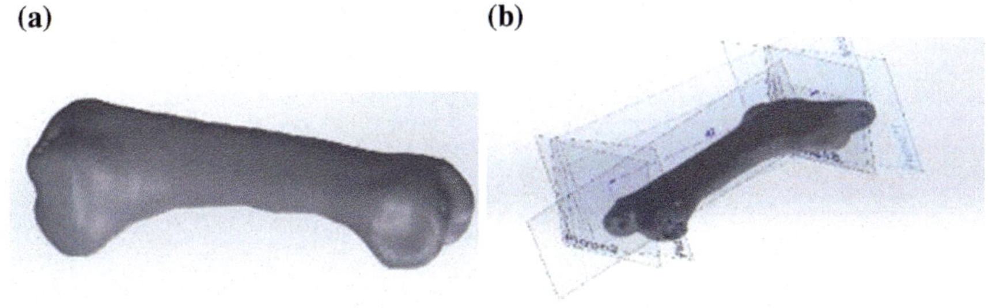

Fig. 18 **a** Proximal phalanx of the middle finger. **b** Proximal phalanx, edited middle finger in solid works

Finally, for the proximal phalanx, eight planes were used in order to manipulate the head and the base of this phalanx. Because of the tomography's contour, several inclined planes (parallel and perpendicular planes) to the phalanx's faces were used (Fig. 18).

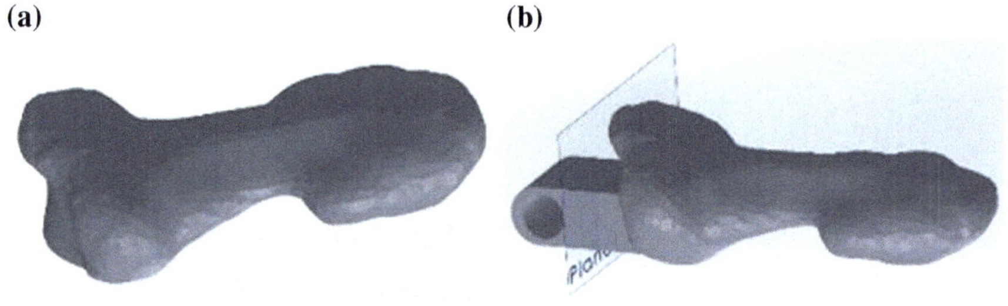

Fig. 19 **a** Distal phalanx of the middle finger. **b** Distal phalanx, edited ring finger in solid works

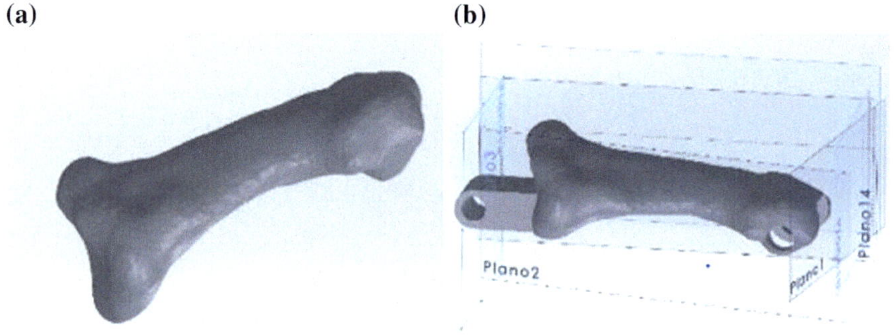

Fig. 20 **a** Medial phalanx of the ring finger. **b** Medial phalanx, edited ring finger in solid works

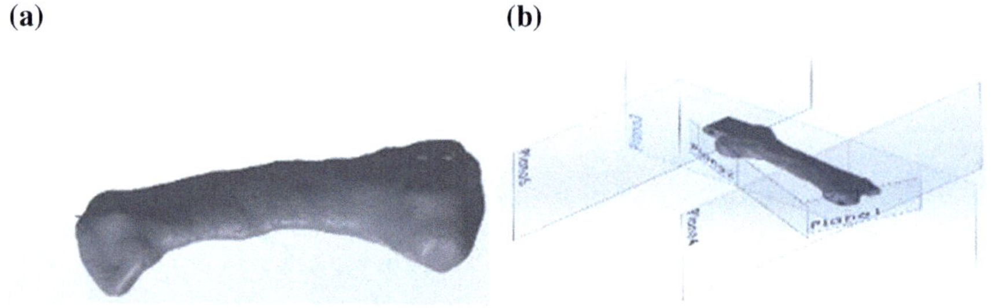

Fig. 21 **a** Proximal phalanx of the ring finger. **b** Proximal phalanx, edited ring finger in solid works

5.3 *Ring Finger*

In regards to the distal phalanx, only one parallel plane to the phalanx's base was required (Fig. 19).

The middle phalanx required six planes for the extrusion, perforation and assembly processes, as shown in Fig. 20.

For the proximal phalanx, five planes were used so that the final piece could have movable extremes. Of these five planes, two had an inclination angle in order to create a functional edition of the piece (Fig. 21).

(a) (b)

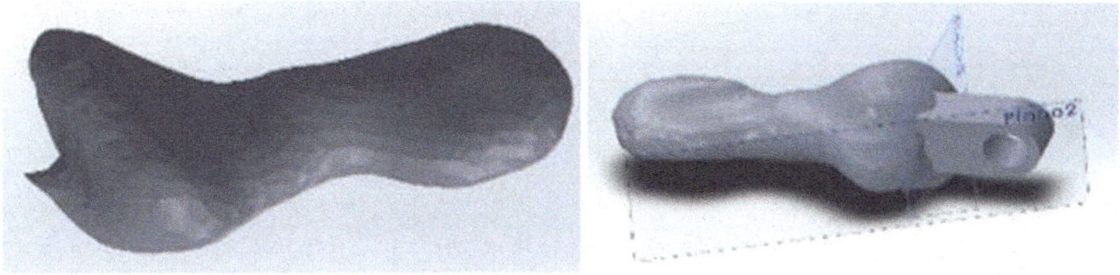

Fig. 22 **a** Distal phalanx of the ring finger. **b** Distal phalanx edited little finger in solid works

(a) (b)

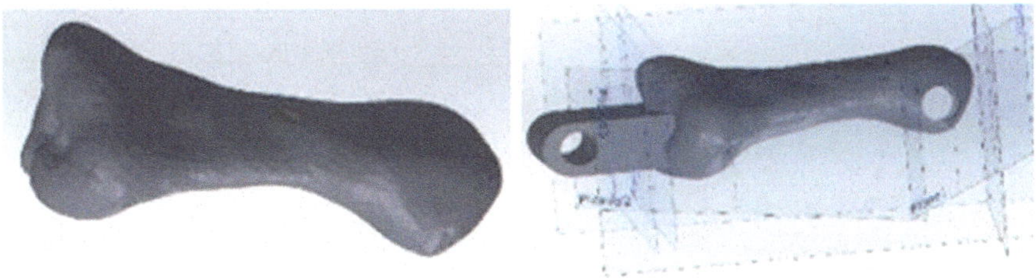

Fig. 23 **a** Medial phalanx of the little finger. **b** Distal phalanx edited little finger in Solid Works

5.4 Little Finger

For the little finger, it must be taken into account that during the final movement the pieces need to resist the displacement forces. If the phalanges are too little, they could easily break.

Figure 22 shows the usage of two planes, one possesses a certain grade of inclination due to the curve position of the little finger with respect of the four remaining fingers of the hand.

For the middle phalanx, eight planes were required; four of them had different grades of inclination (Fig. 23).

For the proximal phalanx, four planes were used. All of them with a precise inclination with respect to the planes of the base (Fig. 24).

Finally, the data segmentation, including the unions and partial assemblies of the joints, which need to be cinematically controlled are shown in Fig. 25.

6 Robotic Prototype for the Flexion-Extension Movement of the Four Fingers

For the mechanical analysis of the natural phalange's movement (flexion-extension) is important to have a wide knowledge of the hand's anatomical constitution. The

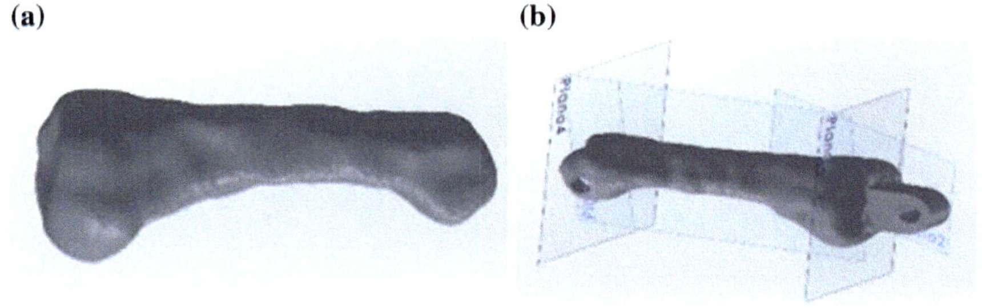

Fig. 24 **a** Proximal phalanx of the little finger. **b** Proximal phalanx, edited little finger in solid works

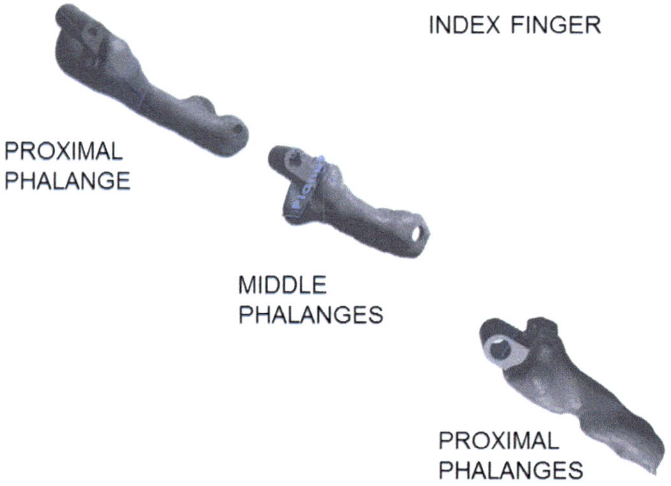

Fig. 25 Assembly of the index phalanx including joints, using PTC Creo Parametric

movement of the 27 hand's bones is performed by four principal groups of articulations shown in Fig. 26, of which conforms the phalanges of the hand, 5 metacarpals, and 8 carpals (Table 7).

6.1 Extension Movement

For the extension movement, the phalanges are wrapped by the extensor mechanism HOOD, which has a principal coating over the metacarpal bone. Two tendons control the movement of the proximal articulations (PIF) and the metacarpophalangeal articulations (MCF). They also have lateral systems which are inserted at the base of the distal phalanx (Fig. 27).

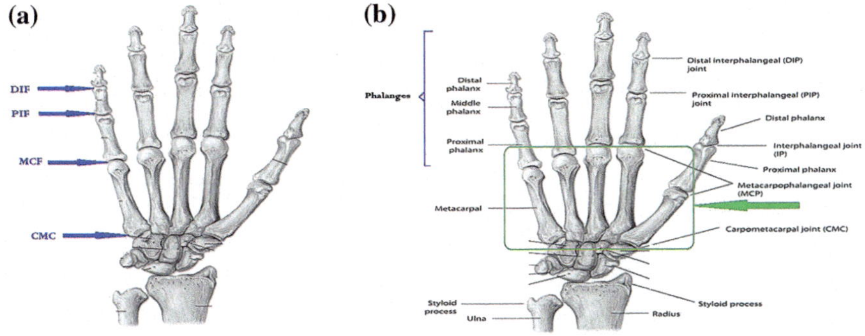

Fig. 26 **a** Distal Interphalanx (DIF), Proximal Interphalanx (PIF), Metacarpal Phalanx (MCF) and Carpal-metacarpal (CMC), **b** distributed conformation of a group of articulations of the hand

Table 7 Properties of the phalange's movements

Articulation	Description	Performed movements
MCF	Allows the angular movement in two directions (biaxial)	Flection–extension, Abduction–adduction
PIF	Allows the movement in only one axis (monoaxial)	Flection–extension
DIF	Allows the movement in only one axis (monoaxial)	Flection–extension

Note Each interphalangeal articulation has two collateral ligaments, which give stability to the articulation during the movement, allowing the performance of the movement in only one axis

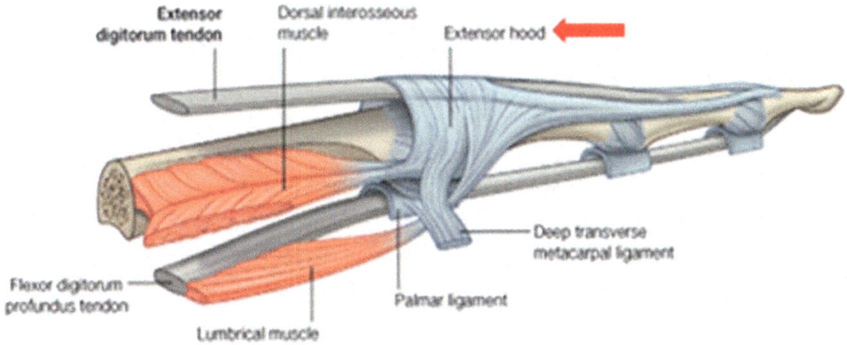

Fig. 27 Extensor mechanism HOOD (dyed in blue) [5]

6.1.1 Bending Movement

In regards to the flexion movement of the metacarpophalangeal articulation (MCF), the sagittal bands are displaced forwards in order to apply force on the extensor. When the interphalangeal articulation is bent, the lateral tendons are displaced from the palm of the hand, which relaxes them and allows the bending if the interphalangeal, distal articulation.

6.1.2 Range of Movement of the Phalanges

The range of movement of the fingers is given by the maximum grades of flexion and extension for each phalanx, in conjunction with the maximum grade of separation between them each time an adduction-abduction movement is performed by the metacarpal bones (Table 8).

7 Cinematic of the Hand

According to the performed movements of the phalanges, in conjunction with the grades of freedom each element possesses, the next step consists of the calculation of the hand's cinematic. All the fingers have the same kind of movement; therefore only one formula will be used for all them based on the "Denavit–Hartenberg" method, which establishes the positioning of each rotation axis in robotic-articulated systems (Fig. 28).

The following table shows the Denavit-Hartenberg parameters of the index finger (Table 9).

Table 8 Maximum movement degree of the phalanges [6, 7]

Articulation	Flexion-extension θ's	Abduction-adduction (φ)
MCF	0 a 30°–45°	0°–20°
PIF	0 a100°	–
DIF	10-0 a 90°	–

Table 9 Denavit-Hartenberg parameters of the index finger

i	α_{i-1}	a_{i-1}	d_i	θ_i
1	$\varphi_1 - \frac{\pi}{8}$	0	0	θ_1
2	*0*	l_1	0	θ_2
3	0	l_2	0	θ_3
4	0	l_3	0	0

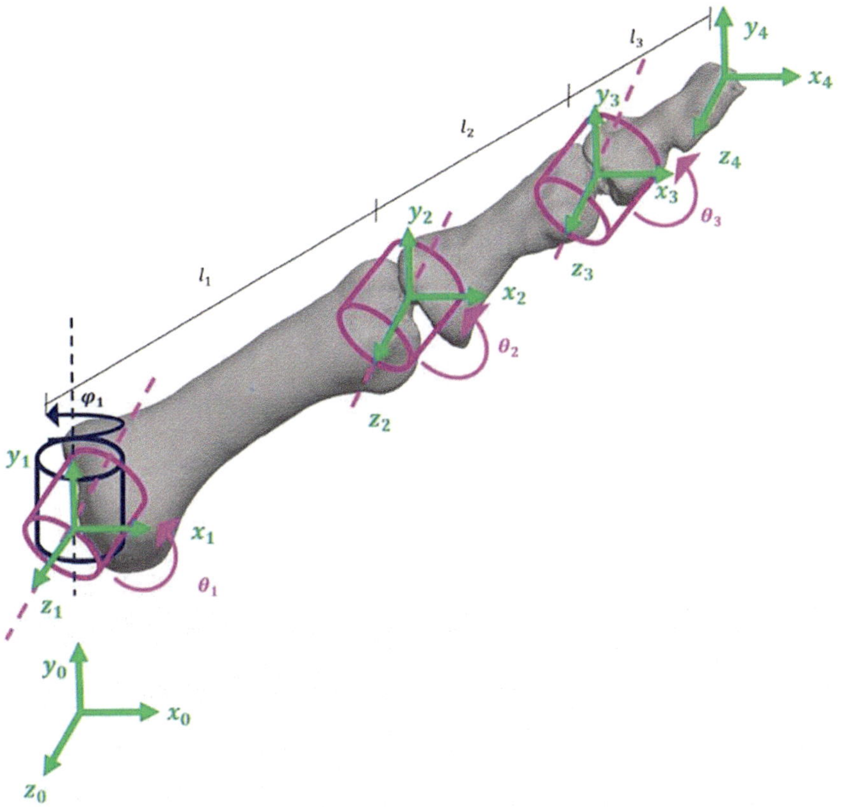

Fig. 28 Axis and angles according to the Denavit-Hartenberg method

7.1 *Analysis of the Hand Movement*

For the bending-extension movement of the hand, the following sequence was developed. For the extension sequence, the model shown in Table 10 was followed.

7.2 *Control System of Open Loop for the Movement Generation*

Based on Table 10, the bending-extension movement can be analyzed from 3 main points:

1. MCF articulation, which performs a −45° movement.
2. PIF articulation, which performs a −90° movement.
3. DIF articulation, which performs a −90° movement.

Table 10 Sequence of the bending-extension movement

1	
2	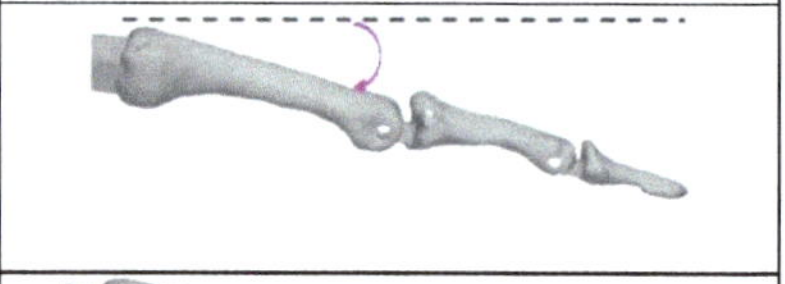
3	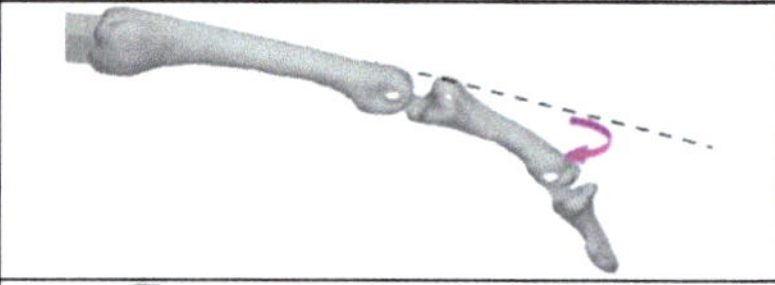
4	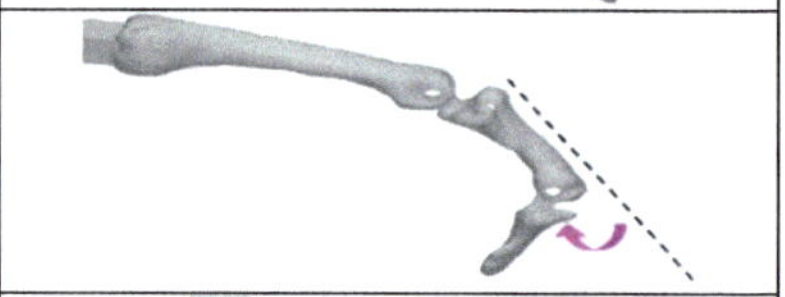
5	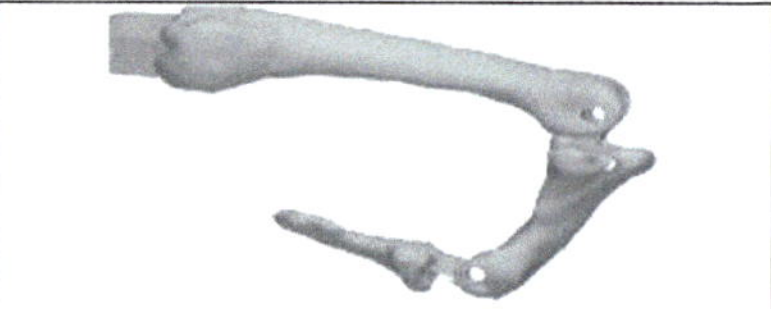

Table 11 PWM angles

Degree	PMW work cycle (ms)
−90	1
0	1.5
90	2

In the case of an external power source, a servomotor model SG90 that uses a PWM pulse, trademark Towerpro was considered for this study. The advantages this device offers are the following: angular movements with a range of [90°, −90°] and a force moment of 2.5 kg/cm [8, 9, 10].

Based on the servomotor's work cycle of 1-2 ms, the degrees of the angles are shown in Table 11.

Taking into account that the hand possesses a movement for the four fingers of the hand and for each finger, three movements need to be considered, 12 PMW independent signals needed to be used. Therefore, a microcontroller with 12 output ports was used in order to configure the movement through the PMW signals. The microcontroller that was chosen is the PIC18F4550 microchip. The data sheet for this device is shown in Fig. 29.

Because each interphalangeal union has independent movements, the microcontrollers were displayed as followed (Table 12).

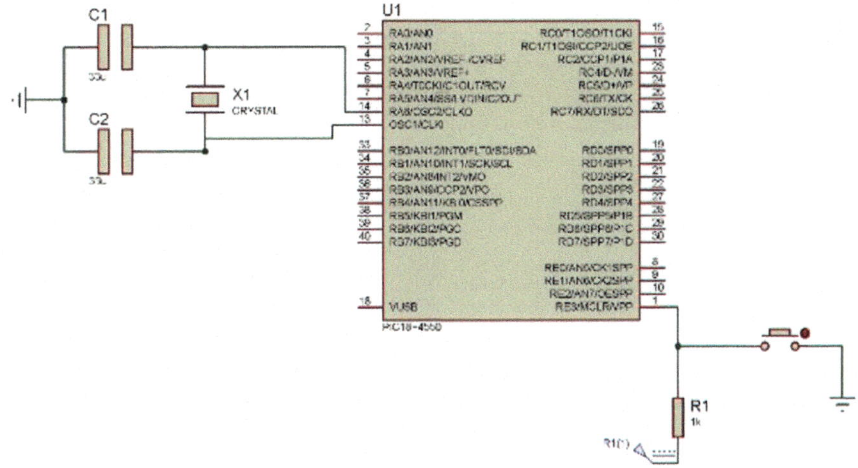

Fig. 29 Datasheet that includes the 12 ports for the PWM signals

Table 12 The relation between the output ports of the microcontroller in relation to the interphalangeal unions

Index finger		Ring finger	
Interphalangical union	PIC18F4550 output port	Interphalangical union	PIC18F4550 output port
DIF	RC0	DIF	RB3
PIF	RC1	PIF	RB4
MCF	RC2	MCF	RB5
Middle finger		Little finger	
Interphalangical union	PIC18F4550 output port	Interphalangical union	PIC18F4550 output port
DIF	RB0	DIF	RB6
PIF	RB1	PIF	RB7
MCF	RB2	MCF	RC4

The program used in order to give movement to the interphalangeal union's capable of performing a flexion-extension movement in only three seconds, one second is used for the flexion, another second for keeping the phalanx extended and one more second is required to extend the phalanx. This specific sequence can be appreciated in Tables 13 and 14.

This same process is going to be repeated for each one of the four fingers.

Table 13 Movement sequence, flexion of a finger

		Union	*Pulse*	*Time*
0		MCF 0°	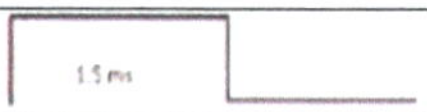	**0 s**
		PIF 0°	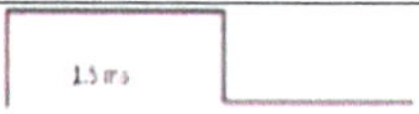	
		DIF 0°		
1	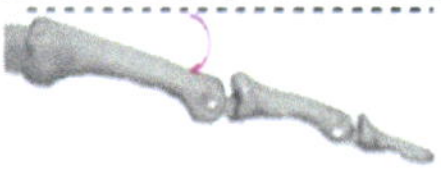	MCF -45°		**0-0.3 s**
		PIF 0°	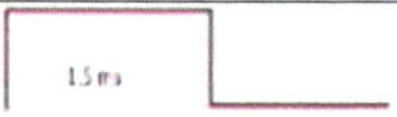	
		DIF 0°		
2	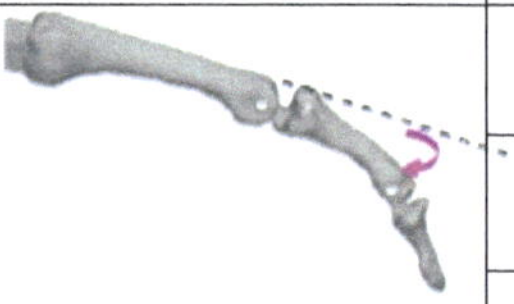	MCF -45°		**0.3-0.6 s**
		PIF -90°		
		DIF 0°		
3	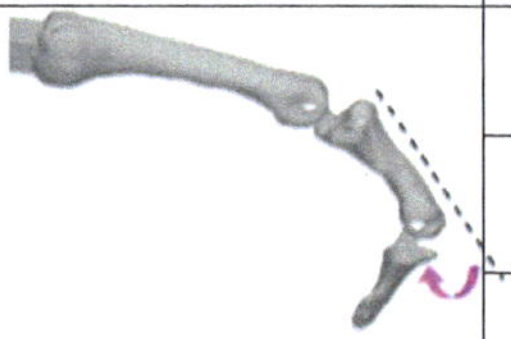	MCF -45°		**0.6-1 s**
		PIF -90°		
		DIF -90°		

8 Generation of the Functional Movement for the Patient Parting from the Tissue's Reconstruction and the Results of the Computational Tomography

Based on the DICOM files that were acquired through the computerized tomography of the patient's hand, a zip file which contains images of each one of the performed cuts during the study was analyzed with the purpose of making a relation of the bones and phalanges that the patient should have.

For the morphological design of the hand, the computational program Scan IP was used, which allows the user to create a rendered model using the tomographical slides as DICOM archives to create the profile of the target model. Using the tools paint, draw and color the contour of the different slides of the CT scans were defined (xy, yz, zx), as shown in Fig. 30.

Using the different planes, the user has the option to manipulate the cut to delimit the form of the min sketch. After defining the contour and selecting the option to

Table 14 Movement sequence, flexion of a finger

		Union	Pulse	Time
0	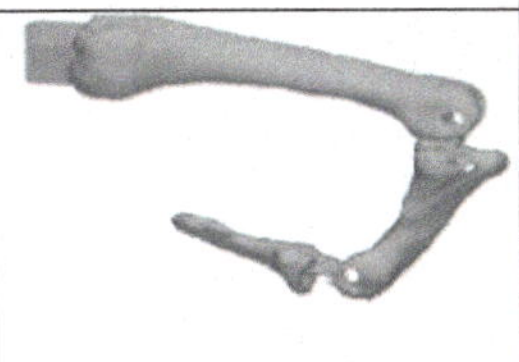	MCF -45°	1.25 ms	1 -2s
		PIF -90°	1 ms	
		DIF -90°	1 ms	
1	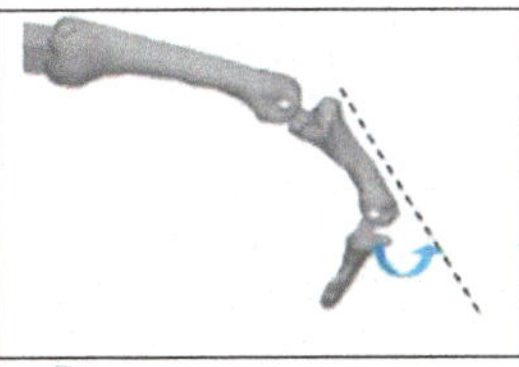	MCF -45°	1.25 ms	2-2.3 s
		PIF -90°	1 ms	
		DIF 0°	1.5 ms	
2	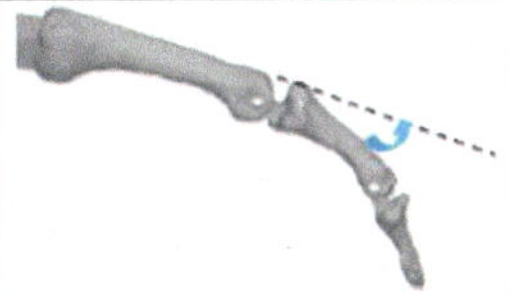	MCF -45°	1.25 ms	2.3-2.6 s
		PIF 0°	1.5 ms	
		DIF 0°	1.5 ms	
3	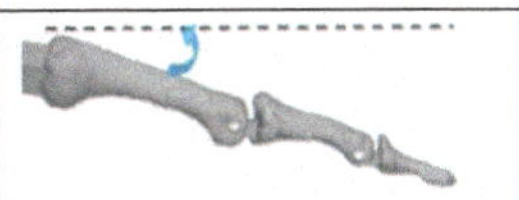	MCF 0°	1.5 ms	2.6-3s
		PIF 0°	1.5 ms	
		DIF 0°	1.5 ms	

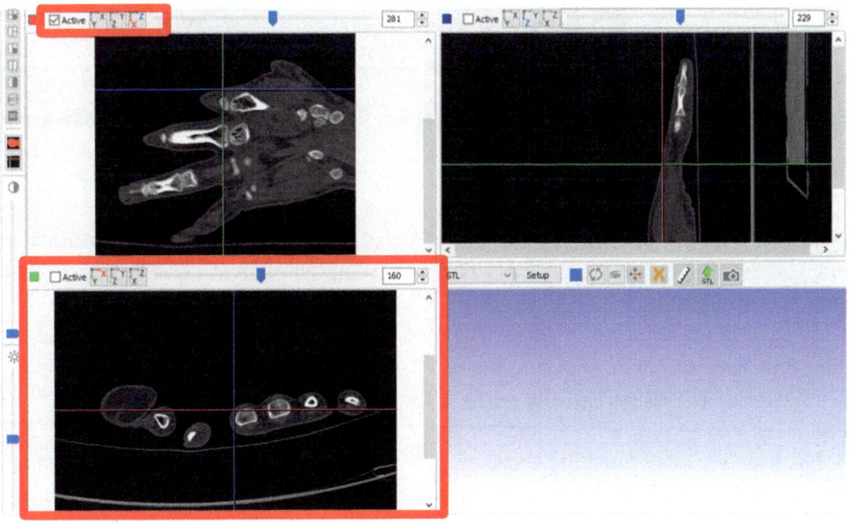

Fig. 30 Use of the Scan IP program in different visualization planes

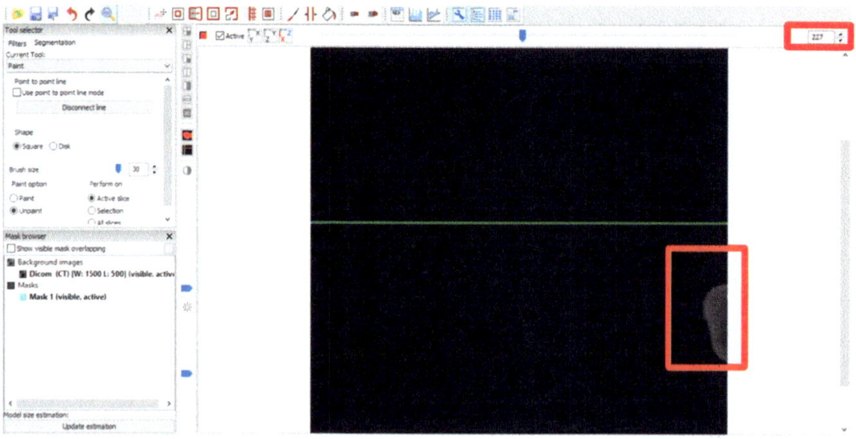

Fig. 31 Example of the cut in the z, x plane indicating the number of the tomographical cut

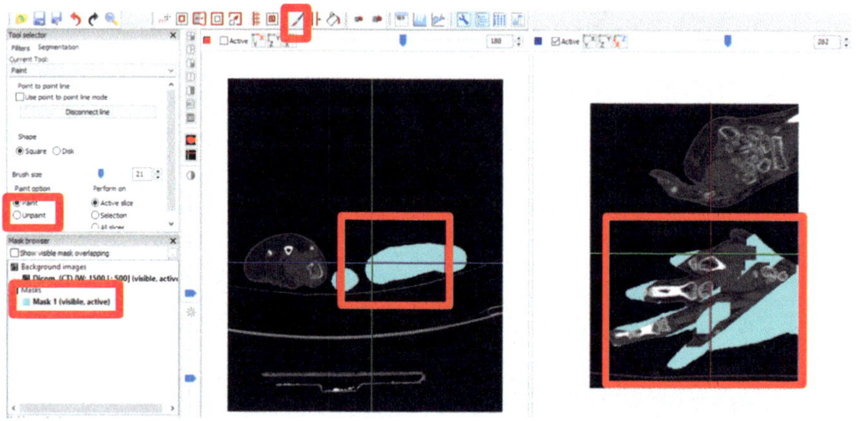

Fig. 32 The window of the SCAN IP program, which shows the painting procedure of the slices, focusing on the scheme of the planes

paint the interior of the defined figure, a mask with the form of the extremity or joint is created, as shown in Fig. 31. The first cut in the plane z,x can be appreciated parting from the image 227.

A mask in the selected window and plane will be drawn at the same time in the other windows and planes, giving the user the possibility to collocate guide cursors to indicate in which part of the surface the joints and extremities will be defined (Fig. 32).

The last step after finishing the sketch of all the cuts is the render of all the slices to create a tridimensional model. Using the Setup button an emergent window will appear, as shown in Fig. 33.

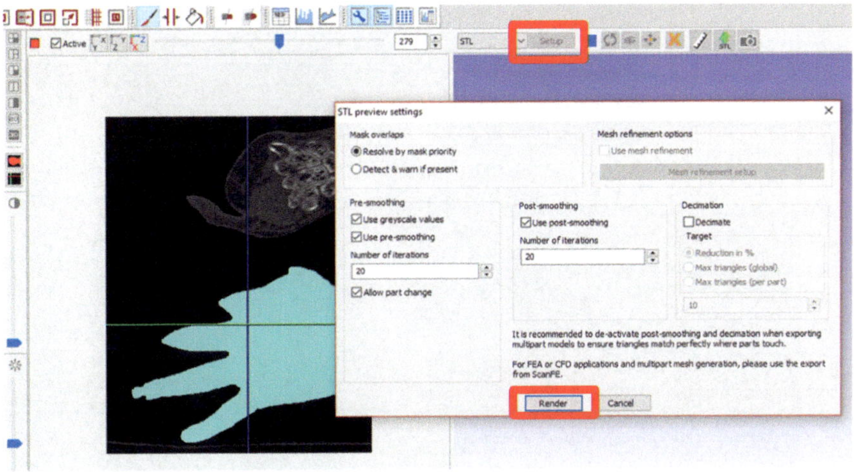

Fig. 33 The window for the rendered SCAN IP program

In the window of Fig. 33, the formulary needed to be filled in with the necessary iteration numbers to heighten the model's quality and avoid possible separations that can be found between the points that compose the model. A major number of iterations slows down the rendering process and takes more time. In order to optimize the process a number of iterations of 20 were chosen because the contour of the structures was meticulously defined.

8.1 Edition of the Solid Model Obtained for the Reconstruction of the 3D Tomography

After obtaining the solid model through the SCAN IP, the archive is saved as an stl. archive to import it as an object and clean the computerized archive of impurities using the program Meshmixer which allows to sculpt, analyze and edit a solid model. First, the obtained archive is imported.

After importing the archive to be analyzed, the program starts searching those points which are not joined, in conjunction with the empty spaces between solid structures automatically using the tool Inspector of the menu Analysis, the toolbar of the program can be found, which detects automatically those points and highlights the points that need to be joined as shown in Fig. 34.

Multiple tools were used in order to sculpt the model, which can be found in the toolbar of the Sculpt menu and were used in order to eliminate imperfections of the model's surface and give it a better texture for its posterior printing, having as result the model.

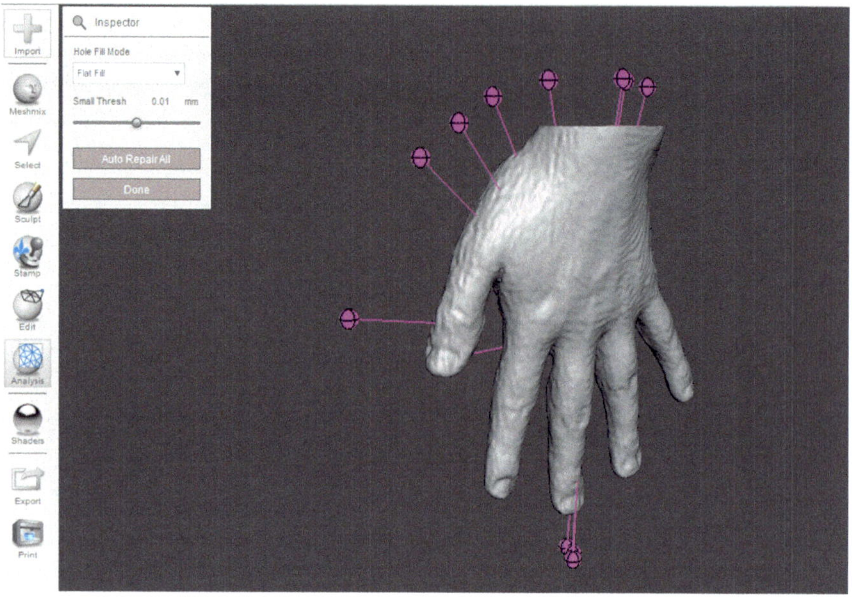

Fig. 34 Inspection of the points and empty spaces of the solid through the Meshmixer program

After sculpting archive is newly saved in stl. format and is imported to the computational program SolidWorks to revise the model's surface and check if there are still empty spaces or imperfections that could cause problems when submitting the model to force simulations to prove its resistance. Also for the development of this particular step, researchers opted to create an osseous like part as a guide for the joint's design. For this purpose, the same steps were followed to create a tridimensional model of the healthy hand, as shown in Fig. 35.

As it can be appreciated in Figs. 35 and 36 a particular cut was made to analyze the structure of the model and make sure that the conjunction of points was able to create a solid model, which lacks empty spaces or imperfections. Posteriorly, the material for further proves was selected and adjusted to perform resistance, force, tension and yielding testing.

8.2 *Angle Association of the Healthy Phalanges Mobility Through the Usage of the Platform Kinovea*

The computational platform Kinovea is a tool that allows the user to take videos or photos of the extremities and joints of the body to track the movement of the body's part and also adjust the location of the point to develop the creation of a path and analyze the movement and angles performed by a specific biomechanical structure.

Fig. 35 Tridimensional model of the hand's osseous structure after completing the previously mentioned steps. The archive was exported to the SolidWorks platform and can be appreciated as the conjunction of thousands of points

Fig. 36 Tridimensional model of the hand's structure after completing the previously mentioned steps. The archive was exported to the SolidWorks platform and can be appreciated as the conjunction of thousands of points

The flexion movements of the hand were obtained as a video archive. For instance, the patient needed to perform flexion movements with each finger of the hand until reaching a grip position. Then, a video which captured each one of the moments when the movements were made and the positions were reached, was opened in the Kinovea program, as it can be appreciated in Fig. 37 to analyze the angles of certain points of interest for the researchers in each phase when the complete movement was performed.

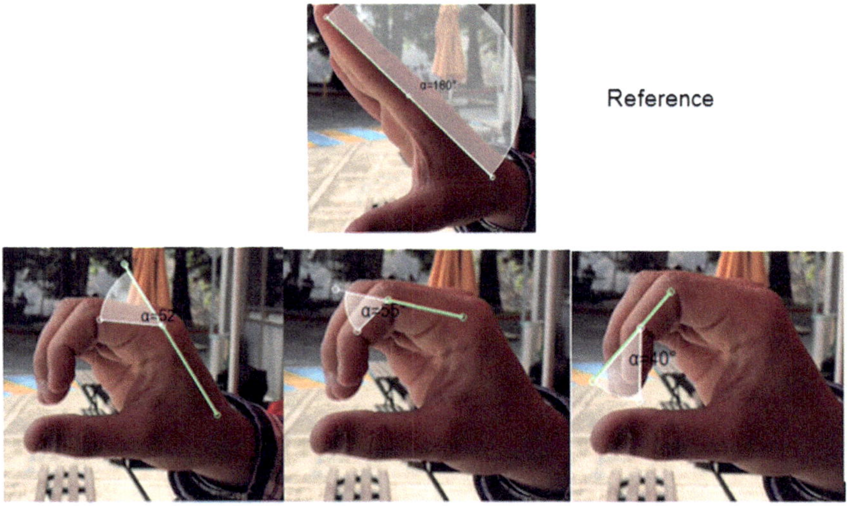

(a) (b) (c)

Fig. 37 Video of the patient's hand movement using the computational platform Kinovea. Measuring bending angles with respect to the anteriorly proximal bone for: **a** phalange metacarpus, **b** falange-phalange, **c** distal phalanx

8.3 Digitalization of the Healthy Patient's Hand

Finally, the healthy patient's hand was scanned with the usage of the ATOS CORE and Artec 3D Scanner in order to adjust the hand-wrist holder (socket) and create a more accurate model. The final archive is composed of a cloud of points called NURBS.

For instance, one of the most important elements of the evaluation process is the fact of achieving a certain immobility of the body part to be evaluated. The patient needs to adopt a static position so that the scanner doesn´t lose the reference point collocated on the patient's skin. This procedure is performed over tridimensional planes and the scanner needs to surround the surface of the object, in this case the patient's extremity, as shown in Fig. 38. Other reference points were collocated on the patient's hand to facilitate the scan process. All points are captured and analyzed by the scanner and multiple layers of the surface are collocated one over the other, only having as guidance the mentioned points. Finally, all the captured layers form a tridimensional model composed of a cloud of multiple points.

Based on the digitalization process, the following morphological parameters were obtained (Table 15).

These results were taken into account to create the mechanism and spherical cover of the final device, which will allow the patient in the development of his principal movement functions.

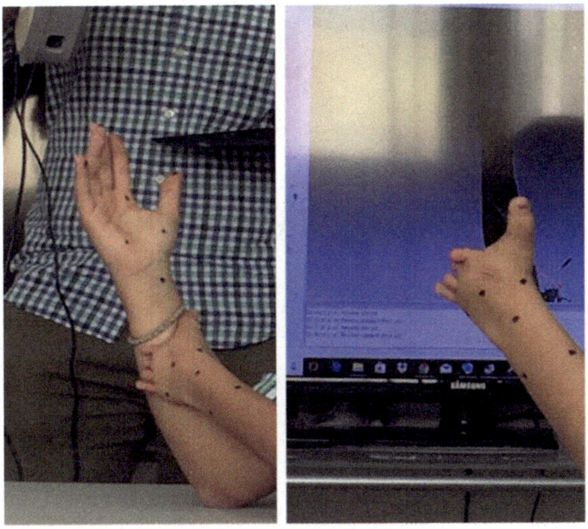

Fig. 38 Scan process of the patient's left hand using ATOS Core and Artec 3D

Table 15 Morphological parameters for each finger

Anthropometric measurements of the patient						
Phalange	Little (mm)	Ring (mm)	Middle (mm)	Index (mm)	Health thumb (mm)	Non health thumb (mm)
Proximal	24.56	41.08	48.30	43.41	30.23	–
Middle	19.88	28.30	27.51	24.54	–	–
Distal	21.27	23.62	23.29	20.05	22.46	–
Proximal-distal	75.78	85.88	93.75	79.53	52.69	37.49

8.4 Selection of the Operational Mechanism

Considering all the performed analysis regarding the advantages and disadvantages of multiple conceptual alternatives and according to the patient's morphology and ergonomic, the costs and assembly facilities.

Considering the analysis of different advantages and disadvantages of alternative concepts and considering the patient's morphology and ergonomic, additionally the prices and facility of the ensemble it was determined that the mechanism to implement on the design it would rather prefer a mechanism with the function of a tensor or tensioner. This is in consequence that the mechanism generated the movement by the forces transmission also by the natural ergonomy and does not it requires a lot of shackles.

On this mechanism, it was used three shackles an a tensioner, which acts like four shackles to produce a mechanism of 4 stick. On the same way, this mechanism

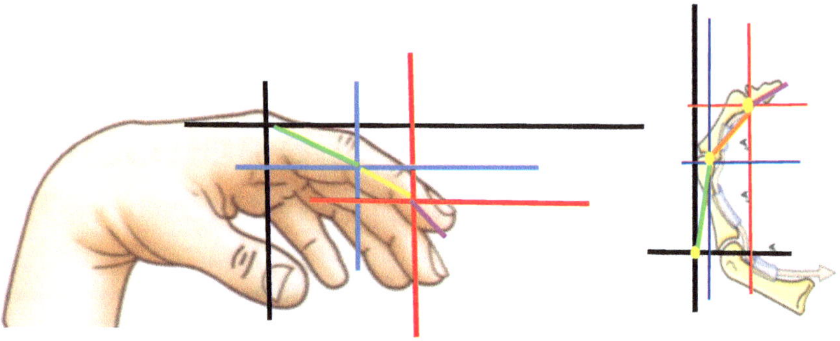

Fig. 39 Kinematic mechanisms planes according to Kutzbatch-Grüber criteria

transfer the movement form the tensioner to the distal shackles of the hand, which transmit the movement to the medium shackles and then successively.

The liberty of a free grade or movement of a mechanism, it could be defined with the number of parameters that must need to fix on this model, in consequence, it could be determined the mechanism position.

Where the mechanisms cinematic plane of the phalanx it is defined by three shackles, one fix shackles and threes cinematic parties (referred to the union if two mechanism members) for the grade of liberty calculation that has the phalanx mechanism (Fig. 39).

Specifically, the Kutzbach-Grübler criteria, explain with the previous concepts, the movement of a flat mechanism. These criteria it was used considering the previously cinematic parameters and just the movements on the plane [11, 12].

After this selection, a prototype of the right hand is made to verify that the operation of the mechanism is adequate and the evolution is shown in Fig. 40.

Finally, the construction of a functional system used a 3D model printing based on 1.8 mm filament of polymers such as ABS (Acrylonitrile Butadiene Styrene) and PLA (Polylactic acid). In particular, the latest one to avoid rejection problems due to allergy and the first one as a low weight material and economic, It is shown in Fig. 41.

9 Structural Behavior in the Articulation of the Distal Phalanx—Index Finger

In order to know the mechanical response of the main joint (Distal-index) a numerical test was carried out, with the NX NASTRAN tool and assigning the loading and boundary conditions, as well as the mechanical properties of the materials (ABS).

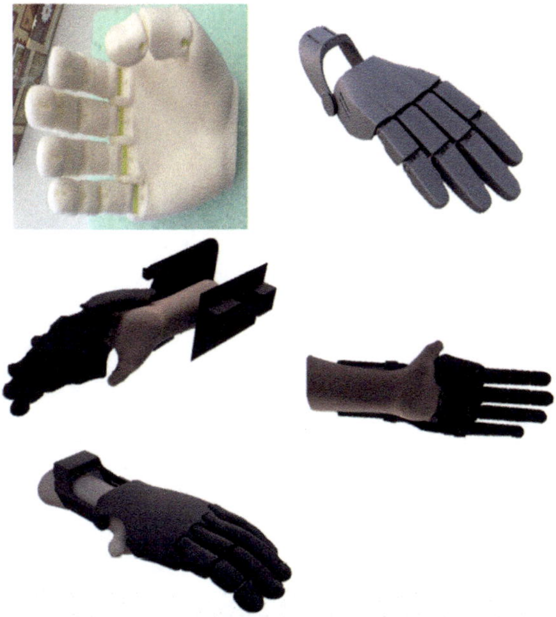

Fig. 40 Evolution of the prototype mechanism of the patient with tensioner and action mechanism

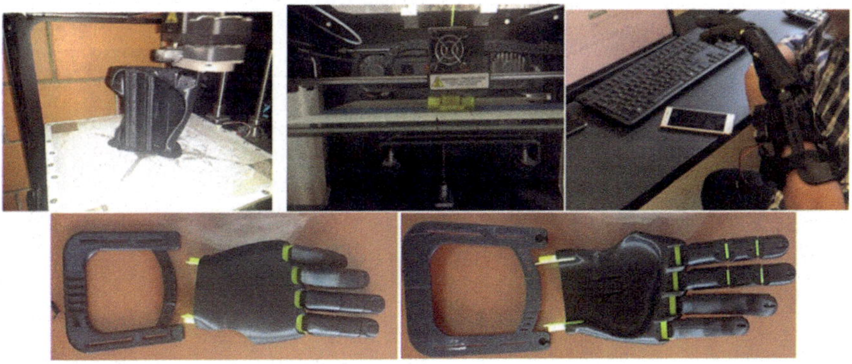

Fig. 41 Additive manufacturing and incorporation of the auxiliary mechanism for the patient

The main assembly considers shared faces created by cuts in the models derived from the tomographic files and converted into faceted solids, in consequence, to obtain contact areas and achieve a correct transfer of the effect of the charges (Fig. 42).

Despite generating a free mesh with tetrahedral elements of ten nodes, only the distal phalanx should analyze the structural effects in the joint base.

In consequence, 13,434 elements and 21,743 nodes were generated for simulation purposes, according to Fig. 43.

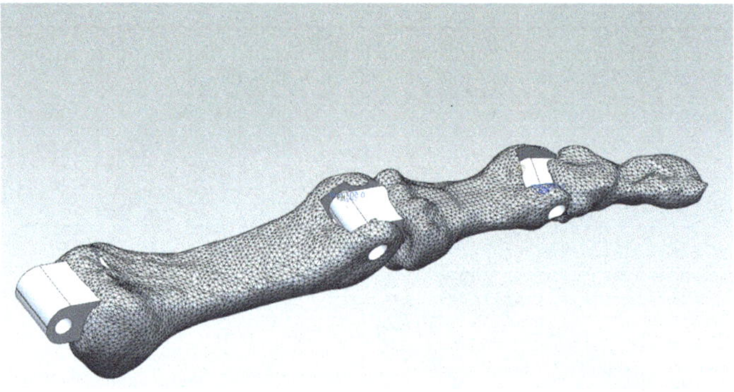

Fig. 42 Assembly of the phalanges if the index finger

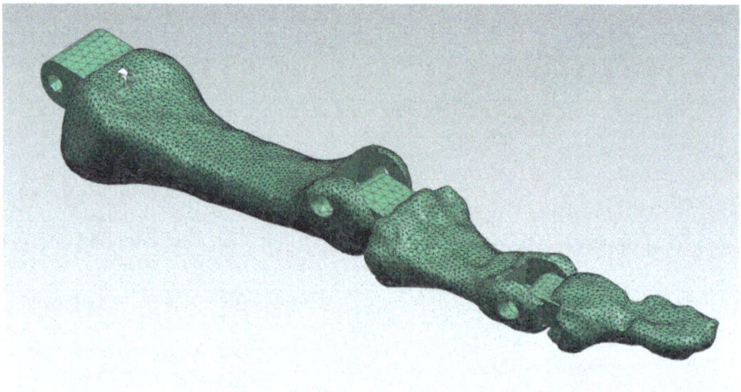

Fig. 43 Mesh assembly of the phalanges of the index finger

It was considered as average load 7 kgf on the distal phalange, simulating a cantilever condition, despite establishing that the device will be based on precision and not on the stiffness [13, 14].

On Fig. 44, the loading and boundary conditions are observed. Movement in the circular joint is restricted to simulate a cantilever condition and demand maximum stress in the surrounding geometric regions.

The final results of the numerical simulation indicate that the maximum displacements in the region near the cantilever in the joint and von Mises stresses do not compromise the structural integrity of the material considered. (ABS—yield stress: MPa, E = MPa and Poisson's relation).

Main results are shown on Figs. 45, 46 and 47.

In accordance to the final stresses, it can be seen that ABS polymer is useful in order to apply common activities in the case of the final prosthesis.

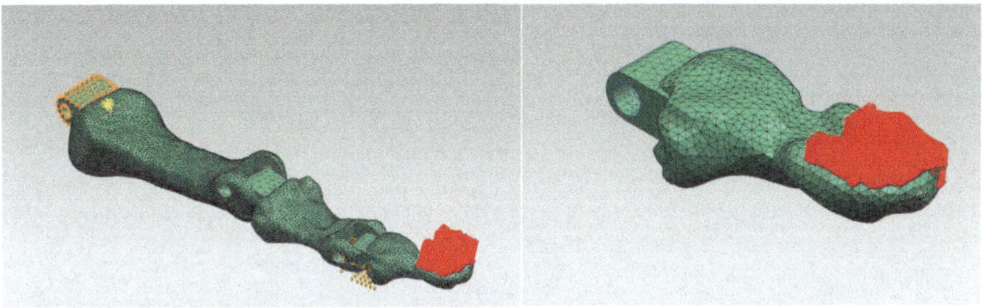

Fig. 44 Loading and boundary conditions in the distal phalanx of the index finger

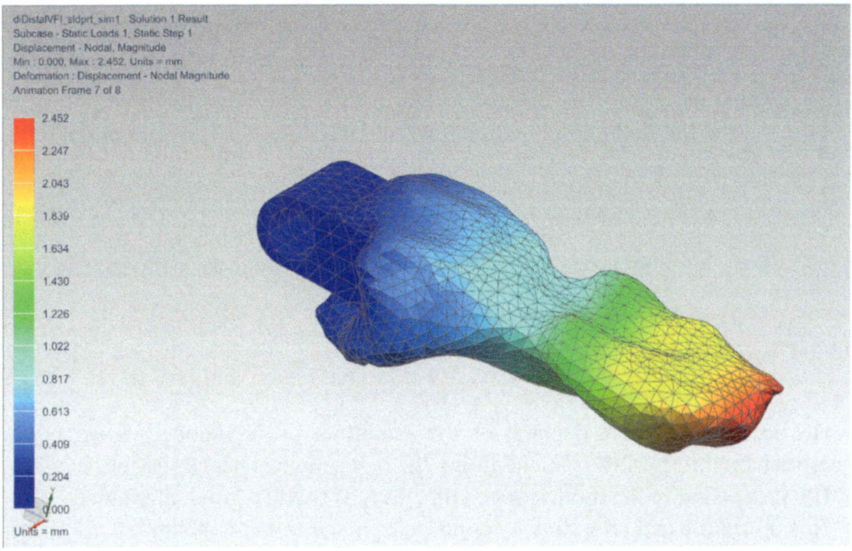

Fig. 45 Displacements resulting from the numerical simulation in the distal phalanx of the index finger

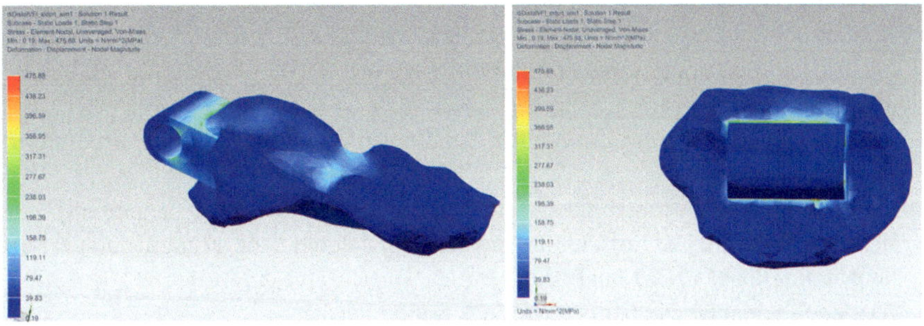

Fig. 46 Von Mises stress from the numerical simulation in the distal phalanx of the index finger

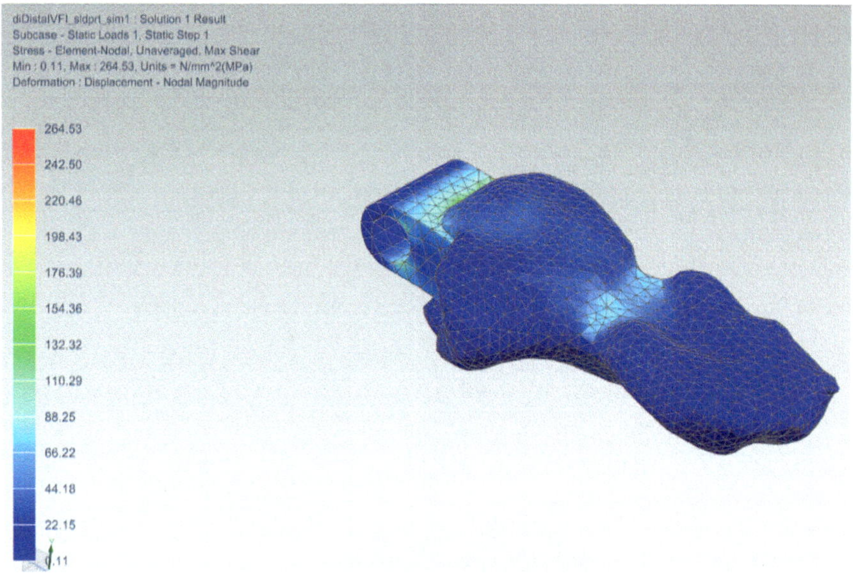

Fig. 47 Shear stress from numerical simulation in the distal phalanx of the index finger

10 Conclusions

- The applied procedure depends on the conditions of the damaged area and the correct alignment of the patient during the magnetic resonance session.
- The processing of the models with DICOM files is still a good alternative.
- This procedure and its simplicity contributes in the solution of similar cases, even though it was a custom system.
- This is one of the most common methodologies that has been approved by the medic specialist.
  The prosthesis needs to be useful.
- Modern engineering evaluation of medical solutions is growing with alternative materials and computational simulations (e.g. orthopedics o maxillofacial areas) [15, 16].
- The knowledge of the biomechanical behavior of the involved bony structures is required.
- The hand prosthesis contributes to the full social integration for the patient.
- The bone structure and muscles are the main for calculate the strength and perform the gripping tasks (FEM analysis)
- Very good advantages of this procedure can be listed:
- Geometrical parameters were obtained from a magnetic resonance and they were adjusted in the user and 3D quick printing models directly.

- Experimental testing on the printed material as well as mechanical analysis of the design before clinical use is essential for the prosthesis to display optimal results and in order to avoid malfunction of the system.
- This is a multidisciplinary job.
- To perform the procedure at the lowest cost
- To simulate with FEM Lattice and Scaffolds structures in order to reduce cost can be a good alternative [17, 18].

Acknowledgements The authors acknowledge the support given by the National Council for Science and Technology, CONACYT, Mexico; the National Polytechnic Institute, IPN, and Hospital La Raza (IMSS) and CILAB (Biomechanics laboratory) for the means and facilities for the development of this research.

References

1. Kikinis R, Pieter S (2011) 3D Slicer as a tool for interactive brain tumor segmentation. In: 33rd Annual international conference of IEEE EMBS, Boston, MA, USA
2. Siddiqi AA, Khawaja A, Zafar M, Husain M (2012) 3D Representation of human skull bearing the tumor using image processing. In: Fourth international conference on computational intelligence, communication systems and networks. IEEE
3. Day PA, Digvijay N, Jeeva JB (2014) Pre-operative brain tumor segmentation using Slicer 3D. 2014 international Conference on green computing communication and electrical engineering (ICGCCEE)
4. Pieter S, Halle M, Kikinis R. (2004) 3D slicer. Surgical Planning Laboratory, Brigham and Women's Hospital, Isomics Inc. IEEE
5. Memorang. [Online]. Available. https://www.memorangapp.com/flashcards/67336/Anatomy+%28UL%29%3A+Extensor+Forearm+and+Hand/
6. Velázquez Sánchez AT (2007) Caracterización Cinemática e Implementación de una mano robótica multiarticulada. Ph.D. thesis, IPN, Instituto Politecnico Nacional ESIME, Mexico City
7. Hollerbach K, Ashby AE, Logan C, Martz H, Bossart PL, Rikard D (1996) High-resolution hand dataset for joint modeling. In: 18th annual international conference of the IEEE Engineering in Medicine and Biology Society, Amsterdam
8. Van De Giessen M, Smitsman H, Strackee SD, van Vliet LJ, Grimbergen KA, Streekstra GJ, Vos FM (2009) A statistical description of the articulating Ulna surface for prosthesis design
9. Wikipedia, Encyclopedia online. [Online]. Available. https://es.wikipedia.org/wiki/Mano
10. Synopsis 2017. SimpleWare. [Online]. Available https://simpleware.com/support/videos/?NM6j8ra2Ikw
11. J. R. (n.d.). Las Diferentes Formas del Criterio de Kutzbach-Grübler [PDF]. Salamanca: Facultad de Ingeniería Mecánica, Eléctrica y Electrónica de la Universidad de Guanajuato
12. Bhagat AP (2012) Medical images: formats, compression techniques and DICOM image retrieval a survey. IEEE
13. Escalona P, Naranjo J, Lagos V, Solís F (2009) Parámetros de Normalidad en Fuerzas de Prensión de Mano en Sujetos de Ambos Sexos de 7 a 17 Años de Edad. Revista chilena de pediatría 80(5):435–443 [Online]. https://dx.doi.org/10.4067/S0370-41062009000500005
14. Servomotor SG90. [Online]. Available. http://www.ee.ic.ac.uk/pcheung/teaching/DE1_EE/stores/sg90_datasheet.pdf
15. Moreno-Garibaldi P, Beltran-Fernandez JA, Hernandez-Gomez LH, Lopez-Saucedo F, Corro-Valdez N, Lopez-Lievano A (2017) Mechanical characterization of composite material focused on maxillofacial applications. DYNA 92(4):461–465

16. Microchip, "PIC18F2455/2559/4455/4550 Data Shett'. Microchip Technology 2009
17. Hernández-Gómez LH et al (2019) Characterization of Scaffold structures for the development of prostheses and biocompatible materials. In: Öchsner A, Altenbach H (eds) Engineering design applications. Advanced structured materials, vol 92. Springer, Cham
18. Spong MW, Hutchinson S, Vidyasagar M (2006) Robot modeling and control. Wiley, London
19. Miyazaki N, Kochi M, Mochimaru M, Kurihara T (2005) Finger joint kinematics from MR images. In: 2005 IEEE/RSJ international conference on intelligent robots and systems (IROS 2005)
20. Fisioterapia, Salud & bienestar. [Online]. Available: https://i2.wp.com/fisiostar.com/wp-content/uploads/2011/05/Huesos-de-la-Mano-y-Mu%C3%B1eca.jpg
21. Battula S, Njus GO (2004) A new method to develop finite element model of the bones in the hand from CT scans. IEEE

Reconstruction and Finite Element Evaluation of a Calcaneous Implant by Stereolithographic 3D Printing Technique

Juan Alfonso Beltrán-Fernández, Juan Carlos Hermida-Ochoa, Adolfo López-Lievano, Luis Héctor Hernández-Gómez, Berenice Uribe-Cortes, Pablo Moreno-Garibaldi and Nefi Pava-Chipol

Abstract Flexible flat foot in infants can cause pain and physical limitation. Surgical treatment is preferred when pain is severe. Evans enlargement of the lateral aspect of the calcaneus has good results but has complications due to the collapse of bone grafts. Tridimensional segmentation of anatomical areas allows to obtain stereolithographic models that can be numerically analyzed. A stereolithographic model of a healthy calcaneus was obtained from a CT and virtually sectioned to simulate Evans osteotomy. A wedge-shaped implant was made and implanted as described by Evans. Both models were analyzed with a finite element software, the calcaneus model comprised cortical and cancellous bone layers (E = 20 GPa and 457 MPa respectively) The wedge-shaped was assigned PMMA properties (E = 2.4GPa). Three phases of human gait were simulated and stress obtained. The high-

J. A. Beltrán-Fernández (✉) · A. López-Lievano · L. H. Hernández-Gómez · P. Moreno-Garibaldi · N. Pava-Chipol
INSTITUTO POLITÉCNICO NACIONAL - Escuela Superior de Ingeniería Mecánica y Eléctrica - Sección de Estudios de Posgrado e Investigación Edificio 5, 2do Piso, Unidad Profesional Adolfo López Mateos "Zacatenco" Col. Lindavista, C.P. 07738 Ciudad de México, Mexico
e-mail: jbeltranf@hotmail.com

A. López-Lievano
e-mail: k_lievano@hotmail.com

L. H. Hernández-Gómez
e-mail: luishector56@hotmail.com

P. Moreno-Garibaldi
e-mail: pmg170588@gmail.com

N. Pava-Chipol
e-mail: sowbran@gmail.com

J. C. Hermida-Ochoa · B. Uribe-Cortes
Centro de Investigación y Laboratorio Biomecánico, Carmen #18, Chimalistac San Ángel, 01070 Ciudad de México, CDMX, Mexico
e-mail: jhermida@gmail.com

B. Uribe-Cortes
e-mail: bereniceuc@gmail.com

A. Öchsner and H. Altenbach (eds.), *Engineering Design Applications II*, Advanced Structured Materials 113, https://doi.org/10.1007/978-3-030-20801-1_27

est von Mises stress occurred at the stance phase (18.25 MPa). None of the stress exceeded the yield stress for all materials in all phases. Reconstruction of anatomic models and implants using a stereolithographic technique allow to predict stress and failure. These models could also be printed 3D to elaborate novel implants.

Keywords PMMA · Flat-foot · Finite element · Stereolithographic models · 3D printing

1 Introduction

Flexible flatfoot (FFF) in infants is a normal condition that often needs no treatment since is asymptomatic and usually corrects spontaneously in the first decade of life [1–3]. Congenital FFF with shortening of the Achilles tendon in six to sixteen years old infants is often painful and therefore disrupts locomotion of these patients.

Conservative treatment can help to relieve pain [4, 5] but, in some cases, surgery is the only course of action to alleviate pain and enhance deambulation. [6–10]. A standard treatment for this condition is the Evans surgical method which aims to correct the longitudinal arch of the foot by surgical lengthening of the sagittal axis of the calcaneus through an osteotomy and placement of a wedge-shaped bone graft to maintain the lengthening [11–13], which is often obtained from the iliac crest of the patient and it has to be cut to a specific geometry as described by Evans et al., and it has to have certain specific measures to effectively preserve the surgical enlargement of the calcaneus bone.

Postoperative collapse and subsidence of the bone graft are frequent [13] and result in structural failure of the calcaneus lengthening probably due to poor technique while harvesting and shaping the bone graft. Also, infection and severe pain at the area of bone harvesting at the iliac crest are rare but are possible complications. Some metal devices are available to replace the need for bone grafting but they are expensive and not readily available in emerging economy countries.

Poly Methyl Methacrylate [PMMA] also known as bone cement is a well-known polymer which is vastly used in orthopedics mainly to bond implants to bone as in the case of joint replacement surgery. PMMA has high tensile and compressive strength and it is also highly biocompatible [14, 15], its readily available worldwide and inexpensive. PMMA has been also used to effectively cover craniofacial bone defects and novel 3D printing techniques are in use to allow an anatomical replacement of bone defects [16–18].

In spite of these advantages the use of PMMA as a stand-alone material to elaborate weight-bearing implants is only anecdotical and has been used in few cases as temporary spacers in a revised joint that suffered an infection, but not as a permanent solution. The advent of 3D printers allows the production of a myriad of geometries using different materials including resins such as PMMA. A CT scan of certain region of a bone defect or whereas created during an accident or due to an osteotomy can be segmented to obtain a precise anatomical representation of the defect and a virtual 3D model can be created by the use of a CAD software and then printed with a suitable material for implantation in the real defect in vivo.

The present study aims to characterize and analyze a numerical model of a PMMA implant to predict sustainable structural support to the calcaneus bone after an Evans calcaneal osteotomy has been performed when treating painful FFF in infants without the need of additional bone grafting. In this study a wedge-shaped PMMA implant was designed using CAD software and based on the local geometry of the calcaneus osteotomy and measurements proposed by Mosca et al.

The calcaneus bone 3D model was obtained by segmentation of CT scan images and an Evans osteotomy was virtually performed. A Finite element analysis (FEA) in the ANSYS program, evaluated stress between the bone interphase of the calcaneus and the implant to account for mechanical performance during the three phases of the gait cycle.

2 Materials Ant Methods

Using a computerized axial tomography of a cadaveric specimen of a healthy 25 years old woman with a cut resolution of 0.6 mm, three bones of the foot: distal tibia segment, talus and calcaneus were obtained using image reconstruction techniques using a segmentation program (ScanIP) A virtual osteotomy was performed on the calcaneus model by sectioning and enlarging the lateral aspect of the bone as described by Evans [11] (Fig. 1a).

A curve obtained by contouring the edge of the calcaneus bone section at the site of the osteotomy was extruded to the proper dimensions to represent wedge shaped implant, and posteriorly, it was 3D printed in high impact polyesthyrene (Zortrax M300 printer) just to have a physical representation and plan a future mechanical

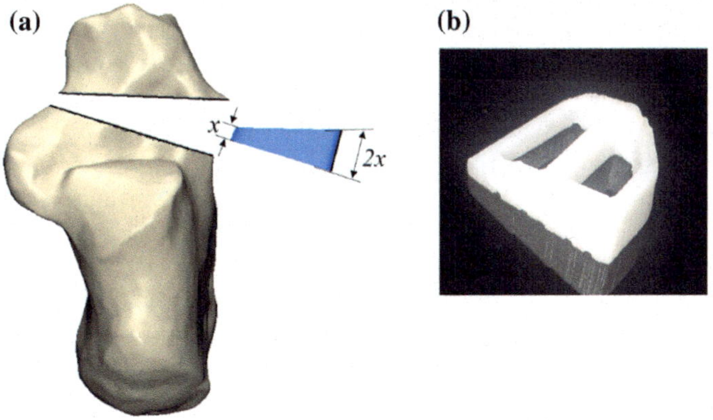

Fig. 1 Evans surgical osteotomy for the calcaneus lengthening (**a**), and wedge-shaped 3D printed implant (**b**)

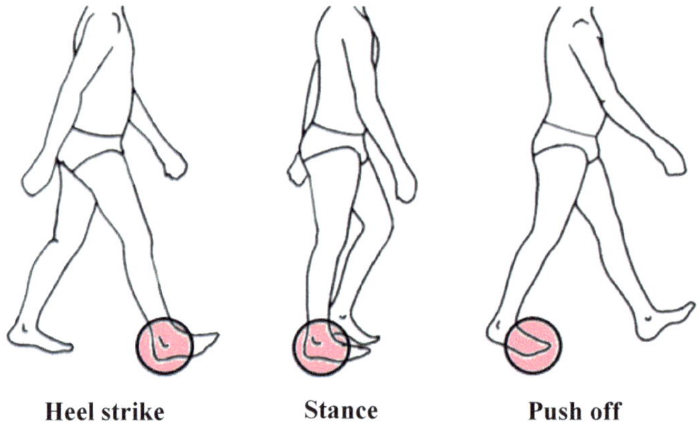

Fig. 2 The three phases of the gait cycle

test on a definitive PMMA printed implant (Fig. 1b). The three main phases of the human gait were simulated (Fig. 2): Heel strike, Stance and Push off, using a Finite element program (ANSYS).

3 Numerical Simulation

The finite element model included cortical and cancellous bone in each of the three bones.

A coefficient of friction of 0.2 [19] was applied between surfaces in interaction: tibia-talus, talus- calcaneus; as well as between the calcaneus and the PMMA implant. Bonded contact was applied to the contact surfaces between the cortical and the cancellous surfaces of each bone.

In all three phases, three times body weight (considering an infant weighing 40 kg) was applied to the tibia in the vertical direction; to represent the position of each of the phases, an angle of 30° was considered for both the Heel strike and Push off phases, and 15° for the Stance phase as shown in Fig. 3.

In both cases, the Heel Strike and Stance phase a restriction was applied in the vertical direction (z) representing the floor, movement was restricted in the x direction on the articular face of the calcaneus to simulate the cuboid restriction. For both talus and calcaneus, rotation was allowed on the x axis to allow greater interaction between the surfaces and the implant; the latter was left free only conditioned by the coefficient of friction to anticipate possible dislocation in any of the prescribed phases (Fig. 3).

The mesh of the model generated in ANSYS, has 367,590 tetrahedral elements and the mesh was refined in the contact zones between the calcaneus and the implant to obtain greater accuracy in the prediction of the stress and displacements, and the quality of the elements was achieved between 0.88 and 1 in those areas.

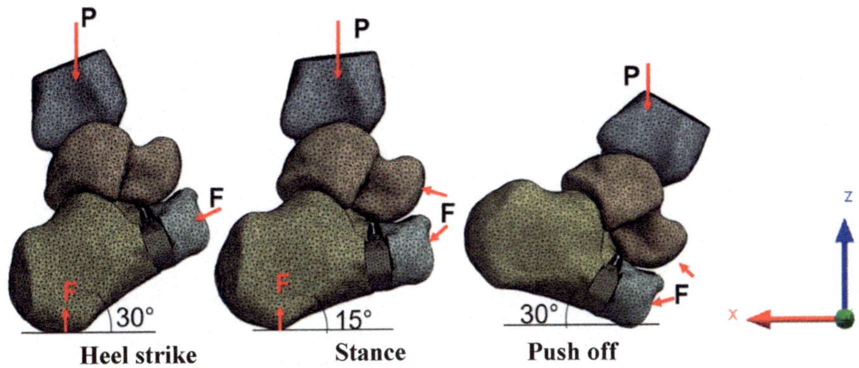

Fig. 3 Boundary conditions of the FEA model, where P = 1177 N (3 three times body weight considering an infant weighing 40 kg), F = fixed

Table 1 Mechanical properties of the materials

Material	Young's Modulus	Poisson' ratio
Cortical bone [20]	20 GPa	0.35
Cancellous bone [21]	457 MPa	0.3
PMMA [22]	2.4 GPa	0.3

To obtain a more accurate prediction of the interaction between the implant and the bone, mechanical properties of both cortical and cancellous bone were included. In Table 1, the mechanical properties used in the finite element model are shown.

4 Results

During Heel strike phase, the equivalent von Mises stress were 3.71 MPa in the implant, at the interphase between the smallest thickness of the implant and the cortical of the anterior and posterior part of the calcaneus, however, in the complete model, the greatest stress was generated in the middle part of the calcaneus (Fig. 4).

During the push-off phase, the highest von Mises stress was mainly generated in the upper and lateral part of the neck of the talus in the complete model; and in the implant were 5.18 MPa, and are located in the same region as in the heel strike phase (Fig. 5).

The stance phase registered the highest stress, concentrated mainly in the calcaneus from the upper facet, passing through the middle part at the peroneal tubercle to the tuberosity.

The wedged-implant had the highest von Mises stress of 18.25 MPa at the interphase between de smallest thickness of the implant and the cortical of the anterior and posterior part of the calcaneus (Fig. 6).

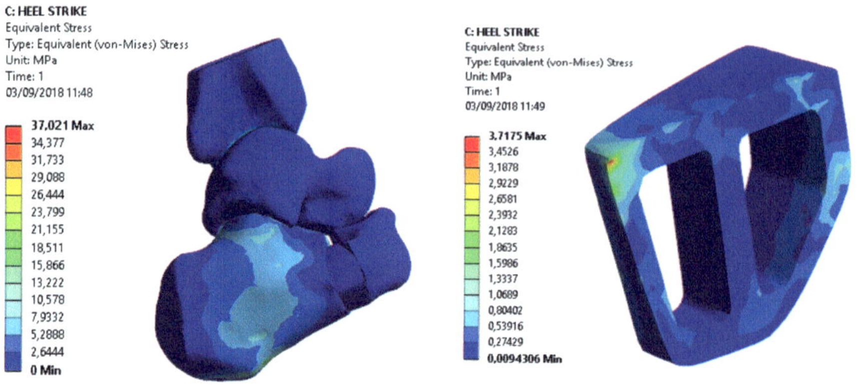

Fig. 4 von Mises stress in Heel Strike phase

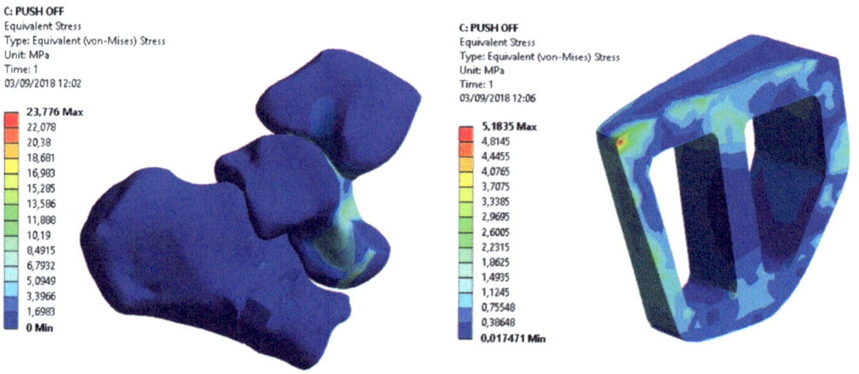

Fig. 5 von Mises stress in Push Off phase

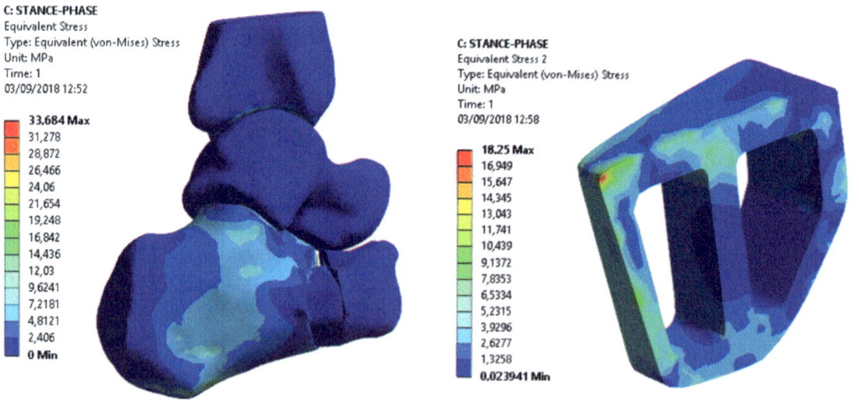

Fig. 6 von Mises stress in Stance phase

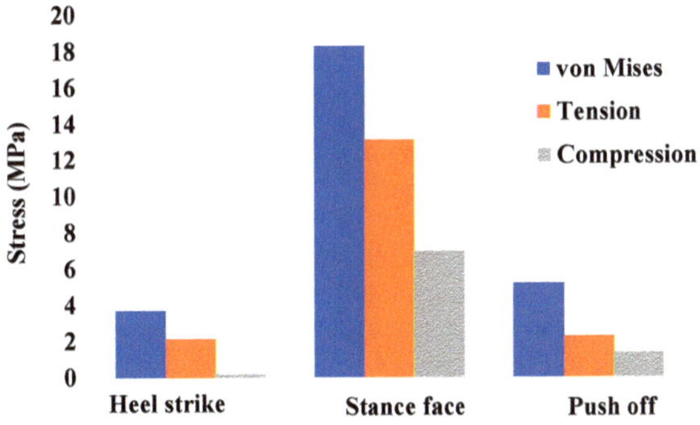

Fig. 7 Maximum principal and equivalent von Mises stress in the implant

The maximum displacement of the implant originated in the Stance phase and was 0.01792 mm, which indicates that there was no dislocation of the implant.

Figure 7 shows the comparison of the tensile and compressive principal stress, and the equivalent von Mises stress.

5 Discussion

Musculoskeletal deformities such as FFF can cause limitation and pain specially in infants decreasing their development and quality of life [3, 4]. Enlargement of the lateral aspect of the calcaneus bone by performing an Evans osteotomy is an effective way to treat FFF in infants and since it corrects the mid foot misalignment inherent to this pathology [4, 5, 14]. Mechanical failure of this enlargement is mainly due to the collapse and subsidence of the wedge-shaped bone graft. Also, since the bone graft is harvested at the patient's iliac crest it could cause severe pain and even infection. Some manufacturers offer metallic implants that could effectively substitute a bone graft, however these are expensive and not readily available worldwide and since their modulus is much higher that the surrounding bone these could potentially subside into the bone, therefore, modifying the lengthening of the calcaneus. Tridimensional printing technology offers a substantial potential to elaborate biological implants but we believe it has not to be limited to only characterize anatomical shapes to correct physical defects, but also yield the defect region with biomechanical properties such as functional weight bearing.

Computerized Segmentation of diverse anatomical through CT scan areas allow a thorough understanding of defects and deformities of the musculoskeletal system. Once an accurate segmentation has been achieved a precise 3D model can be obtained and the design of novel customized implants is highly facilitated. In our study we designed an implant that replicate measurements and geometry of the bone graft described by Evans. A numerical representation of this implant with assigned material properties of PMMA was made and then tested against geometrically accurate models of both segments of the calcaneus. Once the implant has been numerically tested a lithography can be 3D printed in PMMA for further mechanical analysis. PMMA is a 3D printable composite that offers optimal mechanical support to the bone due to its modulus [14, 15], and it has shown through the years of use in orthopedic surgery its high bio compatibility.

Results of our FEA analysis on a wedge-shaped PMMA implant showed that during all three-gait phase simulation stress remained well below its yield stress [23]. The highest peak principal stress occurred during the stance gait at 13 MPa which theoretically will not predict cyclic failure during the integration time of the implant after about 8 weeks or less than 100 K gait cycles. von Mises stress concentrated more on the thinnest edge of the implant (medial side) which might explain why a wedge-shaped bone graft often collapses and fail in the in vivo Evans procedure. Even when the wedge-shaped PMMA implant was not restrained in all displacement and rotation axis during the simulation, no dislodgement occurred during any of the gait phases which might suggest that the geometry could be used as a stand-alone implant without the need of extra fixing features.

6 Conclusions

The final numerical analysis suggested that this kind of implant has the potential to be implanted in vivo and yield optimal mechanical support to the Evans osteotomy instead of the use of an autologous bone graft. It seems to represent a good stand-alone device to allow the surrounding bone to grow and integrate the device between the two divided segments. A Mechanical test of the printed PMMA is still needed to validate the FEA simulation and will be performed in a future experiment.

It is thought that 3D printing and scaffolds offers potential for manufacturing novel implants aiming to solve various skeletal problems. FEA offers an essential tool to evaluate the behavior these implants could have in the skeletal system before they are allowed to be implanted in patients [24, 25].

References

1. Bouchard M, Mosca VS (2014) Flatfoot deformity in children and adolescents: surgical indications and management. JAAOS-J Am Acad Orthop Surg 22(10):623–632
2. Das SP, Das PB, Ganesh S, Sahu MC (2017) Effectiveness of surgically treated symptomatic plano-valgus deformity by the calcaneo stop procedure according to radiological, functional and gait parameters. J Taibah Univ Med Sci 12(2):102–109
3. Mosca VS (2010) Flexible flatfoot in children and adolescents. J Child Orthop 4(2):107–121
4. Evans MJ (1998) Podiatry and paediatrics. Curr Paediatr 8(4):237–241
5. Kelikian A, Mosca V, Schoenhaus HD, Winson I, Weil L Jr (2011) When to operate on pediatric flatfoot. Foot Ankle Spec 4(2):112–119
6. Bauer K, Mosca VS, Zionts LE (2016) What's new in pediatric flatfoot? J Pediatr Orthop 36(8):865–869

7. Mosca VS, Bevan WP (2012) Talocalcaneal tarsal coalitions and the calcaneal lengthening osteotomy: the role of deformity correction. JBJS 94(17):1584–1594
8. Mosca VS (2014) Management of the painful adolescent flatfoot. Tech Foot Ankle Surg 13(1):3–13
9. Patrick N, Lewis GS, Roush EP, Kunselman AR, Cain JD (2016) Effects of medial displacement Calcaneal Osteotomy and Calcaneal Z Osteotomy on subtalar joint pressures: a cadaveric flatfoot model. J Foot Ankle Surg 55(6):1175–1179
10. Ragab AA, Stewart SL, Cooperman DR (2003) Implications of subtalar joint anatomic variation in calcaneal lengthening osteotomy. J Pediatr Orthop 23(1):79–83
11. Evans D (1975) Calcaneo-valgus deformity. J Bone Joint Surg 57(3):270–278 (British volume)
12. DuMontier TA, Falicov A, Mosca V, Sangeorzan B (2005) Calcaneal lengthening: investigation of deformity correction in a cadaver flatfoot model. Foot Ankle Int 26(2):166–170
13. Mosca VS (1995) Calcaneal lengthening for valgus deformity of the hindfoot. Results in children who had severe, symptomatic flatfoot and skewfoot. JBJS 77(4):500–512
14. Yang J, Zhang K, Zhang S, Fan J, Guo X, Dong W, … Yu B (2015) Preparation of calcium phosphate cement and polymethyl methacrylate for biological composite bone cements. Med Sci Monit Int Med J Exp Clin Res 21:1162
15. McCormack BAO, Prendergast PJ (1999) Microdamage accumulation in the cement layer of hip replacements under flexural loading. J Biomech 32(5):467–475
16. De La Peña A, De La Peña-Brambila J, Pérez-De La Torre J, Ochoa M, Gallardo GJ (2018) Low-cost customized cranioplasty using a 3D digital printing model: a case report. 3D Printing Med 4(1):4
17. Kim BJ, Hong KS, Park KJ, Park DH, Chung YG, Kang SH (2012) Customized cranioplasty implants using three-dimensional printers and polymethyl-methacrylate casting. J Korean Neurosurg Soc 52(6):541
18. Fernandes da Silva AL, Borba AM, Simão NR, Pedro FLM, Borges AH, Miloro M (2014) Customized polymethyl methacrylate implants for the reconstruction of craniofacial osseous defects. Case reports in surgery
19. Lennon A, McCormack B, Prendergast P (2003) The relationship between cement fatigue damage and implant surface finish in proximal femoral prostheses. Med Eng Phys 25(10):833–841
20. McCormack B, Prendergast P (2018) Microdamage accumulation in the cement layer of hip replacements under flexural loading
21. Nicholson P, Strelitzki R (1999) On the prediction of Young's modulus in Calcaneal Cancellous bone by ultrasonic bulk and bar velocity measurements. Clin Rheumatol 18(1):10–16
22. Jeffers J, Browne M, Lennon A, Prendergast P, Taylor M (2007) Cement mantle fatigue failure in total hip replacement: experimental and computational testing. J Biomech 40(7):1525–1533
23. Race A, Miller M, Mann K (2008) A modified PMMA cement (sub-cement) for accelerated fatigue testing of cemented implant constructs using cadaveric bone. J Biomech 41(14):3017–3023
24. Hernández-Gómez LH et al (2019) Characterization of scaffold structures for the development of prostheses and biocompatible materials. In: Öchsner A, Altenbach H (eds) Engineering design applications. Advanced structured materials, vol 92. Springer, Cham
25. Beltrán-Fernández JA et al (2019) Design and Manufacturing of a Temporomandibular Joint (TMJ) prosthesis for mandibular bone necrosis using the finite element method. In: Öchsner A, Altenbach H (eds) Engineering design applications. Advanced structured materials, vol 92. Springer, Cham

Development of an Auxiliary Biomechanical Assistance System for Blindness or Partial Vision

L. H. Hernández-Gómez, J. A. Beltrán-Fernández, J. Martínez-Paredes, J. Medel-Ortiz, Itzel Bantle-Chávez, Carolina Alvarado-Moreno, A. Vega-López, D. Villaseñor-Chávez and J. O. Miranda-Peña

Abstract In the present work a proposal for the design of an auxiliary device for people who suffer from blindness or weak vision in form of a diadem, is developed. The device is capable of orientating patients in an urban environment. The main goal of this project is to assist patients when performing displacements and diminish the possible collisions with objects or obstacles at the height of their heads or at a middle height (abdomen height), so that patients can have more freedom when performing a certain movement or displacement. The auxiliary system emits vibrations with certain intensity, according to the distance between the object and the patient. The

L. H. Hernández-Gómez · J. A. Beltrán-Fernández (✉) · I. Bantle-Chávez ·
C. Alvarado-Moreno · A. Vega-López · D. Villaseñor-Chávez · J. O. Miranda-Peña
INSTITUTO POLITÉCNICO NACIONAL - Escuela Superior de Ingeniería Mecánica y Eléctrica - Sección de Estudios de Posgrado e Investigación Edificio 5, 2do Piso, Unidad Profesional Adolfo López Mateos "Zacatenco" Col. Lindavista, C.P. 07738 Ciudad de México, Mexico
e-mail: jbeltranf@hotmail.com

L. H. Hernández-Gómez
e-mail: luishector56@hotmail.com

I. Bantle-Chávez
e-mail: itzi.bantle@gmail.com

C. Alvarado-Moreno
e-mail: aly-oroshiwa@hotmail.com

A. Vega-López
e-mail: quijhua_ceci@hotmail.com

D. Villaseñor-Chávez
e-mail: raiden2500@hotmail.com

J. O. Miranda-Peña
e-mail: orlando_1614@hotmail.com

J. Martínez-Paredes · J. Medel-Ortiz
Universidad Politécnica del Valle de México, Av. Mexiquense s/n esq, Universidad Politécnica, Col. Villa Esmeralda, CP. 54910 Estado de México, Mexico
e-mail: jonmart.biomecanica@gmail.com

J. Medel-Ortiz
e-mail: jaquelinemedelortiz@hotmail.com

A. Öchsner and H. Altenbach (eds.), *Engineering Design Applications II*,
Advanced Structured Materials 113, https://doi.org/10.1007/978-3-030-20801-1_28

vibrations present the tendency to be stronger when the distance between the patient and the object is shorter and rather weak when the distance is longer. This mechanism avoids the usage of sounds or similar tools that normally confuse or disorientate the user.

Keywords Visual incapacity · Blindness · Diadem for blind people · Feedback trough vibration

1 Introduction

Blindness is defined as the total loss of the vision, which cannot be improved through the usage of glasses or contact lenses. This affliction can occur with an unexpected suddenness or through the years, being cataloged as total or partial blindness.

According to a report which was recently published by the World Health Organization (WHO) [1], approximately 81.7% of the 39 million people who suffer of blindness, are in an age of 50 years or older. These patients present an inherent and increasing risk while having trouble when walking or trying to displace in a known environment. Despite the existence of multiple navigation devices, such as guiding sticks and white canes, the assistance they can offer present several limitations and don´t eliminate the risk of getting hurt by colliding against the surrounding objects. The existing electronic devices are capable of resolving that particular problem only at a certain point, but in most of the cases, these auxiliary system can´t detect the so called negative and dangerous obstacles, such as descendant stairs and road borders. In accordance with the National Institute of Statistics and Geography in 2010, the second most important disability in Mexico was the visual impairment, reaching an alarming percentage of 43.24% of the total population (approximately 48 millions), who require personalized healthcare services. Amongst the patients, 1.4 million are under the age of 15 years and suffer of irreversible visual disability [2].

Currently, an increasing number of people with a certain type of disability is growing. However, the visual impairment and total blindness occupy the first or second place worldwide. The disability is classified in five categories of visual deterioration, being the most frequent causes the following ones: noninfectious or progressive processes, usually accompanied of multiple risk factors and nosologic, systematic entities which provoke the so called multiple disability and ocular deficit [3].

The alteration of the visual system presents several consequences, amongst them difficulties in the development of daily activities and the impossibility of acquiring visual information, limitation in the motor development and physical skills, together with cognitive limitations and a crescent tendency of the lack of being able to acquire social and communicative skills. Being this the case, the design and creation of special auxiliary tools that allow the development of the patients with these conditions could help improve their life quality.

2 Problem Statement

The intelligent systems that are capable of aiding blind people have a high cost in the international market. Because of the low acquisitive power in Mexico, most of the people are not able to buy an intelligent stick or an aiding visual device in order to perform daily activities. Even obtaining a conventional stick is complicated for a large number of patients. The social situation creates the need to develop and implement an intelligent prototype for blind people at a lower cost and as a social inclusion project.

3 Antecedents

In 1910 the first telescopes, microscopes and loups were designed and built with the usage of optical systems, very similar in design to the current devices. Unfortunately these systems were heavy, expensive and of limited fabrication. During the Second World War the production of these instruments increased drastically and the use of specialized materials, designed specifically for an optimal function, allowed the massive fabrication of the optical instruments, making them cheaper and with the time better [4].

In regard of the autonomous displacement, before the large, white cane was launched into the market as a mobility aid device, other type of canes were used to assist blind people. Those techniques were unreliable, taught by blind professors and usually possessed a lower security range [5]. After the First World War, guide dogs were started to get trained. The United States was a pioneer in this subject, being the country were several institutions, specialized in dog training like The Seeing Eye Institution were founded. This organization was the first training school for dogs. The founder was Dorothy Harrison Eustis and she opened the doors of the institution in 1929.

The whole picture changed during the 1940s decade (Second World War), when the doctor Richard Hoover realized the fact that the canes were too short, heavy and not as efficient as they were supposed to be when patients needed to evade obstacles. For this reason, he designed a large cane made of aluminum tubes, commonly used to build military planes and also developed an unique method to use it so that injured north American soldiers, who lost their vision during the battles of the Second World War, could avoid obstacles. These techniques were described by Bledsoe, collaborator of the Hoover project, in the Manual for Orienteers (1947) and created so much interest among the blind community that many specialists began to teach them and took courses in the centers of Maryland and Hines to a acquire the needed knowledge.

Even Bledsoe himself (1980) described exhaustively how the whole procedure was developed so that the different techniques for the usage of the mobility cane could be performed. Nowadays, these canes help people with visual impairment to improve their lifestyle, providing their routines more security when moving. Usually patients make use of the assistance of another person, a guide dog or a stick. The original prototype of the cane, which consisted of the "white cane", also known as "Hoover", has been improved in the last years. The devices can be flexible or rigid. The flexible cane presents the disadvantage of having a shorter useful life due to physical wear. It also is not capable of predicting the tactile transmissions in comparison to the rigid stick. It is also important that the patient respects the range of longitude. Otherwise the device won´t help detect the obstacles accurately, forcing the patient to acquire a bad posture and generating with time a physical problem [6].

3.1 Assistance Devices for Blind Patients

Consequently, some of the commonly used devices developed in order to aid visual impaired patients, which also are related to the current project, are mentioned as followed.

The Seeing Aid Device, as shown in Fig. 1: one of the first electronic systems created to assist mobility of people who suffer of blindness. The device is capable to register the objects on the road while emitting infrared signals. The infrared leds of a low cost are set on glasses and produce a certain tone when an object is detected. The system employs a ray of stretch wavelength for the main signal, allowing the user to choose the direction he or she wants to take [7].

A Sonic prototype for blindness aid is shown in Fig. 2. The new technical help, researchers of the Carlos III University (UC3M) in Madrid, have recently developed, works as a sonic guide that informs the user about his or her surroundings. The proposal consists of a stereo visual processor that measures the difference between images taken by two lightly displaced cameras, calculating the distance of each point of the general picture. In order to transmit the acquired information, the user employs a sound code, which informs about the position and distance of the different obstacles. The creators assured that the system aims to be a compliment of the cane or the guide dog, not a substitute. The new prototype has a smaller size in comparison

Fig. 1 Mimic E-2 Eyeglass Seeing Aid, 1972 [7]

Fig. 2 Sonic prototype system for blind patients [8]

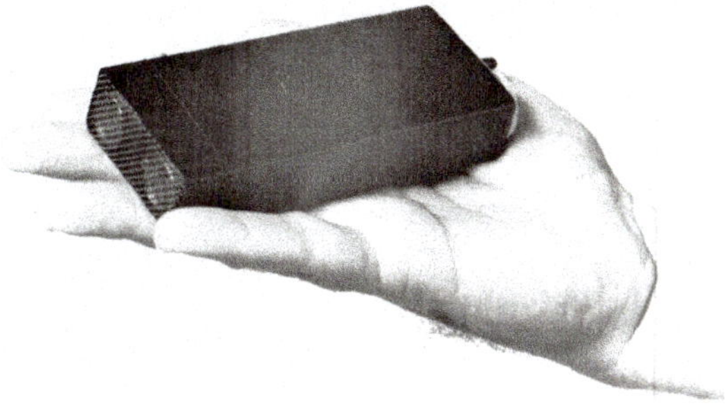

Fig. 3 Mowat Sonar Sensor, 1973 [7]

with the first ones and is composed of glasses with two micro-cameras, a plate of the size of a cigarette package near the battery, a small audio stereo amplifier and helmets for osseous transmission. The system hasn't been launched into the market, because only one of its kind has been built [8].

The Mowat Sensor (Fig. 3) is a device based on ultrasounds, produced by a special artifact, which the patients need to hold in their hands. When an obstacle is near, the distance between the user and the object is detected through vibrations, which frequency is inversely proportional to the distance [7].

Portable, obstacle detector system (Fig. 4): students of the National University of Columbia designed a portable electronic system for blind people, which are able to easily detect obstacles, working with the help of vibrations. The device is composed of two main systems: one for the measures and the second, for the signal processing and announcement. The first one is in charge of detecting the suspended objects and possesses a frontal reach of approximately 80 cm. It is also equipped with a bracelet and two elastic bands, one of which is used for the head and the second for the hand wrist. Both are equipped with an infrared sensor, which is protected by a shell made of acrylic. The second system for signal processing and announcement, measures and detects the obstacles which are on the floor. It contains two boxes of acrylic that store the energy, with which the card of the electronic components that allow

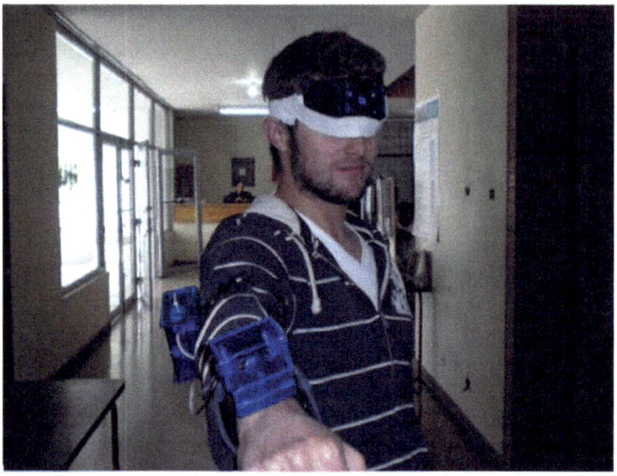

Fig. 4 Portable, obstacle detector [8]

Fig. 5 Sentiri Diadem [9]

the signal processing can work. The signal is acquired through the sensor and the announcement system, composed of an eccentric mass [8].

The Sentiri diadem (Fig. 5), created by the company Chaotic Moon, consists of the usage of multiple infrared sensors that detect the surroundings objects and the height at which these are located. This allows the user to analyze the collision risks [9].

The ARGUS Project (Fig. 6), pretends to work as an artificial retina and assists people who suffer of total blindness due to degenerative illnesses. The system converts the obtained images taken by a micro camera, contained in the glasses. The camera emits electrical impulses to an electrode matrix, located in the user's retina. These impulses stimulate the cells of the retina, generating the corresponding, luminous perceptions in the visual area of the brain. This is how the user is able to partially recover some visual capacity [10].

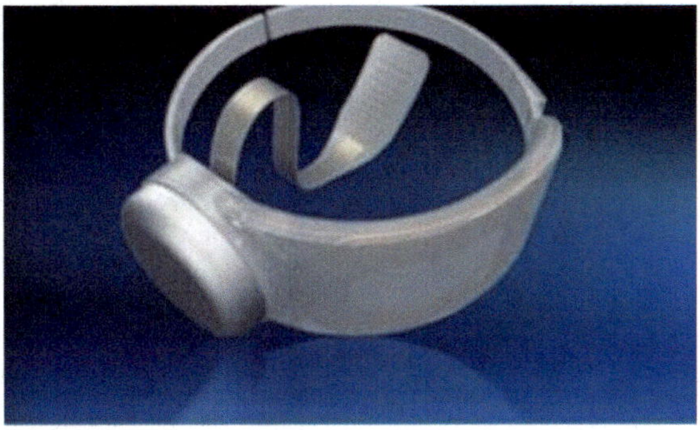

Fig. 6 Artificial retina ARGUS [10]

4 Design and Development

4.1 Requirement Definition

According to the information acquired through recently applied surveys, important information about some people who suffer of visual impairment or are visually weak was recollected. Likewise, through the bibliographic research about the state of the art of the principal aid devices that currently exist in the market, the selection of the specific requirements for the design of the diadem was performed. It is important to emphasize the fact that some of the most important differences this prototype will have in comparison to the previously mentioned ones are the following: a compact size, will be much lighter and cheaper. The device is capable of performing all the necessary functions without having the need of building a system of mayor complexity or of a bigger size. Taking into account that some blind people have also additive problems, the device needed to be capable of emitting vibrations, avoiding distracting the user while producing sounds.

Some of the most important requirements are mentioned in the following list:

- The device has to detect objects in a range of 10–80 cm.
- The sensor needs to indicate the distance of the object through vibrations in front of the user, which will be more or less depending on the distance of the detected object or obstacle.
- Be composed of a circuit and tiny hardware.
- Needs to possess a wide obstacle detection spectra.
- Needs to be light.
- Device of easy collocation and adjustable for different users.
- Ergonomic.
- Esthetic.
- Of a lower cost in comparison to the commercial devices.
- Needs to have an on/off switch.
- Low cost for the maintenance.

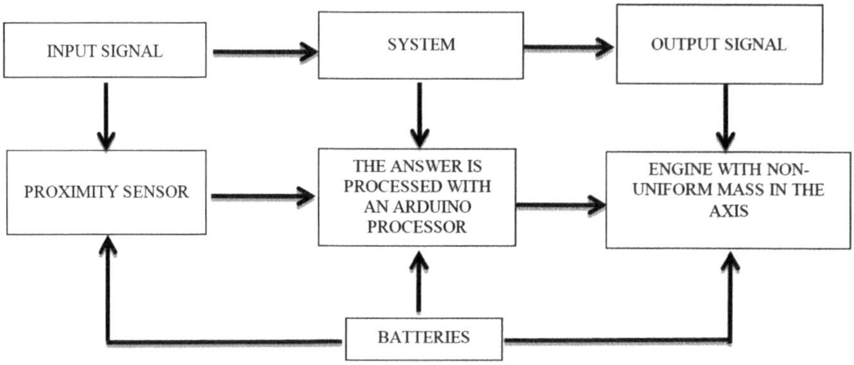

Fig. 7 Flow chart

- The spare parts need to be found in the national commercial market.

After classifying all the economic, functional, spacial, esthetic factors, together with the manufacture and installation specifications, the following system diagram was designed, as shown in Fig. 7.

In the flow chart of the system (Fig. 8), the interaction between the sensors and the vibration motors is illustrated. Depending on the side where the object is detected, it is the motor or the motors the one or those which needs or need to be activated.

This project is specially intended to assist patients with visual impairment and tries to minimize the mobility problems, enhancing the sensorial perception of blind people with the usage of sensors and the indication of these to alert about obstacles. This device was built with materials of good quality of a lower cost, using the stereolithography technique for its tridimensional print. This is in line in order to get lower costs, reduce hours of manufacturing for its elaboration and also reduce weight. The central part of the system is based on distance sensors that are connected to the vibration motors, are in charge of signaling the proximity of an object, producing a mayor vibration when a proximate object is detected and a less stronger vibration when the object is farther.

The control is achieved with the usage of a Nano Arduino and the modulation of a width pulse. Currently, there are other sensorial devices in form of a stick and lenses, designed to allow the user to walk through the streets and be able to detect obstacles. However, there are some obstructions that end to be impossible to detect, because of the poor quantity of sensors they possess.

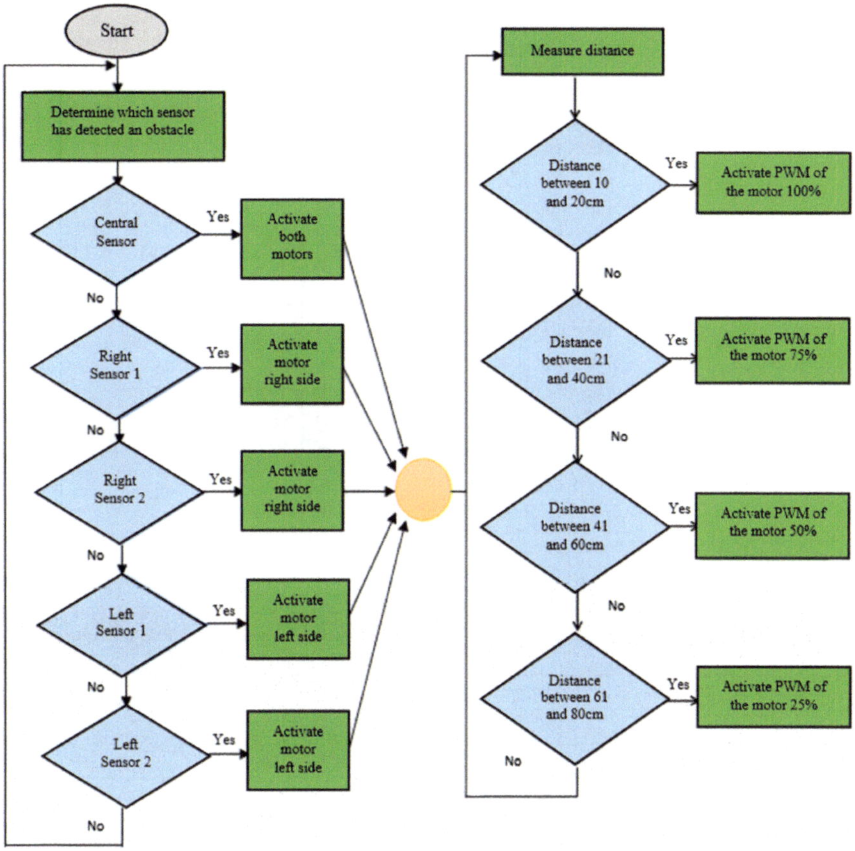

Fig. 8 System flow chart

4.2 *Electronic Subsystem*

4.2.1 Nano Arduino 328

This plate of rapid Prototyping (Fig. 9) controls the whole system with a code that acquires the inputs of the signal and the outputs of the motors, which allows the sensor and the motors to perform their function in the precise moment it is needed.

4.2.2 Vibration Motor

Through the use of the previously mentioned motors, the patient that suffers of blindness is capable of perceiving the signal in form of a vibration when he or she is near an obstacle (due to the stronger vibrations). The motors are located in the

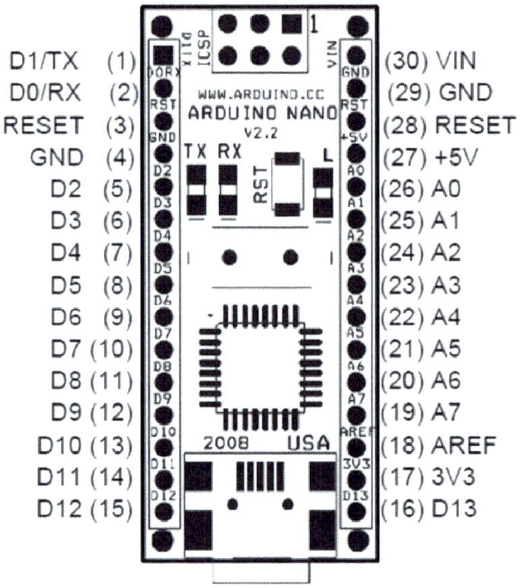

Pin No.	Name	Type	Description
1-2, 5-16	D0-D13	I/O	Digital input/output port 0 to 13
3, 28	RESET	Input	Reset (active low)
4, 29	GND	PWR	Supply ground
17	3V3	Output	+3.3V output (from FTDI)
18	AREF	Input	ADC reference
19-26	A7-A0	Input	Analog input channel 0 to 7
27	+5V	Output or Input	+5V output (from on-board regulator) or +5V (input from external power supply)
30	VIN	PWR	Supply voltage

Fig. 9 Nano arduino pin layout [11]

frontal part and in both sides of the diadem. The motors are DC type, of an uniform mass along the axis that generates the vibration (Fig. 10). The specifications are the following ones:

- Functioning voltage: 1.5–6 VDC
- Velocity: 1500 rpm
- Size of the vibration axis: 18 × 10 × 4 mm (L * W * T)
- Body size: 12 × 25 mm (H * D).

In order to control the velocity variation of the DC motors and therefore, be able to vary their vibration, the modulation of the pulse (PWM) of a square, digital signal, as shown in Fig. 11. Under the PWM control, the motor rotates to a specific velocity according to the average level of the square signal.

Fig. 10 DC motor

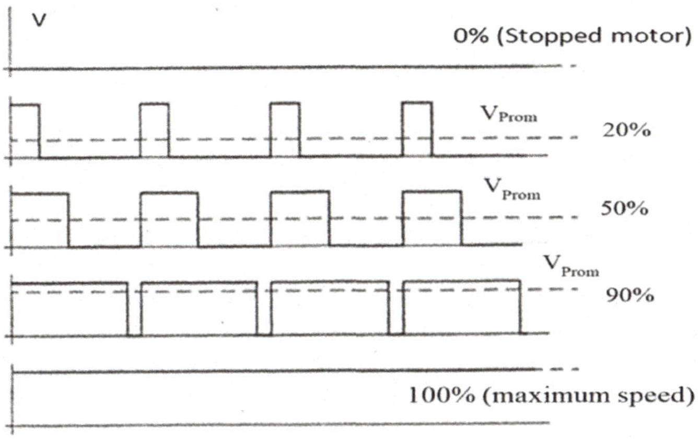

Fig. 11 Graphs of velocity control of a DC motor by using PMW

The average continuous tension of the motor is controlled maintaining a constant frequency and varying the time, the signal stands at the top. This means the work cycle must suffer a variation (Duty Cycle). This is how if the work cycle consists of 50%, the average tension supplied to the motor equals half of the maximum value. With a work cycle of 20%, only 1/5 of the maximum tension is supplied to the charge.

Changing from one work cycle of 50% to another of 20% will diminish the velocity of the motor. The PWM regulation offers an effective method through the usage of a simple, digital, control signal.

With this control method the current quantity stays the same, the tension doesn't present any variation and, as a consequence, the torque of the motor is constant, which is the aim of the whole procedure. Figure 12 illustrates the circuit simulation of the diadem signal with the elements that compose it.

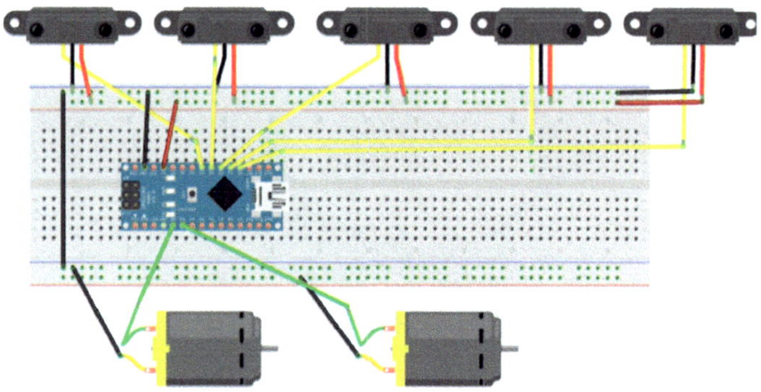

Fig. 12 Simulation of the control system

4.2.3 Distance Sensors GP2Y0A21YK0F

The distance sensor is collocated in a specific position to detect the area of work of 180°, being the most precise points, positioned at 0°, 45°, 90°, 135° and 190° in relation to the transversal plane of the body. This is a unit of distance measure for the sensor, composed of a combination of the PSD (sensitive position sensor), IRED (infrared diode) and a processing circuit for signals. The variety of the objects reflectivity, the environment temperature and the duration of the work cycle, don't have an easily influence in the distance detection due to the triangulation method. The device emits the corresponding voltage of the detected distance. The sensor scans the continuous form of the objects located in front of the user and provides the output using an analogical tension reference. For instance, the analogical inputs of the Arduino need to read the value of the distance. The measure range oscillated between 10 and 80 cm, keeping a high precision grade amongst the whole range. The feeding tension is of 4.5–5.5 V and the current consume of 33 mA.

The Sharp sensor is more precise in comparison to the ultrasonic sensors, such as the HC-SR04 sensor, particularly with short and medium distances, where the ultrasonic sensors can present inaccuracies in their mediations due to the reboots or echoes produced by the geometry of the surroundings. Unfortunately, it can't operate at short distances (less than 10 cm) where the HC-SR04 sensor, is capable of taking measures, even though these are of low precision. By using more than one sensor, some interference can be produced between the measures if both sides interfere. However, the inconveniences are less in comparison to when the sensors are combined, where some interference can be present even if both sides of the sensors do not enter on contact because of the echoes and reboots originated by the sensor HC-SR04. The infrared LED emitter produces an infrared light beam with a wavelength of 850 ± 70 nm. A luminous beam with pulses is sent to minimize the influence of the light and the color of the object in the measurement process.

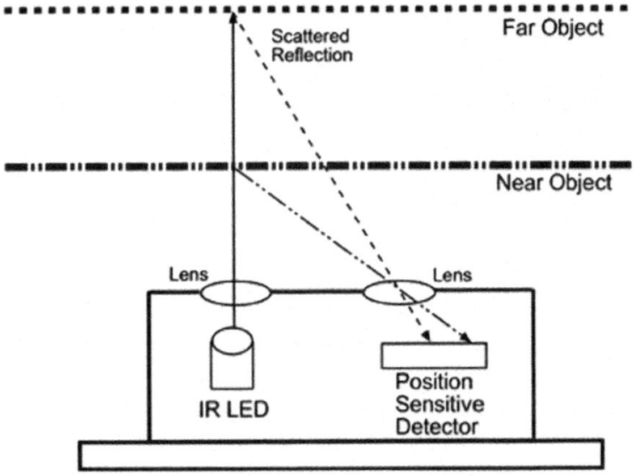

Fig. 13 Detector de posición PSD

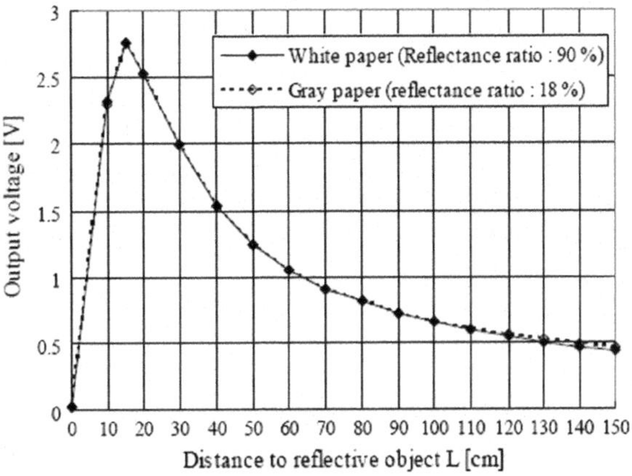

Fig. 14 Response of the analogical signal according to the object´s reflection [12]

The position detector PSD is a tiny linear, CCD sensor that receives the reflected light of any object, caught in the beam's trajectory. The sensor employs the triangulation to determine the distance between the sensor and the detected objects (Fig. 13).

The analogical output has a value of 2.5 V at 20 cm and 0.4 Vat 150 cm. The answer is not linear, which makes it necessary to interpole the value to obtain an adequate precision level (Fig. 14).

4.2.4 Switched off and on Interruptor

The user will be able to switch on and off the device everytime it is not being used to save battery. This will let the system get energized, which also requires a specific charge of 3.5–5 V for its functioning through a rechargeable Litium-Polium battery 5 V 1 A.

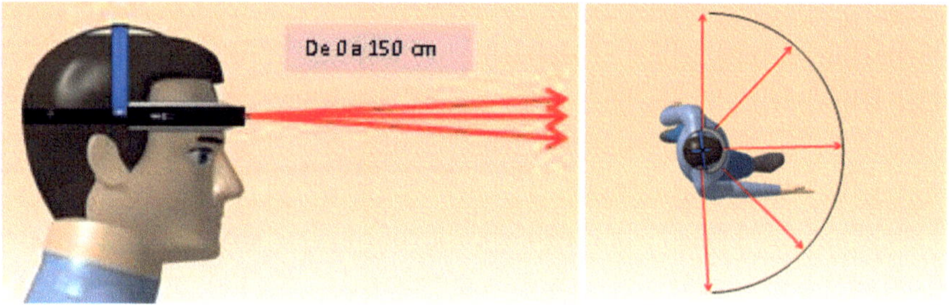

Fig. 15 Detection range of the diadem

4.3 System Characteristics

The linear reach of each sensor is of approximately 10–80 cm. he angular reach of each sensor tends to vary between 20° and 30°, giving as a result a total angular reach for the whole system, reaching the 180° (Fig. 15). The normal visual ranges of a person take into account the design of the distractors in order to avoid an obstruction of the user´s visual field, reaching 62° in their left and right limit, as measured form the standard vision line.

This is how with the usage of the stick, the user is able to detect borders, unevennesses and objects that could collide with the users head (Fig. 15).

4.4 Mechanical Subsystem

The tridimensional model of the diadem (Fig. 16) for the distance sensors, specifying the dimensions of each one of the elements that compose it, was performed. To obtain a more esthetic appearance, most of the cables were collocated in the inside of the device and the distance sensors were integrated to an adjustable structure to make the device more comfortable (Fig. 17).

4.4.1 Material and Measures for the Diadem

It represents the structure where all the pieces of the circuit are contained. It is esthetic, ergonomic and adjustable. The contained elements need to be as light as possible to make the device comfortable for the user.

For the material´s selection of this component, the following characteristics are required:

- Low density material, so that the final piece is as light as possible.
- Resistance to the flexion, because the diadem will be summited to different forces each time the user collocates it over his or her head.

Fig. 16 Representation of the user employing the diadem

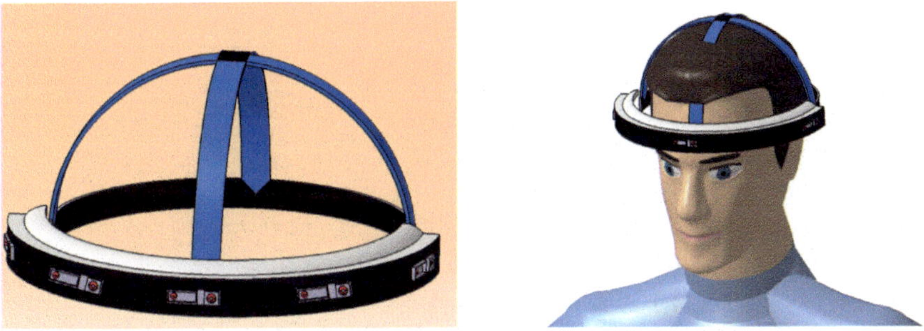

Fig. 17 Tridimensional design of the diadem for patients with visual impairment and visual weakness

- Low fabrication cost.
- Hardness, so that the device can last longer.

The preselected material according to the previously mentioned characteristics are the Acrylonitrile butadiene styrene (ABS) and the polylactic acid (PLA), are two polymers that can be used in the stereolithographic process for three tridimensional print of the prototype (Table 1).

4.4.2 Analysis with Infinitive Elements Method

Subsequently, a finite element analysis is perform to verify if the structure of the device, in PLA material, elastically get deforms when it is subjected to flexion, in addition to the interaction with the head of the user when are using the headband.

Table 1 Anthropometric dimensions of the skull of a young male of 15–17 years old with the purpose to develop the design of an extern maxillary systems [13]

Male
15 to 17 years old

Measures		15 years (n = 91)					16 years (n = 121)					17 years (n = 138)				
		Percentiles					Percentiles					Percentiles				
		$\overline{X}$	D.E.	5	50	95	$\overline{X}$	D.E.	5	50	95	$\overline{X}$	D.E.	5	50	95
33	Head diameter a-p	190	8	177	190	203	292	7	180	191	204	192	7	180	192	205
48	Head perimeter	557	19	526	560	588	558	17	530	558	586	561	18	531	560	591
34	Head width	156	7	146	156	167	156	7	144	156	168	156	6	146	156	166
35	Neck width	109	8	96	109	122	112	8	99	110	125	113	8	100	112	126
36	Face height	130	7	118	130	142	130	8	117	130	143	129	7	119	130	141
37	Face width	134	10	118	135	150	135	9	122	135	150	135	8	122	135	148
38	Interpupillary diameter	54	5	46	55	62	55	3	50	56	60	55	5	47	56	63

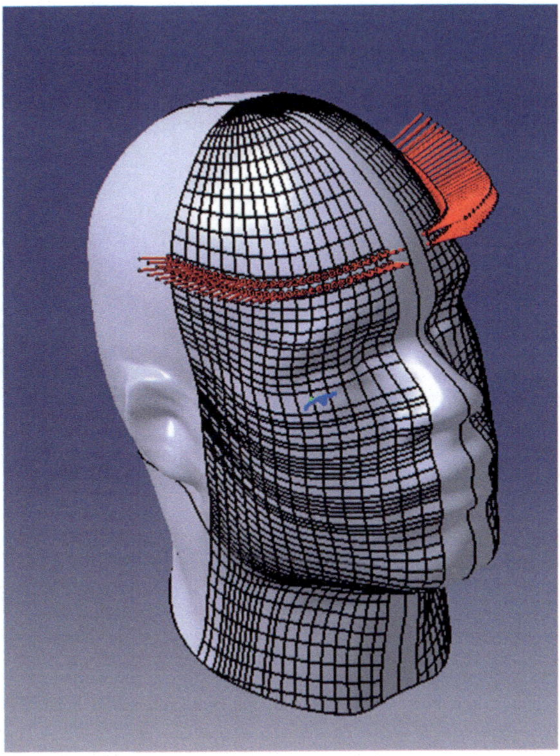

Fig. 18 Border conditions for the analysis of the head and the headband

For this study the base of the head was fixed and a distributed force of 5 N was applied to simulate the tightening by adjusting the headband. On Fig. 18 it can been seen the placement of boundary conditions.

When the mesh is generated and the material with the characteristics of cortical bone for the head is applied. It can be observed that the force exerted on the head of the user is minimal. Therefore, does not represent any danger on its use. The maximum Von Mises stress is 234 Pa (Fig. 19).

In relation to the maximum displacement, it corresponds to 0.52×10^{-8} mm, as can be seen in Fig. 20.

For the front part of the headband it is fixed on its central part and two equal forces are applied, and in reverse way, of 5 N to verify if the geometry with PLA material does not presents plastic deformation (Fig. 21).

A new mesh is generated on the geometrical model and a material are applied with the characteristics of the PLA for 3D printing. It is observed that the Maximum Stress of Von Mises is 1.22 MPa (Fig. 17). Therefore, the material correctly supports the forces to which the headband can be exposed.

According with the maximum displacement, observed on Fig. 22, its obtained 0.384 mm.

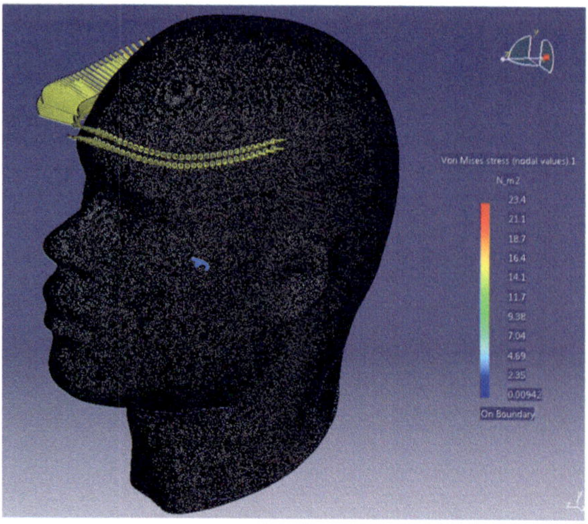

Fig. 19 Stress resulting from Von Mises on the head

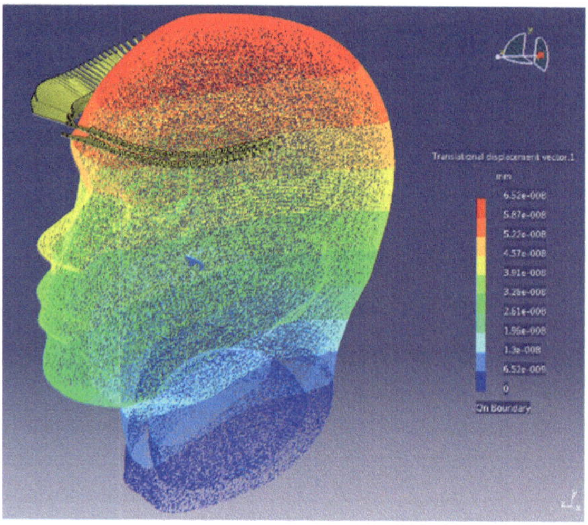

Fig. 20 Maximum deformation on the head

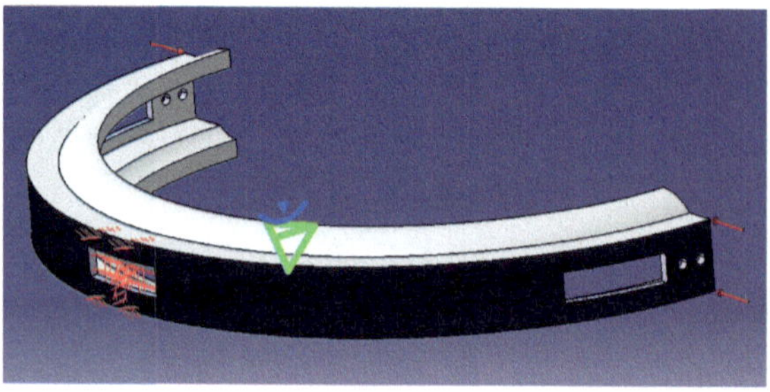

Fig. 21 Border conditions for the headband

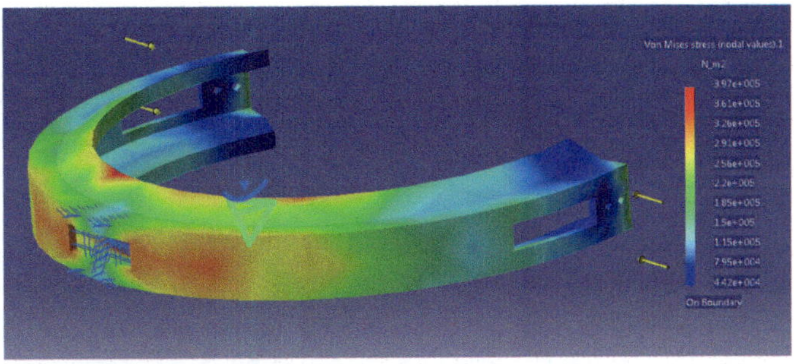

Fig. 22 Efforts resulting for the headband in base of Von Mises

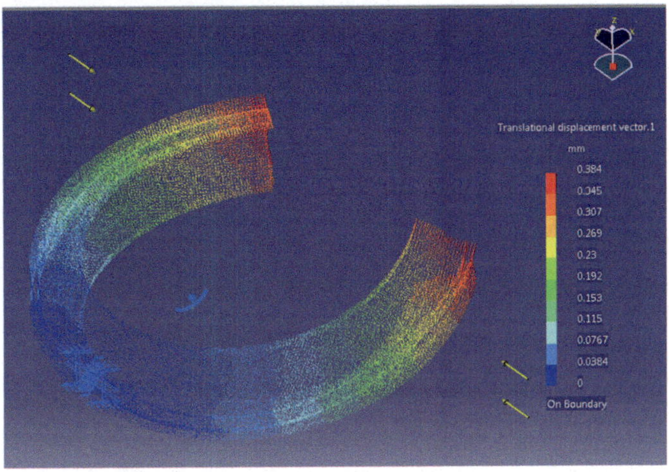

Fig. 23 Maximum deformation on the headband

4.4.3 System Manufacturing

Once the pertinent tests have been development with the device, modifications and adaptations of the distance sensors are carried out, in order to printing of the three-dimensional parts of the headband by stereo-lithography. This technique has been used to produce anatomical grafts models in maxillofacial surgeries [14–16].

The digital model of the pieces saved, is required as a good quality .stl file. These files can be used in the program Cura 14.07® that serves as an interface with the Prusa I3® printer. Firstly, the pieces must be located them on the way that a better print detail can be obtained, less material is required and less printing time (Figs. 23 and 24).

After calibrating the printer and having the file ready, they are sent to print the pieces and then finally obtain the physical prototype (Fig. 25).

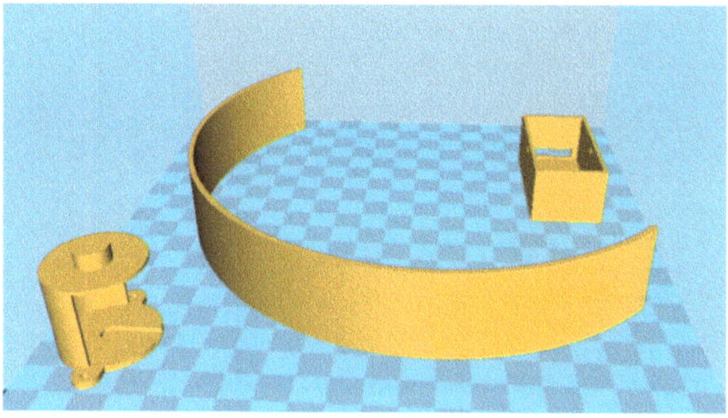

Fig. 24 Imported model on the software Cura 14.07®

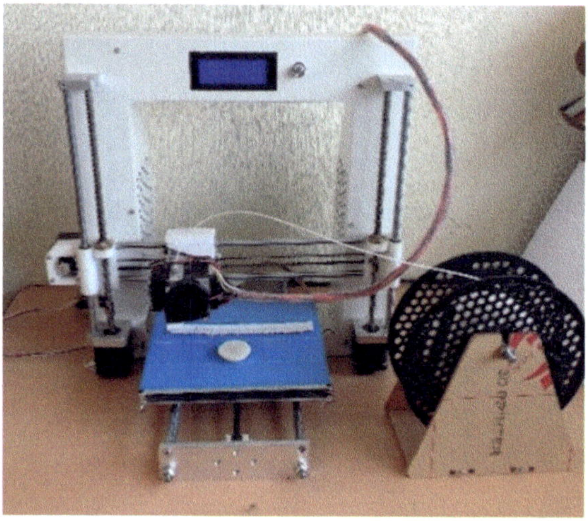

Fig. 25 Printing of the three dimensional pieces

Once the printed pieces are done, it is proceed to clean them and remove the support material (Fig. 26).

With all the printed pieces, the device is assembled and the electronic control system is implemented and connected as shown in Fig. 27.

5 Experimental Tests

Sensor tests were performed with the purpose to demonstrate the error in the measuring range of the device, it was necessary to make several readings using the device and a tape measure as a check pattern. For distances less than 10 cm, the sensor reading could be confused with a large distance, according to its behavior graph. For

Fig. 26 Printing, cleaning and assembly of printed pieces

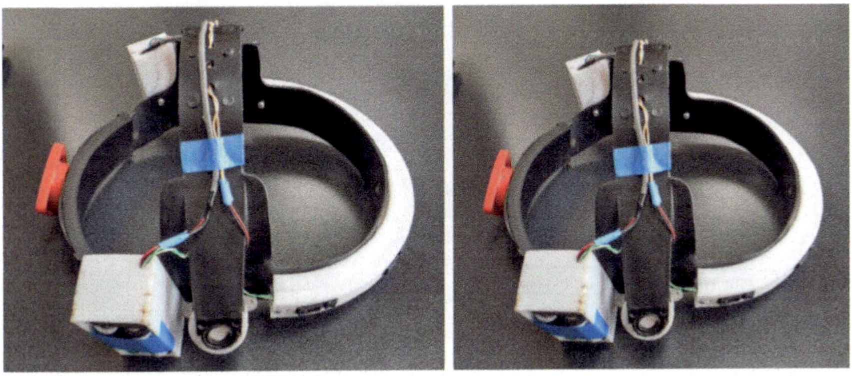

Fig. 27 Constructed headband

this reason, it is recommended to ensure that the object is not too close to the sensor in order to avoid an incorrect reading (Fig. 28).

To perform the sensor tests, a table was used; a solid surface for sensing and a flexometer as a measurement reference (Fig. 24), to verify that the data thrown by the sensor correspond to the actual distance measured.

As the distance is varied at different times, the values thrown by the sensor and interpreted in the arduino program (Fig. 25), should be equivalent to those measured (Figs. 29 and 30).

A sample is taken of the iterations made by the sensor every 2 and 3 ms for the reading time and the measured distance (Fig. 31).

For this system, that uses the GP2Y0A21 sensor, it is necessary to avoid approaching less than 10 cm where is known that the sensor behaves in an unstable manner.

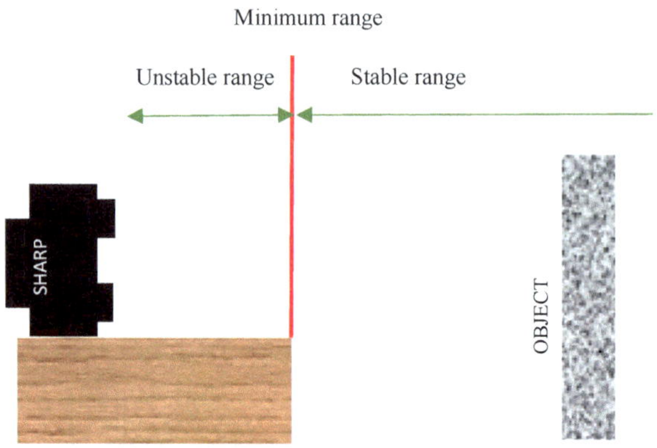

Fig. 28 Sharp 2Y0A21 sensor distance ranges for lecture

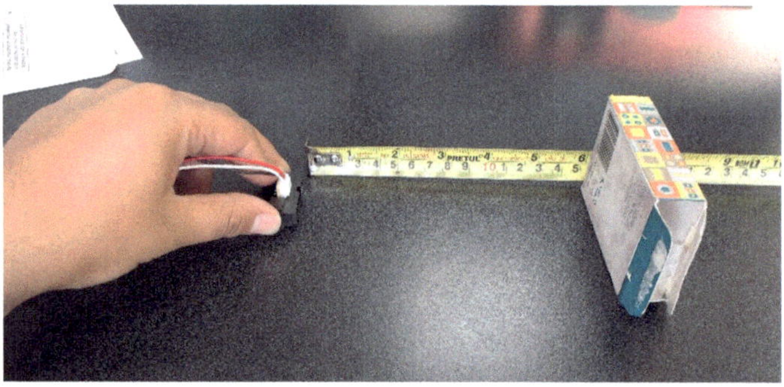

Fig. 29 Sensor measurement test

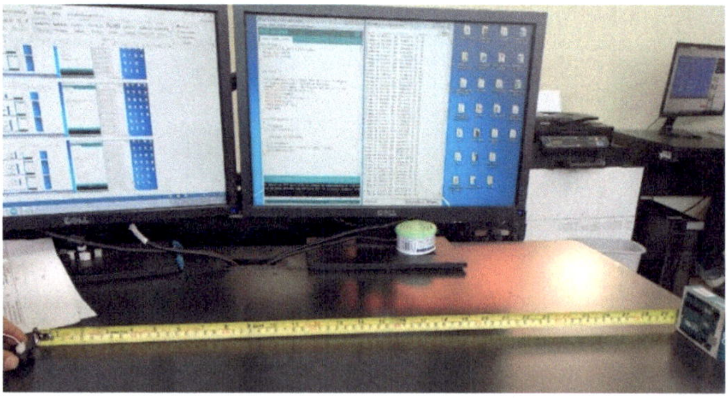

Fig. 30 Data thrown by the sensor

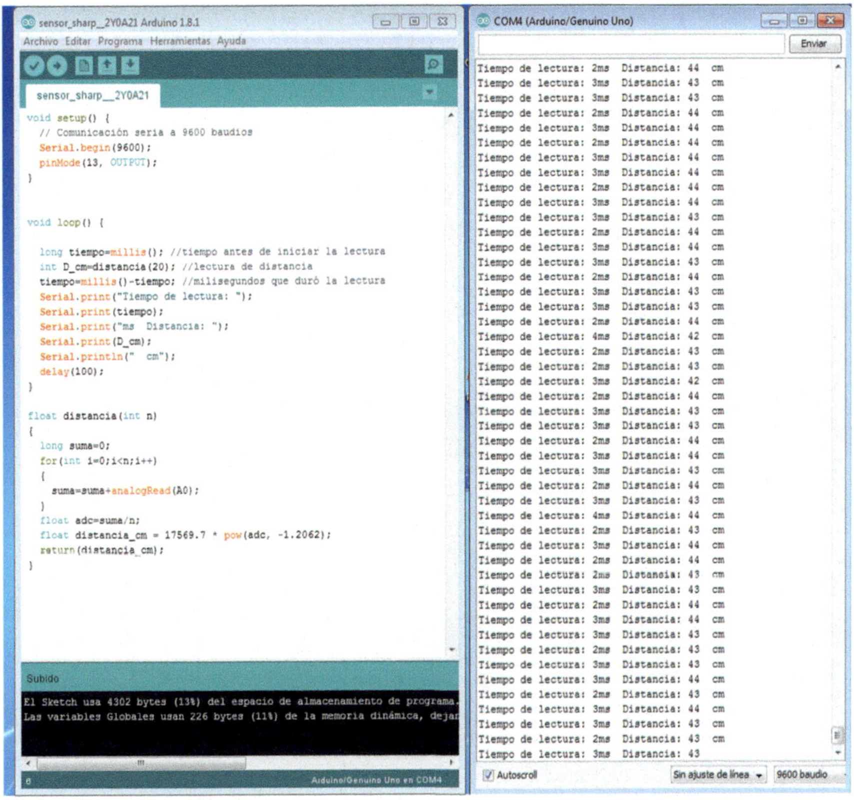

Fig. 31 Sample of data thrown by the sensor

As a result, a device capable of measuring distances in a range of 10–80 cm was obtained. In order to check the error in the measurement range of the device it was necessary to perform a sampling every 3 and 4 ms, with a variation of the measurement for 15, 25, 45, 60 and 75 cm and obtain the average, absolute error, relative error and percentage error as shown in Table 2.

On Fig. 32 it can be seen that the variation of the measured distances in contrast to the real distance that the sensor should detect. It was obtained a very low response and almost linear.

According to the absolute, relative and percentage errors (Fig. 28), it is observed that the relative error is very low and with a linear response, however the percentage error is high for the measured distances of 15 and 25 cm (Fig. 33).

Table 2 Absolute, average and relative values, and errors of the sensor measurements

Real value (cm)	Measured value (cm)										Average	Absolute error	Relative error (cm)	Percentage error (%)
	Measure 1	Measure 2	Measure 3	Measure 4	Measure 5	Measure 6	Measure 7	Measure 8	Measure 9	Measure 10				
15	16	16	15	16	16	15	16	15	15	16	15.6	0.6	0.038	3.846
25	24	24	24	24	25	24	24	25	24	24	24.2	−0.8	−0.033	−3.306
45	45	44	44	45	45	45	45	45	45	45	44.8	−0.2	−0.004	−0.446
60	59	59	59	59	59	59	58	60	59	59	59	−1	−0.017	−1.695
75	72	78	74	75	75	75	74	75	75	77	75	0	0	0
											Average error	−0.28	−0.003	−0.320

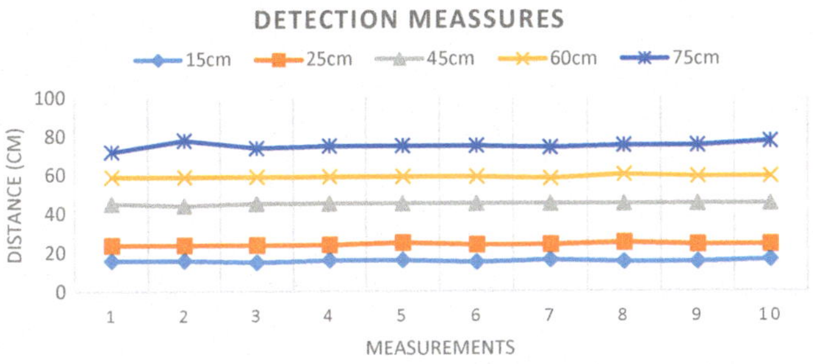

Fig. 32 Measured values against actual values

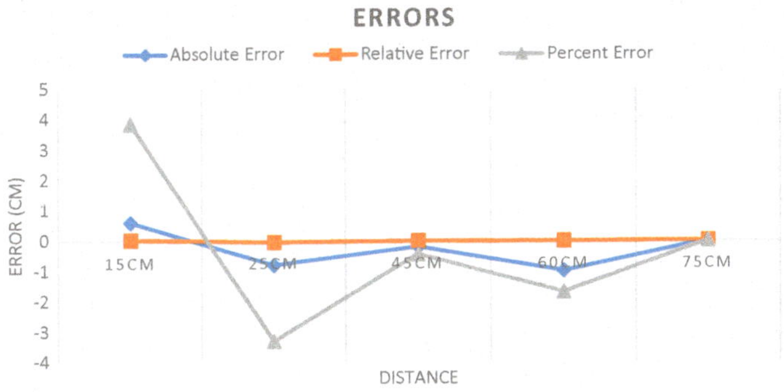

Fig. 33 Failures graph

6 System Test in Everyday Environment

Finally, the physical tests of the device were performed in a daily environment to verify its correct functioning and perform any additional requirement calibration of the sensors. As can be seen in Fig. 29, the white cane was used in conjunction with the headband to obtain greater spatial coverage, also to preventing the user from tripping or hitting any object within the detection area, which is extended with the complementary use of a cane (Fig. 34).

7 Discussion

- The sensors were tested in order to determine the error in the measurement of the distance. Several readings were done. The results showed that measurements bellow 10 cm were confused and they could be considered erroneously as a large distance signal.

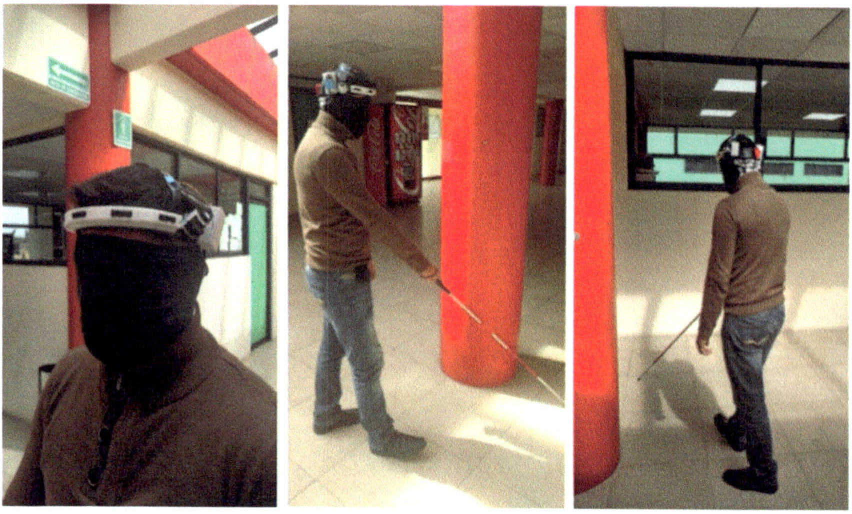

Fig. 34 Performance of physical tests of the headband for visual disabled patient

- The analog output was 0.25 V, when the object was located at 10 cm. On the other hand, it was 0.55 V, when the object was at 80 cm. However, a non-linear response was obtained, so an interpolation was required in order to improve the accuracy of the evaluation.
- Although the sensors were susceptible to ambient light, this factor was corrected, when the sensor did not operate at top of the sensing range.
- The variation of the reflection of the objects, the environmental temperature and the time of operating duration did not influenced on the evaluation. The triangulation method improved the performance of the sensor.

8 Conclusions

- In the final design, the GP2Y0A21YK0F proximity sensors were used. A good performance was observed, with less interference and it was easier to calibrate.
- The Assistance System for Blindness or Partial Vision is an assistance device and it is not a total substitute for the cane, since its main function is to detect objects at the height of the head of the individual or obstacles that are above the ground. It is considered that the use of both devices is ideal to cover more area of detection of objects and thus avoid collisions.
- Greater security for users was obtained and, therefore, an improvement is obtained in their daily activities.
- The originality of the system lies in the use of proximity sensors as a core part of the detection system, increasing the range of total angular range that the system

has. In addition to using stereo-lithography technology for its manufacture at low cost.

Acknowledgements The grants awarded by the *Consejo Nacional de Ciencia y Tecnología, Instituto Politécnico Nacional* and *Universidad Politécnica del Valle de México* are grateful acknowledged by the authors.

References

1. Pascolini D, Mariotti SP (2011) Global estimates of visual impairment: 2010. Br J Ophthalmol 96(5):614–618. https://doi.org/10.1136/bjophthalmol-2011-300539
2. La discapacidad en México, datos al 2014, consultada por internet el 24 de junio del 2018. Dirección de internet: http://conadis.gob.mx/gob.mx/transparencia/transparencia_focalizada/La_Discapacidad_en_Mexico_datos_2014.pdf
3. Ontiveros-Paredes SD, Rojas-Balbuena D, Martínez-Paredes J (2014) Diseño y construcción de un bastón blanco electrónico para personas invidentes. Científica 18(2):63–70
4. Discapacidad visual y autonomía personal (2011) Enfoque práctico de la rehabilitación, Organización Nacional de Ciegos Españoles (ONCE), primera edición 2011, Madrid, España. ISBN: 978-84-484-0277-8
5. Yablonski MS (2000) Functional orientation and mobility. In: Silverstone B, Rosenthal BP, Lang MA, Faye EE (eds) The Lighthouse handbook on vision impairment and vision rehabilitation, vol II. Oxford University Press, Nueva York, pp 855–868
6. Fragoso D, Medleg M, Rivera JA, Santillan E (2018) Casco con sensores de proximidad para invidentes" consultada por internet el 24 de junio de 2018. Dirección de internet: http://vinculacion.dgire.unam.mx/Memoria-Congreso-2016/trabajos-ganadores-02/2.pdf
7. Howard Freiberger AM (2018) Movility aids for the blind, Electronics Engineer, Research for Prosthetics Veterans Administration New York, consultada por internet el 24 de junio de 2018. Dirección de internet: https://www.rehab.research.va.gov/jour/74/11/2/73.pdf
8. Peralta MA, Urmendiz JV (2014) Sistema de asistencia y guía para personas invidentes, Universidad Autónoma de Occidente, Facultad de Ingeniería, Departamento de Ingeniería Biomédica, Santiago de Cali, consultada por internet el 24 de junio de 2018. Dirección de internet: https://red.uao.edu.co/bitstream/10614/6868/1/T05005.pdf
9. Melian G (2018) "Chaotic Moon", consultada por internet el 24 de junio de 2018. Dirección de internet: http://amolatecnologia.com/chaotic-moon-idea-una-diadema-que-avisa-a-los-ciegos-cuando-tienen-un-obstaculo-delante/
10. Martínez AD (2012) Bastón blanco para prevenir obstáculos, Instituto Politécnico Nacional, Escuela Superior de Ingeniería Mecánica y Eléctrica Unidad Zacatenco, Ciudad de México, México, consultada por internet el 24 de junio de 2018. Dirección de internet: http://tesis.ipn.mx/bitstream/handle/123456789/11470/21.pdf?sequence=1
11. Datasheet Arduino nano consultada por internet el 24 de junio de 2018. Dirección de internet: https://www.arduino.cc/en/uploads/Main/ArduinoNanoManual23.pdf
12. Datasheet Sharp GP2Y0A21YK0F consultada por internet el 24 de junio de 2018. Dirección de internet: http://www.sharp-world.com/products/device/lineup/data/pdf/datasheet/gp2y0a21yk_e.pdf
13. Bushby KMD, Cole T, Matthew JNS, Gooship JA (1992) Centiles for adult head circumference. Arch Dis Child 67:1286–1287
14. Beltrán-Fernández JA, Martínez-Paredes J, González-Rebattú M, Hernández-Gómez LH, Ruíz-Muñoz O (2017) Customization and numerical simulation of a cranial distractor using Computed Axial Tomography (CAT). In: Öchsner A, Altenbach H (eds) Properties and characterization of modern materials. Advanced structured materials, vol 33. Springer, Singapore

15. Moreno-Garibaldi P, Beltrán-Fernández J, Hernández-Gómez L, López-Saucedo F, Corro-Valdez N, Martínez-Paredes J (2017) Methodology for modeling and manufacture of a lower jaw prosthesis for its possible treatment by congenital deformation. Journal de Ciencia e Ingeniería 9(4):7–12. ISSN: 2145-2628
16. Martínez-Paredes J, Beltrán-Fernández J, Hernández-Gómez LH, Gonzáles-Rebatú M, Rangel-Elizalde AI (2014) Orbital cavity reconstruction with hydroxyapatite and titanium mesh prosthesis. Journal de Ciencia e Ingeniería 6(1):7–12. ISSN: 2145-2628

Printed by Printforce, the Netherlands